Technology and the Future of European Employment

The researches on which this book is based have benefited from the financing of a TSER project of the EC (Targeted Socio-Economic Research programme of the European Community).

Technology and the Future of European Employment

Edited by

Pascal Petit

Director of Research, CEPREMAP/CNRS, Paris, France

Luc Soete

Maastricht Economic Research Institute on Innovation and Technology (MERIT), Maastricht University, The Netherlands

Edward Elgar

Cheltenham, UK • Northampton, MA, USA

Published by
Edward Elgar Publishing Limited
Glensanda House
Montpellier Parade
Cheltenham
Glos GL50 1UA
UK

Edward Elgar Publishing, Inc.
136 West Street
Suite 202
Northampton
Massachusetts 01060
USA

A catalogue record for this book
is available from the British Library

Library of Congress Cataloging-in-Publication Data
Technology and the future of European employment / edited by Pascal Petit, Luc Soete.
    p. cm.
   Includes bibliographical references and index.
    1. Labor supply—Effect of technological innovations on—Europe. 2. Technological innovations—Economic aspects—Europe. I. Petit, Pascal. II. Soete, Luc.

HD6331.2.E85 T43 2001
331.12'094—dc21

                                00-050289

ISBN 1 84064 517 2

Typeset by Manton Typesetters, Louth, Lincolnshire, UK.
Printed and bound in Great Britain by Bookcraft (Bath) Ltd.

Contents

List of tables   vii
List of figures   xi
List of boxes   xiv
Contributors   xv

Introduction   1
Pascal Petit and Luc Soete

PART I   LONG-TERM STRUCTURAL CHANGES

1   Long-term structural changes: a reappraisal   23
    Christopher Freeman
2   Technology, growth and employment in postwar Europe: short-run dynamics and long-run patterns   46
    G.N. von Tunzelmann and Ü.D. Efendioglu
3   Europe in the triad: growth pattern and structural changes   78
    Pascal Petit

PART II   SECTORAL CHANGES AND DEMAND

4   Structural dynamics and employment in highly industrialized economies   111
    Ronald Schettkat and Giovanni Russo
5   Innovation, demand and employment   142
    Mario Pianta
6   Technical change and employment growth in services: analytical and policy challenges   166
    Pascal Petit and Luc Soete
7   The European unemployment problem: a structural approach   204
    Michael Landesmann and Robert Stehrer

PART III   CHANGES IN ORGANIZATION AND DISTRIBUTION

8   New technologies, organizational change and the skill bias: what
    do we know?   259
    Eve Caroli
9   Unemployment and labour market flexibility: a misplaced
    question?   293
    Donatella Gatti
10 Sweeping the chimney before kindling the fire as a workable
    option for employment policy   317
    Adriaan van Zon, Huub Meijers and Joan Muysken
11 Determinants of sectoral average wage and employment growth
    rates in a specific factors model with production externalities
    and international capital movements   352
    Ivo De Loo and Thomas Ziesemer
12 Modelling the link between skill biases in technical change and
    wage divergence through labour market extensions of
    Krugman's North–South model   387
    Adriaan van Zon, Mark Sanders and Joan Muysken

PART IV   INSTITUTIONAL CHANGE

13 Changing working time patterns   405
    Vincenzo Spiezia and Marco Vivarelli
14 Europe's system(s) of innovation   425
    Bruno Amable and Robert Boyer
15 Employment, unemployment and ageing in the West European
    welfare states   451
    Lars Mjøset
16 Policy conclusions: on the future of European employment   507
    Pascal Petit and Luc Soete

Index   541

Tables

1	Technical change and employment: a reversed relationship?	4
1.1	Short (Juglar) business cycles in long (Kondratieff) waves	26
1.2	Unemployment levels in selected countries, 1980s and 1995	29
1.3	Exports of ICT equipment to major OECD countries, as a percentage share of total manufacturing exports to those countries, from selected countries	30
1.4	Patents taken out in the United States by various countries and regions, 1977–96	33
2A.1	Relationships between per capita real GDP and the rate of investment	63
2A.2	Relationships between per capita real GDP and the rate of investment in producer durables	64
2A.3	Relationships between total R&D intensity and aggregate GDP	65
2A.4	Relationships between total R&D intensity and rate of investment	66
2A.5	Relationships between total R&D intensity and rate of investment in producer durables	67
2A.6	Relationships between higher-education R&D intensity and aggregate GDP	68
2A.7	Relationships between per capita patents and per capita GDP	69
2A.8	Relationships between per capita patents and rate of investment	70
2B.1	Unit root tests on R&D intensity, patents, GDP and rate of investment	71
2C.1	Results from the key variables	76
2C.2	Summary of panel data two-period estimates	76
2D.1	Results of ECM panel data regressions, abstracted	77
3.1	Labour market characteristics in Europe, the USA and Japan	92
3.2	'Good' jobs and 'bad' jobs, by sector and country	94
3.3	Three interdependent structural changes	99
3.4	The three structural changes show up at various levels	103
4.1	Ratios of services to manufacturing in the institutional division and in the final product concept	116

4.2	OLS estimates of the effects of structural dynamics of demand on employment and final product employment	118
4.3	Share of intermediate goods and own value added in gross output	119
4.4	Productivity and employment changes (both final product and industry concepts), USA, 1972–90	122
4.5	Productivity and employment changes (both final product and industry concepts), Germany, 1978–90	123
4.6	Productivity and employment changes (both final product and industry concepts), Japan, 1970–90	124
4.7	Productivity and employment changes (both final product and industry concepts), France, 1972–90	125
5.1	Main variables from the European Community Innovation Survey, country averages	152
5.2	Regression estimates	155
5.3	Performance of product and process innovation-based industries	159
6.1	Tertiarization: the rise of business services in recent decades	184
6.2	Employment ratios by NACE sectors	189
7.1	ISIC classification	212
7.2	Results of spline regressions, sector 3	226
7.3	Results of spline regressions, sector 9	228
7.4	Summary of spline regressions	230
7.5	Sectoral growth and unemployment (single regressions)	233
7.6	Sectoral growth and unemployment (joint regressions)	234
7A.1	Equilibrium structure of output and employment	241
7A.2	Equilibrium structure of output and employment with redistribution	244
7A.3	Effects of technical change and redistribution	247
7B.1	Country and sector-specific dynamics of change	252
9.1	A parametric characterization of the MoOs	305
11.1	Sectors with correct expected signs	364
11.2	Relative importance of explanatory variables in explaining per country wage growth	364
11.3a	Growth rates of explanatory and dependent variables over the estimation period given by sample (SMPL)	365
11.3b	Growth rates of explanatory and dependent variables over the estimation period given by SMPL, 1980s onwards	366
11.4a	Relative importance of dependent and explanatory variables in explaining per sector wage growth	367
11.4b	Relative importance of dependent and explanatory variables in explaining per sector wage growth, 1980s onwards	368

11.5   Sectors with correct expected signs when the roles of labour
          and wages are interchanged   368
11.6   Relative importance of explanatory variables in explaining
          per country wage growth   369
11.7a   Growth rates of explanatory and dependent variables over
          the estimation period given by SMPL   370
11.7b   Growth rates of explanatory and dependent variables over
          the estimation period given by SMPL, 1980s onwards   370
11.8a   Relative importance of dependent and explanatory variables
          in explaining per sector labour growth   371
11.8b   Relative importance of dependent and explanatory variables
          in explaining per sector labour growth, 1980s onwards   371
11.9a   Growth rates of explanatory and dependent variables over
          the estimation period given by SMPL   374
11.9b   Growth rates of explanatory and dependent variables over
          the estimation period given by SMPL, 1980s onwards   374
11.10a   Relative importance of dependent and explanatory variables
          in explaining per sector wage growth   374
11.10b   Relative importance of dependent and explanatory variables
          in explaining per sector wage growth, 1980s onwards   374
11.11a   Growth rates of explanatory and dependent variables over
          the estimation period given by SMPL   375
11.11b   Growth rates of explanatory and dependent variables over
          the estimation period given by SMPL, 1980s onwards   376
11.12a   Relative importance of dependent and explanatory variables
          in explaining per sector labour growth   376
11.12b   Relative importance of dependent and explanatory variables
          in explaining per sector labour growth, 1980s onwards   376
11.13   Sectors that have adverse effects from trade and decreasing
          wage growth   378
11A.1   Sector classification and abbreviations   386
11A.2   Country codes   386
13.1   Long-run changes in total working time, average working
          time and employment   408
13.2   Average annual changes in hourly productivity and in
          average working time, 1960–95   412
13.3   Distribution of the average annual growth rate in hourly
          wage between wage per employed person and average
          working time, 1960–95   413
13.4   Overtime hours and part-time working, 1983–94   417
13.5   Percentage of full-time dependent employees working most
          frequent usual weekly hours (mode), 1985 and 1995   418

13.6	Patterns of changes in working time	420
14.1	Institutions and organizations influencing endogenous growth	430
14.2	Institutional analyses of innovation and production systems: a long way from the 'social capability' approach	434
15.1	The West European unemployment experience	452
15.2	Historical background to Western Europe's present challenges	456
15.3	Growing life expectancy in selected European countries, 19th and 20th centuries	464
15.4	The new risk structure	467
15.5	The three welfare models in the post-Golden Age period	473
15.6	Basic labour market indicators, 1983 and 1996	475
15.7	Comparative tax structures	479
15.8	Service sector employment disaggregated, mid-1990s	480
15.9	Retirement age	485
16.1	Key issues in the institutional changes of the 1980s and 1990s	516
16.2	GDP growth and productivity growth	526
16.3	Diversified demographic structures in European countries, 1995	534

Figures

1.1	Gross domestic investment, Brazil and South Korea, 1973–93	31
1.2	Employment in manufacturing, Brazil and South Korea, 1973–93	32
1.3	Measures of the US high-tech economy	
2.1	'Average' intensities of GERD, BERD and HERD, 23 OECD countries, 1963–95	48
2.2	'Average' per capita GDP and rates of investment across 22 OECD countries, 1950–95	49
3.1	Productivity and output growth, 1960–71	81
3.2	Productivity and output growth, 1981–90	81
3.3	Productivity and output growth, 1991–95	83
3.4	Productivity and output growth, 1995–2000	83
3.5	Unemployment rates	85
3.6	Employment ratios	85
3.7	Worries over job security	93
3.8	Productivity growth, 1980–97	96
3.9	Catching up, 1980–97	100
4.1	Industry-specific iso-employment curves	113
4.2	The relationship between employment and labour productivity (in levels) in different countries and in selected industries	114
4.3	Employment trend and trends in capital stock, capital–labour ratio, labour productivity and total factor productivity across industries in different countries	128
4.4	Total factor productivity and labour productivity across industries in different countries	131
5.1	Key links between innovation, demand and employment	145
5.2	A framework for investigating the impact of innovation on growth and employment	146
5.3	Product innovations in R&D and in sales	154
6.1	ICTs' effects on the tradeability of goods and services	176
6.2(a)	Innovation scheme in manufacturing	181
6.2(b)	Innovation scheme in services	181
6.2(c)	ICTs on tradeability	182

6.3	Employment growth in industry and services	187
6.4	Job gains and losses, by industry, OECD total	187
6.5	OECD employment trends in manufacturing and services industries	188
6.6	Employment growth, by skill level, in manufacturing and services	194
7.1	Shares in sectors I, II and III	214
7.2	Employment shares in manufacturing	216
7.3	Employment shares in services	217
7.4	Dynamics of change, sectors 1 and 3	221
7.5	Dynamics of change, sectors 6–9	222
7.6	Unemployment rates	231
7A.1	Effects of wage subsidy for low-income groups	243
7A.2	Tax burden	245
7A.3	Output and employment with biased technological progress	246
7A.4	Output with biased technological progress and redistribution	247
7A.5	Tax burden with biased technological progress	248
7B.1	Country and sector-specific dynamics of change, sectors 1 and 3	249
7B.2	Country and sector-specific dynamics of change, sectors 6–9	250
8.1	Technology, organization and skills: the black triangle	260
8.2	Organizational change, technical change: a theoretical framework	271
8.3	Organizational change, technical change and skills	284
9.1	Vocational training systems	298
9.2	The matching process	302
9.3	Equilibrium opportunity rate	309
9.4	Equilibrium unemployment rate	312
10.1	*Ex ante* job substitution and *ex post* skill substitution	321
10.2	Expected wage growth and skill switching	323
10.3	*Ex ante* job choices	325
10.4	Parameter estimates of the relative efficiencies	335
10.5	Parameter estimates of the learning costs	336
10.6	Unemployment change due to an increase in low-level jobs	340
10.7	Unemployment change due to an increase in high-level jobs	340
10.8	An increase in low-level jobs	341
10.9	An increase in high-level jobs	342
10.10	Unemployment change due to a decrease in low-skilled wage costs	344
10.11	Employment effects due to a decrease in low-skilled wage costs	345
12.1	Labour market specialization regimes	397

12.2	Parameter constellations and labour market regimes	399
13.1	Average annual changes in total working time, average working time and employment	407
13.2	Estimated contribution of part-time and full-time working to changes in average annual hours of employees, 1983–93	416
14.1	Why institutions shape the growth regime	427
15.1	Long-term implications of the US strategy	471
15.2	Long-term implications of the Nordic strategy	477
15.3	Long-term implications of the continental type	485
16.1	Three interdependent structural changes	524
16.2	Employment and growth: the main issues	536

Boxes

1.1   Greenspan says US is in a 'virtuous cycle'   42
3.1   Unemployment theories and their relevance for the USA, Japan
     and Europe   89

Contributors

Bruno Amable is a professor at the University of Lille 2 and a researcher at CEPREMAP, Paris, France.

Robert Boyer is a CNRS director of research at CEPREMAP and at EHESS, Paris, France.

Eve Caroli is a researcher at INRA/LEA and CEPREMAP, Paris, France.

Ü.D. Efendioglu is a researcher at INTECH, UN University, Maastricht, The Netherlands.

Christopher Freeman is an emeritus professor at SPRU, University of Sussex, Brighton, UK.

Donatella Gatti is a researcher at WZB, Berlin, Germany.

Michael Landesmann is a professor at the University of Linz and scientific director at the Vienna Institute for International Economic Studies, Vienna, Austria.

Ivo De Loo is an assistant professor at the Open University, Heerlen, The Netherlands.

Huub Meijers is a senior researcher at MERIT and in the Department of Economics and Business Administration of Maastricht University, The Netherlands.

Lars Mjøset is a professor in the Department of Sociology and Human Geography, and a researcher at the ARENA project, University of Oslo, Norway.

Joan Muysken is a professor in the Department of Economics and Business Administration of Maastricht University, The Netherlands.

Pascal Petit is a CNRS director of research at CEPREMAP, Paris, France.

Mario Pianta is a researcher at CNR-ISRDS, Rome, and a professor at the University of Urbino, Italy.

Giovanni Russo is an assistant professor in the Department of Economics at Utrecht University, The Netherlands.

Mark Sanders is a researcher at MERIT and in the Department of Economics and Business Administration of Maastricht University, The Netherlands.

Ronald Schettkat is a professor in the Department of Economics at Utrecht University, The Netherlands.

Luc Soete is director of MERIT and a professor in the Department of Economics and Business Administration of Maastricht University, The Netherlands.

Vincenzo Spiezia is a researcher at the International Labour Office in Geneva, Switzerland.

Robert Stehrer is a researcher at the Vienna Institute for International Economic Studies (WIIW), Vienna, Austria.

G.N. von Tunzelmann is a professor at SPRU, University of Sussex, Brighton, UK.

Marco Vivarelli is a professor at the Catholic University of Piacenza, Italy.

Thomas Ziesemer is a senior researcher at MERIT and an associate professor in the Department of Economics and Business Administration of Maastricht University, The Netherlands.

Adriaan van Zon is a senior researcher at MERIT and in the Department of Economics and Business Administration of Maastricht University, The Netherlands.

Introduction

Pascal Petit and Luc Soete

---

1. EMPLOYMENT IN EUROPE: THE FACTS AND THE DEBATES

Over the last decades, technology has come to be regarded as a major force behind international competitiveness, growth and employment creation. At the same time it has also emerged as a central concern behind the rapid rise in unemployment in the late 1970s and 1980s and the persistence of such high unemployment rates in Europe. The rapid introduction of new technologies, spurred by Europe's own integration process and the broader globalization process, has often been singled out as one of the main factors behind widespread employment displacement and job losses. At the aggregate level such employment displacement obviously becomes much more visible in periods of sluggish growth and recession; at the sectoral level it will often be concentrated in particular industries; at the individual job level it will affect some workers with particular routine skills much more than others. New technologies have also been held responsible for increased wage and income disparities and work insecurity more generally. No single country seems to have escaped these pressures, even though some have been much more successful in coping with them than others.

Yet the concerns about the employment and distributional implications of contemporary technological change, as has been emphasized by many economists over the centuries, are not really based on any historical precedents. Concerns about so-called *technological* unemployment have a long track record, predating even the Industrial Revolution. In general terms, it could be argued that those predicting the employment implications of a particular set of technologies have systematically tended to overestimate the visible, direct employment displacement effects of the substitution of labour for capital and to underestimate the various indirect employment compensation effects which are likely to operate in the economy at large and in the longer run. In many ways, this analytical bias towards employment displacement is part of the much wider tendency to assess a particular new technology first and foremost in terms of what it now can do more efficiently and better. Ultimately, this was of course the aim of the research of the inventor. This inventor's bias is often reflected in the first, original name given to the particular new technology: the radio or 'wireless' replacing the wired telegraph; the car replacing the horse and cart … or even today the notion of 'electronic commerce' conveying the impression of replacing first and foremost 'physical' commerce. As many economists, particularly in the crisis years of the 1930s or 1980s illustrated, it is particularly difficult not to come up with significant employment displacement results if account is only taken of these substitution effects, the price and income elasticities being generally too low to bring about such output growth to offset the productivity gains realized thanks to the new technology. It is indeed as difficult for the economist as for the technologist to imagine and estimate the employment impact of possible new applications of the technology in areas far removed from the original technology and bringing about new growth and demand.

From this perspective, the early postwar decades of fast economic growth and strong productivity gains leading to a rapid return, after the end of World War II, to full employment in the 1950s and 1960s, with a more or less perfect institutional mix between the distribution of the gains in productivity into higher wages and a large and widespread unsatisfied demand for standard, homogeneous goods, were rather more an exception than the rule. While such advances helped to clear the suspicion associating technological change with unemployment, they also gave full credit to analyses where output compensation effects depending on price and income elasticities, as well as indirect employment creation effects in the new capital goods-producing sector, and possible substitution effects following wage adjustments, would practically automatically result in full compensation of the initial destruction of jobs and might even have positive net effects.

As a result, the dominant set of economic arguments put forward in the early 1980s, when the employment crisis became visible in practically all OECD countries, emphasized in particular the imperfections in labour and capital markets to explain the fact that compensation effects did not materialize and that various forms of 'technological unemployment' became apparent. Practically by definition, such illustrations of structural unemployment became translated into market frictions: thus the notion of a 'natural' rate of unemployment or of the non-accelerating rate of unemployment (NAIRU) became popular. This view was undoubtedly also the one most clearly reflected in many of the official and less official European and OECD studies carried out in the late 1980s and culminating in the 1994 OECD Jobs Study, which most explicitly blamed the over-regulation of European labour markets as being at the core of the European unemployment problem. The successful activation of labour market policies as introduced in a number of European countries, with in some cases significant reductions in unemployment rates (with Denmark and the Netherlands as the most evident cases), undoubtedly lends strong support to this view. The monitoring of individual European countries' employment policies, as proposed at the Luxembourg summit in 1997, and since then part of an annual evaluation procedure by the European Commission, highlights further the broad current policy consensus on Europe's failure to generate sufficient new employment opportunities.

The approach chosen in this book, however, is a different one. As early as the mid-1970s, in response to the then standard Keynesian view, as most explicitly expressed in the OECD McCracken report, that the oil shock-induced crisis was a temporary phenomenon, a number of economists, including both the present writers (Soete, 1987; Clark *et al.*, 1982; Petit, 1986) emphasized the underlying structural nature of the emerging employment crisis, associated with a different path of technological accumulation and growth. We continue to do so. By focusing on the many structural features of technological change, we will thus depart sometimes from the standard market equilibrium view, to insist on the many underlying learning processes that are associated with a number of new structural changes associated with technological change. These learning processes are rarely smooth and easy. They will involve various coordination problems: how to bring about changes on the shop floor, at the organization level of production and distribution, at the level of consumers' willingness to adjust habits and (re)learn, at the institutional level – all in a mutually reinforcing, synergetic way?

From this perspective, the widespread slowdown of productivity gains over the 1980s and 1990s is for the standard economists maybe a 'paradoxical' phenomenon, often used, at least until recently, to play down the significance of the technological transformation. In our view it is rather a plausible, historically analysed phenomenon. Thus, rather than denoting a slowing down of technical progress or a lack of innovations, as many standard neoclassical growth analyses were claiming as recently as the late 1980s (see, among others, Englander and Mittelstadt, 1988),[1] such a trend appears in our view more indicative of the pervasiveness and diversity of the many mismatches between different learning processes: of agents, of organizations and of institutions. These mismatches can even be found reflected within the economic profession itself, with a burgeoning amount of detailed, micro-based studies, highlighting the many success stories of firms or sets of firms having used new information and communication technologies in productive ways (see, for example, the Washington Conference of the Department of Commerce, March 1999), yet with very little macroeconomic evidence, with the notable exception of the US computer industry (Gordon, 1999), to put forward.

In Table 1 we try in a very approximate and crude fashion to summarize the changes in the perception of the relation between employment and technology. The table highlights how, in the 1980s and 1990s, the lack of a clear,

Table 1   Technical change and employment: a reversed relationship?

	Technical change	Productivity	Growth	Employment
1950s and 1960s	+	+	+	+
1980s and 1990s	+	?	?	?
Question		Paradox of productivity slowdown	Measures in real terms	Arbitrage between unemployment and 'poor workers'

positive relationship between technical change and productivity, the productivity paradox discussed above, is part of a broader set of questions with respect to the relationship of technology and growth and the increasing measurement problems of output, particularly in some of the most rapidly growing new service sectors. The unclear relationship between technical change and employment creation can then be viewed in part at least as depending on a trade-off between downward wage adjustment and the creation of low wage employment which can be characterized as the working poor, or a growth in unemployment as a result of the elimination of routine jobs.

On the surface, Europe, at least in aggregate terms, appears to have followed this second line and the USA the first. The contrast between the European and US experience over the last decade is indeed particularly striking. The USA experienced in the 1990s a surprisingly long economic upswing, unemployment today is at a record low level and a high volume of employment has been and continues to be created every year. Japan, which was the growth and employment success story, very much reflecting the 1950s and 1960s set of positive relationships indicated in Table 1, has now experienced slow growth for over a decade and is facing the problem of unemployment, with registered unemployment approaching 5 per cent in 1999: a totally new problem for a country which enjoyed 'natural' full employment for four decades.

There are of course many reasons which might be invoked to explain these diverging trends over the last decade amongst the so-called 'triad countries'. In this book, as argued above, we focus first and foremost on technology (and the technical and organizational knowledge that goes with it): the most important, if not unique, factor in explaining long-term growth. From the technology perspective, the growth divergence amongst the USA, Europe and Japan in the 1990s appears, at least at first sight, paradoxical, given the increased international access to and diffusion of technology. Most politicians, businessmen, technology experts and economists, whether of US,

European or Japanese origin, have very similar expectations with respect to the long-term growth and employment creation potential of new technologies, particularly information and communication technologies (ICTs). These expectations are built at least partly on the historical record of the overall positive employment impact of previous waves of technological change in the postwar period, as we saw before, and partly on the special features of current ICTs.

Three such features appear of particular relevance today: the dramatic decline in the price of information processing, the convergence between communication and computer technology, and the rapid growth in international electronic networking, all of which are likely to boost innovation in organization, in changing market practices and in transforming work. While different concepts or terms are still used in each country ('electronic highways', the 'knowledge-based economy', the 'global information society'), they all point to a rapid increase in the information and knowledge base of the economy closely associated with electronic networking. The more pessimistic or sceptical views stress rather the problems involved in using these new technologies: problems of organization, of security, of privacy, of taxation, of (intellectual) property that may all, in one way or another, block or delay the growth and expansion of these new uses and activities (see Litan and Niskanen, 1998). In general, the debate ends up in some vague consensus among policy makers, businessmen and academics that there is a potential to reap but that severe traps or deadlocks (or 'lock-in') must be avoided. Here, though, comes the real dividing line between those who believe that such caveats can only be met by 'free' market forces and those who believe that the challenges imply some policy interventions and adequate institutional changes.

From this perspective the comparison between Europe's poor experience and the successful US experience is particularly revealing. Free marketeers praise this example as the proof that no intervention is the rule to follow to take advantage of ICTs. They argue that Hayek's argument that the world is too complex for policy makers to do anything else than messing around in a counter-productive way is more valid than ever, precisely because of the very decentralized and pervasive nature of the new technologies. Interventionists, on the contrary, stress that the US example illustrates rather the opposite. The US economy, or more exactly its ICT sector, continues to enjoy a rent on innovation as leader in an area where the US government and administration have been quite active (large military programmes, spillovers from military networks, semiconductor trade agreements, inspirers of codes of conducts and of intellectual property rights, and so on). Secondly, and more importantly, US growth has been accompanied, at least as revealed by aggregate indicators, by an impoverishment of part of society, to an extent which might question the long-term outcome of the US growth pattern. This debate over the 'dualist' potential of ICTs is all the more relevant since it reflects a general trend in all OECD countries, with the USA being the 'worst practice', most affected case so far.

This more or less sets the scene for the issues addressed in this book. What is the potential of the so-called 'new' technologies, particularly ICTs? What various patterns of growth and employment creation, and what sort of employment can countries expect to be created following the introduction of these new technologies? How should firms behave to adjust to more favourable production and distribution patterns? How should policy makers? How far does one rely on national intervention, or regional policy action? Or does one respond only to global actions?

Of course this book does not pretend to produce all the answers to questions which are in our view still very dependent on the course of action that will be taken at all the levels previously mentioned. But the book seeks to show how to address the questions at each stage, how to disentangle experiences of countries from their specific institutional 'path' advantages and disadvantages. It is very telling in that respect to look at common changes and we stress in that regard the common interrelated structural changes that are specific to the historical phase that developed economies are currently facing. Secondly, we underline the strong sectoral dimension of the problem, because of the role of services both as a pole of employment and as a central logistics base for the new network economy. But the changes have also strong organizational dimensions which concern the firms and the way they articulate reorganization of the workplace and market links. Finally, we shall also address the questions of the institutional context in which these organizational issues and policy choices are being raised.

We now focus in more detail on each of the four sets of structural change issues along the lines of the four parts into which this book is divided, namely aggregate structural change at the world level, sectoral changes, organizational changes and, finally, institutional changes.

2.   CONTEMPORARY STRUCTURAL CHANGE AT THE WORLD LEVEL

We start Part I of this book with Freeman's chapter, which compares the dynamics of growth and employment in the triad countries, and underlines the impact on growth patterns by the financial sector. Freeman compares Europe as a whole with the USA and Japan, and asks whether specificities of service industries in Europe account for its lagging behind in the triad. His analysis underscores the fact that large differentials in the rate of technical change across sectors have led to a wide dispersion in growth, price and quality trends across sectors with, as a consequence, continuous structural shifts in the demand and supply of old and new commodities. The shifts in employment in practically all European economies from agriculture and manufacturing to services, the latter sector employing today some two-thirds of total employment, are a striking illustration of such differential impact. However, Europe, the USA and Japan have experienced rather different trends within this broad development framework (trends which are surveyed in Chapter 3).

The USA seems to have a pre-eminent position with regard to the mastering and diffusion of new technologies, but also in terms of international outreach and business service logistics, all of which appear to have contributed to restoring its hegemonic position. Conversely, Japan and Europe are in positions differing on many grounds, which cannot be easily interpreted in simple terms of catching up. Japan is slowly recovering from the challenge represented by the reorganization of its own system of production under the pressure of internationalization and tertiarization, while Europe, less different as a whole from the USA, may be too hampered by its internal diversity and difficulties to take advantage of the scale and scope economies that should be brought about by the Union, particularly in knowledge-based activities and services.

Freeman, in his contribution to this volume, insists on the specific roles of leaders and followers in this process, as well as on the complexity of the catching-up process. The recent crisis in East Asia is cited as an example of such complexity, resulting from a specific feature of the present phase of tertiarization, the role of financial institutions. If the finance industries play a key role in monitoring the high levels of risks of the present worldwide diffusion of technologies, then countries developing in all areas but with a weak financial sector (of which Japan seems to have given a good example) are effectively running extraordinary risks bound to be fatal at some stage. Furthermore, even the leader country can be forced by competition from followers to engage in risky activities. The actual level of risk associated with a new economy, which has many of the hallmarks of a bubble economy, makes it all the more compulsory to have a central coordination of macroeconomic policies, acting as a G7 lender of last resort to rescue the world economy from systemic crises.

Because services are, by their development and history, very much country-specific, the emergence of a new growth regime is marked by national specificities, although one might have expected the opposite in times of increased internationalization. In fact, there will be both factors of international differentiation and factors of international harmonization in the strong structural changes that are manifested by the globalization and the tertiarization trends depicted above.

The dynamics of employment will then depend on two central issues: first, how firms will organize work internally, and, second, how welfare policies and local customs will influence the participation rates that remain so widely different in most of Europe and in the triad. These two issues will be of central importance in the next two sections of the present chapter. Regarding participation rates, the low level of the employment content of economic growth in most European countries is significantly linked with their relatively low employment ratios (percentages of large age groups, such as 25–55, in the labour force, are on average lower than in the USA or in Japan).

This points to a major characteristic of Europe as being a zone where people work less, both because there are more unemployed and because their levels of activity are low. The reasons for low levels of activity are relatively straightforward, however. On the one hand, the pursuit of studies tends to last longer for men and women alike, and, on the other hand, Europeans enjoy a much earlier retirement age. On top of that, one could point to longer holidays than in the USA or Japan. Some try to stress the bleak side of these figures, arguing that earlier retirement is due to a lack of jobs, as is the later entrance into the active population. This, however, may be looked upon as a transitory phenomenon linked with the mass unemployment rates of the 1980s. Others may look at it in a more positive way, as a voluntary choice in favour of a certain quality of life. However, low participation represents (as Mjøset argues in Chapter 15 of this volume) a welfare state luxury. Either it is wanted and perceived as a priority or it becomes increasingly unbearable and undermines the whole welfare fabric of European society. Increased participation, in so far as it monetizes non-economic activities, represents a double dividend advantage to Europe's sophisticated welfare states: fewer people to cater for and more social revenues.

The relative weight of consumption and leisure in our highly developed economies has, however to be kept in mind on at least two grounds. First, it is a favourable condition, if not an opportunity, when questions of education and training are raised, as it presents advantages in terms of lifelong learning, alternative schemes of formal and on-the-job training and other themes which are recurrent in the basic requirements of a new economy. Second, it opens up opportunities for the organization of service activities, at a time when user–producer relationships are becoming more intense.

These issues will have to be borne in mind when considering the outcomes of the choices of firms regarding their internal organization of work, or the perspectives given to welfare policies. Having stressed that the new role of services will go hand in hand with the modernization of the other activities in a new mix, where both relations between firms and work organization within firms will have been thoroughly revised, we turn towards the key issue of work organization.

3. SECTORAL GROWTH PATTERNS AND THE DYNAMICS OF EMPLOYMENT

Manufacturing has long been considered as the engine of growth, because of its capacity to organize and restructure production in ways allowing steady productivity gains (see, for example, Cornwall, 1977, and Fagerberg and Verspagen, 1999, for a more recent empirical assessment). Static and dynamic economies of scale, such as replication on a larger scale of production processes, or dynamic learning effects from cumulated experience and incremental innovation, have been the way to sustain this process. This went together with the old classical Smithian principle that large markets allow for a more extensive division of labour. Young (1928) insisted on the fact that such division occurred both within firms and between firms, that it stimulated (incremental) technological change, which in turn stimulated demand so that economic growth propagated itself in cumulative ways. This was also basically the mechanism referred to by Kaldor and later post-Keynesian scholars when putting forward the notion of manufacturing as an engine of growth (see McCombie and Thirlwall, 1994).

Fagerberg and Verspagen (1999) conclude that the role of manufacturing in overall growth of the developed countries has declined over the last decades, while it remained strong in some of the newly industrializing countries. In light of the discussion in the previous section, which stresses the increasing role of services as suppliers of high-quality inputs into the manufacturing process and other parts of the economy, one might be tempted to ask whether (business) services can be the new engine of growth for the whole economy. The industrialized economies have effectively experienced a significant shift away from manufacturing to services, as the increasing shares of the latter in nominal gross domestic product (GDP) and in employment show. Still, the general slowdown in productivity and in output growth does not allow one to interpret this shift towards services (as yet) as the source of a new engine of growth. Moreover, the forces underlying these sectoral trends remain a matter for debate (for example, see Pianta's contribution in Chapter 5 of this volume, as well as von Tunzelmann and Efendioglu in Chapter 2).

There are two sides from which one may look at the shift from manufacturing towards services: deindustrialization or tertiarization. Landesmann and Stehrer in Chapter 7 show that a sizeable group of continental European countries displayed a significant speeding-up of the deindustrialization process during the mid-1980s and early 1990s, a move that did not show up in the UK or in the non-European OECD economies. In addition, a majority of European countries experienced a sharp slowdown in the rate of employment absorption in the community and social services sector (the sector with the largest employment share), which contrasts with the high employment rates of this sector in continental Europe over the period 1975–85.[2] Thus, according to this interpretation, changes in rates of unemployment in Europe could be explained mainly by the lagged deindustrialization of most continental countries (completed earlier in the UK and the USA, and more recently in the Netherlands) and by a slowdown in the growth of employment in welfare services. Overall, these results point to a process of adjustment of the dynamics of employment of continental Europe to the dynamics observed in other OECD countries. This adjustment concerns changes in the strategies of both manufacturing firms and welfare policies. More than a process driven by the expansion of services, it manifests itself in a reduction of the room for manoeuvre of continental European countries, or in a convergence of continental Europe to global trends.

However, the process of deindustrialization may be overrated because in fact what has happened is a growing outsourcing of tertiary tasks. To check this effect, Schetkatt and Russo in their contribution to this volume (Chapter 4) implemented the concept of vertically integrated product sectors (which in fact corresponds to aggregating all the stages of production into the lines of final products), by using an internationally comparable input–output (I/O) database. Their analysis decomposes the changes in output into several effects, of which the first quantifies shifts in the final demand structure of the economy. In this respect, a shift towards services industries is most pronounced in final household consumption, partly compensated by reverse trends in exports. This effect turns out to be rather sizeable, but other variables are important as well. Outsourcing does appear to have gained in importance over the period of the 1980s and early 1990s examined, and contributed to the shift towards services industries. However, it amounted to only a minor part of the overall shift. Moreover, there are some important and surprising differences between countries when looking at the shares of intermediate inputs of services (relative to gross output). Overall, the conclusion from this analysis of I/O statistics is that technological developments, rather than shifts in the inter-industry division of labour, drive productivity increases.

What this points to is that a simple 'accounting' approach to the process of tertiarization and the growth of business services does not suffice. Beyond the real problem of measurement of these services activities (for example, see Griliches, 1994) it seems that their impact has to be analysed in terms of a 'deeper' level of technology and organizational dynamics.

4.   SKILL AND WORK ORGANIZATION ISSUES

Technical change, that is, the emergence of ICTs, as well as the other structural changes that accompany it, have challenged the work organization of firms. This organizational issue is crucial to understanding both the slowdown in productivity gains and the potential of a new growth regime that would rely on ICTs and on the exploitation and accumulation of knowledge that they allow. This major question of organization, which concerns consumption activities as well as production activities, has led to heated societal debates, connected to a process of trial and error, which has more than ever been left to private agents. It contrasts with previous changes in the technological system, whose implementation required public interventions to establish frameworks, norms and infrastructures. Never has such a major transformation been left to the coordination of market forces, which in addition are taking place worldwide. That is basically why organizational issues are so crucial in the appreciation of the emergence of a knowledge economy.

Debates on this issue have been linked to the demand by firms for employment that is more flexible, more skilled and more adaptable to new qualification requirements. The significance and importance of this skill-biased technical change have been the subject of numerous econometric studies, aimed at disentangling effects so closely linked as those of trade and technology. The underlying causal relationships at work have been addressed much less. This is a main reason why the great majority of the contributions to this project have focused on the complexity underlying the demand for various categories of qualifications, technological change and international competition. Theoretical modelling has been one way to track some of the effects that these organizational issues may bring to the surface. Here again, implications in terms of policy may have to be differentiated to account for the diversity in the organization principles prevailing in European countries.

In general, continuous organizational changes take the form of a move towards more 'organic' forms of workplace organization. This includes more responsibility being awarded to workers at the lower layers of the hierarchy, a more collective organization of work, more horizontal communication and a greater variety of tasks being performed by each single employee. Moreover, new forms of workplace organization tend to be associated with a high technological intensity. Technical change and organizational change go hand in hand.

With the help of studies coming from the economic literature, but also from sociology and management, Caroli in her contribution to this volume (Chapter 8) highlights some possible sequences of choices between technical and organizational change. Recent evidence regarding the spreading of information technologies and new work practices suggests that organizational changes have largely taken place in response to technological evolution. This is due to the combination of ICTs calling for, as well as permitting, new forms of workplace organization and strong organizational inertia. In a dynamic perspective, this ends up in technical change providing a crucial impulse to organizational evolution. However, organizational change is far from being an optimal response to technical change. The implementation of new HRM (human resource management) practices heavily depends on labour–management relations and on cultural as well as institutional factors. This leads Caroli to develop a rather evolutionary approach whereby technical and organizational changes are shown to coevolve. Debates over which of the two – technological change or organizational change – determines the other then turn out to be irrelevant.

Disentangling the roles of technical and organizational changes in continuing microeconomic evolution is particularly important in the current debate about the so-called 'skill bias'. Recent econometric studies indicate that organizational change is actually as much skill-biased as technical change. The analysis suggests that organizational changes call for more general education, whereas technological changes could be coped with by relying on specific training. In short, organizational change would be knowledge-biased, while technical change is essentially skill-biased. Not surprisingly, the room for manoeuvre left to human resources management by firms thus depends heavily on the supply of educated and skilled labour provided by the educational system and the labour market. But the balance and ability to substitute one kind of resources for another may not be the same in the new context. Interactions between the conditions of labour supply and work organization by firms have, for this reason, been central to the questions raised in most of the contributions in this book.

Gatti in Chapter 9 of this volume has taken a broad perspective in modelling how different microinstitutional conditions at the firm level would interact with various kinds of labour markets, leading to a possible explanation for differential rates of unemployment observed in Europe, Japan and the USA. The definition of these different microinstitutional conditions encompasses (a) the structure of the firm, (b) the system of vocational training, and (c) the nature of labour markets, making it possible to build a model of equilibrium rates of unemployment in an efficiency wage framework. The results of this mainly theoretical exercise point to the important role that can be played by the mode of organization of the firm and the nature of workers' competencies in determining the rates of unemployment, as compared with the relatively minor role played by the flexibility of the labour market, so often referred to as the central cause of unemployment in Europe.

Again, this explanation does not rule out the fact that characteristics of labour supply may have a sizeable impact. In particular, tertiarization boosted by a widespread increase in the level of education of the labour force (as manifested in the average number of years spent at school or at university) is certainly a key issue in determining both the kind of work organization used by firms and the kind of labour market that will develop. Van Zon, Meijers and Muysken focus their contribution to this volume (Chapter 10) on this issue of skill allocation. Their model investigates the influence of asymmetries in substitution possibilities between skills, but also of asymmetries in learning capabilities between workers of different skill levels. Labour demand is defined in terms of jobs, which have to be matched with the heterogeneous supply of skills. The putty-clay[3] model of allocation is neoclassical in nature, except for the asymmetries mentioned above, and takes into account the relative efficiency of the labour force on the jobs (meaning that a skilled worker in a low classified job has, nevertheless, a greater efficiency).[4] The model also describes how a 'chimney' effect allocates labour supply to meet the demand of firms.

Empirically, the model is implemented using a skill allocation model estimated for Germany and the Netherlands. The analysis suggests that 'upgrading' job requirement in the Netherlands is rather successful in the manufacturing sectors, but much less so in the tertiary sector. This implies that speaking of a skill bias would be justified in manufacturing activities, but much less in services in the Netherlands. The results are much less obvious in Germany, although results are hinting at similar phenomena. These are important qualifications in the debate over the skill bias nature of technical change, which would be worth checking for other countries. One has to keep in mind, however, that measures of relative efficiency are entirely based on wage differentials, which may also reflect a general practice of differentiating wages in favour of more educated workers, whatever their occupation is.

Following up these analyses of the effect of structural changes in the labour supply, it is important to check how the three facets of structural change affect wage formation. In that respect, the above analysis left out the issue of internationalization, while it is often claimed that increasing trade and foreign direct investment (FDI) with low-wage countries had a detrimental impact on the relative wage of low-skilled workers. De Loo and Ziesemer have in the present book precisely tried to estimate all the relative effects of trade, technical change and shifts in labour supply on employment (see Chapter 11). They also accounted for the impact of changes in interest rates conditioning investment flows. The estimation of their perfect competition model for 67 combinations of countries and sectors yields the result that technical change explains a higher percentage of both wage and employment growth than changes in the terms of trade do before the 1980s. From the 1980s onwards, international trade is slightly more influential than technical progress. Much more important than these two, however, are changes in the sector-specific labour supply in all countries but the UK (where terms of trade changes matter most).

These results (following an assumption of perfect competition but apparently robust to an assumption of increasing returns) seem to re-enforce the

idea that, over the 1980s and 1990s, which is the period we are concerned
with, much of the assumed skill bias is linked with a shift in the supply of
labour. It is consistent with the emphasis placed in the project on an aspect of
structural change that also has major policy implications. An important issue
is then to know whether the prevalence of this labour supply effect will lead
to an increasing differential between wages of high- and low-skilled workers.
In other words, is the growing divergence that one may observe in the USA
an inevitable part of the process that European countries are going to face?

The contribution to this volume by van Zon, Sanders and Muysken (Chapter 12) suggests that there is nothing inevitable in these evolutions. Using a
model similar to Krugman's North–South model of international trade, but
with two sectors within one country, they show that, if technological change
can be the cause of wage divergence between high-skilled and low-skilled
workers, different constellations of technology and labour supply characteristics by country may induce different labour market regimes regarding the
evolution of relative wages. This theoretical model recalls that countries are
not 'condemned' by contemporary technological change to wage divergence
and increasing income inequalities.

Summarizing the work in this area of work organization and skills, the
analysis suggests on various accounts that education or, more broadly, issues
of labour supply, have a major impact on employment trends. This leads us to
analyse the impact of institutional change implied by the transformation of
educational systems and the broader institutional context influencing the
quality of the labour force and its participation rates.

5.   INSTITUTIONS THAT MATTER

We have stressed above that the diversity in organization principles between
sub-sectors of services or between firms is borne by the institutional variety
of European countries. The previous section also insisted on the role of
labour supply that hinted strongly at the institutions monitoring this supply.
We also mentioned that this diversity was a real challenge for Europe as it
was much more difficult to design any policy of institutional change for such
a fragmented institutional fabric. However, diversity may also have some
advantages in offering institutional frameworks closer to issues that have
strong local links such as education, labour supply and inter-firm linkages.
When new synergies are required to take full advantage of local assets, such
diversity may help, although diversity does not vanish with the imperative
need to adapt the old structures to the new situations.

We shall illustrate some of the problems raised through three different
perspectives. One is concerned with the social organization of working time,

which is obviously a principal dimension in the organization of labour markets. A second has to do with the evolution of national systems of innovations, a field that has given way to numerous comparative analyses of institutional structures. We have taken advantage of these studies, attempting to illustrate the broad kind of challenge that lies in the provision of the proper human resources, the importance of which was stressed above. Third, we have tried to link our preoccupations on institutional matters with the question of the future of the welfare state. Not only does the welfare state influence and encompass as a general framework the systems of education and innovation which we have just mentioned, it also conditions two big issues raised earlier, namely the dynamics of social and community services, which remain a cornerstone for the dynamics of employment, and the redistributive issue in an era where divergence in wages and incomes tend to increase inequalities. Organization of working time, systems of education and innovation, and the evolution of the welfare state are not the only institutions that matter in the shaping of contemporary new growth regimes, but they are clearly the issues closest to the points made in the previous sections of this chapter.

The reorganization and the reduction of working time are currently on the policy agenda of most European countries and of the European Union as a whole. Working time, once all its dimensions are accounted for (legal full-time, all-year-long distribution of work periods, conditions of part-time work, and determinants of participation rates) is an essential indicator of welfare and of the functioning of labour markets. Not only is the total amount of working time an appropriate measure of the employment possibilities existing in a country, but per capita working time is also a measure of quality of life of its workers. It is also a central issue in the organization of services, and the use of ICTs in those activities where time is a more directly binding constraint.

Spiezia and Vivarelli in Chapter 13 draw up a balance sheet of the main changes in the patterns of working time in Europe over the last 35 years, with respect to three important aspects. The first is the interrelation between the dynamics of total working time and per capita working time in determining the dynamics of employment. It shows that, on the one hand, if total working time is given, shortening working hours is a necessary condition for an increase in employment, but the increase in hourly labour cost that it carries with it may reduce the total working time required. On the other hand, if more free time leads to an increasing time spent on some activities, especially connected to services (be it leisure, education or health), then this may counteract the negative effect of a rise in labour cost, not to mention the increase in productivity brought about by a reorganization of work done in the process. One may think that there are few productivity gains of the sort to be gained by further reduction in working time (although the issue is more open in the service sectors, where reorganization can spread out a lot of part-time jobs or take advantage of shorter full-time jobs to reorganize front-office activities if not the back-office division of labour). More promising, however, seem to be the opportunities presented by the uses of free time. Much depends in that case on the institutional framework and on the specific policies that each country will be able to launch. Policies so far have not been very imaginative and the focus on competitiveness has led to a too large focus on the cost effect of a reduction in working time. Very little has been said about the advantages and limits of an increase in non-working time. The idea that it will simply disappear into the sands of such inert occupations as watching TV is quite common, although this has been challenged, if only because of the attraction of the more interactive Internet. As we stress in Chapter 6, there is plenty of room for policies to stimulate productive uses of free time (for oneself and for the common good).

The fact that a good number of European countries favour clear improvements in urban life styles may even be considered as a comparative advantage. The potential for coordinated adjustments in time budgets is a central issue. The relative inertia, if not reluctance, to change them is in large part tied to the threat to low incomes presented by the evolution of labour markets. This is deeply counter-productive and cumulative processes in the form of incentive plans to get out of such traps should be initiated. Again, the evolution of labour markets under the spell of ICTs may help in that respect.

National systems of innovation, which form the second institutional perspective taken here, encompass a broad range of issues. In a way one could say that the concept aims to capture all the institutions that one way or the other contribute to improving and organizing the human resources in a country, in order to promote economic growth and welfare in societies where these resources have obviously become the central asset of development (a way to paraphrase what is currently qualified as knowledge-based economies). In that respect, these national systems shape the various aspects of labour supply discussed above. Not only does the concept stress similarities and divergences between countries but, as these systems are very much policy-oriented in essence, they help to question what kind of tools could be considered and which policy target to select.

Such advantages come clearly out of the contribution of Amable and Boyer in Chapter 14 of the present volume. The authors undertake to cluster scientific, technological, social and financial institutions in order to derive configurations of these institutions that can be interpreted in terms of some key principles of adjustment: the market, the social negotiation process, the leading role of large companies, or a complete set of public interventions. The diversity of systems of innovation within Europe emerges from this analysis as a potential problem for the cohesion of the Union and the efficiency of European policy, whether in the area of science and technology or more broadly. At the microeconomic level, the institutional arrangements present in the different countries provide different sets of incentives to agents in terms of opportunism or trust, short-term or long-term planning horizons, flexibility and so on. Therefore, one may expect not only very different patterns of investment, types of specialization and forms of innovation, which is what is found, but also very different patterns of response to structural policy impulses.[5] To give an example, a set of European research policy measures based on market arrangements is more likely to have strong responses in the market-based countries than in other countries. Or, to go one step further, a competition process based on price competitiveness is likely to be more detrimental to countries which have highly elaborated sets of non-market mechanisms. Conversely, large old-style European technological programmes may be of benefit mainly to the public institution-based model, to the detriment of other systems with different features. The challenge to any European policy is therefore to promote this institutional diversity, while creating more and more complementarity among economic specialization patterns of the individual countries.

From this perspective, one should not underestimate the impact of the European monetary integration process on the dynamism of innovation, as a method for reconciling dynamic efficiency and the preservation of an extended welfare system. Back in the 1970s, an economy that experienced an erosion of its competitiveness could always adjust the exchange rate. During the last two decades, the objective of stable exchange rates among European countries has been the target of various policy strategies, the most frequent being a movement towards an increase in labour market flexibility. But an unexpected currency crisis could always, within one day, erode the structural competitiveness built up by a country over a decade or more. The Amsterdam Treaty is clearly putting an end to such possibilities. If a pure defensive strategy in terms of wage reduction and slimming down of the welfare state is to be prevented, a strong and dynamic innovation policy has to be promoted, both at the national and European level.

This discussion on the importance of the coherence and dynamic efficiency of institutional arrangements points to another set of policy recommendations in addition to those aimed at industrial specialization. This is why a differentiated set of policy measures adapted to the particularities of the different systems of innovation may be more apt for the present European situation. One may object that such a differentiation would neglect the process of integration within Europe. It is true that EU scientific programmes and measures favouring human capital mobility have taken steps in the direction of a unified Europe, but the integration is stronger in the goods market and in the area of monetary policy, and this has created, so far, few echoes in the institutional fields relevant to systems of innovation. There is no unified financial system for Europe, in spite of a very strong movement of financial liberalization. Besides, this pattern of unification towards a more financial market-based system may not be the improvement it is meant to be, particularly in terms of financial fragility and investment project-monitoring abilities.

Large European industrial and technological programmes have been, so far, more successful when they corresponded to a public action organized around some large projects than in the promotion of networking and more decentralized industrial and technological integration. But this problem is not specific to the Union and is present at the national level too. This is more a matter of adapting public policies to changes in the forms of innovation and technological competition, at whatever level, than a strictly European problem. The lack of coordination between member states and their respective policies may exacerbate the problem, but it did not create it.

However, the most telling examples of institutional changes that have been implemented following a policy programme are given by the emergence of fully-fledged welfare states in the immediate postwar era. This brings us to the final point of this section. Obviously, these welfare systems have not been created overnight in the aftermath of the war. They are all rooted in the past history of each nation. How they took advantage of this past to develop new forms and how they are under strain in present times and have to reform themselves are crucial issues for the contemporary changes and their coordination at the European level. This context also strongly conditions the dynamics of employment in the sector of personal and community services. Mjøset investigates in Chapter 15 the various forms taken by these welfare states in Europe. Esping-Andersen's (1990) tripartite typology of welfare states (Scandinavian, continental and southern European) is found to obfuscate central issues in the current context. For example, it does not help to distinguish between France and Germany, while the trajectories of these countries are obviously specific. Still such broad characterizations are useful. They broaden the range of 'ideal types' of welfare states to which we can refer to target global policies. A detailed comparison between the French and German welfare systems shows how different these systems are, especially if one wants to reform them. Policies will have to find precisely differentiated ways. Moreover, central issues in the evolution of the welfare state go beyond the changes in education and health provision systems, the changes in pensions systems and more generally all the changes accompanying an ageing of the European populations. This issue is clearly linked to those resulting from changes in participation rates. Projects on the future of societies are clearly as important as the drive to ensure the competitiveness of countries in defining the future structural policies of a unified Europe.

To conclude, it is clear that the old Taylorist organizational model, whether within the context of private firms or public administrations, within traditional manufacturing or services or within the organization of work or non-work activities, is undergoing a deep transformation. Organizational and institutional changes, as underlined in the various sections above, are an essentially man-made and policy-based feature of the technology, growth and job creation dynamics. While Europe, probably more than any other region in the world, has been characterized by major institutional changes, these institutional changes appear, we suggest in Chapter 16, to have by and large side-stepped a lot of the issues raised above and in this book. The policy orientations to be drawn from our work should therefore aim to fill this gap and propose measures which would alleviate, if not take advantage of, some of the diversities which have been so neglected in the process of European integration. But that is a task, and subject, we leave to the similar diversity of European, national and regional policy makers.

NOTES

1. It might be argued that the debate about the more recent Asian miracle and the apparently small part technological change played in the rapid industrialization of the Asian NICs, as claimed by some more traditional growth economists (Young, 1995), is a modern, Asian version of the OECD productivity paradox story.
2. With some differing experiences for the Scandinavian countries at the turn of the 1990s.
3. Putty-clay meaning that, once work organization in terms of job skills is chosen by the firm, it cannot be altered.
4. The skill allocation model retained is part of a more general production structure that has an explicit vintage production structure (see Master, 1997, for more details).
5. One should recall at this stage that the 'old systems' may well have entered a phase of decline, as suggested by some of the project's contributions that diagnosed decreasing returns of global R&D expenditures in the 1980s and onwards; see Chapters 2 and 5, for example.

BIBLIOGRAPHY

Clark, J., C. Freeman and L. Soete (1982), *Unemployment and Technical Innovation*, London: Frances Pinter.

Cornwall, J. (1977), *Modern Capitalism: its Growth and Transformation*, London: Martin Robertson.

Elfring, T. (1989), 'Evidence on the Expansion of Service Employment in Advanced Economies', *Review of Income and Wealth*, 35(4), December.

Englander, S. and A. Mittelstadt (1988), 'La productivité totale des facteurs: aspects macro-économiques et structurels de son ralentissement', *Revue Economique de l'OCDE*, 10, Paris.

Esping-Andersen, G. (1990), *The Three Worlds of Welfare Capitalism*, Cambridge: Cambridge University Press.

Fagerberg, J. and B. Verspagen (1999), 'Modern Capitalism in the 1970s and 1980s', in M. Setterfield (ed.), *Growth, Employment and Inflation*, London: Macmillan.

Fagerberg J., P. Guerrieri and B. Verspagen (eds) (1999), *The Economic Challenge for Europe: Adapting to Innovation-based Growth*, Cheltenham, UK and Northampton, MA, USA: Edward Elgar.

Freeman, C. and L. Soete (eds) (1987), *Technical Change and Full Employment*, Oxford: Basil Blackwell.

Gordon, R. (1999), 'Has the New Economy Rendered the Productivity Slowdown Obsolete?', Working paper, North Western University, May.

Griliches, Z. (1994), 'Productivity, R&D and the Data Constraint', *American Economic Review*, 84(1), 1–23.

Litan, R.E. and W.A. Niskanen (1998), *Going Digital*, Washington, DC: Brookings Institution Press.

McCombie, J.S.L. and A.P. Thirlwall (1994), *Economic Growth and the Balance of Payments Constraint*, London: Macmillan.

MASTER (1997), 'Final Report to the Commission of the EC', MERIT, Maastricht.

OECD (1997), *Employment Outlook*, Paris.

Petit, P. (1986), *Slow Growth and the Service Economy*, London: Frances Pinter.

Petit, P. and L. Soete (1997), 'Is a Biased Technological Change Fuelling Dualism?', working paper, CEPREMAP, Paris, December.

Setterfield, M. (ed.) (1999), *Growth , Employment and Inflation*, London: Macmillan.

Soete, L. (1987), 'Employment, Unemployment and Technical Change: A Review of the Economic Debate', in C. Freeman and L. Soete (eds), *Technical Change and Full Employment*, Oxford: Basil Blackwell.

Young, A. (1928), 'Increasing Returns', *Economic Journal*, December.

Young, A. (1995), 'The Tyranny of Numbers: Confronting the Statistical Realities of the East Asian Growth Experience', *Quarterly Journal of Economics*, 110(3), August, 641–68.

PART I

Long-term Structural Changes

1. Long-term structural changes: a reappraisal

Christopher Freeman

1. UNEMPLOYMENT AND 'LONG-WAVE' THEORIES OF STRUCTURAL CHANGE

Economists have conventionally classified unemployment into several different categories, and policies have typically been devised to deal specifically with one or other of these types. From the 1950s until the 1970s the aggregate level of all types of unemployment combined was generally less than 3 per cent of the labour force in most European countries, with the important exception of Italy. This rather low level of unemployment was often defined as 'full employment' and although short-term cyclical unemployment sometimes pushed the level temporarily above 3 per cent, the labour market situation in Europe in the 1950s and 1960s could fairly be characterized as one of 'full employment'.

All this changed in the 1970s, and, in the 1980s and 1990s, levels of unemployment were typically higher than 3 per cent in almost all European countries, often far above this level. As early as 1977, the OECD McCracken Report showed some anxiety about this situation, which it attributed to the OPEC crisis of 1973 and its destabilizing effect on prices. The majority report somewhat complacently assumed that the OECD countries could return in a relatively straightforward way to the halcyon days of the 1960s, with full employment and high rates of economic growth. However, a minority report already raised the issue of structural unemployment, which since then became a major focus of the European debate and increasingly of the world-wide debate, together with Keynesian unemployment.

This chapter will seek to relate the problems of structural unemployment to the nature of technical change. Ever since Ricardo's famous remarks in 1821 and the ensuing debate, economists have recognized the two-edged nature of technical change: that it both destroys old jobs and creates new ones. In general, economists have argued that the job creation effects have in the long run outstripped the job destruction effects, albeit accompanied by a steady reduction in working hours throughout the 19th and 20th centuries. Nobody has claimed, however, that 'compensation' is automatic, painless or instantaneous; as Ricardo pointed out, the new jobs may not match the old ones either with respect to skill or to location. Where the mismatch is severe and/or prolonged, economists speak of 'structural unemployment' and the problems of 'structural adjustment', although the precise borderline between 'structural' and the more usual everyday 'frictional' unemployment is not always easy to define precisely. Nevertheless, the existence of some fairly severe problems of structural unemployment from the 1970s to the 1990s is now universally recognized, as became obvious from the rapid increase and high rates of 'long-term' unemployment or male 'non-employment' in Europe.[1]

Schumpeter (1939) gave a new twist to the whole debate with his conception of 'successive industrial revolutions' when new technologies were diffusing through the productive system. In a recent issue of *The Economist* magazine (23 September 2000), Pam Woodall, the Economics Editor, compared information technology with earlier technological revolutions following this Schumpeterian conception. *The Economist* survey of the 'New Economy' argued that 'A period of pervasive structural change lies ahead. Economies will enjoy big gains overall, but these will not be evenly spread. Many existing jobs and firms will disappear' ('The New Economy Survey', p. 9). Whether or not they accept Schumpeter's long-wave ideas, few economists or engineers today would deny the enormous worldwide impact of information and communication technology (ICT). In fact, many commentators go even further and suggest that ICT is ushering in an entirely new era or 'post-industrial' society. Everyone would today accept that the extraordinary reduction in costs associated with microelectronics in successive generations of integrated circuits, of telecommunications and of electronic computers is having great effects on almost every branch of the economy, whether in primary, secondary or tertiary sectors. As *The Economist* survey argued, earlier new technology systems, such as steam power or electricity, had similar pervasive effects. However, ICT is unique in affecting every function within the firm as well as every industry and service. Scientific and market research, design and development, machinery, instruments and process plant, production systems and delivery systems, marketing, distribution and general administration are all deeply affected by this revolutionary technology (Freeman and Louçã, 2001).

This chapter discusses the contemporary problems of unemployment and computerization in the context of the Schumpeterian theory of long waves in the economy and in technology. Following Schumpeter (1939) and Perez (1983), the long-wave concept is interpreted as successive structural transformations in the economy, rather than simply cyclical variations in gross domestic product (GDP) growth. Schumpeter himself argued that aggregate measures, such as gross national product (GNP), obscured more than they revealed. Since in the real economy some firms, products and industries were increasing their output very rapidly, whilst others were stagnating or declining, statistical averages or compound index measures as in GDP, or even industrial production, could not convey any satisfactory picture of these contradictory trends. Schumpeter therefore advocated the study of company reports, technical journals and business histories which, he thought, could reveal far more about the processes of qualitative change in which he was interested.

If we look at long waves as successive qualitative transformations of the economic and social system, rather than a purely statistical phenomenon of alternating periods of faster and slower growth of some aggregate indices, attention should be concentrated on the loss of old types of skill and employment and the growth of new employment. Obviously, this is closely connected to new investment. Ever since the Industrial Revolution, the fluctuations in the nature, rate and direction of new investment have generated the possibility of a serious mismatch with the available numbers and skills of the workforce.

Keynesian (short 'Juglar' cycle unemployment) interacts with Schumpeterian (long 'Kondratieff' cycle unemployment). During the periods of intense turbulence when a cluster of new technologies are entering the economy, as with information technologies today, mass production technologies in the 1920s and 1930s, heavy and electrical engineering in the 1880s and 1890s, the downturns of the ordinary Juglar business cycle are accompanied by much more serious levels of unemployment than in those periods when those new technologies are firmly established as the dominant technological regime, as in the periods 1848–73 (iron, steam power and railways), 1895–1914 (steel and electrification) and 1948–73 (oil, mass production of cars and consumer durables). Christopher Dow's (1998) magnum opus on the major recessions, completed just before his death, showed big differences between the business cycles of 1950–72 and those of the inter-war period and the 1972–93 period (Table 1.1). Keynes showed no explicit interest in long waves and was sharply criticized by Schumpeter for confining his analysis entirely to the shorter business cycles. However, in his *Treatise on Money*, in 1930, Keynes did actually accept Schumpeter's idea that waves of technical innovation were one of the principal determinants of investment behaviour, with the rate of interest playing a permissive role:

> In the case of fixed capital it is easy to understand why fluctuations should occur in the rate of investment. Entrepreneurs are induced to embark on the production of fixed capital or deterred from doing so by their expectations of the profit to be made. Apart from the many minor reasons why these should fluctuate in a changing world, Professor Schumpeter's explanation of the major movements may be unreservedly accepted. (Vol. 2, p. 86)

Table 1.1   Short (Juglar) business cycles in long (Kondratieff) waves

Output trend and deviations	Long-wave upswing (1950–72)		Long-wave downswings (1920–38 and 1972–93)	
	4 'fast' phases	5 'slow' phases	4 'fast' phases	5 major recessions
% deviation of output from previous trend, height/depth of recession	2.4	–3.8	5.6	–10.9
Length of period (years)	2	2¾	2¾	2½
% of output deviation reflected in unemployment	13	16	55	51

Source: Adapted from Dow (1998).

Unfortunately, Keynes never followed up this line of thinking. Had he done so he might have extended his original theory of investment behaviour beyond the Juglar (short) business cycle. Keynes rightly pointed to the impossibility of accurate mathematical calculations of the future rate of return on innovation investments and to the role of confidence and 'animal spirits' in business behaviour. As Siegenthaler put it in his outline of Keynesian theory:

> Actors get confident not on the basis of adequate knowledge, not as a result of procedures leading to objectively superior forecasting methods, not as an outcome of individual optimising strategies of selecting and handling information ... but they *do* get confident despite uncertainty. (Siegenthaler, 1986)

In Schumpeter's theory this confidence derives from waves of technical change, when entrepreneurs and investors perceive many new opportunities for profit making from the growth of new industries. Only when numerous bandwagons begin to roll does a general climate of confidence develop and a long-wave boom take off. One of the main criticisms of 'Business Cycles' made by Simon Kuznets in 1940 was that Schumpeter did not really show how any single innovation or cluster of innovations could possibly be big enough to cause long-term fluctuations throughout the economy. As early as 1948, the Canadian economist Keirstead, in his exposition of Schumpeterian analysis, went some way to answering this criticism with his demonstration of *economic* as well as technological linkages in many *clusters* of innovations which he called 'constellations'. Nelson and Winter (1977) took the argument one stage further with their concept of 'technological trajectories' extending over long periods and embracing many innovations. Although they did not elaborate on their idea of 'generalized natural trajectories' or relate it to long waves, they did indicate mechanization, electrification and scaling up as phenomena with very long-term economic consequences. Giovanni Dosi (1982) built on this with his concept of 'technological paradigms', whilst Freeman *et al.* (1982) developed the idea of 'new technology systems' common to most historians of technology. They pointed to the fact that new technologies combined numerous innovations in consumer goods and materials with innovations in capital goods. Such 'technology systems' took decades rather than years to develop and diffuse, thus offering a possible explanation of long-wave fluctuations.

However, it was Carlota Perez (1983) who went beyond all these partial and embryonic approaches with her concept of pervasive 'technoeconomic paradigms'. She pointed out that such changes as steam-powered mechanization, electrification, motorization or computerization affected many sectors of the economy and not just a few leading industries. Furthermore, they led to a new style of management since organizational innovations were necessary to implement their numerous applications. It was a question not just of a cluster of discrete innovations in any particular sector but of a combination of interrelated innovations which had matured over a considerable period. Her theory corresponds best to the Nelson and Winter concept of 'generalized natural trajectories', which, once established as a powerful influence on engineers, designers and managers, can become a 'technological regime' with strong lock-in effects for several decades. A new technoeconomic paradigm develops initially within the old, showing its decisive technical and economic advantages already during the 'downswing' phase of the previous Kondratieff cycle. However, it becomes the dominant paradigm only after a prolonged crisis of structural adjustment, since many social and economic institutions are still geared to the old dominant paradigm. Perez thus provided an explanation for what Schumpeter had described as the 'pathological' features of the depression phase of the long wave and also indicated one of the main reasons for the persistence of high levels of unemployment during such a period of structural adjustment: new industries are growing very rapidly but the job-creating effects of investment in the new areas are not yet strong enough to create a sustained boom, while the labour-saving efforts of employers to cut their costs and maintain profits continue apace.

It is this Perez theory of paradigm change which underlies the current (1998–2000) debate on the prospects for the US economy. The optimistic ('new paradigm, new economy') school of analysts argue that the huge growth of the computer industry and related sectors, together with the radically new Internet and telecommunication infrastructure, go far to explain the relatively successful performance of the US economy in the 1990s and its relatively low levels of unemployment, and augur well for sustained future growth. The sceptics, such as Paul Krugman ('Requiem for the New Economy', 1997) and *The Economist* magazine ('Once upon a time on Wall Street', 24 October 1998), prefer a more conventional analysis, taking into account such phenomena as the US 'financial bubble' and the 'irrational exuberance' of the US stock market (in Alan Greenspan's famous phrase). This debate is obviously an extremely important one both for the USA and for Europe and for the rest of the world. The concluding section of this chapter will therefore return to it.

2. THE 'EAST ASIAN MIRACLE' AND THE EAST ASIAN CRISIS: THE INSTABILITY OF INVESTMENT AND SURPLUS CAPACITY

Any satisfactory account of structural change and employment in the late 1990s must pay special attention to the dramatic changes in East Asia, including, of course, Japan. This book is primarily concerned with Europe, but no one would deny that events in Europe are profoundly influenced by the world economy and especially by the evolution of the economies of East Asia and America. For this reason, not only this chapter but Chapters 2 and 3 also place the analysis of the European economy firmly in the context of the global economy.

This chapter concentrates on events in Asia (this section) and in the USA (section 3). The East Asian 'Tigers' are of special interest because they were until the mid-1990s some of the fastest 'catch-up' economies in the world and were characterized by very rapid structural and technical change. These countries also had rather low unemployment (Table 1.2), which was unusual for any group of developing countries, and until recently contrasted also with the EU countries, which in the 1980s and 1990s had far higher levels of unemployment. It is important to understand this East Asian transition from relatively full employment in the 1970s and 1980s to increasingly serious unemployment in the late 1990s. Finally, all this is equally relevant to the case of Japan, now experiencing levels of unemployment higher than at any time since the immediate aftermath of World War II. As the second largest economy in the world, Japan is also obviously a major direct influence on the

Table 1.2   Unemployment levels in selected countries, 1980s and 1995

	1982–92 average	1995
Germany	7.4*	9.5
France	9.5	11.6
Italy	10.9	12.1
Spain	19.0	22.9
UK	9.7	8.2
Belgium	11.3	9.9
Netherlands	9.8	7.3
Sweden	2.3	9.2
S. Korea	2.5	2.3
Singapore	2.9	2.6
Japan	2.5	3.1

Note: * German Federal Republic.

Source: EU, OECD.

growth of world aggregate demand (or lack of it) and on world trade and investment flows. Schumpeter stressed that his theory of long waves required analysis of the special characteristics of each one and the East Asian crisis is clearly one such characteristic of the present worldwide recession.

The divergence between rates of economic growth in various parts of the world was very wide in the 1980s and 1990s. Per capita growth in the 1980s was actually *negative* on the average both in Africa and in Latin America, yet in the 1960s and 1970s it was Latin American countries which were regarded as 'miracle' economies, attracting a huge inflow of foreign investment. The leading Latin American countries (Argentina, Brazil, Mexico, Chile and Venezuela) had been well ahead of both South and East Asian countries in levels of industrialization and per capita incomes in the 1950s. Yet, by 1992, the four 'Tiger' economies (South Korea, Taiwan, Singapore and Hong Kong) had per capita incomes much greater than the larger Latin American countries and even Thailand had surpassed Brazil in per capita income, estimated on a purchasing power parity (PPP) basis. It was these huge divergences in growth rates which the World Bank Report on the 'East Asian miracle' (1993) sought to explain.

It was surely right to recognize some common features in the performance of both East and South-East Asian countries, such as their export performance in the 1980s and 1990s and their relatively high rates of investment. Ever since the classic work of Maizels (1963) on *Industrial Growth and*

World Trade, economists have accepted the empirical evidence for a close link between export performance and achieved rates of economic growth. Maizels and his colleagues explained this link not only in terms of the export share in final demand but also in terms of change in competitive power associated with export growth. The World Bank Report, in emphasizing the latter point, relates it to 'openness' in access to international experience and international technology. However, while Maizels and his colleagues analysed the changing commodity composition of exports, the World Bank Report paid no attention to this point.

Maizels analysed the world trade data from 1899 to 1959, but the evidence for the most recent period is even more compelling. The growth of exports from the Asian countries in the 1980s was by far the most rapid of any part of the world, amounting to over 15 per cent per annum for Taiwan and South Korea. In the leading Latin American countries (Brazil and Mexico), it was less than half this rate. The attainment of these extraordinarily high rates of export growth was made possible by a massive *structural* change in the composition of output, investment *and* exports.

On a world scale, exports of *manufactures* have been growing much more rapidly than primary commodities for a long time, but, within the manufacturing sector, some commodity groups have been characterized by exceptionally rapid growth, especially machinery, cars and electronics. The group of com-

Table 1.3   Exports of ICT equipment* to major OECD countries, as a percentage share of total manufacturing exports to those countries, from selected countries

	1970	1980	1992
Japan	21	17	29
Germany	7	5	6
USA	14	15	22
France	6	5	6
Netherlands	6	7	8
UK	6	7	13
Taiwan	17	16	28
S. Korea	7	13	26
Singapore	20	36	65

Note: * Computers, office machines, telecommunications equipment and other electronic equipment.

Source: OECD Trade Data Base.

modities classified as 'ICT goods' (computers, telecommunication equipment, office machinery, electronic consumer goods and electronic components) was growing at 13 per cent per annum in the 1980s, much faster than any other commodity group.

Clearly, those countries which succeeded in increasing their share of world trade in the fastest-growing commodity groups could do exceptionally well in the growth of exports. Not only did the East Asian countries succeed in increasing the share of manufactures within their total merchandise exports but, within the class of manufactures, they were especially successful in the fastest-growing categories of manufactures in the markets of the industrialized countries (Table 1.3).

The share of exports within the total output of the East Asian countries was increasing in many manufacturing sectors, but, of course, the growth of total output, including exports, depended on a very high rate of new investment in plant and equipment, and there is scarcely any disagreement among economists that these high rates of investment were another important feature of the overall high rates of growth of the Asian countries. In the 1980s and early 1990s, in South Korea, the *average* share of investment in GDP was over 30 per cent from 1975 to 1993 (Figure 1.1). This was significantly higher than the share in other Third World countries or in the fastest-growing European economies, such as Germany, Italy or the USSR during their phase of rapid growth after World War II.

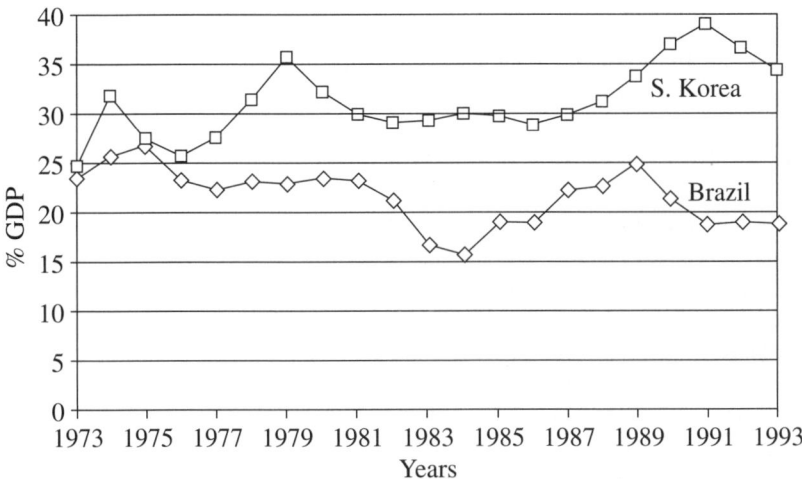

Source: Viotti (1997).

Figure 1.1   Gross domestic investment, Brazil and South Korea, 1973–93

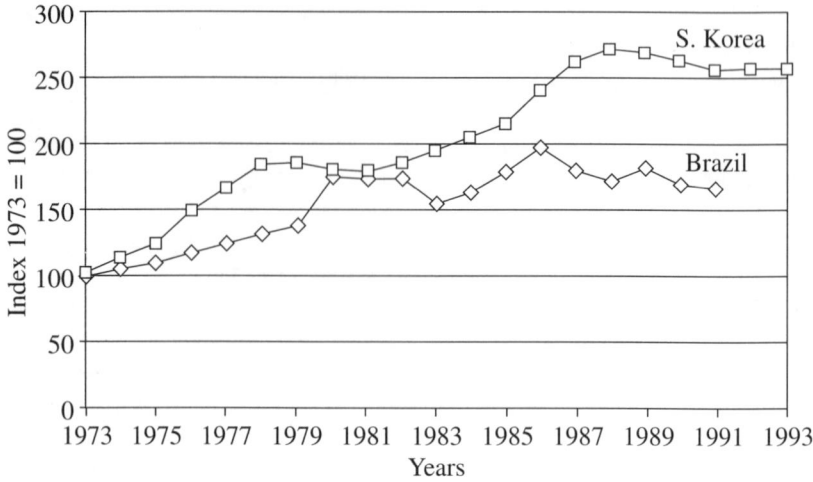

Source: Viotti (1997).

*Figure 1.2   Employment in manufacturing, Brazil and South Korea,
                1973–93*

In fact, the growth of the traditional factor inputs (capital and labour) was so great (Figures 1.1 and 1.2) that some economists have explained Asian growth almost entirely in these terms (see, for example, Krugman, 1994; Young, 1995). This emphasis on growth of the labour force (Figure 1.2) and accumulation of physical capital is very much in the tradition of neoclassical growth theory and of some 'old' growth models. But, whilst the pioneers of growth modelling, such as Abramovitz (1956) and Solow (1957), pointed to the 'residuals' as evidence of the great importance of social capability for technical change, some of the economists who have worked on mathematical models of East Asian growth seem, on the contrary, to belittle the importance of technical and institutional change. The World Bank Report itself makes little or no reference to research and development or other scientific and technical services, yet it would have been impossible for firms in Taiwan, South Korea and Singapore to design, manufacture and export many new fast-growing products and processes in the 1980s and 1990s without intensive in-house research and development activities, especially, but by no means exclusively, in the electronic industries.

In-house R&D activities in Korean *firms* had hardly begun before 1970, but since that time they have grown with extraordinary rapidity. In most other developing countries, R&D activities are still concentrated in government and university laboratories and this is still the main contrasting feature

Table 1.4   Patents taken out in the United States by various countries and regions, 1977–96

	1977–82	1983–89	1990–96
Taiwan	382	2 292	11 040
S. Korea	70	580	5 970
Hong Kong	272	633	1 416
Singapore	17	66	337
4 'Tigers'	741	3 571	18 763
Mexico	245	289	314
Brazil	144	212	413
Argentina	130	135	187
Venezuela	51	122	192
Top 4 Latin America	570	758	1 106
East & Central Europe	3 444	2 417	1 317
Total, all countries	**393 629**	**565 739**	**772 927**

Source: Kumar (1997).

between Latin American R&D and East Asian R&D. Nowhere is this more evident than in the patenting activities of the four 'Tigers' between 1977 and 1997 (Table 1.4). The relatively slow growth of Latin American patents – doubling over 20 years – and the headlong decline of East European patenting are in stark contrast with the explosive growth of patenting from the four 'Tigers'. Since the statistics are based not on domestic patents (which are biased towards the home country for many reasons) but on patents taken out *in the USA*, they provide clear-cut evidence of a substantial increase in the technical capability of East Asian engineers and scientists and of the capacity of East Asian firms to compete in increasingly complex product markets and processes (see also Jae-Yong Choung, 1995). 'Catch-up' in technology is an extraordinarily difficult process for any latecomer country, since the leading countries have very sophisticated science and technology systems ('national systems of innovation') and are themselves constantly improving and changing their range of products, processes and systems. Theories of simple 'exposure' to foreign technology ignore all the real-world complexities of this patient build-up of autonomous capability, and continuously changing forms of technological competition. Historically, even within Europe catch-up was a very prolonged and difficult social transformation.

If the East Asian countries, and to a lesser extent the South-East Asian countries, were making good progress in technological competition, why then the collapse of their economies? Many of the comments on the Asian crisis of 1997–8 are characterized by emphasis on the supposed sins of the Asian governments. In particular, they have blamed corruption of governments for some of the unwise and inept investment decisions of the 1990s. Of course, there has been corruption in many Asian countries, and in some, especially Indonesia, it has been on a very large scale. There has also been corruption in European countries and in the USA. But it is fanciful to put the whole blame for the collapse on corruption and to ignore the misallocation of private investment and the instability of foreign investment flows. As Jeffrey Sachs (1997) has put it:

> It is somehow comforting, as in a good morality tale, to blame corruption and mismanagement in Asia for the crisis. Yes, these exist, and they weaken economic life. But the crisis itself is more pedestrian. No economy can easily weather a panicked withdrawal of confidence, especially if the money was flooding in just months before.
> (…)
> The problems emerged in the private sector. In all of the countries, international money-market managers and investment banks went on a lending binge from 1993 to 1996. To a varying extent in all of the countries, the short-term borrowing from abroad was used, unwisely, to support long-term investments in real estate and other non-exporting sectors.

The general instability of investment in a market economy has long been recognized by almost all business cycle theories (Samuelson, 1980). The liberalization of capital flows and foreign exchange transactions which was the predominant trend in the last 20 years has heightened this instability. The collapse of investment in several Asian economies in 1997–8 arose from two main sources: (a) an outflow of foreign funds based on loss of confidence in particular currencies, political regimes and domestic policies; and (b) a slowdown in exports and domestic investment based on the emergence of surplus capacity and declining profitability in several of the key industries which had been among the fastest-growing in the previous decades.

It was the first of these two problems which triggered the panic in financial markets, but the second may prove to be the more fundamental long-term problem. The success of the economies of South Korea, Taiwan and Singapore and, to a lesser degree, of several other South-East Asian countries was based to a considerable extent in the 1980s and 1990s on the extraordinarily rapid growth of investment, output and exports of the *electronic* industries. The major structural change in the world economy in the last quarter of a century, based on the worldwide diffusion of information and communication technology (Perez, 1985), meant that many Asian firms were able to position themselves well to take advantage of this situation by developing the capability to manufacture at first components and later increasingly complex electronic products and systems. They adopted a variety of strategies to acquire the necessary technologies – inward investment, licensing agreements, joint ventures and ultimately, in the most advanced cases, autonomous design, research and development (Hobday, 1995, 1998). The most frequent partners in such arrangements were Japanese, although American and European firms have also played a significant role. Following the collapse of the Japanese 'bubble economy' in the late 1980s, the scale of Japanese investment in other Asian countries greatly increased as a result of the intensification of worldwide competition and the efforts of Japanese firms to acquire lower-cost manufacturing locations for the simpler products and components. At the same time, China was very rapidly increasing its exports both to Japan and to the USA.

The longer-term problem which confronted the Asian economies in the 1990s was thus how to sustain the growth of their leading sectors in the face of a growing crisis of overcapacity in manufacturing and more difficult 'catch-up' as they approached the world frontiers in technology. The crucial problem of overcapacity was identified by the Bank for International Settlements in its Annual Report for 1998:

> As noted in last year's Annual Report, indications of excessive investment in particular sectors had already emerged in 1996. In that year, the massive investment in Asia's electronic industry contributed to conditions of oversupply and a resulting price collapse in world markets. But investment has sharply increased in other areas as well (such as automobile construction, household appliances and electricity generation) at the risk of flooding local and foreign markets ... Overinvestment in particular sectors has tended to erode the rates of return on capital in recent years.

As ICT has become established as the dominant leading world technology, the role of *software* has become more and more important in most products and systems. Japanese firms and, later, other Asian firms were often highly successful in overtaking American firms in *hardware* manufacture, as for example in colour televisions or in videocassette recorders. They have been much less successful in software products and systems. Of course, they do design and use software, and have some very strong in-house software groups in both manufacturing and service industries, but the market both for standard software products and for complex systems has been dominated by American firms worldwide, because of their technological and organizational leadership in new development and applications.

The establishment of the Internet and the World Wide Web has at least temporarily enhanced this leadership, based on English language dominance, Microsoft dominance in operating systems (a problem for other American firms as well as non-American firms), American leadership in Browsers, and so forth. In all other industries, but especially in service industries, competition increasingly depends on capability in the fast-changing software market as well as the latest hardware. Moreover, successful American and European *manufacturing* firms, such as GE, Ericsson or Benetton, increasingly base their operations on *service* activities such as design, worldwide marketing and financial services, which require complex software. These close links between financial strength, R&D, worldwide marketing, hardware and advanced software systems present a renewed source of hegemony for the leading US multinationals and to a lesser extent for European multinational corporations (MNCs). The later stages of 'catch-up' by Asian countries are thus encountering some severe difficulties, as several close observers, such as Hobday (1995), Ernst (1995) and Ran Kim and Cawson (1997), have demonstrated. Ernst often pointed out that the achievement of autonomous design and product development capability was the most difficult stage of all.

The problems of catch-up in technology were aggravated by the acute social tensions engendered by the investment crisis and the International Monetary Fund (IMF) medicine. In a region previously characterized by rather high levels of employment and a strong demand for labour, unemployment became a social problem. The president of the World Bank, James Wolfensohn, was one of the first to recognize that the social problems associated with high unemployment would now require major policy attention, including World Bank programmes:

> The region must tackle social issues if it is to foster sustainable economic recovery and East Asia's financial crisis risks undermining one of the most remarkable economic and social achievements of modern history. What began as a financial crisis has spilled over into the real economy, severely hitting both production and employment.
> In that case, was the miracle a mirage? Emphatically not. No other group of countries in the world has produced more rapid economic growth and dramatic reductions in poverty. (*Financial Times*, 29 January 1998)

As recent events have shown, dependence on the global economy, and on the IMF and World Bank, can be a mixed blessing. If these institutions were operating as Keynes had originally envisaged, there would be much greater reason for optimism. In the present state of globalization, the instability of the system and the repercussions of events such as the Asian crisis in Eastern Europe, Latin America and, ultimately, in Western Europe and the USA itself seem more likely to aggravate the turbulence. This means that the reform of the IMF is now becoming an urgent question for the management of the global economy. Again, as Jeffrey Sachs (1998) pointed out, the lack of accountability and transparency in the operations of the IMF means that disagreement with its advice is now often regarded as synonymous with a sinful rejection of financial rectitude punishable by the markets. Yet its advice has often been mistaken, and not only in East Asia. It forecast growth of 1.5 per cent in Mexico in 1995 but actual growth was minus 6.1 per cent and again in Argentina forecast growth was 2 per cent and actual growth minus 4.6 per cent. Its handling of successive crises, in Latin America in 1995, in Bulgaria in 1996 and East Asia in 1997, calls into question its ability to handle the volatility of the private capital market in a way which does not damage future growth in countries in which the 'fundamentals' for sustained growth are relatively favourable.

All of these problems were compounded by the slowdown, stagnation and, later, absolute decline of the Japanese economy. Indeed, it was in Japan long before the other East Asian countries that the problem of surplus capacity in the electronic and other industries first became apparent, as well as the huge financial 'bubble' associated with asset inflation. The problems of the Japanese economy have given rise to an interesting debate among economists in which a Japanese minister entered into the correspondence columns of the *Financial Times*. Paul Krugman had pointed to the severe problems of surplus capacity not only in East Asia but also in Japan. This provoked a response from the Japanese Vice-Minister for Finance, Mr Sakakibara. The frustration of the Japanese government, after seven successive 'packages' had failed to produce the desired result of stimulating the economy, is clearly apparent in his response:

> In his analysis of the Japanese economy ('Even worse than you think', October 27) Paul Krugman states the obvious in saying that Japan's central problem is inadequate aggregate demand. Everybody, including those of us in the Ministry of Finance, knows that much, at least ... Although I am not as powerful an economist as Prof. Krugman, I once studied 'orthodox' economic theory and understand the basic logic that underlies the discipline. However, having been involved in actual policy making and the interactions with markets, I came to believe that current market capitalism is inherently unstable. (*Financial Times*, 30 October 1998)

3.   THE UNITED STATES, EUROPE AND THE WORLD ECONOMY

Sakakibara's comment on market capitalism is reminiscent of Schumpeter's work on the instability of capitalism and his theory of business cycles, discussed in section 1 of this chapter. The instability of investment behaviour is primarily due to three types of uncertainty:

1.   general uncertainty about future events, such as wars, revolutions, politi-

cal crises and natural catastrophes: most of these cannot be accurately predicted or foreseen, yet they are certainly important influences on the system's behaviour;

2. technical uncertainty: scientific discoveries, inventions, innovations and their diffusion can be foreseen only to a limited extent and are indeed by definition to some degree unpredictable, except for diffusion. The prevalence of technological uncertainty is particularly apparent from the history of computers and semiconductors since 1940;

3. market uncertainty: no firm can accurately predict its future market share or that of its competitors. Investment decisions are often taken in the expectation or hope of enlarging future market share, but it is quite obvious that not all these hopes can be fulfilled, whether in the case of simultaneous leaps of investment leading to overcapacity, or changes in consumer behaviour or new entrants into markets, or changes in profitability or other causes.

The work of Mariana Mazzucato (1998) illustrates all three of these points well. In particular, her study of long-term changes in the market shares of US car producers from 1909 to 1995 shows the high degree of instability which prevailed from 1909 to 1935. Before this, *technical* uncertainty predominated. It was by no means clear whether electric cars, steam cars or the internal combustion engine would become the dominant technology and there were more firms in the first two categories than in the third (Klein, 1977; Freeman and Soete, 1997). In fact, it was not until the 1950s that a relatively stable pattern of market shares emerged and, at the international level, this more stable pattern was of course itself disrupted by the dramatic entry of the Japanese producers into the US market in the 1970s and 1980s.

Even greater instability in market shares and in dominant technologies has been characteristic of the computer industry (see, for example, the entertaining account in Cringely's (1994) *Accidental Empires*). The present uncertainties about satellite TV, cable TV and digital TV are a further vivid illustration of this point. The case of East Asia also illustrates all these points particularly well. Many firms took an optimistic view in the 1970s and 1980s of their prospects for enlarging their product markets both through exports and home demand. They invested very heavily in the hope of taking advantage of these expanding markets and often, by a combination of good judgement and good fortune, they succeeded in fulfilling their expectations. Yet, from the mid-1990s, many of them confronted problems of surplus capacity and falling market shares, leading to falling profits, withdrawal of much foreign investment and great instability of domestic investment.

The special features of the East Asian crisis characterize it as a crisis of 'Gerschenkronian uncertainty'. Gerschenkron (1962) in his analysis of latecomer 'catch-up' pointed to many advantages of latecomers, notably access to new technologies without all the trials, errors and costs incurred by the pioneers and access to markets originally developed also by the pioneers. In his study of 'catch-up' by the German and Russian steel industries, he also pointed out that latecomers could take advantage of such economies by building plants larger than the established leaders', although this might require social and financial innovations to facilitate the mobilization of large amounts of capital at low rates of interest.

Jang-Sup Shin (1996) in his 'Gerschenkronian' analysis of South Korean 'catch-up', illustrated his argument with the cases of the Korean steel industry and the Korean semiconductor industry. In both cases, Korean firms took advantage of access to foreign technology and foreign skills and established markets to build large plants for successive generations of equipment, with the aid of low-cost finance from the banking system. In these ways they were able to avoid or reduce many of the costs incurred by the earlier pioneers and greatly to reduce both the technological and the market uncertainty which the pioneers had faced.

However, their great success did not eliminate uncertainty altogether, but in some ways actually amplified it on a global scale. Since they followed in the wake of Japan, which had pursued a similar strategy in many industrial sectors, and since their example was followed in turn by other Asian countries and, above all, by China, it led ultimately to a crisis of Gerschenkronian uncertainty in which very large tranches of new investment were coming on stream in several different countries almost simultaneously. This led to the widespread crisis of overcapacity noted by the Bank for International Settlements (1998) which applied not only to such new sectors as memory chips and consumer electronics but also to older industries, such as steel and cars. This crisis of overcapacity was felt most acutely in world export markets and was of course aggravated by competitive devaluation of currencies.

In principle, the uncertainty which accompanies this Gerschenkronian crisis could ultimately be resolved, as Krugman and Sakakibara agree, by a renewed expansion of worldwide aggregate demand, giving firms a breathing space to adjust their capacity, their technology and their marketing strategies to the new realities of the world economy. Whether such a hopeful outcome can be achieved, and how quickly, depends above all on the evolution of the economies of the USA and Europe in the early years of the new millennium. The remainder of this book is dedicated primarily to the analysis of the European situation, while the rest of this chapter concentrates on the situation in the USA.

The points which have been made about general business uncertainty, technological uncertainty and market uncertainty apply just as much to the US economy as to the East Asian economies, but, of course, with the major difference that the USA is the leader in most of the new technologies and not a latecomer 'catch-up' economy. Whether or not the US economy can sustain a relatively stable growth trajectory into the coming decade depends therefore not only on developments in the rest of the world economy but especially on the ability of the USA to continue its successful pattern of structural change in developing and using information and communication technology.

Opinions differ strongly on the likelihood of the continuation of the USA on its 1990s path of relatively stable low unemployment, low inflation, high growth through 1999 and into the 21st century. For the sake of simplicity, this chapter will polarize this complex debate into two main schools: 'optimists' and 'pessimists'. Both the optimists and the pessimists draw upon Schumpeterian arguments about structural change. The optimists argue that the years of relative stagnation, instability and high unemployment (1973–92) were a period of structural adjustment, which the US economy has now successfully completed. They refer frequently to the triumph of the 'new paradigm' by which they mean not only a paradigm of management of the economy on a non-inflationary stable trajectory but also a new paradigm in the neo-Schumpeterian sense of a new technological and management style in which ICT rules. The dramatic expansion of the Internet and of Internet-related business is for them conclusive evidence of the triumph of the new paradigm, with its new computer-based infrastructure. The US stock market in late 1998 and early 1999 clearly reflected this type of thinking. Under the headline, 'Return of Irrational Exuberance? Investors catch Internet Wave', K.G. Gilpin reported in the *International Herald Tribune* of 2 January 1999 on the booming New York Stock market:

> the excitement generated among investors by the Internet's potential is justified, analysts said. The Internet is the most exciting business phenomenon since the airplane or television, Mr. Wien of Morgan Stanly said. 'It is truly a dramatic life-changing event, an open-ended situation investors have not been confronted with for some time.'

The report showed that for the fourth quarter of 1998 the 50 stocks that make up Interactive Week's Internet Index rose by 71.4 per cent.

The optimists argue that the 'high-tech' industries in the USA now have sufficient weight to drive the entire economy into a sustained boom. Their capacity to generate new investment and new employment and to counteract any inflationary tendencies has now been adequately demonstrated, according to the special features in *Business Week* (24 August 1998) and to comentators such as Walter Eltis (1998). (See Figure 1.3.)

Allan Greenspan, at the head of the Federal Reserve, hovers uneasily between accepting the optimistic view that the new paradigm has indeed triumphed and his earlier pronouncements that the US stock market was characterized, in his famous phrase, by 'irrational exuberance'. Since he first

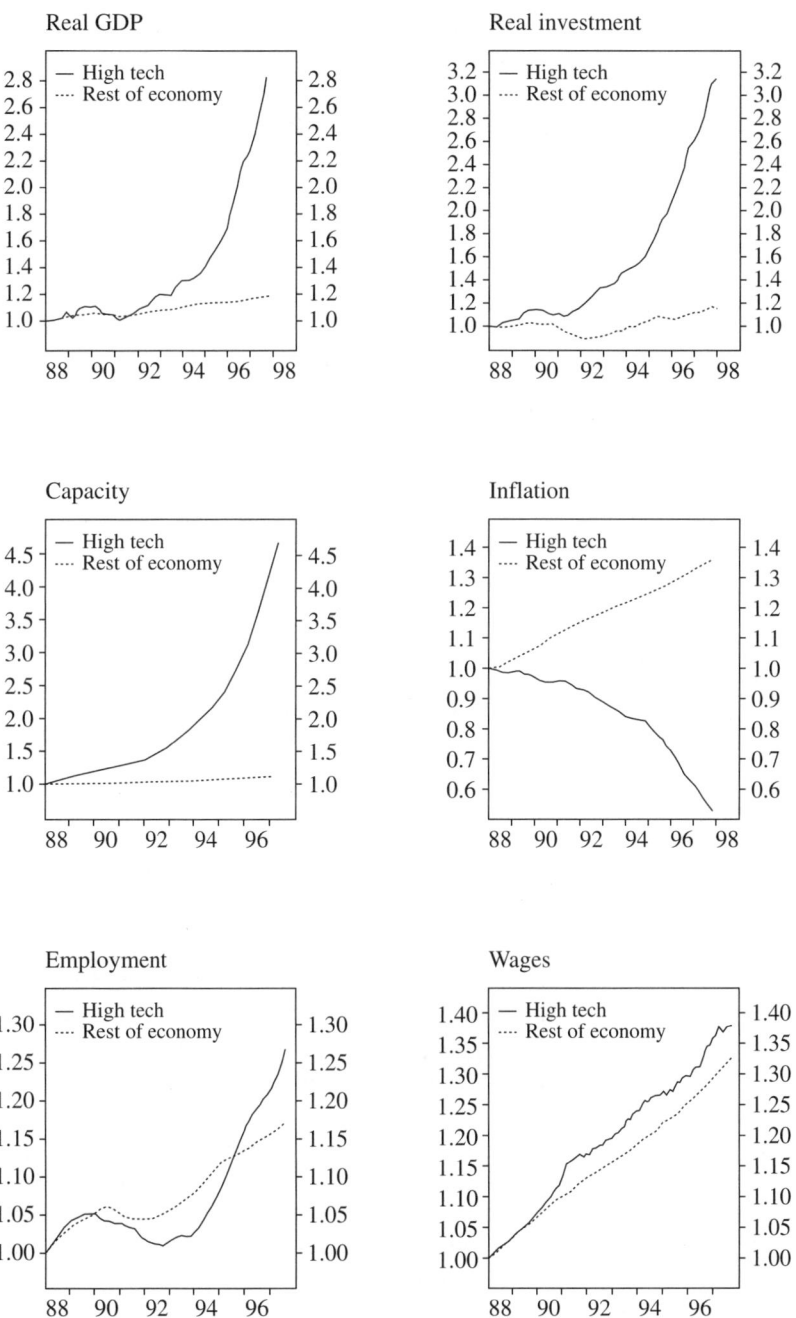

Source: Eltis (1998).

Figure 1.3   Measures of the US high-tech economy

used this expression the stock market advanced by a further 30 per cent, so that the pessimists can argue that there is a high probability that 'historical relationships will eventually re-assert themselves' (Box 1.1).

The Economist magazine has persistently cast doubt on this optimistic view of the US stock market, most notably in the piece entitled 'Once upon a time on Wall Street' (24 October 1998):

> The truth is that financial bubbles, unlike fairy tales, never have happy endings. Mr. Greenspan cannot prevent a bear market with a mere wave of his wand. From the 17th-century Dutch tulip mania through to emerging-market madness and today's Wall Street, bubbles always burst eventually – and the result is always painful. Yet in truth, America's bubble was inflated at home. For over two years the Fed all but ignored the explosion in share prices and lax credit conditions. And to the extent that America's rapid economic growth in recent years has been fuelled by a splurge of credit and by large capital gains, both its economic and its financial boom remain unsustainable.

The Economist leader argued that American monetary policy was at fault in 1998 in failing to deal with the asset inflation: 'Central Banks cannot and should not prevent an overdue correction in share prices. They should only cushion the economy from the worst consequences' (page 16). This points to one of the differences among pessimists: those who forecast a 'soft landing' and those who believe in a 'hard landing', that is, a slump. Those who believe in a soft landing point to the example of the fall in share prices in 1987 which did not have very drastic consequences for the 'real' economy. Typical of those who took a more pessimistic view was Andrew Smithers, who in his article for *The Economist* review, 'The world in 1999' on 'Why Markets will Fall' predicted a serious slump. He justified this view by a comparison between the American financial bubble, the earlier Japanese bubble and the low level of personal savings in the USA: 'As the stockmarket falls, personal

---

BOX 1.1   GREENSPAN SAYS US IS IN A 'VIRTUOUS CYCLE'. ECONOMY WEATHERS ASIAN THREAT

This 'is not what historical relationships would have led us to expect at this point in the business expansion, and while it is possible that we have in some way moved beyond history, we also have to be alert to the possibility that less favourable historical relationships will eventually re-assert themselves' (*Herald Tribune*, 11 June 1998).

savings will rise. This is not really surprising, given they fell to only 0.2% of disposable incomes in June 1998. This compares with an average of around 6% over the past 20 years and 15% in Japan' (Eltis, 1998).

A similar warning came from Robert Reich (1999), the former secretary for labour in the Clinton Administration. While pointing out that the USA had its lowest rate of unemployment for 30 years in 1998, he said, in a warning to Europe not to follow the US example:

> What Europeans don't know is that America's economic ebullience rests on a house of cards. It's not sound macro-economics and a flexible labour market that's put everyone here in such a good mood. It's a stock market that has soared into the stratosphere combined with plummeting world commodity prices ... The ratio of prices at which shares of stock are now selling compared with how much the companies are earning continues to rise to levels not seen since just before the Great Crash of 1929 ... But when the euphoria ends as it must many Americans will be a lot poorer than they feel right now. (*Observer*, 3 January 1999)

Paul Krugman in his 'Requiem for the New Economy' also suggested that the Fed's monetary policy in the 1990s had made a soft landing impossible. Whilst many 'pessimists' would accept that the US economy has indeed been in much better shape in the 1990s than in the 1980s or 1970s and that the diffusion of the new information and communication technology is a positive feature of the situation, they would nevertheless point out that this does not eliminate the instability of the US economy. Indeed, financial bubbles may be blown up by the exceptional performance of a few leading firms in new technology so that the Internet financial bubble of 1998–9 may be compared with the British 'canal mania' of the 1790s, the 'railway mania' of the 1840s, or the later US car and real estate mania of the 1920s. In each case the bubble was followed by a severe crash, although ultimately there was indeed an enormous expansion of the relevant industries. This pattern could be considered as rather typical of waves of new technology, in terms of the theory outlined in section 1. Only when a new institutional and social framework has been firmly established on a global basis, and when the new paradigm has clearly spread from a few leading high-growth sectors to many other industries and services, will a more stable pattern of investment, growth and prosperity become a realistic scenario.

It was not the purpose of this chapter to make a forecast for the future of the US or world economy. On the contrary, the purpose was to indicate the high degree of uncertainty which attends any forecast and to present a few of the key arguments for alternative views of the future. By the time this book is being read, some of these uncertainties will have been resolved and readers will be able to decide more easily for themselves between an 'optimistic' and a 'pessimistic' scenario. In any case, the lesson for Europe is clear: it is to hope and prepare for the best, but also to prepare for the worst. The rest of this book shows many ways in which Europe may hope to prepare for the best. To avoid the worst outcomes, Europe must be prepared to cooperate closely with Japan and the USA in a joint strategy to sustain world aggregate demand and to reform the relevant international institutions so that they are able to carry out a global Keynesian strategy in the hard world of the new millennium.

NOTE

1. The latter rate, which is less subject to definitional differences in the measurement of unemployment between countries, is to some extent a more correct measure of the 'unused' labour potential in an economy.

REFERENCES

Abramovitz, M. (1956), 'Resource and output trends in the United States since 1870', *American Economic Association Papers*, 46(2), 5–23.

Bank for International Settlements (1998), *Annual Report*, Basel.

Business Week (1998), 24 August.

Choung, Jae-Yong (1995), 'Technological capabilities of Korea and Taiwan: an analysis using US patenting statistics', Steep Discussion Paper no. 26, SPRU, University of Sussex.

Cringely, R. (1994), *Accidental Empires*, Harmondsworth: Penguin.

Dosi, G. (1982), 'Technological paradigms and technological trajectories – a suggested interpretation of the determinants and direction of technical change', *Research Policy*, 11(3), 147–62.

Dow, J.C.R. (1998), *The Major Recessions, 1920–1995*, Oxford: Oxford University Press.

The Economist (1998), 'Once upon a time on Wall Street', 24 October.

Eltis, W. (1998), *High Tech Industries in the US Economy*, London: Foundation for Manufacturing.

Ernst, D. (1995), 'Berkeley Round Table on the International Economy', working paper.

Freeman, C. and F. Louçã (2001), *As Time Goes By: From the Industrial Revolutions to the Information Revolution*, Oxford: Oxford University Press.

Freeman, C. and L. Soete (1997), *The Economics of Industrial Innovation*, 3rd edn, London: Pinter/Cassell.

Freeman, C., J. Clark and L. Soete (1982), *Unemployment and Technical Innovation: a study of long waves and economic development*, London: Pinter.

Gerschenkron, A. (1962), *Economic Backwardness in Historical Perspective*, Cambridge, MA: Harvard University Press.

Gilpin, K.G. (1999), 'Return of Irrational Exuberance?', *International Herald Tribune*, 2 January.

Hobday, M. (1995), *Innovation in East Asia: the Challenge to Japan*, Aldershot, UK and Brookfield, USA: Edward Elgar.

Hobday, M. (1998), 'Crisis and Recovery in Pacific Asia: Insights from the Electronics Industry', conference paper for ESRC Pacific Asia Programme, November.

Keynes, J.M. (1930), *Treatise on Money*, 2 vols, London: Macmillan.

Klein, B.M. (1977), *Dynamic Economics*, Cambridge, MA: Harvard University Press.

Krugman, P. (1994), 'The Myth of Asia's Miracle', *Foreign Affairs*, 71(6), 62–78.

Krugman, P. (1997), 'Requiem for the New Economy: Millennial optimism confronts reality', Internet, 10 November.

Krugman, P. (1998), 'Even Worse than you Think', *Financial Times*, 27 October.

Kumar, N. (1997), 'Technology Generation and Technology Transfer in the World Economy: recent trends and implications for developing countries', UNU, INTECH, # 9702.

Kuznets, S. (1940), 'Schumpeter's Business Cycles', *American Economic Review*, 30, 257–71.

Maddison, A. (1991), *Dynamic Forces in Capitalist Development: a long-run comparative view*, Oxford: Oxford University Press.

Maizels, A. (1963), *Industrial Growth and World Trade*, Cambridge: Cambridge University Press and NIESR.

'McCracken Report' (1977), *Towards Full Employment and Price Stability*, Paris: OECD.

Nelson, R. and S.G. Winter (1977), 'In search of a useful theory of innovation', *Research Policy*, 6(1), 36–76.

OECD (1993, 1994, 1995), *Employment Outlook*, Paris: OECD.

Perez, C. (1983), 'Structural change and the assimilation of new technologies in the economic and social system', *Futures*, 15(3), 357–75.

Perez, C. (1985), 'Micro-electronics, long waves and world structural change', *World Development*, 13(3), 441–63.

Ran Kim, S. and A. Cawson (1997), 'The Korean electronics industry: from semiconductors to multi-medias', *InfoWin Bulletin*, May.

Reich, R. (1999), *Observer*, 3 January.

Sachs, J.D. (1997), 'IMF orthodoxy isn't what Southeast Asia needs', *International Herald Tribune*, 4 November.

Sachs, J.D. (1998), 'Out of the Frying Pan into the IMF Fire', *Observer*, 8 February.

Sakakibara, E. (1998), Letter, *Financial Times*, 30 October.

Samuelson, P. (1980), *Economics*, New York: McGraw-Hill.

Schumpeter, J.A. (1939), *Business Cycles*, 2 vols, New York: McGraw-Hill.

Shin, Jang-Sup (1996), *The Economics of Latecomers*, London: Routledge.

Siegenthaler, H. (1986), 'The state of confidence in the '30s and '70s', in I.T. Behrend and K. Borchardt (eds), *Papers of Section 5 of International Economic History Congress*, Berne: International Economic History Association.

Smithers, A. (1998), 'The World in 1999', *The Economist*, Annual Survey, p. 137.

Solow, R.M. (1957), 'Technical Progress and the Aggregate Production Function', *Review of Economics and Statistics*, 39, 312–20.

Viotti, E.B. (1997), 'Passive and Active National Learning Systems', PhD dissertation, New School for Social Research, New York.

Young, A. (1995), 'The Tyranny of Numbers', *Quarterly Journal of Economics*, 110 (3), 641–80.

Wolfensohn, J. (1998), 'Asia, the long view', *Financial Times*, 29 January.

2. Technology, growth and employment in postwar Europe: short-run dynamics and long-run patterns

G.N. von Tunzelmann and Ü.D. Efendioglu

1. INTRODUCTION

The purpose of this contribution to the TSER programme is to present data that we see as relevant to establishing the correlates of technological and growth performance, viewed in a macroeconomic context. The main focus is the dynamic interactions between economic and technological variables, assessed at the macro level. Section 2 of the study draws heavily on previous findings and presents summary graphs of some key trends. Section 3 presents estimates from a set of simple regressions, estimating bivariate relationships in both directions for each country separately, in order to determine leads and lags. Section 4 develops a dynamic analytical framework in which to estimate the interrelationships, while section 5 extends the results relating to technology, investment and growth to employment issues. Some brief conclusions for theory, empirical results and policy are provided at the end of the chapter.

The limitation here to macroeconomic performance and to country-wide data should not be taken as implying that we view such an approach as sufficient unto itself. On the contrary – we believe that any such macro perspective needs to be complemented as far as possible by the micro-foundations of macroeconomic behaviour, not only at the theoretical but also at the empirical level. The obstacles to presenting cross-country empirical micro data with the degree of diversity of the macro data are, unfortunately, enormous, if not insuperable. In the data presented here we do not descend below the broad sectoral (SIC one-digit) level in terms of disaggregation of the production (and so on) data, while the empirical estimation confines itself to macro data.

The data at the time of writing cover 18 countries of Europe other than eastern Europe, for as much as possible of the 46-year postwar period from 1950 to 1995. This limitation was imposed by questions of data availability, relying as we do particularly on OECD and other international sources for data of the requisite depth. These results have been extended to the non-European OECD countries (Canada, USA, Japan, Australia, New Zealand), but are not reported here. For the remaining two OECD countries, Yugoslavia and Turkey, the data are meagre even from the OECD publications, and for other countries the prospects of obtaining the panel data for sufficiently long tracts of time at the required levels of disaggregation do not seem particularly promising. This means that the data are relevant at best to describing just one tier of those established by Baumol (1986) and other writers in the 'catching up' literature. Without labouring the point unduly, it may be pointed out that even this curtailed list of countries involves 18×46 (or 23×46) matrices on each of the many included variables, and a huge job of collating, disaggregating or reaggregating, estimating incomplete series, and above all reindexing the data to provide complete (or relatively complete) time series. We regard this data accumulation and processing stage as one of our main contributions to the TSER programme.

2. AVERAGE MACROECONOMIC PERFORMANCE

This section also builds on our previously published work (von Tunzelmann 1992, revised 1999), which may be consulted for details about sources (mostly OECD data sets) and data procedures. Here we report a more specific set of newer results.

To cope with what is a very patchy set of time series, the figures in the chart are computations using least-squares dummy variables (LSDV). Both countries and years are dummied, and the results show the outcomes of the latter. That is, they represent a kind of 'average' across countries for each year considered separately. These averages are based on the intercept dummies for each country, which serve to offset the impact of missing data. In effect, an interpolated series for each country can be developed, based on the country intercepts, with the interpolation derived from what was happening in the other OECD countries in the specific years.

Figure 2.1 shows a marked acceleration in the ratio of GERD (Gross Expenditure on R&D) and BERD (Business Expenditure on R&D) to GDP (respectively GERD% and BERD% in the figure) in the first half of the 1980s. Some part of the acceleration in R&D intensities is the result of a falling denominator, that is, the pause in GDP growth across countries at the beginning of the 1980s, but it is still noteworthy that R&D intensities are counter-cyclical in the recession of the early 1980s but more procyclical in that of the early 1990s. Recession or no recession, countries in general diverted quite a lot more of their GDP to formal R&D during the 1980s. The

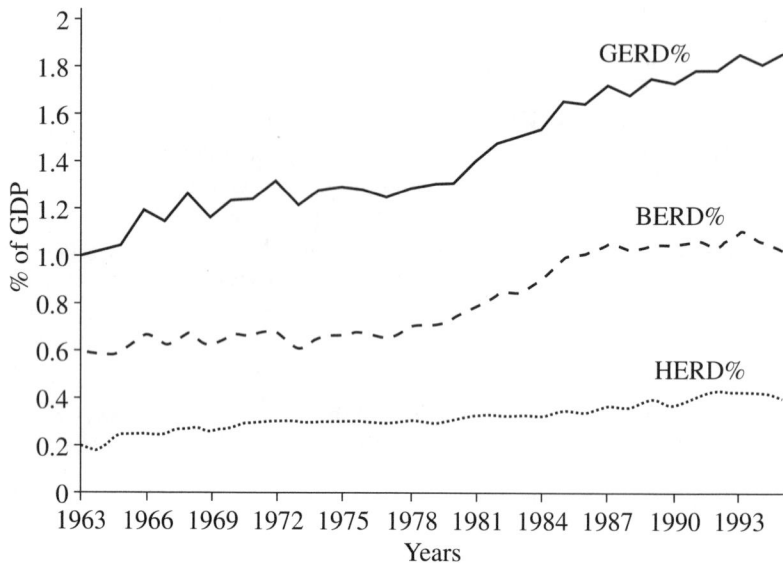

Figure 2.1 '*Average' intensities of GERD, BERD and HERD, 23 OECD countries (base USA), 1963–95*

ratio of Higher Education Expenditure on R&D to GDP (HERD% in the figure) grows more slowly and steadily.

Corresponding results for per capita GDP and rates of investment are shown in Figure 2.2. GDP per head rises fairly linearly in arithmetical terms (the vertical scale is measured in thousands of 1990 US dollars), though of course with some slowdown from the early 1970s in growth rates. The ratio of gross fixed capital formation to GDP ('rate of invest-ment') reaches a peak in the early 1960s and another about a decade later, but then goes into decline with only temporary reversals. One possibility is that the fall in the aggregate rate of investment is responsible – at least in part – for the declining rate of growth of GDP from about this time, for example being transmitted through a declining share of manufacturing in total output and employment. Also charted is the 'producer durables rate': this is included as an approximation of the investment in machinery and equipment, which De Long and Summers (1991) believe to be closely correlated to economic growth across countries.[1] This ratio appears by eye to fluctuate procyclically with real GDP, which supports some linkage to growth, but over the longer term it too shows a later decline. The relation-ship between investment, or machinery investment, and GDP growth can go in either direction, according to standard theorizing. 'Autonomous' increases

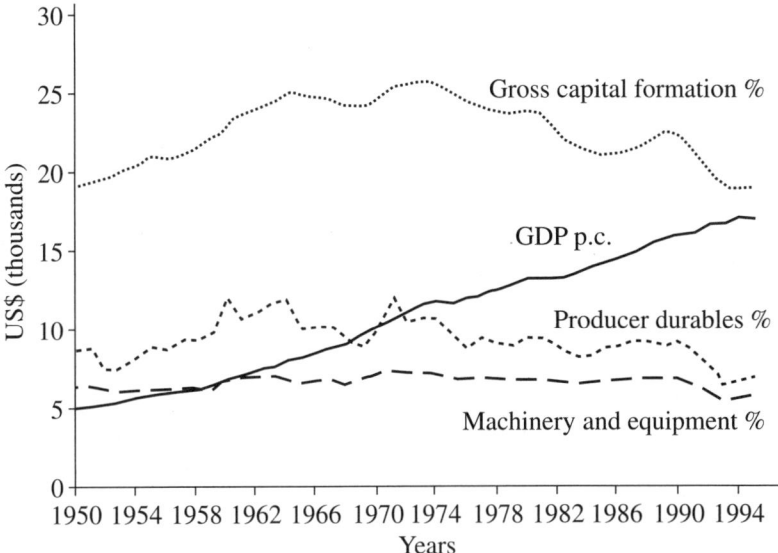

Figure 2.2   *'Average' per capita GDP and rates of investment across 22 OECD countries (base USA), 1950–95*

in investment can generate greater output and growth, for example according to orthodox Keynesian multiplier thinking. Alternatively, higher output can 'induce' larger investment, such as through the accelerator.

The role of investment declines in bringing about falling growth will be assessed more rigorously in the next section, where we also assess some alternative explanations for the stuttering pattern of growth over the last quarter of the 20th century. In particular, we examine some indicators of technological activity, based on the hypothesis of a change in the longer-term pattern away from capital-intensive activities to the knowledge-intensive, which had rather negative medium-term effects but opened up possibilities of restoring *secular* growth rates (von Tunzelmann and Anderson, 1999). Again, standard theorizing can lead to hypotheses that run in either direction. According to the 'linear model' of technology, for instance, growth in science and thereby innovation can result in increased economic performance. The opposite is the 'demand–pull model' suggested by the work of Schmookler (1966), in which investment in the American economy appeared to precede patenting. Finally, investment (in physical assets) may be positively linked with R&D (investment in intangible assets), for example by 'embodiment' of the innovations in physical capital, or negatively linked, through 'crowding out' between these alternative forms of investment, for instance.

Comparing Figures 2.1 and 2.2, the results indicate a period of technological development in these technologies, especially in the 1980s, which did not
bring commensurate benefits at the time in terms of growth (GDP) or other
macro indicators such as production, investment or employment. GDP growth
in fact wavered and sometimes declined. The outcome is one way of concep-
tualizing the well-known 'productivity paradox' identified by Solow and
others. Put in more traditional terms, the expansion of R&D outlays in this
era betokened 'diminishing returns' in respect of the failure of output to grow
commensurately (for similar findings on a much smaller group of countries,
see Jones, 1995a, 1995b).

This seems to run directly counter to the nostrums of the 'new growth
theory', that R&D should generate some gains that are non-rivalrous and
non-excludable and thus represent positive externalities, albeit allowing for
some gains to be privately captured by patenters and so on. There are also
some important implications for public policy which will be raised at the end
of the chapter.

3.   SINGLE-COUNTRY ESTIMATES

It follows that the key to understanding macroeconomic change comes from
the interactions between investment (in physical capital), technology (intangible investment in R&D or output of patents) and growth. The lead–lag
interrelations give some insight into the likely causal patterns, but they are
not easy to assess from graphical treatment. We revert to regression analysis
to do this, allowing us to consider the situation in individual countries.

The first set of regressions (results reported in Tables 2A.1–2A.8) between
pairs of variables is conducted country by country, and is undertaken both in
levels of the variables (including ratio variables such as the rate of investment
to GDP) and in changes in the same variables. The variables included in this
first set are GDP, either in aggregate (GDPP, for purchasing power parity
calculations) or in per capita terms (GDPC), gross capital formation relative
to GDP either for all investment (GFKR) or for investment in producer
durables (GIPD), R&D relative to GDP in total (GERD) and subdivided into
business R&D (BERD) and higher-education R&D (HERD), and lastly patents per capita (PATM). Each variable is regressed against each of the others
in this first set, taking either as the left-hand-side variable.

The pattern of leads and lags (precedence) is established from the pattern
of *t*-values assessed across the range of lags estimated, which here run up to
two years in both directions. The procedure is to look for 'curvature' in the
pattern of *t*-values, to assess which represents the peak lead or lag (this makes
it similar in principle to obtaining a Wald statistic). Where this peak occurs

with a one- or two-year lag there is no problem with identifying which variable leads. When the peak occurs at zero, things are slightly more difficult; here we look at which side of zero the higher t-values occur to establish which variable leads and which lags. Note that it is possible for a particular variable both to lead and to lag in a bivariate relationship, if the pattern of t-values is bimodal. Apart from the lead–lag pattern, it is crucial to assess the sign of the relationship, to help indicate whether underlying hypotheses can be supported.

The data in Tables 2A.1–2A.8 are arranged in geographical order so as to make identification of regional patterns more evident, though we do not have space to develop this issue here.

Table 2A.1

In the relationship between per capita GDP (GDPC) and the rate of investment (GFKR), for countries where GFKR precedes GDPC it is consistently positive, with a two-year lag; that is, higher levels of the *rate* of investment lead to higher levels of per capita income two years later. This suggests the existence of a Keynesian-style positive impact of autonomous investment on growth, or more precisely the negative impact of the observed decline in investment. However, it is more common to find GDP leading investment, and in these cases the relationship also involves a two-year lag, but equally consistently negative. Hence there is no evident long-term 'induced' effect of GDP on investment. This is in keeping with the findings from Figure 2.2, with the longer-run decline of investment rates. The columns showing changes do, on the contrary, suggest a positive short-run 'induced' impact, generally occurring in the same year (zero lag), but a more ambiguous short-run 'autonomous' effect. It may be concluded that the fall in investment is more consistently the result of a preceding fall in GDP than its cause.

Table 2A.2

The results obtained by using GIPD (investment in producer durables) rather than GFKR (all investment) are generally similar. The 'autonomous' effect of exogenous increases in such investment, measured in levels, is not quite so strong as for all investment (Norway shows a negative correlation); on the other hand, the short-run impact (measured in changes) is more consistently positive. This provides the strongest support that we have for the De Long and Summers viewpoint, though their argument addresses levels rather than changes. Their conclusion that machinery investment is a better predictor of growth than all investment does not seem to be supported.

Table 2A.3

When we turn from investment to R&D, for the relationship between GERD and GDP, the relationships in levels are almost all *positive* (only the UK shows a weakly significant negative relationship). They are, however, split fairly evenly between GERD leading, as in the 'linear model' (supply–push), and GDP leading, as in the demand–pull model. Either way, there is generally a two-year lag between them (the values for other OECD countries follow similar patterns). In terms of changes, again the causality is divided fairly evenly between GERD changes leading and GDP changes leading. However, the majority of significant relationships are negative, contrary to the underlying theories, and especially so for the supply–push side. We conclude that GDP growth is associated with the upward shift in R&D intensities in the longer run, with the faster increases in R&D associated with periods of slower growth in GDP (for example, during the 1980s depression), but the direction of causation is unclear.

Table 2A.4

The relationship in levels of GERD and GFKR (rate of investment) is fairly similar to that between GERD and GDP – a split between R&D preceding investment and investment preceding R&D, and generally a one- or two-year lag at work. The big difference is that the relationships here are uniformly *negative*, again with the UK as the sole exception. That is, as already noted, physical investment fell as intangible investment rose. However, this does not seem to have been a 'crowding out' effect, as in the short run the relationships were often *positive*, especially when investment preceded R&D.

Table 2A.5

Much the same applies to the link between GERD and GIPD (investment in producer durables). When machinery investment precedes R&D, the relationship is invariably a two-year lag with negative sign. Again, in the short run this relationship is most often positive, giving some support to the 'embodiment' notion.

Relationships between BERD (business expenditure on R&D) and these three variables are generally similar, as is to be expected since it is movement in BERD that accounts for most of the variation in GERD in this period (these results are not shown explicitly here). However the relationship between levels of BERD and levels of GIPD (machinery investment) are less consistently negative than for GERD, while conversely that between changes in BERD and changes in GIPD is less consistently positive, possibly indicating a degree of 'crowding out' at the industry level.

Table 2A.6

The relationship between the three economic variables and HERD is somewhat different. The table shows the significant correlations between HERD and GDP. First, in levels, GDP is more often seen as preceding HERD than GERD or BERD. This suggests that greater prosperity permits greater intangible investment in higher education research. A similar picture emerges in the short run: in contrast to GERD and BERD, where these links were usually negative, in the case of HERD they are mostly positive. Overall this suggests a degree of 'circular causation': that greater wealth allows countries to invest for the long term in upstream research. The relationships between HERD and both GFKR and GIPD are, however similar to those for GERD and BERD.

Table 2A.7

For patents, the comparison has been undertaken with GDP per capita (GDPC) rather than aggregate GDP. In the same vein, patents are measured per million of population (PATM). The period covered is now generally the full extent of the data, that is, 1950 to 1995 (46 observations), since patents data for most countries except Finland are available throughout. Since both are trending upwards, the relationships in levels are always positive for these countries. As would be implied by the 'linear model', patents precede GDP levels, for the bulk of the countries, generally by two years. Strikingly, however, in terms of changes the precedence pattern goes the other way, with changes in per capita GDP fairly consistently leading changes in per capita patents by one year. However, we need to take into account here that the patents figures are for patents granted; if we allow for the fact that the gap between patent applications and patent grants at the US Patent Office is about 18 months, then we should probably conclude that per capita patents and GDP moved contemporaneously in the short run. Where patents do show more unequivocal signs of leading in the short run, the relationship is sometimes negative (this is also true for Australia and Canada, not shown here).

Table 2A.8

A rather different picture emerges when we look at the relationship between patenting and the rate of investment (GFKR). In levels, when patents lead investment the relationship is always negative, contrary to the 'linear model'.

However, unlike GDP, investment leads patenting as often as the reverse, and here the relationship in levels is generally positive, so supporting a version of the Schmookler argument about the demand–pull from investment to patenting. This is also true in the short run, which is more precisely equivalent to what Schmookler (1966) actually assessed, though of course his work was at the micro level of individual industries. But in the short run the 'linear model' relationship of patents leading investment is also mostly positive for those countries that show this as significant, contrary to Schmookler's result. For the relationship between PATM and GIPD (producer durables investment), the patterns in both levels and changes are similar to those for PATM and GFKR, except that fewer of the short-run relationships in either direction are statistically significant.

4. A MODEL FOR ESTIMATION

The simple single-equation assessments used to this point are obviously crude, and it would be desirable to build a more convincing case for causality patterns than just relying on single-country simple lead–lags. The 'error-correction' (ECM) model is used, again on pairs of variables, for this purpose. First, we have to establish the order of integration of the variables, in order to defend the use of the ECM approach, which will evidently be most convincing where the variables are I(1). It is not self-evident that all our variables, many of which are ratio variables, are such. In Table 2B.1, we carry out unit root tests on the variables used in the previous set of tables.

The unit root tests show that most of the variables for most European countries indeed display evidence of a unit root. The autoregression coefficient is usually significantly different from zero, and often not significantly different from unity. The Dickey–Fuller and Augmented Dickey–Fuller tests on the variables measured in levels are usually non-significant. On the other hand, the Dickey–Fuller estimated on changes in the variables is usually very significant. Thus, in the right-hand column, the order of integration is usually assessed to be I(1), though there are a few possible exceptions. This reinforces the case for drawing the key conclusions from ECM calculations, allowing the identification of both short- and long-term relationships.

The unit root tests are known to be influenced by the effect of any structural change in the variables and their interrelationships within the data set. There is good a priori evidence for a structural break in the early 1970s in many of these European countries (von Tunzelmann, 1992, 1999). Accordingly, we have divided the whole period into two and re-estimated the Dickey–Fuller and so on for 1950–73 and 1973–95. The results – not reported explicitly here for lack of space, but available from the authors – are even more convincing evidence of cointegrated series that, within these sub-periods, experience first-order integration.

Our more refined specification is based on a dynamic classical model of supply–demand interaction, drawn from our earlier research. This model is appropriate here, for several crucial reasons. First, it places heavy emphasis on diminishing (and, in reverse, increasing) returns, though in its dynamic version not just in the static context to which diminishing (as opposed to increasing) returns are so often consigned. Second, it emphasizes the dynamic interlinkages which characterize macroeconomic systems, which can lead on to produce vicious circles, virtuous circles, and so forth, depending on how the parameters shape up. Thirdly, it gives some attention to uncertainty and to learning processes in the development of macroeconomic behaviour (von Tunzelmann, 1991).

Fourthly, this model lends itself directly to some versions of modern time-series analysis, which were of course developed for quite different reasons – essentially for reasons of statistical purity. In the formulations below, we use the error-correcting mechanism (ECM) version of time-series estimation. This is usually estimated in levels of the variables, but in our cases that procedure generated consistently high correlation coefficients in which time patterns predominated, running the risk of 'spurious correlations'. Instead, we are reporting results from the strict ECM transformation of the equations in levels. That is, if one takes an equation in levels such as (Hendry and Doornik, 1996):

$$y_t = \alpha_1 y_{t-1} + \beta_0 x_t + \beta_1 x_{t-1} + \varepsilon_t \text{ with } \varepsilon_t \sim \text{IN}(0, \sigma_\varepsilon^2) \tag{2.1}$$

then this is estimated here in the form:

$$\Delta y_t = (\alpha_1 - 1)y_{t-1} + \beta_0 \Delta x_t + (\beta_1 + \beta_0)x_{t-1} + \varepsilon_t. \tag{2.1a}$$

This equally transforms into the form:

$$\Delta y_t = \beta_0 \Delta x_t + (\alpha_1 - 1)(y_{t-1} K x_{t-1}) + \varepsilon_t \tag{2.1b}$$

in which β_0 is the 'impact effect' of x on y, $(1 - \alpha_1)$ is the 'feedback effect', and K is the long-run response. The term $(y - Kx)_{t-1}$ is the ECM, and K corresponds to the cointegration between y and x if the series are I(1). The dynamic classical model we are utilizing suggests various extents of error correction, but also indicates more richly than the econometrics alone why such error correction may come about and what it signifies. It also indicates the importance of simultaneous relationships between the ys and the xs.

We adopt a general panel data approach, in which the data for all countries and all (available) periods are combined. Consonant with equation (2.1a), typically the change in a variable is regressed against its level in the previous period, the change in the RHS variable, and the previous-period level of the RHS variable. This is then switched around, by putting the RHS variable on the LHS, to test for reverse causation and the interactive effects noted above. In contrast to the usual panel data approaches (for example, Hsiao, 1986; Baltagi, 1995; and references therein), we estimate dummies not just for the country intercepts but also for country slopes. This results in the estimation of large numbers of RHS variables in each equation. We follow general-to-specific modelling as recommended by adherents of the ECM approach, so that all dummies are first included, and then decisions are taken about which to omit.

For reasons of space, the full tables are suppressed here, but Table 2C.1 summarizes the conclusions reached as regards the variables which we have focused upon to date. Some patterns appear to emerge from this matrix. An obvious point is that, quite often, the short-run (SR) impacts ran in the opposite direction (sign) to the long-run ones (LR) obtained from the cointegration procedure. In this section we shall pay most attention to the long-run relationships. Causal patterns have been inferred from the relative strength of the set of correlations (comparing explanatory power in one direction with that from the opposite direction). Further work is proceeding on the exact time lags involved.

Demand Factors

The long-run impact of GDP on the technology variables (GERD, BERD, HERD and PATM) appears to be greater than the reverse. It may be concluded that GDP generally leads technology rather than vice versa. Demand factors would seem to explain why GDP appears to be leading. This suggests a Schmookler-like mechanism of demand–pull operating at the macro level. In the short run, however, growth in GDP often has a negative impact on R&D, perhaps partly because it is the denominator of R&D intensity (so if R&D is relatively stable, intensity will fall when GDP rises). In the long run, they move up together. But one should beware of simple conclusions of the Schmookler kind: using a series for total patents, instead of patents per capita, we found that patents appeared to lead GDP.

Supply Factors

While R&D intensities are positively linked to GDP in the long run, capital intensities (here capital–output ratios) are negatively linked, following on from Figure 2.1 above. Conversely, capital intensities are positively and quite strongly related to GDP in the short run. In the long run, GFKR and to a lesser degree GIPD are also negatively related to GERD, BERD and HERD. It will be noted that there are some apparently inconsistent results; for example, if causation was running from PATM to GFKR and GIPD, the long-run relationships were negative (see the last row in Table 2C.1), but if the opposite were true (as in the last column of the table) our results indicate a positive long-term association. By default, we assume that supply-side factors lowering capital intensities or the incentives to accumulate capital underlay these comparative trends.

Figures 2.1 and 2.2, as well as these econometric results, suggest that the 45-year period from 1950 to 1995 ought to be split into two, the period from 1950 to 1973 representing faster growth and rising capital intensities, while the reverse held thereafter. The results of panel data estimations are summarized in Table 2C.2. Note that R&D figures at the cross-country level are not available before 1963 and the data become more abundant only from the 1970s and 1980s, so relationships involved with R&D are not re-estimated. Thus the preceding R&D results should be taken as applying to the second (more recent) sub-period. Unit root tests have been reassessed to accommodate this 'break in trend', but as already noted they tend to strengthen the consistent I(1) pattern of most of the series for most European countries.

The main shift observable in Table 2C.2 is evidently the long-term relationship of GFKR to both GDPC (GDP per capita) and PATM, which is positive if not very strong in the first sub-period but negative and also rather weak in the second sub-period. This of course is what Figures 2.1 and 2.2 also suggest. Although it may be a reasonable finding in the light of the underlying data, it nevertheless can be used to draw some strong conclusions, which we come back to later.

5. RELATIONSHIP TO EMPLOYMENT AND COSTS

This section presents results to date on extending the above interrelated results to a wider set of macro-level variables, including employment and factor costs. These results have been prepared in panel data ECM form. The full country-by-country tables are too extensive to list here, but were given in an earlier version of this chapter, obtainable from the authors.

To date we have not been able to compile entirely adequate cross-country series for total employment since 1950. The data on agriculture and services are rather limited before the early 1960s for many countries, and even later for some. Although the data are now about as complete as it seems possible to get, the figures used in the tables hereafter use employment in manufacturing alone (its share in total employment). Obviously, there is a problem with using this figure, because by the mid-1960s many of the leading countries were beginning to show evident signs of deindustrialization, as already implied; but this may be precisely the kind of problem that it is necessary to identify.

Table 2D.1 summarizes the main results. Some of the panel data were estimated for the full set of 23 countries, to which data like the R-bar squared thus pertain, but the number of positive and negative signs is computed for the 17 or 18 European countries alone.

Growth of employment in manufacturing could influence economic growth from the supply side, for instance by providing the inputs into a sector (manufacturing) deemed critical for overall development. Conversely, the growth of output could have a demand-side impact on the growth of employment. From Table 2D.1, row 1, it can be seen that the change in output had a generally positive impact on the growth of manufacturing employment once the levels of the variables are controlled for. The complete results show the absolute impact being highest in the base country, Germany. At the same time, the level of output (GDP in PPP terms) had a consistently negative impact on the change in manufacturing employment, so that, as output levels rose over the period, the gain in manufacturing employment weakened; in practice, as noted above, the gains turned into losses as deindustrialization proceeded.

The supply-side impact being tested for in row 2 gives an even higher R-bar squared (= 0.549 measured in changes of GDP). Despite the deindustrialization, the impact of changes in manufacturing employment on changes in GDP was consistently positive. The level of manufacturing employment also had a generally positive impact on output. The time trend in the overall relationship, moreover, was positive. All of these were constrained by the negative relationship between the level of GDP and changes in GDP, which masks the supply-side relationship in a simple correlation between output and manufacturing employment. We conclude that 'manufacturing matters', at least in terms of the employment impact on overall growth. The technology effect in this relationship will be assessed below.

The direct links between investment and manufacturing employment are assessed in the next two rows. On a priori grounds, there are a number of plausible links possible, including substitution between them as well as complementarity. The figures for employment here are the share of manufacturing employment in total employment, since the rate of investment is also a ratio variable; this differs from the level of manufacturing employment used in some of the preceding tables. The effect of deindustrialization is shown in the strong negative time trend in row 3.

On the whole, complementarity seems to predominate in row 3, with a positive relationship existing between changes in the manufacturing employment share (of total employment) and changes in the investment share (of total output), for most but not all countries. The levels of the variables have little impact here. However, in row 4, examining the reverse causation, there is greater evidence favouring the substitutability notion – though, again, positive relationships outnumber negative ones. Evidently, greater investigation of the relationship between the factor inputs is called for; including of course the extension to other sectors of the economy.

The comparisons between employment and technological variables give stronger correlations when the technological variables are interpreted as the 'independent' variables (compare rows 5 and 6 for BERD). The clearest pattern is for per capita patents and employment, where both short-run and long-run impacts seem to be positive. This is presumably because manufacturing undertakes a high proportion of total patenting activity. It is perhaps more surprising that the same effect does not arise in regard to R&D: this may have been the result of unusually large increases in R&D within a shrinking manufacturing sector, or because some R&D shifted to the rising services sectors.

Turning to the links of factor supplies to factor costs, we would expect to find (from classical assumptions) a negative impact of employment growth (or population growth) on real wages but a positive impact of wage increases on employment. In these tables, we use data on real aggregate 'employee compensation' per head as our measure of average real wages (RWR).

The coefficients on changes in the real wage rate in row 9 of Table 2D.1 are very mixed across countries: for Germany as the base country, there is a strong positive association as the classical model would predict, but there are many with a negative sign, including the UK and France. We might use these results to doubt the view that changes in the cost of labour negatively influenced employment. On the other hand, the *level* of the real wage rate does indeed appear to have exerted a negative influence on changes in manufacturing employment, as shown in the final columns. The level of manufacturing employment (in period $t-1$) almost always has a negative impact on changes in that employment, as is to be expected from the deindustrialization in later years of virtually all these countries.

The reverse relationships in row 10 are much weaker. Only Ireland shows an apparently significant negative relationship between changes in manufacturing employment and consequential changes in the real wage rate. The level of the real wage rate seems to have exerted a negative but often non-significant effect on changes in real wages, and it may be noted that catching-up countries experienced this as much as the more advanced.

Nor did real interest rates (RRI) behave over time in the way that might be expected by economic theory; instead, they fell when capital formation (the rate of investment, that is the ratio of gross domestic fixed capital formation to GDP) itself declined, especially in the 1970s. Row 11 shows that there is a strong negative time trend in the rate of investment. However, none of the country dummies of either intercepts or slopes, and for any of the variables, are statistically significant. Similar results emerge from the reverse causation possibility assessed in row 12, except that here the level of previous-period real interest rates had some (negative) impact upon the change. We tentatively conclude that the 'supply price of finance', which has been stressed not just in the financial literature but even in the rather limited public debates over links between macroeconomics and technology (like the 1995 EU *Green Paper on Innovation*, section II.4), played little part in determining the longer-term course of capital formation.

It may be concluded that technology data can be linked effectively to economic data, but the patterns of interrelationship are complex, and simple one-period dynamics are unlikely to get us very far. It seems probable that we will need to understand more of the micro-level determinants and impacts of technological change in so developing more substantive macro dynamics. However, by the same token, there is even less support for classical economic models in which quantities depend on prices.

6. CONCLUSIONS

Theoretical Implications

Probably the most consistent finding from the empirical as well as theoretical work is the prevalence of diminishing returns observable in both static and (here) dynamic terms. Whereas the 'new growth theory' has focused upon increasing returns to technology or human capital inputs, we find diminishing returns to these, as well as to conventional factor inputs. No convincing assessment can be made until the whole macroeconomic system is put together, which we have not yet done, but we do feel that the 'new growth theory' has missed a crucial point.

Our emphasis thus falls upon the greater cost of technology or human capital inputs as levels of output rise. We suspect that this is the main cause of the long-term shift in postwar growth, from the 'Golden Age' of the 1950s and 1960s to the more disturbed conditions of the 1970s and 1980s. There was, however, a positive side to this, in the rise of new pervasive technologies, of which information technology is the most conspicuous example, which had the potential to unleash a new phase of growth (and possibly increasing returns). In another paper (von Tunzelmann, 1999) we have found that extra R&D in ICTs account for between 40 per cent and 60 per cent (depending on the base country chosen) of the pronounced overall increase in R&D across OECD countries between the late 1970s and the late 1980s.

A second major theoretical finding is that disequilibrium models which allow for (a) dynamic as well as static conceptions of demand and supply and (b) a conceptual separation between the demand and supply factors (before trying to interrelate them) have considerable power in accounting for short-period changes in complex technoeconomic systems, and pave the way towards better specification and understanding of the way longer-period interrelationships function.

Empirical Findings

The most substantial empirical findings include the following. First, cost-based supply factors like prices and interest rates appear to have had a rather limited role to play in promoting growth and capital formation, though there are some circumstances in which their contribution cannot yet be ruled out, and in any case they probably constitute parts of a much more elaborate system in which prices are outcomes as much as inputs into the whole process.

Second, demand factors need equal consideration with supply factors in interrelating economic, demographic and technological variables; in some cases, demand factors appear more capable of explaining the pattern of time-series and cross-country relationships than do those from the supply side.

Third, technology appears to have effects on employment that are on balance beneficial, after allowing for extraneous factors; but further thought needs to be given to ways in which technology factors should be interrelated with economic and other variables, which may involve more detailed use of micro as well as macro data, and certainly more careful consideration of short-, medium- and long-term factors.

Policy Implications

The finding of pervasive diminishing returns does not in our view reduce the importance which policy makers should place on any of these variables. Cynics have pointed out that, if education is negatively related to growth, the obvious policy inference is to reduce education. Our finding that this is reversed when levels of education and growth (and so on for other variables) are allowed for gives rise to the opposite conclusion: that more and more effort has to be given to technology or human capital as growth proceeds, simply to ensure that growth continues. However, in our recent work we have also argued that the 'diminishing returns' impact of additional R&D may be fading away, as the advanced countries move towards reaping the benefits of applying these new infrastructural technologies across an ever-widening range of users (von Tunzelmann, 1999).

The second main policy finding is that demand factors have to be given similar weight to supply factors, and the emphasis that has been placed rather exclusively on the latter for the last two decades may have been seriously misplaced. It is early days to claim too much, but it may well turn out to be the case that the striving for 'stabilization' on the supply side has actually been to the detriment of innovation and ultimately economic growth. It is little comfort to find that Europe continues to see the way ahead at the macro level as consisting almost exclusively of supply-side policies, and that even its *Green Paper on Innovation* (1995) pays practically no attention to the role of demand. The postwar development of the European economy points to a need for fundamental rethinking of present-day policy making.

NOTE

1. Producer durables include transport equipment as well as 'machinery and equipment', but the series is available for more countries and over longer periods than that for the latter.

REFERENCES

Baltagi, B.H. (1995), *Econometric Analysis of Panel Data*, Chichester: Wiley.
Baumol, W.J. (1986), 'Productivity growth, convergence and welfare: what the long-run data show', *American Economic Review*, 76, 1072–85.
De Long, J.B. and L.H. Summers (1991), 'Equipment investment and economic growth', *Quarterly Journal of Economics*, 106, 445–502.
European Commission (1995), *Green Paper on Innovation* (E. Cresson and M. Bangemann), Brussels/Luxembourg: EC DG XIII/D.
Hendry, D.F. and J.A. Doornik (1996), *Empirical Econometric Modelling Using PcGive 9.0 for Windows*, London: International Thomson.
Hsiao, C. (1986), *Analysis of Panel Data*, Cambridge: Cambridge University Press.
Jones, C.I. (1995a), 'Time series tests of endogenous growth models', *Quarterly Journal of Economics*, 110, 495–525.
Jones, C.I. (1995b), 'R&D-based models of economic growth', *Journal of Political Economy*, 103, 759–84.
Schmookler, J. (1966), *Invention and Economic Growth*, Cambridge MA, Harvard University Press.
von Tunzelmann, G.N. (1991), 'Malthus's evolutionary model, expectations and innovation', *Journal of Evolutionary Economics*, 1, 273–91.
von Tunzelmann, G.N. (1992), 'The main trends in European economic history since the Second World War', in D. Dyker (ed.), *The European Economy*, London/New York: Longman, pp. 15–50.
von Tunzelmann, G.N. (1999), 'Growth and supply in Europe since the Second World War', in D. Dyker (ed.), *The European Economy*, revised edn, Harlow: Addison-Wesley/Longman, pp. 11–42.
von Tunzelmann, G.N. and E. Anderson (1999), 'Technologies and skills in long-run perspective', mimeo, SPRU/IDS, University of Sussex.

APPENDIX A

Table 2A.1 *Relationships between per capita real GDP and the rate of investment*

GDPC/ GFKR	Obs.	GDPC leads		GFKR leads		DGDPC leads		DGFKR leads	
Neths	46	−2	****						
Belgium	46	−2	*			+0	**		
Luxem.	46			+2	*				
France	46			+2	***			+0	***
Germany	46	−2	****					+0	****
Austria	46	−2	*					+0	*
Switz.	46			+2	***	+0	****		
Denmark	46	−2	****			+0	****		
Finland	43	−2	****			+1	****	−2	**
Sweden	46	−2	****			+0	****		
Norway	46	−2	****					−2	*
Iceland	46	−2	***					−1	***
Ireland	46			+2	**				
UK	46			+2	***	+0	****		
Spain	42							+0	****
Portugal	44			+2	****	+1	**		
Italy	44	−2	****			+0	***		
Greece	46			+2	****	+1	**	−2	***

Notes to all tables in Appendix A: obs. = maximum no. of observations; numbers in lead/lag columns represent years of lead; signs in front of these numbers indicate positive/negative correlation; significance levels: * = 20% level; ** = 10% level; *** = 5% level; **** = 1% level.

Table 2A.2   *Relationships between per capita real GDP and the rate of investment in producer durables*

GDPC/ GIPD	Obs.	GDPC leads		GIPD leads		DGDPC leads		DGIPD leads	
Neths	46	−2	****			−2	**	−2	*
Belgium	46	−2	****					+2	***
Luxem.	40	−2	****						
France	46			+2	****			+1	**
Germany	45	−2	****						
Austria	46	−2	****						
Switz.	25			+2	****	+1	**		
Denmark	46	−2	*			+0	***		
Finland	43					+0	****		
Sweden	46							+0	****
Norway	42	−2	****	−2	****			−2	***
Iceland	45					−2	*	+0	*
Ireland	46			+2	*				
UK	42			+0	**	+0	****		
Spain	39	−2	****			−2	**	+0	***
Portugal	42			+2	****				
Italy	44							+0	****
Greece	45			+2	****	+1	***		

Table 2A.3 Relationships between total R&D intensity and aggregate GDP

GERD/ GDPP	Obs.	GERD leads		GDPP leads		DGERD leads		DGDPP leads	
Neths	26					−0	***		
Belgium	18			+2	****	−1	*		
France	32	+1	****	+2	****			−1	***
Germany	31	+2	****			−0	*		
Austria	19			+0	****			−2	*
Switz.	22	+2	****					−0	***
Denmark	24	+2	****			−0	*		
Finland	20			+2	****	−0	***		
Sweden	16	+1	****			+1	***		
Norway	22	+2	****						
Iceland	18			+2	****	−1	*		
Ireland	21	+1	****					+1	***
UK	20	−1	*	−2	**				
Spain	28	+1	****			+1	***	+2	***
Portugal	13			+0	****			+1	*
Italy	31	+2	****						
Greece	11	+2	****	+1	****			−2	***

Table 2A.4   Relationships between total R&D intensity and rate of investment

GERD/ GKFR	Obs.	GERD leads		GKFR leads		DGERD leads		DGKFR leads	
Neths	26					+2	*		
Belgium	18	−1	****	−2	****	−2	*		
France	32	−1	****					+1	**
Germany	31			−2	****				
Austria	19			−2	****	−2	****		
Switz.	22			−1	*			−0	*
Denmark	24			−1	****				
Finland	20	−2	****			−0	**		
Sweden	16			−1	****	+1	***		
Norway	22	−1	**					+1	**
Iceland	18	−1	****			−1	*		
Ireland	21	−1	****			−1	***		
UK	20	+2	**						
Spain	28	−1	****					+2	**
Portugal	13	−1	*						
Italy	31	−2	****	−2	****	−1	*	+1	**
Greece	11			−2	****			+0	**

Table 2A.5   Relationships between total R&D intensity and rate of investment in producer durables

GERD/GIPD	Obs.	GERD leads		GIPD leads		DGERD leads		DGIPD leads	
Neths	26	−2	****					+1	***
Belgium	18			−2	****			−2	*
France	32			−2	*				
Germany	30			−2	****				
Austria	19	−0	****	−2	****	−2	****		
Switz.	12			−2	*			−2	***
Denmark	24					−0	*		
Finland	20	−2	****			−0	*		
Sweden	16					+1	****		
Norway	20			−2	****				
Iceland	18	−1	****			−1	****	+1	**
Ireland	21	−0	***						
UK	17					−1	**		
Spain	28	−0	****					+2	***
Portugal	13	−2	**			−2	*		
Italy	31							+1	**
Greece	11			−2	*			+0	**

Table 2A.6   Relationships between higher-education R&D intensity and aggregate GDP

HERD/ GDPP	Obs.	HERD leads		GDPP leads		DHERD leads		DGDPP leads	
Neths	26	+2	**	+2	***	+2	**		
Belgium	17								
France	32	+1	****	+2	****				
Germany	31			+1	****			+2	*
Austria	7	+0	****					+1	**
Switz.	24			+0	****				
Denmark	23	+2	****	+2	****	−0	**		
Finland	20			+2	****	−0	****	+2	***
Sweden	15	+1	****			+1	*		
Norway	22			+0	****				
Iceland	16			+2	****				
Ireland	22	+1	****					+1	*
UK	26	+2	****						
Spain	28			+2	****	−1	****	+1	***
Portugal	13	+2	****			+1	**		
Italy	31			+1	****				
Greece	8			+1	****	+2	*	−2	***

Table 2A.7   Relationships between per capita patents and per capita GDP

PATM/ GDPC	Obs.	PATM leads		GDPC leads		DPATM leads		DGDPC leads	
Neths	46	+2	****						
Belgium	46	+0	****						
Luxem.	46	+2	****						
France	46	+2	****			+0	*		
Germany	46	+2	****						
Austria	46	+2	****			+2	***	+1	*
Switz.	46	+2	****					+1	***
Denmark	46			+0	****			+1	*
Finland	28			+2	****	−1	**	+2	***
Sweden	46	+2	****					+1	**
Norway	46			+2	****			+1	*
Iceland	46	+2	**						
Ireland	46	+2	****						
UK	46	+2	****					+1	**
Spain	46	+0	****			+0	***		
Portugal	44	+2	****						
Italy	44	+1	****					+1	*
Greece	46	+1	****						

Table 2A.8   *Relationships between per capita patents and rate of investment*

PATM/ GFKR	Obs.	PATM leads		GFKR leads		DPATM leads		DGFKR leads	
Neths	46	−2	****			−2	*	+1	*
Belgium	46					+2	*		
Luxem.	46					−2	**		
France	46			+2	****			+0	***
Germany	46	−2	****						
Austria	46								
Switz.	46			+2	****			+1	*
Denmark	46	−2	*			+2	**		
Finland	31	−2	****			−1	*	+1	*
Sweden	46	−2	***			+2	*		
Norway	46			−2	***	+1	*	−2	*
Iceland	46					−1	*		
Ireland	46			+2	*	+1	*		
UK	46			+1	****	+2	***	+1	****
Spain	42							+1	**
Portugal	44			+0	***	−2	***		
Italy	46	−2	****						
Greece	46			+1	****			+1	***

APPENDIX B

Table 2B.1 Unit root tests on R&D intensity, patents, GDP and rate of investment

Country	Variable	Obs.	Autoreg. Coefft.	Sig.	DF	Sig.	ADF	Sig.	DDF	Sig.	Int.
Neths	GERD	26	0.812	**	-1.50		-1.93		-3.17	*	1
	PATM	26	0.517	**	-3.29	*	-1.98		-8.11	**	1
	GDP	26	0.919	**	-2.27		-1.98		-4.67	**	1
	GFKR	26	0.863	**	-1.83		-2.91		-3.64	*	1
	GIPD	26	0.743	**	-2.28		-2.59		-4.08	**	1
	BERD	25	0.863	**	-1.21		-2.54	*	-2.50		~1
	HERD	26	0.258		-3.95	**	-3.44		-5.96	**	0
Belgium	GERD	18	0.806	**	-2.31		-1.21		-3.95	***	1
	PATM	18	0.571	**	-2.85	*	-2.26		-4.89	**	1
	GDP	18	0.935	**	-3.10		-2.55		-2.63		0
	GFKR	18	0.813	**	-1.27		-2.13		-3.02		~1
	GIPD	18	0.785	**	-1.45		-2.10		-2.68		~1
	BERD	17	0.895	**	-2.13		-2.50		-1.90		?
	HERD	17	0.519	*	-2.47		-2.73		-3.62	*	1
France	GERD	32	0.922	**	-1.38		-0.79		-3.79	***	1
	PATM	32	0.774	**	-2.55		-1.84		-7.81	***	1
	GDP	32	0.966	**	-2.31		-1.79		-5.44	***	1
	GFKR	32	1.011	**	+0.17		-0.73	**	-3.81	***	1
	GIPD	32	0.775	**	-1.86		-3.80		-4.01	***	~1
	BERD	32	0.934	**	-1.64		0.07		-3.85	**	1
	HERD	32	0.894	**	-1.54		-0.51		-5.27	**	1

Table 2B.1   continued

Country	Variable	Obs.	Autoreg. Coefft.	Sig.	DF	Sig.	ADF	Sig.	DDF	Sig.	Int.
Germany	GERD	31	0.882	**	-2.78		-1.85		-3.77	**	1
	PATM	31	0.777	**	-2.62		-1.66		-7.42	**	1
	GDP	31	0.994	**	-0.27		-0.38		-5.51	**	1
	GFKR	31	0.884	**	-1.62		-1.64		-2.91		~1
	GIPD	30	0.783	**	-1.82		-2.49		-3.19	*	1
	BERD	31	0.908	**	-2.11		-1.71		-3.39	*	1
	HERD	31	0.788	**	-2.45		-2.61		-4.57	**	1
Austria	GERD	19	0.753	**	-4.74	**	-5.97	**	-7.21	**	0
	PATM	19	0.647	**	-2.98		-2.64		-4.04	**	1
	GDP	19	0.877	**	-3.00		-2.16		-4.07	**	1
	GFKR	19	0.815	**	-2.08		-3.34	*	-2.77		0?
	GIPD	19	0.588	**	-2.33		-3.34	*	-3.22		1
Switz.	GERD	22	0.576	*	-2.10		-0.41		-5.65	**	1
	PATM	23	0.695	**	-2.27		-1.81		-4.75	**	1
	GDP	23	1.034	**	+0.70		-0.10		-2.84		~1
	GKFR	23	0.837	**	-1.34		-1.73		-2.35		~1
	GIPD	12	0.615	*	-1.80		-2.42		-2.22		?
	BERD	22	0.399		-2.95		-1.46		-5.48	**	1
	HERD	24	0.994	**	-0.08		+0.40		-6.52	**	1
Denmark	GERD	24	1.024	**	+0.86		-0.17		-2.07		~1
	PATM	24	0.314		-3.29	*	-1.38		-11.14	**	0?
	GDP	24	0.942	**	-1.64		+0.42		-3.97	**	1
	GFKR	24	0.886	**	-1.28		-1.93		-3.95	**	1

Country	Variable	N	(1)	(2)	(3)	(4)	I
Finland	GIPD	24	0.524*	-2.57	-2.88	-4.70**	1
	BERD	23	1.039***	+1.30	+1.59	-2.84	~1
	HERD	23	1.020***	+0.27	+0.36	-3.20*	1
	GERD	20	0.982***	-0.69	-0.81	-3.27*	1
	PATM	20	0.919***	-0.88	-0.60	-4.93**	?
	GDP	20	0.884***	-2.61	-1.70	-2.27	~1
	GFKR	20	0.940***	-0.43	-0.87	-2.49	1
	GIPD	20	0.794***	-1.20	-1.39	-3.56*	1
	BERD	20	0.990***	-0.30	-0.18	-4.17***	1
	HERD	20	0.922***	-1.08	-1.02	-3.48*	1
Sweden	GERD	16	0.950***	-0.72*	-1.01*	-3.87**	1
	PATM	16	0.275	-3.44	-3.23	-4.27**	0
	GDP	16	0.918***	-1.46	-1.41	-3.13*	1
	GKFR	16	0.858***	-0.79	-0.89	-3.43*	1
	GIPD	16	0.298	-2.65	-2.72	-3.18*	1
	BERD	15	1.060***	+0.72	+0.46	-2.53	~1
	HERD	15	0.905***	-1.17	-1.07	-2.87	~1
Norway	GERD	22	0.889***	-1.84*	-1.47	-2.48	~1
	PATM	22	0.710***	-2.09**	-2.08**	-4.62***	~1
	GDP	22	1.021***	+0.71*	+0.81*	-3.28*	1
	GKFR	22	0.718***	-1.65	-1.37	-4.14***	1
	GIPD	20	0.702***	-1.79	-1.00	-4.04***	1
	BERD	21	0.970***	-0.54	-0.49	-2.76	~1
	HERD	22	0.664**	-3.04	-2.31	-3.13*	0
Iceland	GERD	18	0.963**	-0.38	+1.57**	-4.84***	1
	PATM	18	-0.135	-6.18	-3.97	-7.79***	0
	GDP	18	0.893**	-3.22	-2.50	-2.96	0
	GFKR	18	0.927**	-0.63	-0.64	-4.86**	1

Table 2B.1 *continued*

Country	Variable	Obs.	Autoreg. Coefft.	Sig.	DF	Sig.	ADF	Sig.	DDF	Sig.	Int.
	GIPD	18	0.618	**	-2.80		-1.56		-4.32	**	1
	BERD	16	1.052	**	+0.69		+1.82		-3.45	*	1
	HERD	16	0.843	**	-1.18		-0.37		-4.34	**	1
Ireland	GERD	21	1.042	**	+0.45		+1.06		-1.45		?
	PATM	21	0.763	**	-1.80		-1.01		-6.02	**	1
	GDP	21	1.000	**	+0.01		+0.39		-3.74	*	1
	GKFR	21	0.846	**	-0.99		-0.72		-3.02		~1
	GIPD	21	0.905	**	-0.76		-0.72		-3.21		1
	BERD	21	1.141	**	+2.40		+2.03		-0.69		?
	HERD	22	0.954	**	-0.71		+0.13		-3.22	*	1
	GERD	20	0.429		-2.68		-2.99		-4.98	**	1
UK	PATM	20	0.430		-2.75		-1.99		-6.20	**	1
	GDP	20	0.954	**	-1.15		-1.48		-5.26	**	1
	GFKR	20	0.834	**	-1.00		-1.58		-2.16		~1
	GIPD	17	0.237		-2.57		-1.55		-2.46		?
	BERD	20	0.487	*	-2.42		-1.74		-4.64	**	1
	HERD	26	0.950	**	-0.77		-0.40		-4.61	**	1
Spain	GERD	28	0.991	**	-0.36		-0.38		-4.91	**	1
	PATM	28	0.841	**	-1.65		-1.03		-6.74		?
	GDP	28	0.953	**	-2.32		-0.46		-2.74	*	1
	GFKR	28	0.893	**	-1.12		-1.79		-3.27		1
	GIPD	28	0.896	**	-1.30		-1.77		-3.84	**	1
	BERD	28	0.971	**	-0.81		-0.54		-5.50	**	1

		Obs.	Autoreg. coefft.	DF	ADF	DDF	Int.
Portugal	HERD	28	1.021 **	+0.49	−0.01	−3.73 **	1
	GERD	13	1.133 **	+0.64	+3.37	−2.45	~1
	PATM	13	−0.162	−3.60 *	−2.39	−5.15 **	0
	GDP	13	1.001 **	+0.01	+0.69	−1.80	?
	GFKR	13	0.132	−2.77	−1.58	−3.77 *	1
	GIPD	13	−0.056	−3.45 *	−2.75	−3.92 **	0
	BERD	13	0.722 *	−1.07	+0.35	−3.82 *	1
	HERD	13	1.310 **	+2.41	+6.14	−1.39	?
Italy	GERD	31	0.949 **	−1.38	−1.50	−3.87 **	1
	PATM	31	0.861 **	−1.86	−1.45	−5.49 **	1
	GDP	31	0.965 **	−1.94	−1.91	−7.03 **	1
	GFKR	31	0.826 **	−2.06	−0.21	−4.99 **	1
	GIPD	31	0.638 **	−2.53	−2.05	−5.73 **	1
	BERD	31	0.950 **	−1.24	−1.43	−3.60 *	1
	HERD	31	0.905 **	−1.22	−1.18	−5.28 **	1
Greece	GERD	11	1.137 **	+0.83	+1.40	−3.78 **	1
	PATM	11	−0.092	−3.17	−2.13	−7.05 **	1
	GIPD	11	1.290 **	+2.24	+2.22	−1.44	?
	GFKR	11	0.330	−3.86 *	−3.79 *	−3.77 *	0

Notes:

Obs. = no. of observations.

Autoreg. coefft. = autoregression coefficient (i.e. value in *t* regressed on value in *t*−1).

DF = Dickey–Fuller statistic.

ADF = Augmented Dickey–Fuller statistic (allowing for higher-order lags).

DDF = Dickey–Fuller estimated on changes in the variable.

Int. = estimated order of integration (~ = approximate, ? = ambiguous).

* = significant at 5% level; ** = significant at 1% level.

APPENDIX C

Table 2C.1   Results from the key variables

From \ To		GDP	GFKR	GIPD	GERD	BERD	HERD	PATM
GDP	SR		+H	+H	−M/L	~M/L	−M/L	+L
	LR		−H	−M/L	+M/L	+M/L	+M	+H
GFKR	SR	+H			~L/M	~L	−L/M	+L
	LR	−M			−L/M	−M/L	−M	+M
GIPD	SR	+H			~L/M	~L/M	−M	+L
	LR	−M/H			~M/L	~L/M	−M	+M
GERD	SR	−M/L	~L/M	~L/M				
	LR	+L/M	−L	−L/M				
BERD	SR	~M/L	~L/M	~M/L				
	LR	+M/L	−L/M	~L/M				
HERD	SR	−M/L	−L/M	−M				
	LR	+M	−M/L	−L/M				
PATM	SR	+L	+L	+L				
	LR	+M	−M	−M				

Notes: H = high proportions of significant correlations; M = medium; L = Low (etc.); + = mainly positive correlations; - = mainly negative (cases where there were roughly equal numbers are shown as ~); SR = short run, LR = long run.

Table 2C.2   Summary of panel data two-period estimates

From \ To		PATM		GDPC		GFKR	
		50/73	73/95	50/73	73/95	50/73	73/95
PATM	SR			−L	+L	~L	+L
	LR			+L	+L	+L	+L
GDPC	SR	−L	+L			+L	+M
	LR	+M	+M			+M/L	−L
GFKR	SR	+L	+L	+L	+M/L		
	LR	+L/M	−M/L	+L/M	−L		

Notes: See Table 2C.1.

APPENDIX D

Table 2D.1 Results of ECM panel data regressions, abstracted

Row	Dep./indep. var. (change in) i	R-bar sq. ii	Time trend iii	+/− coeffts lagged dep. var. iv	+/− coeffts change in indep. var. v	+/− coeffts lagged indep. var. vi	Short-run impact vii	Long-run impact viii
1	EMPM/GDP	0.362	+1.32	4 / 14	16 / 2	0 / 18	+	−
2	GDP/EMPM	0.549	+0.864 ***	0 / 18	17 / 1	13 / 5	+	+
3	EMPM/GFKR	0.266	−0.028 ***	13 / 5	15 / 3	10 / 8	+	~
4	GFKR/EMPM	0.170	−0.013	0 / 18	9 / 9	11 / 7	~	+
5	BERD/EMPM	0.154	+0.002	3 / 14	9 / 8	8 / 9	~	~
6	EMPM/BERD	0.357	−0.065	5 / 12	10 / 7	13 / 4	~	+
7	EMPM/PATM	0.242	−0.041 ***	2 / 16	16 / 2	13 / 5	+	+
8	EMPM/GERD	0.322	−0.055	3 / 14	8 / 9	14 / 3	~	+
9	EMPM/RWR	0.376	+2.26	2 / 16	9 / 9	0 / 18	~	−
10	RWR/EMPM	0.089	0.063 ***	0 / 18	10 / 8	15 / 3	~	+
11	GFKR/RRI	0.105	−0.029 ***	0 / 17	8 / 9	8 / 9	~	~
12	RRI/GFKR	0.351	0.042 ***	0 / 17	10 / 7	0 / 17	~	−

Notes:

Col. i: 'dependent' variable followed by 'independent' variable.

Col. ii: R-bar squared.

Col. iii: time trend and significance (*** = 1% level).

Col. iv: no. of positive/negative coefficients on the lagged 'dependent' variable.

Col. v: no. of positive/negative coefficients on the change in the 'independent' variable.

Col. vi: no. of positive/negative coefficients on the lagged 'independent' variable.

Col. vii: predicted sign of 'short-run' impact from col. v (~ = indeterminate).

Col. viii: predicted sign of 'long-run' impact from cols iv and vi (~ = indeterminate).

3. Europe in the triad: growth pattern and structural changes

Pascal Petit[1]

This chapter aims to assess the main features of structural changes, economic growth and employment in Europe as a whole when compared with economies of comparable magnitude such as the USA and Japan. All three entities of this triad had very contrasting experiences, whether during the two decades of sustained growth of the postwar period or during the period of slower growth of the last two decades.

In the postwar period, Europe and Japan featured as successful models of economic growth enjoying growth and full employment. Meanwhile, the US economy appeared as a more slowly growing economy with a relatively high level of unemployment. In fact, the USA was also experiencing during this period relatively high rates of economic growth, if considered in a historical perspective (see Maddison, 1991), while conversely economic growth in Europe and Japan was at the time especially boosted by a process of catching up.

The picture changed markedly in the course of the last two decades. In the context of a general slowdown of the OECD economies, and after the erratic ups and downs of a period of transition in the 1980s, the USA experienced in the late 1990s the highest economic growth rate among the large economies as well as the longest upswing phase in peacetime in its history. Meanwhile, Europe and, more recently, Japan experienced a marked slow growth with mass unemployment in the case of Europe and a new meaningful level of unemployment in Japan.

Moreover, during these last two decades the technological gap between Europe and Japan, on the one hand, and the USA, on the other, seemed to widen rather than to close. These features may be transitory and represent a phase in the continuing reorganization of the world economy and in the diffusion of a new technological system centred upon the technologies of information and communication. The patterns of cyclical moves seem to have changed with time. Business cycles in the OECD economies first appeared in the 1980s to have increased their simultaneity while internationalization was increasing. This move reversed in the early 1990s, when Europe and Japan went into a long and marked recession, and when the US economy bounced back rapidly and entered a lasting upswing phase. Europe and Japan thus seemed in the late 1990s to be back in the position where they have to catch up with a technological progress led by the US economy. Differences in levels of productivity by kind of activities help to assess this gap but they also point to some difficulties of measurement when output combines to an unprecedented extent tangible and intangible characteristics, the value of which is conditioned by the limited information and know-how the users may have of them.

How this dynamic of technical change and economic growth is linked with the dynamics of employment is another major question. The USA has become the country of full employment, having been for decades the country with a slack labour market. The notion of a high natural rate of unemployment attached to the US economy until the early 1990s is an indication of this past situation. High unemployment has in the meantime become a European phenomenon. It is clear that much of this difference is tied up with the labour content of economic growth. Differences in employment ratios, for example in the shares of the population between 15 and 65 years of age who are employed, are thus very telling. These ratios increased constantly in the last decades in the USA and in Japan, while they declined strongly in Europe on average.

How important and lasting are these differences in the dynamics of technical change, growth and employment in the 'triad' of Europe, the USA and Japan? These are the broad questions that this chapter wants to address in order to recall the origins, timing, magnitudes and perspectives of the major structural changes at work in these economies and which constitute the major background of all the issues on structural change addressed in this book. For this purpose three structural changes, technological change, internationalization and tertiarization, are considered in particular.

Our goal is to be comprehensive and synthetic in comparing the growth patterns of Europe, Japan and the USA. We shall try to see Europe in the first place as an entity and not as a set of different countries, though we shall not neglect the fact that this heterogeneity of Europe and the process of integration which has been going on for the past four decades are important factors in explaining the economic performances of Europe. Moreover, we shall focus our questioning upon an issue which concerns more especially Europe: that of unemployment. In fact it is in Europe that the general commitment in favour of full employment received the broadest legitimacy and marked most strongly the radical change towards 'modern capitalism' in the postwar period – all of which strongly contrasts with the overall persistence of massive unemployment which seems until now to be the plague of Europe.

Our comparative analysis of growth patterns thus runs as follows. Section 1 presents the main features which characterize the growth of productivity and employment in the three 'economies' during the two periods of the 'Golden Age' and of the last two decades. It provides us with a set of stylized facts on the growth patterns. Section 2 then presents, with the help of a survey of unemployment theories, the range of explanations for unemployment and their relative relevance to the economies under view. This helps us to feature the status of unemployment and employment in each country and how it may be evolving in the current period.

Section 3 tries to assess the extent to which the main structural changes that we have indicated transform the inner fabric of the 'economies' of the triad. It goes on to emphasize the importance of the sectoral dimension in the dynamics of employment. The links between the overall sectoral evolution and the relations with the changes in technology and internationalization are investigated. Section 4 draws some preliminary conclusions on the growth patterns and trends and questions the relative capacity of Europe as a whole to catch up and to reverse the adverse trends of employment.

1. PRODUCTIVITY AND GROWTH IN THE TRIAD: FROM THE 'GOLDEN AGE' TO THE YEARS OF STAGNATION

To compare the trajectories of Europe, the USA and Japan, we shall first make use of simple figures illustrating the relative growth of productivity (output per person employed) and output over long periods of time covering respectively the 1960s and the 1980s and 1990s. To contrast the period of rapid economic growth for most OECD countries of the 1950s and 1960s with the period of slow economic growth, we shall select two distinct periods: 1960–70 and 1980–2000. This skips the 1970s, a period of troubled transition, with the first oil shock, the sharp rise in price inflation and the large ensuing swings in exchange rates.

The first series of four figures map the growth rates of productivity versus the growth rates of output for the three economies as well as for a sub-set of European countries which indicates the heterogeneity of European experiences. The figures give a rapid account of the dynamics of productivity and output which was at the core of the rapid growth of the 1960s. The conclusions are straightforward and stress the magnitudes of the changes between the two periods. Both growth rates of productivity and output over the 1960s (Figure 3.1) recall the exceptional experience of Japan and the relative slow growth in the USA, while Europe presents an average performance, with all its members states ranking somewhere between the USA and Japan. The

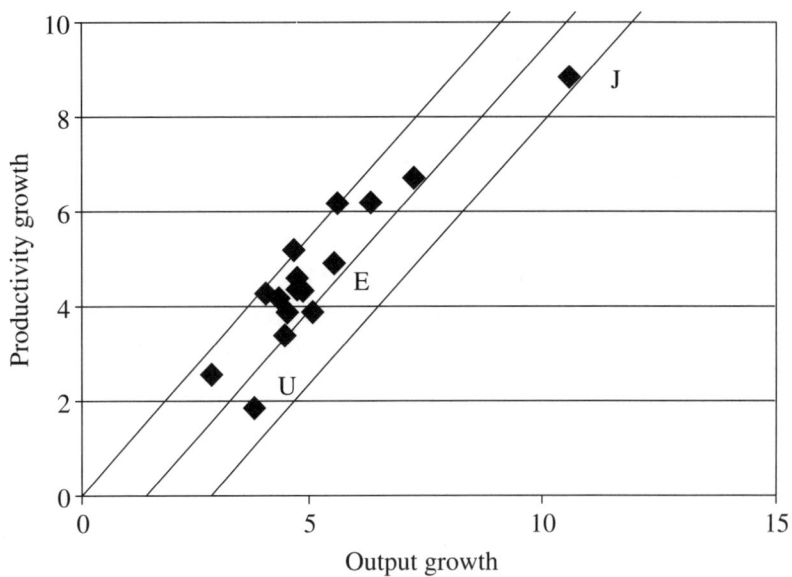

Figure 3.1   Productivity and output growth, 1960–71

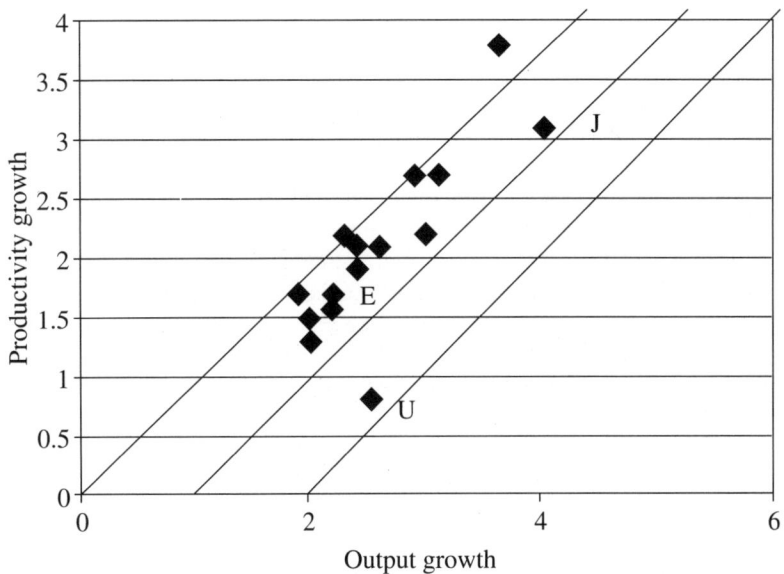

Figure 3.2   Productivity and output growth, 1981–90

whole of Figure 3.1 confirms the idea of a linear relationship between productivity and output growth (*à la* Kaldor and Verdoorn). In the 1980s (Figure
3.2) the growth rates have been halved, still preserving the overall linearity of
the relation, with a noticeable outlier, the USA, where overall productivity
growth has been reduced to a greater extent than output (which already gives
an insight on the origin of US employment growth). The relative positions of
the three economies of the triad have also changed slightly, with the USA
overtaking Europe, by a small margin, in terms of output growth (but not in
terms of productivity growth) while Japan has kept its leadership.

The first half of the 1990s (Figure 3.3) shows a drastic change in relative
positions, as well as in the general feature of the graph. Growth rates were
low in both productivity and output (with no linear relationship between
them) and Japan for the first time shared the poor fate of most countries, the
only outlier in the set of European countries that we retained being Ireland.
Only in the late 1990s did we again find a seemingly linear relation between
productivity and output growth rates (Figure 3.4), with Japan at the lower end
of the achievements and the USA being slightly ahead of the performances of
Europe. The greater heterogeneity of the results within Europe showed in the
better results of countries like Ireland, Portugal and Finland, as opposed to
the poor results of Italy, France or Germany. It should be noticed that the
return of productivity growth in the USA in the 1990s is still relatively low if
compared with other countries in the 1960s. It follows that the strong dynamics of employment in this country from the mid-1990s onwards is largely
linked with low productivity achievements. We can demonstrate the more or
less labour-intensive nature of the various growth patterns by looking at the
relation between output and employment. For this purpose we have used in
the Figures 3.1 to 3.4 three lines corresponding to iso-employment growth
curves of, respectively, 0, 1 and 2 per cent, which show that in all periods the
USA has effectively been closer than expected to the highest employment
line, while Europe was, on the contrary, the most distant.

In section 2 we investigate the reasons behind these differences, in particular in terms of the working of the labour markets. We will return in section 3
to the issues regarding the changes in the catching-up processes that have
been raised in the introduction and that are reflected in the changes in the
ranking of countries shown by the Figures 3.1 to 3.4.

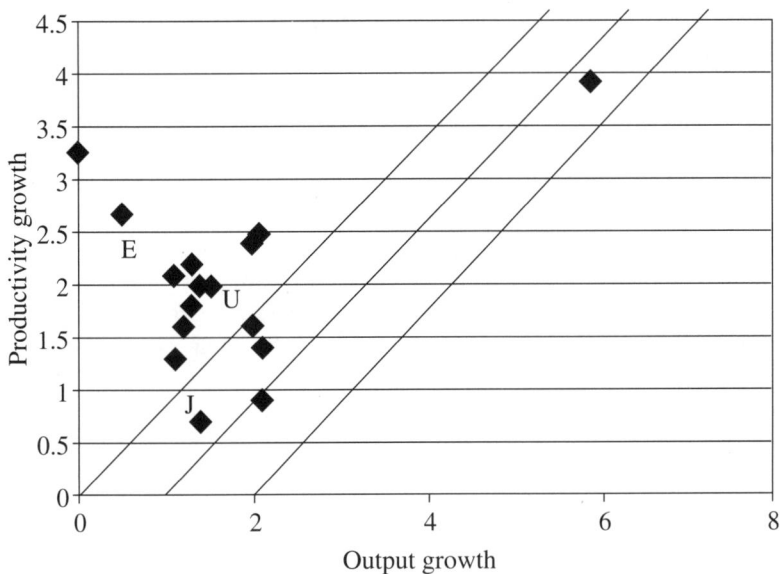

Figure 3.3   Productivity and output growth, 1991–95

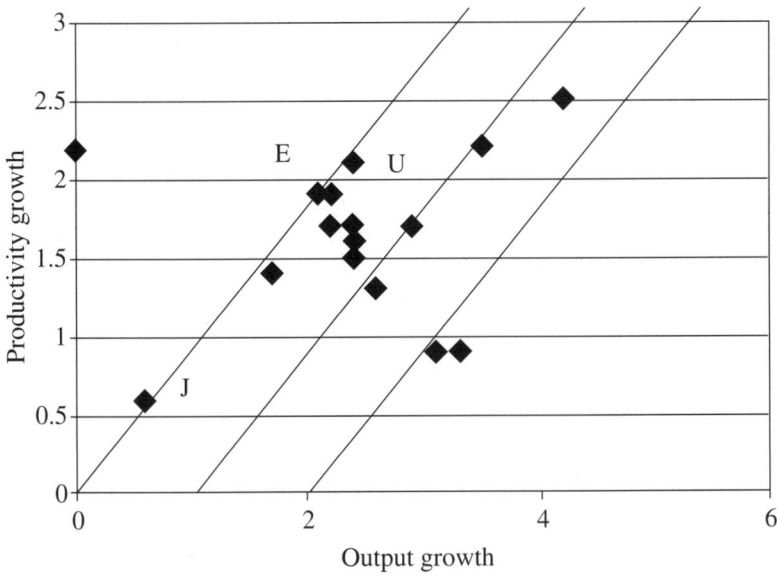

Figure 3.4   Productivity and output growth, 1995–2000

2. UNEMPLOYMENT AND PARTICIPATION: COMPARING COUNTRIES' PERSPECTIVES

Unemployment is not a straightforward category that can be compared without caution among countries with rather different social and occupational structures. Certainly this caveat is familiar when comparing countries where the importance of the agricultural sector differs strongly. Even in economies where the number of people engaged in agriculture has been largely reduced, employment ratios and demographic structures still display strong differences, which imply different status for the wage labour and for unemployment. On top of this, countries also differ strongly in the composition of employment, in terms of part-time, temporary and tenure jobs, and so on. If we add that the simple registration of unemployment depends on the institutional context (on the rules of entitlement and on the benefits allocated to the unemployed) it is not straightforward to interpret differences observed across countries in unemployment rates.

Figure 3.5 illustrates the rising trend of unemployment in Europe beyond its cyclical variations of the last three decades.[2] Conversely, the cyclical variations of unemployment in the USA seem to follow a downward trend, while in Japan unemployment remained low and steady, though slightly rising at the end of the period. Figure 3.6, looking at the employment ratios, such as the proportion of the 15–65 age group that is employed in each period, displays quite opposite trends. While these ratios have been steadily rising in the USA and in Japan over the last three decades, they slowly declined in the EU.

Again this result mixes together very different experiences in Europe,[3] but still the contrast is striking, with Europe combining rising unemployment and decreasing participation when the USA is having the exact opposite. How to explain unemployment in such a diversified context? We shall appreciate hereafter how relevant are the various explanations that are usually given to account for involuntary unemployment. Let us consider first the theories about the ways in which the labour market works as a cause of unemployment.

Unemployment may, for instance, be interpreted as a period of search for a better job. The acceptance of information imperfections in labour markets (together with Stigler's 'optimal search': Stigler, 1962) provided the basis for a revival of voluntary 'unemployment', which now was voluntary search unemployment. Search theories explain rising unemployment by anti-cyclical job search (it is assumed that workers search more efficiently while unemployed).[4] Within this framework, replacement rates, unemployment insurance in general, and so on, will lead to lower search costs, more search and thus higher unemployment. High unemployment is, according to search theories,

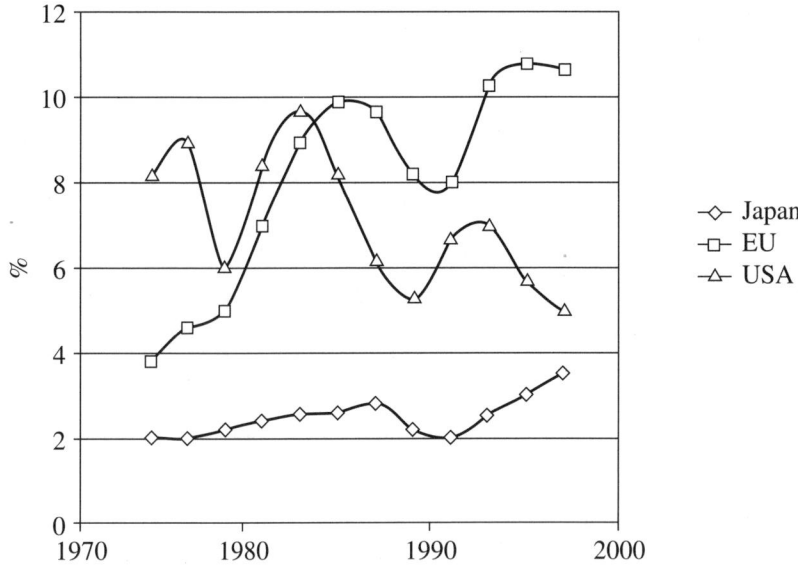

Figure 3.5   Unemployment rates

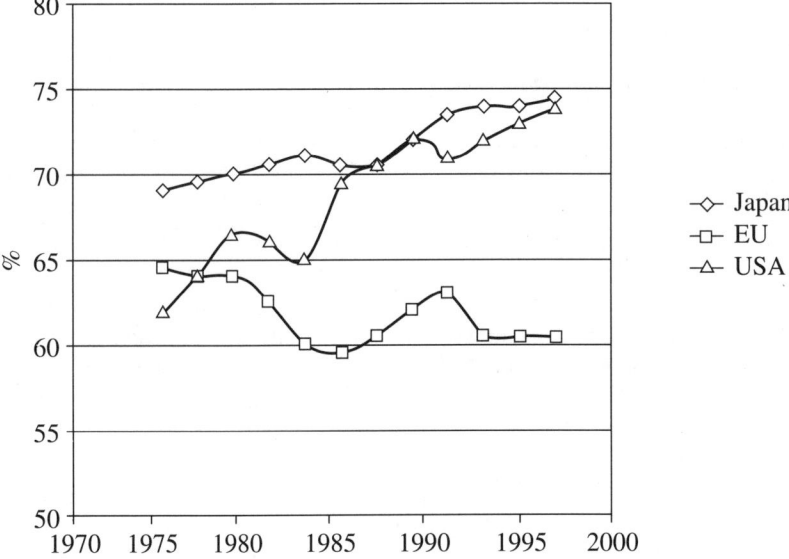

Figure 3.6   Employment ratios

caused by distorted incentives for workers' mobility. Search costs are too low as the effect of unemployment insurance.[5]

It is also claimed, however, that incentives for mobility are, especially in the European welfare states, too low. The common view is that employment protection, low wage differentials and so on have destroyed mobility incentives.[6] 'Eurosclerosis' approaches usually regard external shocks in connection with rigidities as the main cause of European unemployment. It seems that frictions and rigidities have been identified in labour markets only: insiders exclude outsiders, unions use monopoly power to push up wages above the market-clearing level, unemployment benefits raise the inflow into and the duration of unemployment, employment protection and generous social security measures destroy workers' mobility.[7] These trends may have caused the natural rate (or the NAIRU, although it is not necessarily identical) to move upwards. In other words, reducing unemployment requires a revitalization of labour market mechanisms, that is, a deregulation of the (European) labour markets.

No doubt unemployment is structured; it is concentrated among specific groups: the elderly and the less skilled; but this may be the result of various trends, of which institutional phenomena are only one. Excess supply of labour for a longer period may lead to selection processes in labour markets, which in the end may leave certain negatively selected groups to be concentrated among the unemployed (Reder, 1964). But the cause may very well be lack of aggregate demand, for example. Structured unemployment may result from high unemployment itself, as has been emphasized in 'hysteresis' models.[8]

Both lines of arguments, 'Eurosclerosis' and search theories, although contradicting each other, are based on the assumption that social welfare state measures bias microeconomic incentives and thus lead to an incompatibility of individual and social optima. Social welfare states shifted the 'natural rate of unemployment' upwards.[9] 'Eurosclerosis' as well as search theories would be a convenient explanation for rising unemployment if either the pace of structural change increased over time or institutional changes made search more attractive (search theories) or slowed mobility (Eurosclerosis). But, if anything, structural change slowed compared to the 'Golden Age' and analyses of the functioning of labour markets based on the flow approach do not provide evidence for increasing malfunctioning (for a summary of international studies, see Schettkat, 1996).

Another popular supply-side hypothesis emphasizes demographic and labour supply factors as an explanation for high, persistent unemployment. The puzzle that proponents of this hypothesis have to solve, however, is that low increases in unemployment and a tremendous growth in employment-to-population ratios occurred in countries in which the population grew and/or in which labour force participation increased (see Figures 3.5 and 3.6). Sweden and the USA both experienced low increases in unemployment from the 1960s to 1980s, but at the same time the population and/or labour force participation rose substantially. Germany, France and the UK, on the other hand, did not experience these trends but unemployment rose in these countries substantially.

Changes in the 'natural rate' itself instead of fluctuations of the actual unemployment rate around the 'natural rate' have been proposed by David Lilien (1982), who argued that cyclical variations in unemployment should be interpreted as variations in the 'natural rate' caused by industry-specific productivity shocks following restructuring. Unemployment would be especially affected if restructuring is high but aggregate employment stagnates. In this situation workers have to move from shrinking to expanding industries and with frictions unemployment should be higher than in an economy experiencing steady-state growth, that is, an economy in which the structures remain unchanged (for a more detailed discussion, see Schettkat, 1992, 1996). Without doubt this process can partly explain Europe's unemployment problems.[10]

Advanced macroeconomic models allow for shifts between demand constraint and supply constraint situations (for example, Drèze and Bean, 1990). In these models capital shortage (supply constraint) may follow a period of low economic activity caused by deficient demand. Deficient demand leads to underinvestment, which then may cause supply constraints, a sort of capital hysteresis. Aside from solving capacity constraints, increasing investments may also create income effect through the multiplier mechanism. This way increasing investment can stimulate overall economic activity.[11]

Rowthorn (1995) identifies investment at the industry level as the most important source for industry-specific employment growth because investment and employment are positively correlated in his regressions. The positive correlation, however, may be caused by a third variable, demand, which may stimulate both employment and investments. Malinvaud (1994) points out that the investment motive is not only capacity expansion but also capital–labour substitution, that is, rationalization (capital–labour substitution may be distinguished from actual efficiency gains as in concepts of total factor productivity). He stresses factor–price relations as the driving force for investment in rationalization, which is certainly a main motive.

Moreover, if demand is constrained in product markets (for example, for product-specific reasons, as in Schumpeter, 1934; Freeman and Soete, 1994; Appelbaum and Schettkat, 1997) one expects the rationalization motive to dominate investments. In other words, the market context is important for the motive and the effect of investment. Malinvaud's rationalization investments together with product market trends may lead to specific structural economic dynamics. For example, if product markets for manufacturing goods are saturated, rationalization investments (caused by factor–price relations or other variables) will enforce the trend of declining employment.

'Jobless growth' is another, although less theoretically sound, but popular, explanation for stagnating employment. Economic growth, according to this approach, does not lead to employment growth any more.[12] But productivity growth is nowadays lower than in the 1960s, when full employment was the rule rather than the exception. Furthermore, some countries, like the USA, experience continuously high rates of employment growth with roughly similar GDP growth to that in the European countries (Schettkat, 1992). There is more general evidence that countries with high rates of employment growth experienced low productivity growth, and vice versa (Freeman, 1988). In addition, employment creation takes place in 'low-productivity' rather than in 'high-productivity' industries (Appelbaum and Schettkat, 1995). These studies suggest that inter-country differences in the industry structure, as well as a changing industry structure over time, may be an important explanation for diverging employment performance.

Although the industrial structure changed enormously, and nobody disputes this fact as far as changes in nominal output and employment are concerned, the interpretation of this change and its causation are not clear at all. It may be caused by changes in final demand patterns, as hierarchy-of-needs theories suggest. But it may also be that final demand patterns did not change at all but that differences in productivity growth led to diverging trends in the nominal variables. Technological change affects industries unevenly and this may be the main reason behind structural change. Changes in the inter-industry division of labour and outsourcing are other sources of structural change, which are compatible with unchanged final demand patterns. Factor costs may also influence the division of labour between industries, and trade may influence the structure of the labour market.[13]

Finally, changes in demand patterns and trends, as emphasized in Schumpeter (1934) are largely neglected as a factor influencing growth and employment. Keynesian models emphasize a lack of aggregate effective demand but this is not related to products available. It is solely the effect of insufficient income or too high interest rates.[14] Schumpeter (1934) explained the long-run variations in economic activity by the creation of new products and their finite life cycle. Demand dynamics for old products slow down over time and lead to economic stagnation until new products start a new dynamic demand cycle. The Schumpeterian long-run cycle theory is, contrary to the theories discussed above, an endogenous explanation for employment stagnation. Economic development leads to saturation in markets for 'old' products. The labour-saving effect of productivity growth dominates the market expansion effect and employment declines in saturated markets (Appelbaum and Schettkat, 1995).

Box 3.1 gives a rough overview of the unemployment theories discussed above and assesses their relevance for the USA, Japan and Europe.

BOX 3.1 UNEMPLOYMENT THEORIES AND THEIR RELEVANCE FOR THE USA, JAPAN AND EUROPE

Unemployment caused by imperfect labour markets

Search processes are strongly monitored by a network of institutions in each country. Practices are different and result in very different speed and efficiency in the search for employment. Still this mechanism cannot account for substantial levels of unemployment: it accounts for only a fraction of current levels. Search is conditioned by the institutional context of the working of external labour markets (from employment agencies' actions to local hiring practices via the use of the media). The search hypothesis has greater appeal as a representation of unemployment as residual or functional in the USA than in Europe (where public agencies facilitate the search) or Japan (where local hiring practices clearly define the paths or the queues to employment).

The relatively low levels of long-term unemployment or of youth unemployment in the USA, a country where unemployment was very low at the end of the 1990s (see Table 3.1), confirms that this search hypothesis does not lead to substantial 'natural rates of unemployment'. Explanations for massive unemployment refer more to the influences of general institutional contexts affecting both internal and external labour markets. The problem is that, often, what one sees as the cause of unemployment on one side may also be the cause of economic growth on another side. Thus a labour market giving a strong premium to insiders (people already in a job) may improve the efficiency of the labour force (through a greater commitment) and so lead to greater competitiveness and economic growth.

Therefore all the theories which refer to institutional influences may well be partially contradictory. This may well be the case in Europe, where institutions differ and countries have different combinations. So to identify and cure any institutional 'Eurosclerosis' requires as a prerequisite that we assess the overall impact of all the systems of institutions influencing the labour market. The debate prompted by the 'Eurosclerosis' argument should not thus be oversimplified. It has extended to the issue of poor jobs versus no jobs when comparing the USA and Europe.

The segmentation of labour markets (grounding the insider–outsider arguments) and the more or less unequal structure of wages, as well as the importance of unemployment benefits and of employment protection, are all major characteristics of the institutional contexts under review. An indicator of labour market rigidity in OECD (1999) (see Table 3.1) indicates that, by the end of the 1990s, European labour markets still appeared less flexible than the US labour market (though with large differences among member countries) but chiefly this OECD study underlines that this rigidity has no clear impact on the level of unemployment, but clearly affects its structure by age, sex and level of formation.

Unemployment caused by aggregate demand deficiency

Low effective demand may stem from underconsumption and from a poor competitiveness on external markets. Underconsumption may follow (a) from an erosion of redistributive policies, once the fiscal levy has reached a particular level and has begun to be resisted, or (b) from an excess of savings to cope with the uncertainty of what appears as a transition period. Both phenomena are exacerbated when income inequalities rise. The poor have greater problems in facing current expenditures; the rich have greater incentives to save as social insecurity rises and the financial income rolls over with growing speculation. All these reasons should favour a large expansion of unemployment in the USA and a reduction in Europe, as inequalities are overall more pronounced in the USA than in Europe (see the figures on distribution and redistribution at the bottom of Table 3.1). As a matter of fact this explanation could account for growth differentials, but less for employment, where the number of poor jobs can be increased dramatically in countries with strong inequalities and low wages. Even so, the explanation could work for the 1970s and the 1980s, but not for the 1990s, when the growth of output in the USA overtook growth in Europe. The case of Japan, where distribution is rather concentrated and transfers are low, helps also to underline the fact that the relationship between distribution and growth is not as straightforward as suggested above.

Technological unemployment

The fear that technological change could create unemployment in substituting automated equipment for labour in production processes has come back. It was a constant worry debated in the economic literature until the end of the 19th century and resurfaced in the 1930s (see Petit, 1995), but the boom of the postwar period seemed to have done with this fear. It is a characteristic of the diffusion of the present technological system that it raised fears of job insecurity among workers (see Figure 3.7), even if the feared massive destruction of jobs in many trades or services in the early 1980s by the diffusion of computers did not occur.

Changes are long term and not so deterministic. It may end up with a negative effect on employment, depending on the choices of organization, but finally it depends on the ability of the economies to develop innovative goods and services.

Unemployment caused by increasing mismatches in product markets

Lack of innovative products may come from satiation of the rich, or from the budget difficulties of the poor. The satiation of the rich does not come from excess consumption but from the fact that the new products go hand in hand with new services (either marketed or self serviced) which both take time; therefore satiation is very much a time budget constraint. Data are scarce to give a rough estimation of the potential magnitude of this drift, though statistical works to correct the likely underestimation of the consumer price index (see Boskin, 1996) do show that mismeasurement is occurring especially over matters such as health and education, where the know-how of consumers is a key factor of efficiency.

The emphasis in this chapter of featuring the growth pattern of Europe *vis-à-vis* that of the USA and of Japan over recent decades tends to mask the fact that, within Europe, the paths of member countries have been somehow diverging in contrast to what had happened in the 1950s and 1960s. In the last decade, when the new technological system had already gone through the first phase of diffusion, unemployment rates clearly diverged between European countries. From a high reached in the mid-1990s, unemployment rates

Table 3.1   Labour market characteristics in Europe, the USA and Japan

	EU	Mini EU	Maxi EU	USA	Japan
Employment ratios (1997)	60.5	48.6	77.5	73.4	74.6
Unemployment rates (1997)	10.6	3.7	20.8	4.9	3.4
Youth unemployment (1997)	21.0	6.7	38.8	11.3	6.7
Long-term unemployment (1997)	5.2	0.6	10.8	0.4	0.7
Unemployment benefits	0.24	0.05	0.6	0.05	0.13
Labour market rigidity (late 1990s)	2.5	0.9	3.7	0.7	2.3
Employees receiving professional training	36	20	55	50	–
Gini coeff. 1990 income distribution	29	23	36	40.1	–
Poverty rate (1990)	12	4	37	14	4
Transfers to households as % of GDP (1997)	20.5	13	25	13.4	14.5

Sources: Rows 1 to 4, 'Employment in Europe' (EC, 1999); rows 5 to 7, *Employment Outlook* (OECD, 1999: row 5 is the amount of unemployment benefits in GDP divided by the rate of unemployment, row 6 is the second version proposed by OECD (1999) as an indicator of labour market legal restrictions, row 7 gives the rate of employees aged 25–64 receiving some kind of professional training in relation to their jobs (Europe for this indicator – see OECD, chapter 3, 1999, p. 156 – is only represented by six northern member countries); line 8, ILO (1999); line 9, percentage of population below US$14.4 (1985 PPP) a day per person, ILO (1999).

had declined by 1998 in all countries, with the exception of Germany, Greece and Austria. Still, unemployment rates remained well above 7 per cent for nine countries out of 15, with four of these countries (France, Finland, Italy and Spain) over 10 per cent. At the same time, unemployment in four other countries (Luxembourg, the Netherlands, Austria and Denmark) was under 5 per cent by the end of the 1990s, that is, at a level close to the one observed in the USA or in Japan. This perspective is somewhat at odds with the idea that unemployment is a European disease. It does affect a good number of European countries, but by the end of the 1990s it was far from a common phenomenon (even if such may have been the case during the mid-1980s). Looking at regional unemployment rates confirms the finding that unemployment is rather unequally distributed over geographic space, with some regions showing rates exceeding 20 per cent (in southern Spain, Italy, and eastern Germany: see Eurostat, harmonized regional rates of unemployment). This finally may be a trait shared with the US labour market.

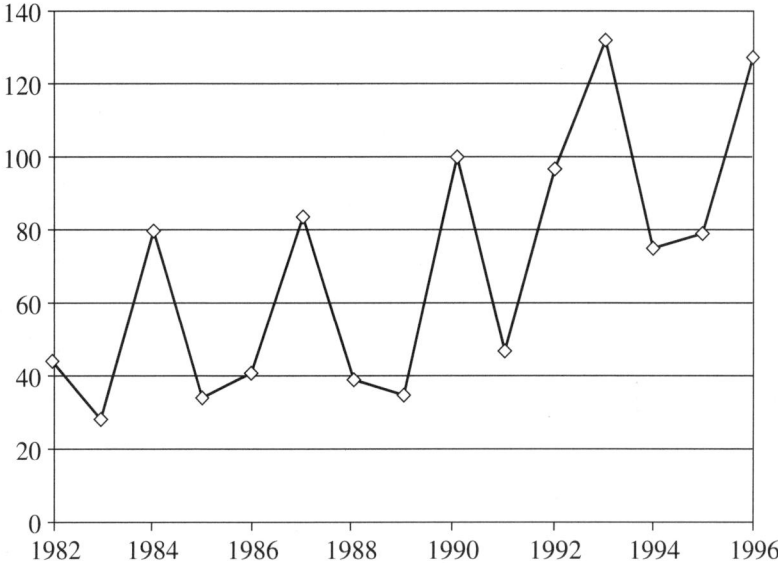

Source: OECD (1997), Employment Outlook.

Figure 3.7   Worries over job security

Conversely, a common trend of the period has been seen in the relative rise of the number of poor jobs. In effect what was supposed to be the plague of the US economy only, that is, the rise in 'poor jobs' (low-paid and precarious) now seems to appear to a significant extent in some European countries (such as the UK) as well. For example, low-wage services represented respectively 17.7 per cent and 13.9 per cent of service jobs in the USA and in the UK in the early 1990s (see Table 3.2, taken from Freeman, 1999). The data for Germany, France and Italy are respectively 11.3 per cent, 9.8 per cent and 10.6 per cent. At the same time, the percentage of jobs with high wages in services is also higher in the USA and the UK (respectively 17.9 per cent and 14.7 per cent) than it is in the other countries.

The widespread insecurity (see Figure 3.7) felt by workers may be rooted not only in the rise in unemployment (which is limited in some countries) but also in this widespread increase in the number of precarious jobs. A key issue in this respect may be the mobility of people in unemployment or in poor jobs. In effect, the outcome in the long run may be quite different between countries where low-wage jobs will have been a short experience for many, or those where it will have been a lasting experience for a smaller number. The equalizing effect of mobility is above average for some countries (such as

Table 3.2   'Good' jobs and 'bad' jobs, by sector and country (%)

	Manufacturing		Services	
	Low-wage jobs	High-wage jobs	Low-wage jobs	High-wage jobs
USA	4.2	4.9	17.7	17.9
Germany	2.3	7.0	11.3	9.7
France	2.5	5.2	9.8	11.1
UK	2.7	5.4	13.9	14.7
Italy	3.3	3.8	10.6	8.0

Source: Freeman (1999).

Germany) and some categories of workers (female and low-education workers) (see OECD, 1997, ch. 2). Many of these effects are clearly tied to certain institutions, such as apprenticeship in Germany, or to specific working of the labour market. In this last case the implications are not always straightforward. The same source (OECD, 1997) thus underlines that countries with more liberalized labour and product markets, such as the UK and the USA, do not show higher mobility which would offset their higher levels of cross-sectional inequality. Taking into account individual mobility may thus soften or enhance the differences observed in the unemployment rates of the European countries.

Whether a larger mobility of low-paid jobs is preferable for the long-run development of a country remains disputable. To share the burden of unemployment or the increase in poor jobs, a larger mobility seems to be preferable. However, in the long run this may well spread widely a detrimental general feeling of insecurity. An obvious advantage of higher mobility is that it may avoid workers in disadvantageous positions becoming definitively unfit for work. Again this has to be viewed in the light of what skills are now required from workers – and of consumers as well.

Thus, there may not be straightforward answers to the dilemma of unemployment or poor jobs. The answer depends on the dynamics of growth and, more precisely, on the learning processes which support it, in the organization of work within and between firms as well as in the organization of consumption, leisure and the domestic sphere. We shall discuss below some of these interrelated learning processes.

3. ADJUSTING TO WIDESPREAD STRUCTURAL CHANGES

The Exhaustion of the Old Engine of Growth

We have already noticed that, despite the broad transformation that had obviously occurred in the production process, the overall labour productivity gains were stagnant in all countries of the triad. Altogether, with the changes in the labour markets, this suggests that the old regime of economic growth is under strain. Changes in the sectoral composition of employment are a sign of this structural transformation. By the end of the 1990s, the service sector had grown to well over two-thirds of employment. In such a situation, the dynamism of the manufacturing sector may be a necessary condition for the economic growth of the developed economies, but it can no longer be considered as a sufficient condition. Comparing the growth of labour productivity in manufacturing and in the economy as a whole for a sample of ten European countries altogether with the experiences of the USA and Japan (see Figure 3.8) does not effectively support the idea that productivity growth in manufacturing promotes productivity growth in the economy as a whole.[15] The efficiency of the manufacturing sector has ceased to be the omnipotent factor of economic growth. It is clear that the efficiency of the manufacturing sector has to be efficiently relayed by the set of domestic services and/or by an efficient use of international markets and production opportunities. Efficient interactions with the 'logistics' of services and with the international environment have emerged at the core of the new engine of growth of nations. The development of the new technological system centred around the technologies of information and telecommunication reinforces this nexus, being itself at the crossroad of these interfaces.

We now crudely explore some of the dimensions of this nexus in investigating parts of the linkages between the three structural changes that represent the diffusion of technology, the spread of services and the growing internationalization of the economies.

Three Major Structural Changes

Three factors of change (the present phases of technological change, internationalization and tertiarization) affect every industry and nation, but in different ways and to a different extent. In other words, each of the factors is industry-specific and interacts with specific national institutional arrangements. Our purpose is to appreciate how they combine and what is their relative importance in the dynamics of employment. Let us first clarify what these major factors consist of.

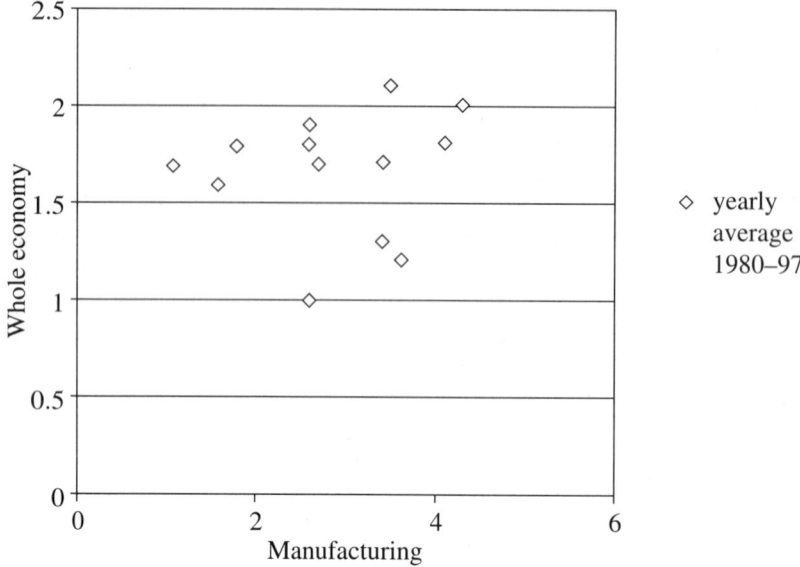

Figure 3.8   Productivity growth, 1980–97

Technological change in the form of information and communication technologies (ICTs) is universal but the speed of diffusion and its specific
application differ between industries. ICTs have formed a new technological
system, which has a pervasive impact throughout the world economy (see,
for example, Freeman and Perez, 1988). But the (beneficial) effects of ICTs
are not immediate and depend on various learning processes. In the short or
medium run, the introduction of ICTs may well thus lead to mismatches and
drawbacks in the economy, resulting in slow productivity growth (David,
1991) and deadlocks (Arthur, 1988).

The wide-ranging diffusion of ICTs is accounted for by two broad factors:
the dramatic decline in the price of information processing which goes along
with the miniaturization of microprocessors, and the convergence between
communication and computer technology which helps to link all uses and
users, locally and internationally. Different concepts or terms are used (electronic highways, the knowledge-based economy, the global information
society), which all point to a rapid increase in the information and knowledge
base of the economy closely associated with the above developments.

Internationalization takes at least three different forms. In its most basic
form, it means the growing importance of world trade relative to world
output. Second, trade flows can change in nature, concern a wider range of
countries and become more complex with the development of intra trade,

while all flows can also be complemented or replaced by foreign direct investment (FDI). Finally, the increased liberalization of financial markets has given to financial capital a new control over all these trade and FDI flows which has transformed the scene of the internationalization process. The present stage of internationalization can be qualified as one of globalization, not so much because trade flows and FDI flows have reached unprecedented levels, but chiefly because trade and FDI patterns involve an unprecedented range of countries and concern a widely diversified range of products in all destinations, with a new importance given to technologically more sophisticated products. Also the present phase of globalization is characterized by the fact that most economic transactions now take place as part of strategies that account for international conditions and opportunities. This is made possible by the diffusion of information on international opportunities, as well as by the existence of intermediaries, whether in the form of international partnerships, through the internal organization of multinational firms or via international networks of services. The large flows of information rapidly available on all sorts of topics, as well as cooperation with academics and other non-market institutions, clearly also contribute to the building up of these logistics of international mediation.

The development of advanced telecommunications has closely accompanied and enabled this internationalization of firm strategies. Information networks have been gradually developing, slowly leading to the recent explosion of the Internet, which again offers new potential for international transactions. This significant enlargement of the scope of actions may not appear in the balance of payments, as long as it does not lead to effective transactions. Still the new scope to act at international levels and its effects on national economies is a key feature of the present phase of globalization. It leads to a larger capacity for a larger number of (small) firms and even households to develop international transactions.

While the new characteristics of this phase of internationalization have been widely described, the phenomenon of *tertiarization* of activities may not have been emphasized enough as a central element in the development of knowledge-based economies. Tertiarization in a broad sense is not a new phenomenon, and the growth of the share of employment in the services sector follows a very long trend, not altered by the relative slowdown in economic growth which has affected most developed economies since the mid-1970s. Still the tertiarization trend experienced over the last two decades has some distinctive features, of which the most marked is the unprecedented growth of business services. Their share in total employment rose from some few percentage points to nearly 10 per cent on average (see Chapter 6 in the present volume) with a net lead for the USA with over 15 per cent of employment in this category in the early 1990s.

The business services sector constitutes a new phenomenon, characteristic of a new fabric of relations between firms, and consists of a wide variety of specialized services to business (from audit and research labs to cleaning and surveying). These new service activities mix very diverse types of activities, some using highly qualified labour and others requiring really low-qualified labour. The development of the highly qualified service activities in many new market niches is clearly linked with the diffusion of the new ICTs, directly for those dealing with the implementation of the technologies (all the industries linked with hardware and software businesses) and indirectly for those helping to exploit the possibilities of the new ICTs in terms of connections, accumulation and exploitation of data, and communications. The elaboration and diffusion of ICTs are thus channelled by all kind of specialized and highly qualified business services (from production of software to all kinds of consultancy).

Parallel to this new expansion of good jobs and bad jobs in business services, the long-term upward trend in personal and social services went on, fuelled by steadily increasing urbanization and participation of women in the labour force.[16] This new phase of tertiarization can be partly seen as a consequence of the others (by means of trade specialization in intangible goods, on the one hand and automation of blue-collar jobs in manufacturing activities on the other). But it also develops in its own right as an important effect of largely more educated labour forces, of increased participation of women and of the net increase in the wealth of the populations under review.

The relations between the three structural changes underlined above help to assess the new context of service and industrial activities, at national and international levels, which are bearing the growth and innovative process allowed by the diffusion of the new technologies. Table 3.3 summarizes some of these linkages between the diffusion of ICTs, the development of highly qualified business services and a greater internationalization of the strategies of firms. This multidimensional perspective indicates the way in which the old growth model where manufacturing industry acted as the main engine of growth has been transformed into a multisectoral nexus governing the specific combination of uses of ICTs, forms of internationalization and employment of knowledgeable workers which developed in each country.

Looking at these structural changes helps to highlight the broad organizational challenge with which our economies are confronted. It concerns firms, how they organize their work in-house or take advantage of external suppliers (especially services), use ICTs and develop international strategies. It also concerns households, who can also fundamentally reorganize their domestic activities, making use of services and ICTs, with a greater attention to the opportunities presented by easier international access.

Table 3.3   Three interdependent structural changes

Effects of \ Effects on	Diffusion of ICT equipment	Internationalization	Tertiarization
Diffusion of ICT equipment	Imitation & networking cumulative effect	Open the way to multinational transactions and market access to smaller firms	Capacity to store, treat and communicate information spurs the development of specialized service activities
International-ization	Demand of hardware and software for intra-MNEs transactions	Networking cumulative effects, synergies at home and abroad in foreign affairs	MNEs require the development abroad of their home service suppliers
Tertiarization	Bookkeeping, information management, customer filing, intranet	Networks of specialized international services, enlarging the reach of local firms	Development of skilled professionals helping cumulative development

The assessment of the interactions between the three factors also turns out to be strongly industry-specific. Whether one considers two industries, let us say banking and car industries, how they use ICTs, organize work within and between firms or go international to sell or to produce will differ strongly not only in one country but also between countries. This is obvious when it concerns service activities which tend to be highly country-specific,[17] but it is also the case between high-tech and low-tech manufacturing industries which react differently to international competition. It may be a transitory phase, but this set of structural changes has transformed the process of catching up which allowed relatively less developed countries (still within the club of industrialized European economies) to grow more quickly. The end of the old model of catching up is rather clearly shown by the experiences of the European countries over the last two decades. Growth rates over the period 1980–96 have not been determined by the initial growth levels, as shown in Figure 3.9.[18]

Conversely, the most advanced countries may well over this period have widened the gap, enhancing their leadership by means of an appropriate dealing with the contemporary trilogy of structural changes. We have already stressed that, even if the odds were in favour of the USA, it did not show in

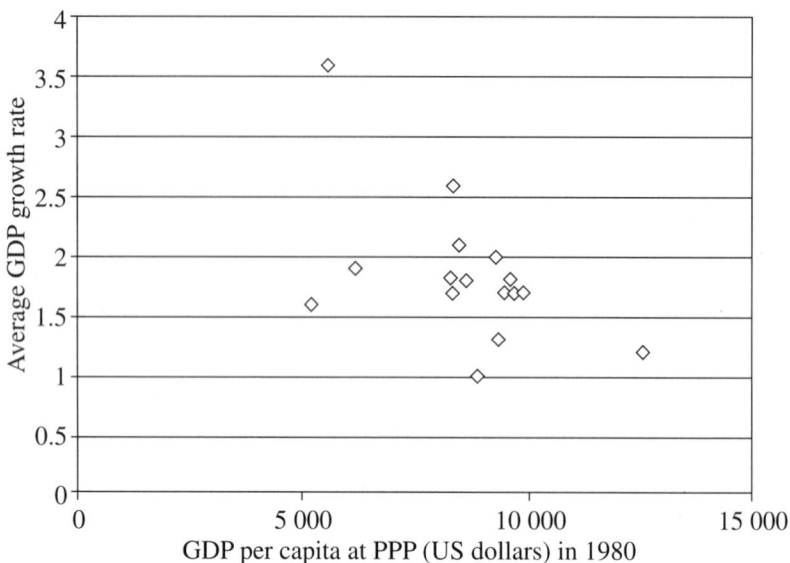

Figure 3.9   Catching up, 1980–97

the statistics of overall productivity growth. An assessment of the virtues of
the various knots of structural change is therefore premature. Still it is clearly
in the working of these trilogies that one is bound to find the characteristics
of the new growth pattern. A central reason for this is that this combination of
structural changes conveys most of the interdependence between the dynamics of production and demand which is at the heart of the cumulative causation
mechanism *à la* Kaldor on which economic growth relies.

The question is then to identify how the interactions between the three
structural factors (technological change, internationalization and tertiarization)
actually shape positive interactions between demand and supply. Technology
has an impact on the supply and demand side of markets. On the supply side,
we stressed that it blurs the borders between activities. The services sector is
generally considered to be a main user of technology and their combined uses
can upgrade the technological content of all activities, blurring in turn the
distinction between high-tech and low-tech activities, a definition which
relies heavily on the direct R&D content of activities when indirect R&D
content or knowledge content of activities are as important today.

On the demand side, IC technology has an immediate impact on income
elasticities (generally higher for more advanced or luxury-type goods) and
the intensity of price competition as against non-price competition. Productivity growth may result in falling employment, if price elasticities are low.

But substantial product innovations, implying changes in the content but also in the provision of goods and services, can compensate by stirring demand. Product innovation and price trends are both influencing competitiveness of the industry (quality competition, price competition) and thus affect the exports and imports. In mature markets with standardized products and only small variations in product quality, the share of world demand that a specific country can attract depends on the price and quality, whereas the world demand will depend on world income. In the extreme, imports may reduce demand for domestic production to zero, for example, if quality-controlled price differences are substantial (that is, if the cross-price elasticity between domestic products and imported products is high), especially in saturated markets.

The price elasticity of demand is related to product differentiation, of which innovation is one source. In the 'new trade literature', product differentiation (or 'imperfect markets') typically reduces the price elasticity of goods, by creating 'small monopolies' (for example, for different brands of cars). This is generally seen as an explanation for the existence of intra-industry trade between developed countries.

The idea of intermediation logistics refers to the effects related to the sectoral linkages in the economy in the broadest way. One issue here relates to the concept of externalities as, for example, developed in the recent 'new growth literature'. Following insights by Kaldor (1972) and chiefly expanding on Romer's and Lucas's exploitation of the endogenous growth paradigm, many studies have been devoted to the impact that some activities could have on economic growth beyond their direct contribution in terms of input–output relationships or in terms of income-generating sources. It is postulated that these 'logistics' have external effects on the efficiency of all economic activities (for example, Ashauer, 1989, and Munell, 1992, on infrastructures, Romer, 1990, on research and development). These studies are broad-brush accounts of effects which are likely to be strongly differentiated between sectors. It is important to underline that all such effects occur a priori simultaneously (which does not mean instantaneously) and interact with each other (public infrastructure, telecommunications, finance, distribution, transport and so on).

Outsourcing also figures under this heading of intermediation logistics. At the industry level, insourcing and outsourcing of production will affect labour demand as well. With outsourcing of specific indirect production activities such as marketing, labour demand in the particular industry declines, but this may just be a 'statistical artifact' because these activities are now performed in another industry. However, we expect reallocation of production (insourcing and outsourcing) to improve the efficiency of the 'vertically integrated' (Pasinetti) production process: that is to say, that the productivity of the vertically integrated production process increases through the changing division of labour between industries.

Looking at the effect of trade on the dynamics of sectors is a more common practice, even if the issue is not much simpler. In a rather extreme attempt to stylize matters for analytical purposes, we distinguish between two different sources of competition challenges: low wages and technological knowledge. Typically, within the Europe–America–Asia triad, technologically advanced nations are some parts of Europe, the USA and Canada, Japan and, recently, some of the first-round Asian NICs (Singapore, South Korea,[19] Taiwan, Hong Kong). Most of the other Asian countries, and also to a large extent (up to the mid-1980s) the aforementioned Asian NICs, largely rely on low wages in combination with specific activities such as textiles or shipbuilding.

Competition with these foreign producers takes place in global (domestic and foreign) markets under the influence of market and supply-side characteristics. We consider three interrelated 'vehicles' for such competition: exports, imports and foreign direct investment (FDI). Increased competition from foreign producers may lead to increased imports or decreased exports, and in the somewhat longer run to FDI. The complex relationship between FDI and trade (see, for example, Narula and Wakelin, 1997) allows a multitude of effects. To the extent that exports and FDI are complementary (for example because the exporting firm wants to set up production facilities closer to its foreign markets), the employment-replacing effects of imports (as stressed, for example, by Wood, 1994) may be (partly) offset by increased inward FDI after a while.

Input–output-type relationships between sectors further complicate the causal framework. Increased competition from foreign producers may open up markets for supplying firms,[20] giving rise to a link between increasing imports and exports. In the even longer run, increasing per capita income in the competing country may lead to increased demand for imports.

The exact (quantitative) effect of all of these links depends, as indicated by the sectoral production and market characteristics, on the nature of the product and its market, as well as on government policy. For example, competition from countries who compete mainly on low wages is likely to be concentrated in a few sectors only, and similarly for very technology-intensive products. Thus the sectoral mix of the domestic economy, as well as directly related government actions such as trade policy, matters.

To break down some of these interactions between structural changes, it is also useful to take into account the fact that they occur along different time horizons. Some are directly embedded in input–output flows, others are linked with the building up of stocks of tangible or intangible capital and, finally, some are setting the overall long-term context of actions, as exemplified by

Table 3.4   The three structural changes show up at various levels

Structural changes	Technological change	Internationalization	Tertiarization
Level 1 Input–output relations	Diffusion of new equipment, information and codified knowledge	Export and import trade flows	Intermediation and business services
Level 2 Tangible and intangible capital stocks	R&D investment, building up of scientific knowledge and technological expertise	FDI and financial capital investments	Learning processes in consumption and production behaviours
Level 3 Institutional contexts	Institutions for education and training, as well as institutions supporting research activities	National and international institutions governing international relations	Institutions governing the laws of market competition as well as redistribution

the regulatory frameworks or the institutional bases of various national systems.

Table 3.4 tries to summarize some of the main points of the discussion. It stresses that linkages between the three structural changes are positioned in various time horizons and that interactions could be broken down as occurring vertically, between various time horizons, and horizontally, between various kinds of structural changes.

4.   CONCLUSIONS

What to conclude regarding growth patterns in the triad? We have stressed that there has been a period, starting with the 1980s, where patterns of economic growth did not seem to operate as neatly as they did in the 1950s and 1960s. The catching-up process may have been reversed while the manufacturing sector, to say the least, had to share its role with business and intermediation services as an engine of growth. This does not mean that one can track the ingredients of the growth dynamics of countries through a mix of indicators on tertiarization (as measured by the share of some business and intermediation services), internationalization (as measured by an indicator of success in exports, for instance) and technological diffusion (as measured by the share of ICT equipment in total investment).

The combination is apparently rather complex, being country- and industry-specific, and changing over time. It may stem from the transitory nature of the present period, with a progressive convergence, initiated at level 3 of the institutional contexts (Table 3.4), unifying partly the dynamics of growth. Still the transition period is long because of the nature of the structural changes under consideration and the paths followed in this period by the various economies may turn out to be decisive in shaping their long-term future.

The idea of a new American leadership being established for the decades to come is behind such a question. But the answer may not be as certain for the long run as it looks in the medium run. If a great part of the potential of the structural changes under way depends on the abilities of countries to spread their dynamics across the board of economic activities and social groups, then the followers of the USA – Europe and Japan – can draw useful lessons and take advantage of lagging a short way behind to take a better route. The US road remains very bumpy so far. This turn of the century is crucial in that respect. Of course, the roads to success of Japan and Europe are bound to be different: while Japan has to adapt a unique national model to the present context, Europe is facing the challenge of being very diverse in its institutional roots and customs, which may be either a drawback or an advantage, something which European structural policies should consider very closely.

This raises many issues, one of them having to do with the evolution of the technological competitiveness of Europe, a challenge which has been dealt with extensively in Fagerberg *et al.* (1999a). The aspect of the question we address in this book is more oriented towards the evolution of the general growth pattern of the economies, such as the dynamics of employment, of change in production and consumption processes, as spurred by a dynamics of innovation closely dependent upon agents' interactions.

NOTES

1. We thank Ronald Schettkat (Utrecht University, Department of Social and Institutional Economics) and Bart Verspagen (MERIT, Maastricht University and Eindhoven University) for their contributions to a preliminary version of this chapter.
2. Figures 3.5 and 3.6 are taken from *Employment in Europe*, 1998, EC-DGV.
3. The situation of Europe is an average which masks, for instance, very different positions between the UK and Denmark or Sweden on the one side and Italy or France on the other.
4. Charles Holt, who developed flow-oriented, dynamic labour market analysis in the 1960s (see Holt, 1970, 1996) did not follow the main interpretation of 'search unemployment' but argued that job search is actually done on the job. Job search does not require unemployment; it is pro-cyclical rather than anti-cyclical (see also Tobin, 1972; Schettkat, 1996).
5. Recent analysis shows that the impact of replacement rates on unemployment may be insignificant. There is some indication, however, that benefit duration matters (Madsen, 1998; Hunt, 1995).

6. Layard *et al.* (1991) focus on institutions preventing labour markets from clearing. In these models product market and/or labour market demand-side constraints are neglected. (For an interesting test of the Layard/Nickell and other models, see Madsen, 1998.)

7. 'Exogenous shocks theories' are attractive because they seem to suggest that economic development like that in the 1960s is possible once the economies return to equilibrium.

8. Hysteresis processes of unemployment may be caused by various channels. One channel is wage setting by insiders together with decreasing marginal productivity of labour (Blanchard and Summers, 1986); another channel may be the human capital deterioration of the (long-term) unemployed (Schettkat, 1992, 1996).

9. Milton Friedman (1968) defined the 'natural rate of unemployment' as the level that would come out of the Walrasian system of general equilibrium, provided that the actual structural characteristics of the labour and commodity markets are imbedded in the model, including market imperfections, stochastic variability in demand and supplies, and the cost of gathering information about job vacancies.

10. Most European countries experienced stagnating employment but restructured their economies during the 1970s and 1980s. However, countries like Germany – a typical example of a European economy – had unemployment rates lower than in the USA although the restructuring took place with stagnating employment. Higher unemployment rates with more favourable growth conditions are hardly evidence for the superiority of the US labour market institutions (Appelbaum and Schettkat, 1990).

11. The usual income multiplier mechanism may apply in any case, but this is unlikely to affect specific industries.

12. 'Jobless growth' in the sense of a causal relation requires that economic growth creates more productivity growth than in the past. Actually this so-called 'Kaldor–Verdoorn relation' weakened rather than strengthened.

13. See Freeman (1999) for evidence on the impact of international trade on labour markets.

14. Keynes himself mentions that there may be periods of 'saturated investment', that is, economic situations in which investment opportunities are exhausted (Keynes, 1943).

15. The case of Finland (not represented in Figure 3.8) would be the only counter-example, but this case is too specific (as is the growth of Ireland in the 1990s) to support the thesis of manufacturing being still in itself an engine of growth.

16. This massive component of employment, which had already benefited from the rise of public budgets in the 1950s and 1960s, continued to grow over the period of slow growth of the 1980s and 1990s. It corresponds to a wide range of good and 'disadvantageous' jobs, and the combination and relative importance of sub-sectors in the field is very much country-specific.

17. The regulatory framework of the services industries plays a major role in this differentiation, as shown in the case of the USA where the Glass–Steagal Act, which severely limited the reorganization of the banking industry, was only cancelled in late 1999.

18. See also Petit (1999) on this slowing down of the catching-up process within the new context of technological change.

19. South Korea is now the second most R&D-intensive country in the world, after Sweden, and just ahead of Japan.

20. For example, the Dutch multinational firm, Philips, manufactures integrated circuits in the Netherlands which are used in video-recorders and TVs, goods which are no longer produced in the Netherlands, but for which Philips has production facilities in other EU countries, such as Belgium and Austria, as well as in the Far East.

BIBLIOGRAPHY

Appelbaum, E. and R. Schettkat (1990), 'Determinants of employment developments: A comparison of the United States and the Federal German economies', *Labour and Society*, 15, 13–32.

Appelbaum, E. and R. Schettkat (1995), 'Employment and productivity in industrialized countries', *International Labour Review*, 134, 605–24.

Appelbaum, E. and R. Schettkat (1997), 'Are Prices Unimportant? The Changing Structure of the Industrialised Economies', AWSB Discussion Paper, 97/10.

Arthur, B. (1988), 'Competing Technologies, Increasing Revenues and 'Lock-in' by Small Historical Events', *Economic Journal*, March.

Ashauer, D.A. (1989), 'Is Public Expenditure Productive?', *Journal of Monetary Economics*, March.

Blanchard, O.J. and L. Summers (1986), 'Hysteresis and the European Unemployment Problem', in S. Fisher (ed.), *NBER Macroeconomic Annual 1986*, pp. 15–78.

Boskin, M. (ed.) (1996), 'Towards a more Accurate Measure of the Cost of Living', Final Report to the Senate Finance Committee, December.

David, P.A. (1991), 'Computer and Dynamo. The Modern Productivity Paradox in a Not Too Distant Mirror', in OECD, *Technology and Productivity*, Paris.

Drèze, J.H. and C.R. Bean (eds) (1990), *Europe's Unemployment Problem*, Cambridge, MA: MIT Press.

EC (1999), *Employment in Europe 1998*, Brussels: DGV.

Fagerberg, J. and B. Verspagen (1999), 'Modern Capitalism in the 1970s and the 1980s', in M. Setterfield (ed.), *Growth, Employment and Inflation*, London: Macmillan.

Fagerberg, J., P. Guerrieri and B. Verspagen (1999), *The Economic Challenge for Europe: Adapting to Innovation-based Growth*, Cheltenham, UK and Northampton, MA, USA: Edward Elgar.

Freeman, C. and C. Perez (1988), 'Structural Crises of Adjustment, Business Cycles and Investment Behaviour', in G. Dosi, C. Freeman, R. Nelson, G. Silverberg and L. Soete (eds), *Technology and Economic Theory*, London: Pinter.

Freeman, C. and L. Soete (1994), *Work for all or mass unemployment? Computerized technical change into the twenty-first century?*, London: Pinter.

Freeman, R. (1988), 'Labor market institutions and economic performance', *Economic Policy*, 6, 63–80.

Freeman, R. (1999), 'Wages, Employment and Unemployment: An Overview. Can the EU pass the Job Test?', Wages and Employment, EC/DGV-OECD/DEELSA seminar, EC, Brussels.

Friedman, M. (1968), 'The Role of Monetary Policy', *American Economic Review*, 58, 1–17.

Holt, C.C. (1970), 'Job search, Phillips' wage relation and union influence, theory and evidence', in E. Phelps (ed.), *The Microeconomic Foundations of Employment and Inflation Theory*, New York: W.W. Norton.

Holt, C.C. (1996), 'Flow analysis of labor markets: origins and policy relevance', in R. Schettkat (ed.), *The Flow Analysis of Labour Markets*, London: Routledge.

Hunt, J. (1995), 'The Effect of unemployment compensation on unemployment duration in Germany', *Journal of Labor Economics*, 13, 88–120.

ILO (1999), *Key Indicators of the Labour Market*, Geneva: International Labour Office.

Kaldor, N. (1972), 'The Irrelevance of Economic Equilibrium', *Economic Journal*, 82, 172–201.

Keynes, J.M. (1943), 'The long-term problem of full employment', reprinted in J.M. Keynes (1980), *Collected Writings*, London/Basingstoke: Macmillan, pp. 320–25.

Layard, R., S. Nickell and R. Jackman (1991), *Unemployment: Macroeconomic Performance and the Labour Market*, Oxford: Oxford University Press.

Lilien, D.M. (1982), 'Sectoral shifts and cyclical unemployment', *Journal of Political Economy*, 90, 777–92.

Maddison, A. (1991), *Dynamic Forces in Capitalist Development: a long-run comparative view*, Oxford: Oxford University Press.

Madsen, J. (1998), 'General Equilibrium Macroeconomic Models of Unemployment: Can they Explain the Unemployment Path in the OECD?', *Economic Journal*, 108, 850–97.

Malinvaud, E. (1994), *Diagnosing Unemployment*, Cambridge: Cambridge University Press.

Munell, A. (1992), 'Infrastructure Investment and Economic Growth', *Journal of Economic Perspectives*, Fall.

Narula, R. and K. Wakelin (1998), 'Technological Competitiveness, Trade and Foreign Direct Investment', *Structural Change and Economic Dynamics*, 9, 373–87.

OECD (1997), *Employment Outlook*, Paris, June.

OECD (1999), *Employment Outlook*, Paris, June.

Petit, P. (1995), 'Employment and Technical Change', in P. Stoneman (ed.), *The Economics of Innovation and Technical Change*, handbook, Oxford: Basil Blackwell.

Petit, P. (1999), 'Integration and Convergence in the European Union', in M. Setterfield (ed.), *Growth, Employment and Inflation*, London: Macmillan.

Reder, M.W. (1964), 'Wage Structure and Structural Unemployment', *Review of Economic Studies*, 31, 309–22.

Romer, P. (1990), 'Endogenous Technical Change', *Journal of Political Economy*, 98 (5), October, pp. 71–102.

Rowthorn, R. (1995), 'Capital formation and unemployment', *Oxford Review of Economic Policy*, 26–39.

Schettkat, R. (1992), 'The Labor Market Dynamics of Economic Restructuring: the United States and Germany in transition', New York: Praeger.

Schettkat, R. (1996), 'Labor Market Flows Over the Business Cycle: An Asymmetric Hiring Cost Explanation', *Journal of Institutional and Theoretical Economics*, 152, 639–55.

Schumpeter, J. (1934), *Theorie der wirtschaftlichen Entwicklung*, Berlin: Duncker & Humbolt.

Stigler, G.J. (1962), 'Information in the labor market', *Journal of Political Economy*, 70, 94–105.

Tobin, J. (1972), 'Inflation and unemployment', *American Economic Review*, 62, 1–18.

Wood, A. (1994), *North–South Trade: Employment and Inequality*, Clarendon Press: Oxford.

PART II

Sectoral Changes and Demand

4. Structural dynamics and employment in highly industrialized economies

Ronald Schettkat and Giovanni Russo

1. INTRODUCTION

The demand side of the labour market is significantly underrepresented in economic analysis and most explanations of high and persistent unemployment in the industrialized countries focus on labour supply rigidities caused by labour laws, unions, insider power and so on. Explanations of unemployment which take demand into account usually focus on the aggregate level but abstract from structure. However, industry structure may well be important in explaining international as well as intertemporal differences in economic performance. Without doubt, industry structure as measured in employment and nominal output has changed dramatically over recent decades. Industrial economies have been transformed into service economies. Manufacturing, after its peak at the end of the 1960s or early 1970s (Singh, 1995), has declined almost everywhere. In the meantime, service industries have expanded. However, it is not clear how this shift should be interpreted since the institutional division of national accounts data leaves the sources of change to speculation. Structural change may be caused by changes in final demand patterns, as theories of the hierarchy of needs suggest, but it may also be the result of unbalanced productivity growth and constant final demand patterns measured in real terms. Finally, a changing industry structure may also be caused by variation in the inter-industry division of labour caused either by the movement of tasks to specialized firms and industries (outsourcing) or by capital–labour substitution.

In this chapter we analyse not only the pace of structural change but also its direction, both of which may affect economic development (Appelbaum and Schettkat, 1995). We analyse the latter using a new international comparable input–output database made available by the OECD in combination with the International Structural Database (ISDB).

2.   CHANGING ECONOMIC STRUCTURES

Employment is the result of the interaction of demand and supply forces in product markets. In general, product demand in any market[1] depends on income, the price of the product, its quality and its novelty. The impact of productivity growth[2] on labour demand in a specific industry depends on the distribution of productivity gains among wage earners, employers and consumers. If prices do not respond to increasing productivity (for example, if profit and/or wage growth compensate for the cost reduction made possible by productivity gains) employment in this industry will probably decline.[3] If prices decline, that is if productivity gains are passed on to buyers, the responsiveness of product demand will be important. If product demand reacts strongly to price reductions, that is if product demand is highly price-elastic, the labour-saving effect of productivity growth may be compensated or overcompensated by the expansion of product demand, and labour demand may therefore increase. If, on the other hand, product demand reacts only weakly to price reductions, that is if product demand is price-inelastic, the labour-saving effect of productivity growth will dominate the expansionary effect (Appelbaum and Schettkat, 1995).

The relationship between demand and supply forces may be summarized in the so-called 'iso-employment curve' (see Figure 4.1). The iso-employment curve is the locus of labour demand per unit of output and product demand per head. It depicts the interaction of demand and supply factors. Its negative slope indicates that changes in productivity must be compensated by changes in product demand to keep employment unchanged. Figure 4.1 summarizes the variables affecting employment, which include insourcing and outsourcing, that is, changes in the inter-industry division of labour, in addition to the productivity and general demand effects in the case of a specific industry.

Generally, domestic demand will depend on domestic income, price trends and product innovations. If substantial product innovations occur, these may compensate for saturation effects resulting from rising product penetration. However, even declining rates of productivity growth – as observed in the industrialized countries – may result in falling employment if demand responses to price reductions are weak; that is, if the price elasticities in the market are low.[4] Product innovation and price trends both influence the competitiveness of industry (quality competition, price competition) and thus affect exports and imports. In mature markets with standardized products and only small variations in product quality, the share of world demand that a specific country can attract will depend on price and quality, whereas world demand will depend on world income.

In the extreme case, imports may reduce demand for domestic production to zero, for example, if quality-controlled price differences are substantial

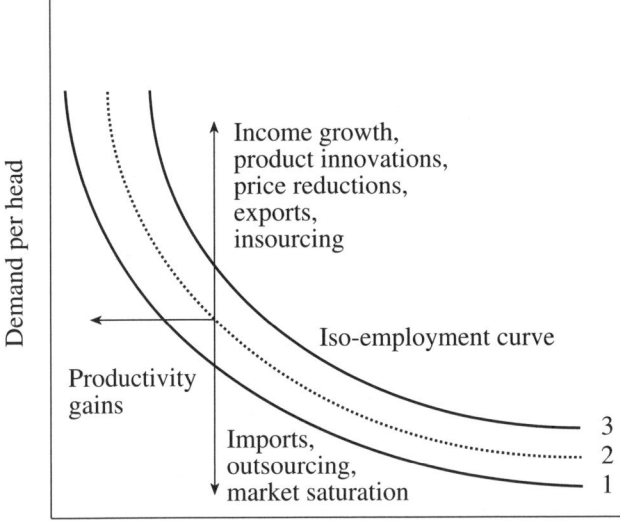

Figure 4.1 Industry-specific iso-employment curves

(that is, if the cross-price elasticity between domestic products and imported products is high because imported products can easily replace domestic products). Saturated markets (where the price elasticity of market demand is low) may be especially vulnerable.

Similar considerations apply to labour demand. The expansion of product demand (as well as labour demand) depends on the individual price of the product (own price elasticity), prices of other goods – especially imports (cross-price elasticity) – income, quality and the novelty of products. At the industry level, insourcing and outsourcing of production will also affect labour demand. If outsourcing takes place, for example of specific indirect production activities such as marketing, labour demand in the particular industry will decline, but this may just be a 'statistical artifact' since the activities concerned will now be performed in another industry. However, we expect reallocation of production (insourcing and outsourcing) to improve the efficiency of the 'vertically integrated' production process (Pasinetti, 1981): that is, the productivity of the vertically integrated production process will increase as a result of the changing division of labour between industries. Labour demand, measured as number of workers per unit of output, is, of course, influenced not only by technology (productivity) but also by average working time per worker. Improvements in technology can therefore be offset by reductions in average working time (ibid.)

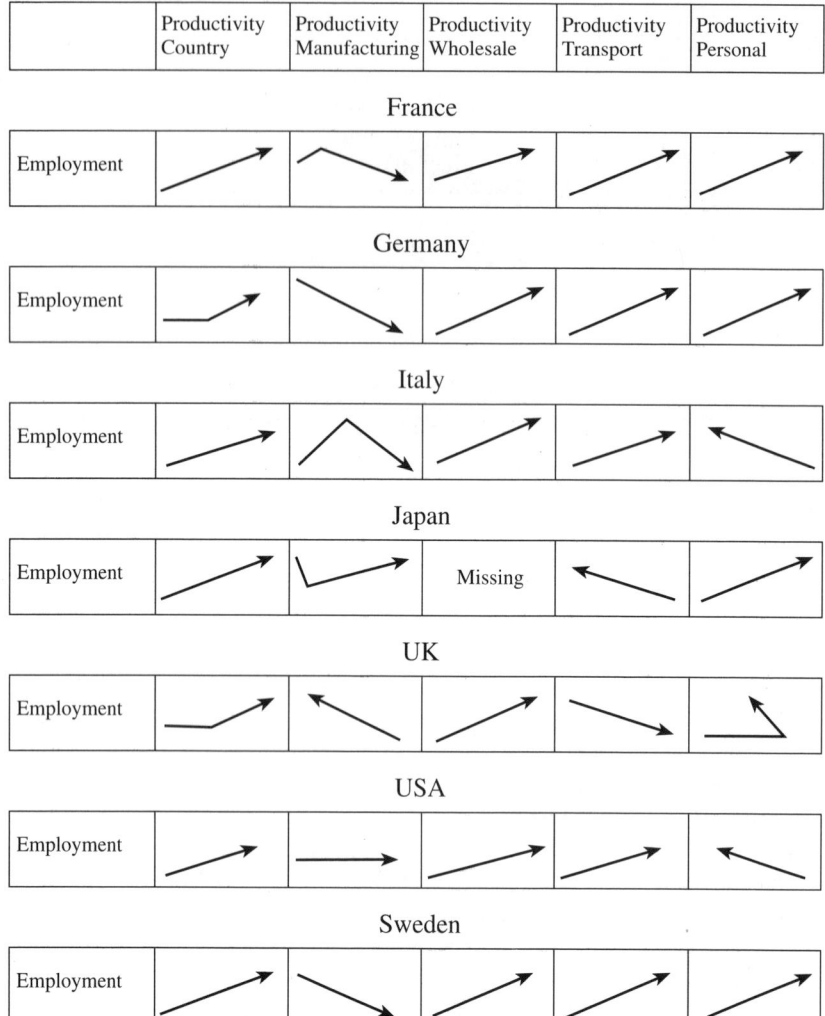

Notes: Productivity is labour productivity (value added per worker); the arrow indicates the direction of time (period 1970–90); Manufacturing (ISIC 3), Wholesale (Wholesale Retail and Trade, Restaurants and Hotels [ISIC 6]), Transport (Transport and Communication [ISIC 7]), Personal (Community, Social and Personal [ISIC 9]).

Source: OECD: ISDB (period 1970–90).

Figure 4.2 The relationship between employment and labour productivity (in levels) in different countries and in selected industries

To summarize, constant industry-specific employment requires the effects of a multitude of variables to balance. This will obviously only occur under very specific conditions (Appelbaum and Schettkat, 1997; Petit *et al.*, 1997). In practice, it is very unlikely that these effects will balance at the individual industry level, but employment decline in one industry may be offset by employment expansion in other industries. Again, this is unlikely to happen as automatically as is implicitly assumed in many economic models.

Figure 4.2 summarizes the empirical relationships between productivity and employment (in levels) both within the overall economy and within selected industries. The pattern obviously differs between industries. For manufacturing – with the exception of Japanese manufacturing industry, where exports were very important – we see a negative correlation between productivity and employment. In other words, the labour-saving effect of productivity growth dominates the market expansion effect. In the transport and communication industry, we find a positive relationship between employment and productivity, showing that compensatory mechanisms are at work. Finally, we observe a mixed pattern for the community, social and personal services industry, where a positive relationship is found in some countries (France, Germany, Japan and Sweden) and a negative relationship (productivity declining as employment rises over time) in Italy, the UK and the USA.

The analysis presented in the following section attempts to disentangle the causes of changing industry structures. These are (a) structure of final demand, (b) changes in the inter-industry division of labour (outsourcing and insourcing) and unbalanced productivity growth, and (c) capital–labour substitution. We investigate the importance of these variables in the following sections.

3. CAUSES OF CHANGING STRUCTURES

Shifts in Final Demand

Table 4.1 (upper panel) displays ratios of value added and employment in services over that in manufacturing in the usual institutional division.[5] In general, these ratios indicate a shift from manufacturing to services, although the level of the ratios differs substantially between the USA and the other countries. However, this is the familiar result obtained from national accounts statistics and may or may not reflect changes in the demand structure of the economy. Indeed, with regard to the final demand concept (lower panel of Table 4.1) the picture is less clear. In the case of overall final demand (comprising private consumption, government consumption, investment and exports), the shift to services is still present in the USA but is much less strong than the

*Table 4.1   Ratios of services to manufacturing in the institutional division
and in the final product concept (ratios times 100)*

	USA	Japan	France	Germany
		Institutional Division		
		Value added		
1970–1978	242	303	227	172
1990	281	207	315	211
		Employment		
1970–1978	277	155	188	149
1990	444	162	323	181
		Final Product Concept		
		Final demand		
1970–1978	164	138	111	89
1990	174	130	114	83
		Consumption		
1970–1978	199	218	129	121
1990	299	247	169	147
		Exports		
1970–1978	50	38	23	15
1990	31	17	31	14
		Gross output		
1970–1978	117	87	97	82
1990	159	99	145	98

Notes: Exact years: USA, 1972; Japan, 1970; France, 1972; Germany, 1978. Services: Wholesale and Retail Trade, Restaurants and Hotels (ISIC 6); Transport, Storage and Communication (ISIC 7); Financial, Insurance, Real Estate and Business Services (ISIC 8); Community, Social and Personal Services (ISIC 9); Manufacturing: Manufacturing (ISIC 3).

Sources: Computations based on OECD: ISDB (upper panel), OECD, Input–Output DataBase (lower panel).

institutional division suggests.[6] In addition, in the other countries the picture is varied. This suggests that outsourcing of service activities may have contributed to the changing industry structures observed in national accounts statistics. However, in the case of private consumption, the trend is very clear: private consumption has shifted away from manufacturing to services. This is especially true in the USA, but similar trends hold for all countries.

The ratio of value added in services to value added in manufacturing rose everywhere (except in Japan, where it actually fell, reflecting the rising share of own value added created in the manufacturing sector). But the differences in the level of these ratios are remarkable, with values of about 210 in Germany and Japan but around 300 in the USA and France. Trends and levels are even more exaggerated in the case of employment (number of workers), where the figures are of course influenced not only by intersectoral productivity differences but also by differences in the average number of hours worked.

Changes in final demand may also be brought about by international competition (Archibugi and Michie, 1998; Ben-David and Papell, 1997). Probably surprisingly, in view of the debate on trade in services, there is no uniform picture in relation to service to manufacturing ratios in exports. France experienced increasing ratios, whereas Japan and the USA saw a declining trend and Germany's ratio remained constant at a very low level. Thus, although shifts in international trade contributed to the increasing share of services, they can hardly explain it. In short, Table 4.1 shows that the trend towards services is not a pure statistical artifact but is a real change, even though outsourcing and other trends may exaggerate the shift in conventional national accounts statistics.

To gain an impression of the impact of the final demand structure on the employment structure, we regressed industries' share in overall employment[7] on the industries' share in capital stock and in total final demand, and on the ratio of industry-specific final demand to industry-specific gross output.[8] The regressions (Table 4.2) confirm the results of the previous analysis.[9] Changes in the final demand structure are an important factor in changes in the employment structure. The regressions also suggest that investment (increase in the industry share in overall capital stock) correlates positively with employment.[10] In other words, investment is mainly complementary to employment rather than simply replacing labour.[11]

Changes in the Inter-industry Division of Labour

The growth of employment in the service sector[12] and the relative (but in Europe also the absolute) decline of employment in manufacturing industries has been explained by reference to the reorganization of the production process. According to this argument, manufacturing industries have outsourced part of their service activities into specialized firms, which are then counted as part of the service sector. As a result, the national account statistics show the service sector growing and the manufacturing sector declining (Appelbaum and Schettkat, 1990). By the same reasoning, USA–Europe differences in industry composition are likewise said to be a statistical artifact. In Europe,

Table 4.2 *OLS estimates of the effects of structural dynamics of demand on employment and final product employment*

	USA			Japan			France			Germany		
	Number of obs = 20			Number of obs = 16			Number of obs = 18			Number of obs = 20		
	$F_{(3, 17)}$ = 10.04			$F_{(3, 13)}$ = 26.16			$F_{(3, 15)}$ = 11.25			$F_{(3, 17)}$ = 11.48		
	Prob > F = 0.0004			Prob > F = 0.0000			Prob > F = 0.0004			Prob > F = 0.0003		
	R-squared = 0.64			R-squared = 0.86			R-squared = 0.69			R-squared = 0.68		
	Adj R-squared = 0.58			Adj R-squared = 0.83			Adj R-squared = 0.63			Adj R-squared = 0.62		
	Root MSE = 1.38			Root MSE = 0.87			Root MSE = 1.32			Root MSE = 1.19		
	Coefficient	Std. err.	*t*-value	Coefficient	Std. err.	*t*-value	Coefficient	Std. err.	*t*-value	Coefficient	Std. err.	*t*-value
RK	0.04	0.19	0.2	RK 0.28	0.09	3.03	RK 0.85	0.22	3.86	RK 0.5	0.18	2.78
RFD	0.79	0.25	3.12	RFD 0.58	0.09	6.4	RFD 0.42	0.14	3	RFD 0.08	0.24	0.33
FDGO	−0.89	0.2	−4.52	FDGO −0.64	0.28	−2.29	FDGO −0.39	0.2	−1.95	FDGO −0.45	0.26	−1.73

Notes:
The dependent variable is annual compound growth rate of employment share (employment in industry h/employment in the country).
The independent variables are RK, annual compound growth rate of capital share (ratio of capital stock in industry h to total capital stock in the country); RFD, annual compound growth rate of share in final demand (ratio of final demand of industry h to total final demand in the country); FDGO, annual compound growth rate of share in final demand in gross output.
The annual compound growth rates are calculated for the period 1970–90 for Japan, 1972–90 for the USA and France, and 1978–90 for Germany.

as compared to the USA, the production process is, according to this argument, less specialized. In Europe a larger share of service activities is performed within manufacturing firms rather than being outsourced to specialized service firms. In national accounts, the industry definition follows the institutional criteria of the main economic activity, and similar service activities are counted as manufacturing in Europe but as services in the USA. In other words, USA–Europe differences in employment structure are not real but a statistical artifact.[13]

Given these arguments, one would expect that the share of intermediate products delivered from service to manufacturing industries in manufacturing's gross output would have increased over time and would be higher in the USA than in Europe. Table 4.3 (columns 1 and 2) shows that this is not the

Table 4.3   Share of intermediate goods and own value added in gross output

	Intermediate products in gross output		Own value added in gross output	
	Services in manufacturing 1	Manufacturing in services 2	Manufacturing 3	Services 4
Japan				
1970	14.22	10.47	20.19	67.10
1990	14.84	10.51	31.00	62.18
USA				
1972	11.77	8.90	34.31	66.33
1990	12.76	7.57	38.32	64.14
Germany				
1978	12.97	11.26	33.18	63.56
1990	16.73	9.24	30.96	62.02
France				
1972	10.82	8.27	33.46	72.30
1990	16.77	6.98	32.19	63.85

Notes:
Services: Wholesale and Retail Trade, Restaurants and Hotels (ISIC 6); Transport, Storage and Communication (ISIC 7); Financial, Insurance, Real Estate and Business Services (ISIC 8); Community, Social and Personal Services (ISIC 9).
Manufacturing: Manufacturing (ISIC 3).

Source: OECD Input–Output DataBase.

case. First of all, it is striking how stable the shares are and how little they vary across countries. Over two decades, the share of intermediate services in manufacturing industry's gross output (the value of all products produced in a sector: final demand plus intermediate goods minus imports) did not change at all in Japan and rose only modestly in the USA. In Germany and France, the share rose slightly more, but the increase is far from the explosion we might have expected in the light of popular discussion.

Most importantly, in the USA – which is usually regarded as the economy with the highest degree of specialization – the share of intermediate services in gross manufacturing output is lower than in France and Germany, usually cited as examples of backward economies. Not only is the share in the USA lower, but the increase is also much less than in the European economies.

The Europe–USA comparison produces similar results if relations are examined from the other side; that is, if one asks how much of the gross output is value added produced within the sector itself. These figures are displayed in columns 3 and 4 of Table 4.3. Over time there is, if anything, evidence for insourcing by the manufacturing sector and outsourcing by the service sector. An exceptional case with respect to the trend of own value added in gross output is manufacturing in Japan. Here the share of own value added increased substantially between 1979 and 1990. This may reflect both the dramatic catching-up process that Japan has achieved and the higher share of innovative products (see Petit *et al.*, 1997).

To conclude, the amount of actual outsourcing appears to be very different from what is commonly assumed. It has not increased dramatically and thus can hardly have caused the observed trend in shrinking manufacturing employment. The common assumption of a highly specialized US economy and backward, non-specialized European economies needs to be revised in the light of the new evidence.

The Impact of Changes in the Inter-industry Division of Labour on Productivity

Outsourcing may improve industry-specific productivity. For example, if relatively unproductive activities are outsourced from manufacturing industries, productivity in those manufacturing industries will increase. If there is no efficiency gain from specialization, outsourcing may improve industry-specific productivity but productivity in relation to the final product (the vertically integrated sector) may remain unchanged. We label the latter final product productivity 'FPP' and define it broadly as the productivity of the vertically integrated product sector (Pasinetti, 1981) comprising all stages of the production process (for a more specific definition, see Appendix 2).

To gauge the importance of outsourcing for productivity we decompose changes in FPP into a structural component (due to reshuffling of employment among industries belonging to the same vertically integrated product sector) and an industry component which measures the effects of changes in terms of industry-specific productivity. The change in industry-specific productivity may or may not be due to outsourcing itself, but technical progress may also play an important role. The results of the decomposition analysis are presented in Tables 4.4 to 4.7 for the USA, Germany, Japan and France.

The structure of the tables is identical: the first column shows the relative change in FPP from the base year (1972 for the USA and France, 1978 for Germany, 1970 for Japan) and the next three columns display the three components (structural, productivity and interaction). Columns 5 and 6 display the annual compound growth rate FPP and industry-specific productivity (value added/employment). Columns 7 and 8 show the annual compound growth rates in final product employment (FPE: employment in the vertically integrated sector) and industry-specific employment. Columns 9 and 10 show the ratio of direct employment (E_{hh}) to indirect employment (FPE_h-E_{hh}), and column 11 displays the annual compound growth rate in this ratio.

The signs of the three components (columns 2 to 4) offer a useful insight into the transformation of the economy. A positive structural effect implies that, within the vertically integrated product sector, employment shifts to high (that is, above-average) productivity industries have outweighed employment shifts to low productivity industries. In other words, high productivity industries have gained importance in the vertically integrated production process. Negative signs indicate that employment shifts to low productivity industries have outweighed employment shifts to high productivity industries. Positive industry effects signal that weighted industry-specific productivity gains have offset weighted industry-specific productivity losses. If industry-specific productivity gains were caused solely by outsourcing of low productive activities the negative structural effect would compensate the positive industry-specific effect.[14]

What emerges clearly from the analysis of the first four columns is that industry-specific effects are the main cause of variation in final product productivity. Structural effects are present, but they are generally much smaller. This indicates that, while outsourcing is certainly taking place, its impact on FPP is limited and the main effect comes from productivity growth within industries. In other words, technology seems to improve the production process.

This result seems to be consistent with what has been found in other analyses on outsourcing. Siegel and Griliches (1992) found only a modest increase in services purchased by US manufacturing industries during the 1977–82 period. Gouyette and Perelman (1997), examining OECD countries,

Table 4.4 Productivity and employment changes (both final product and industry concepts), USA, 1972–90

Industry	Overall change 1 FPP(90)-FPP(72)	Decomposition 2 Structural	3 Productivity	4 Interaction	Compound annual growth rate of: 5 FPP	6 PROD	7 FPE	8 EMP	Ratio (dir emp)/(ind emp) 9 1972	10 1990	11 Growth
Agriculture, forestry & fishing	0.25	-0.08	0.31	0.02	1.26	2.76	0.94	-0.11	2.71	3.15	0.84
Mining & quarrying	-0.19	-0.05	-0.15	0.01	-1.17	-0.92	9.25	0.72	8.32	5.22	-2.55
Food, beverages & tobacco	0.19	-0.01	0.21	-0.01	0.97	1.07	0.16	-0.24	1.87	1.76	-0.34
Textiles, apparel & leather	0.54	-0.05	0.58	0.01	2.43	4.79	0.20	-1.66	2.82	3.40	1.05
Wood products & furniture	0.31	0.00	0.33	-0.01	1.50	2.28	1.81	0.25	2.15	2.13	-0.05
Paper, paper products & printing	0.06	0.00	0.05	0.01	0.32	0.43	1.97	1.46	3.18	2.74	-0.81
Industrial chemicals	-0.05	-0.10	0.02	0.02	-0.31	1.81	1.00	0.81	1.77	2.03	0.75
Non-metallic mineral products	0.12	0.09	0.05	-0.02	0.63	0.99	0.94	-0.75	2.31	2.22	-0.22
Basic metal industries	-0.04	0.06	-0.11	0.00	-0.26	-0.82	-0.11	-2.17	2.79	2.02	-1.79
Metal products machinery equipment	0.50	-0.02	0.48	0.04	2.27	3.22	1.53	0.44	2.79	3.73	1.62
Other manufacturing	0.20	-0.01	0.20	0.02	1.03	1.66	1.53	-0.20	1.92	1.99	0.20
Electricity, gas & water	-0.25	-0.04	-0.23	0.03	-1.56	-1.80	3.25	1.50	3.12	2.06	-2.28
Construction	-0.07	0.02	-0.08	-0.01	-0.42	-1.24	1.67	1.85	2.02	1.87	-0.43
Wholesale & retail trade	0.09	-0.03	0.11	0.01	0.46	1.74	2.57	2.31	3.87	3.96	0.12
Restaurants & hotels	-0.10	0.00	-0.09	0.00	-0.56	-0.76	3.43	3.02	1.99	1.93	-0.18
Transport & storage & communication	0.18	-0.04	0.23	-0.01	0.91	1.60	1.48	1.36	2.90	3.33	0.77
Finance & insurance	0.04	0.00	0.05	-0.01	0.23	0.98	3.83	2.94	3.88	3.96	0.11
Real estate & business services	-0.36	-0.01	-0.36	0.01	-2.47	-2.69	5.56	5.47	6.35	5.95	-0.36
Community, social & personal services	0.02	0.05	0.00	-0.03	0.09	0.21	3.65	2.99	2.84	2.51	-0.69
Producers of government services	0.07	0.00	0.07	0.00	0.38	0.38	1.34	1.21	0.00	0.00	—
Other producers	-1.75	0.00	-1.75	0.00	—	7.23	—	1.23	0.00	0.00	—

Notes:
FPP(90)–FPP(72): ((final product productivity 1990) – (final product productivity 1972))/(final product productivity 1972); FPP: final product productivity (thousands of 1982 US $); PROD: productivity in the industry (thousands of 1982 US $); FPE: final product employment activated from the industry (number of persons); EMP: employment in industry (number of persons); Differences: differences in annual compound growth rates of the variables indicated; Growth: annual compound growth rate of the ratio direct employment to indirect employment in the period 1972–90.
Column 1 = Column 2 + Column 3 + Column 4.

Table 4.5   Productivity and employment changes (both final product and industry concepts), Germany, 1978–90

Industry	Overall change 1 FPP(90)–FPP(78)	Decomposition 2 Structural	Decomposition 3 Productivity	Decomposition 4 Interaction	Compound annual growth rate of: 5 FPP	6 PROD	7 FPE	8 EMP	Ratio (dir emp)/(ind emp) 9 1972	10 1990	11 Growth
Agriculture, forestry & fishing	0.44	-0.08	0.49	0.02	3.06	6.99	-1.93	-3.33	1.95	2.31	1.40
Mining & quarrying	-0.20	-0.05	-0.18	0.03	-1.85	-2.36	-8.70	-2.15	3.23	2.25	-2.99
Food, beverages & tobacco	0.07	-0.02	0.09	0.00	0.57	-0.11	0.28	-0.79	2.05	1.81	-1.02
Textiles, apparel & leather	0.18	0.00	0.18	0.01	1.42	1.21	-0.97	-3.48	3.42	2.67	-2.03
Wood products & furniture	0.05	0.00	0.04	0.01	0.43	-0.95	1.18	-0.84	2.36	1.94	-1.61
Paper, paper products & printing	0.09	-0.01	0.10	0.00	0.71	0.63	4.55	0.71	3.14	2.78	-1.01
Industrial chemicals	0.17	-0.04	0.17	0.03	1.28	2.24	1.25	1.08	2.38	2.85	1.50
Non-metallic mineral products	0.17	-0.03	0.20	0.00	1.34	1.89	0.39	-1.46	2.37	2.20	-0.62
Basic metal industries	0.34	-0.04	0.36	0.02	2.44	5.71	2.22	-1.11	3.51	3.28	-0.54
Metal products machinery equipment	0.15	-0.02	0.18	0.00	1.20	1.20	2.22	0.85	3.22	2.89	-0.90
Other manufacturing	0.26	-0.03	0.29	-0.01	1.92	2.17	-0.42	-1.19	1.98	1.79	-0.81
Electricity, gas & water	-0.17	-0.06	-0.12	0.01	-1.52	-1.07	2.39	0.93	2.68	1.99	-2.45
Construction	0.13	0.00	0.13	0.01	1.02	0.37	-0.36	-0.43	1.99	1.79	-0.85
Wholesale & retail trade	0.18	-0.02	0.20	0.00	1.36	1.58	0.97	0.74	2.92	2.87	-0.16
Restaurants & hotels	-0.10	-0.05	-0.04	0.00	-0.87	-1.74	2.34	2.21	1.70	1.60	-0.53
Transport & storage & communication	0.32	-0.05	0.36	0.00	2.33	3.21	1.16	0.56	2.53	2.64	0.34
Finance & insurance	0.31	-0.02	0.33	-0.01	2.25	2.45	2.58	1.78	4.30	4.04	-0.52
Real estate & business services	0.00	0.00	0.00	0.00	—	—	—	—	—	—	—
Community, social & personal services	-0.01	-0.03	0.01	0.00	-0.09	-1.45	3.24	3.66	2.54	2.40	-0.45
Producers of government services	-0.01	0.00	-0.01	0.00	-0.06	-0.06	1.08	1.13	—	—	—
Other producers	0.12	-0.05	0.17	0.00	0.96	1.43	3.32	3.25	3.58	3.53	-0.11

Notes:
FPP(90)–FPP(78): ((final product productivity 1990) – (final product productivity 1978))/(final product productivity 1978); FPP: final product productivity (thousands of 1985 DM); PROD: productivity in the industry (thousands of 1985 DM); FPE: final product employment activated from the industry (number of persons); EMP: employment in industry (number of persons); Differences: refers to differences in annual compound growth rates of the variables indicated; Growth: annual compound growth rate of the ratio direct employment to indirect employment in the period 1978–90.
Column 1 = Column 2 + Column 3 + Column 4.

123

Table 4.6 Productivity and employment changes (both final product and industry concepts), Japan, 1970–90

Industry	Overall change 1 FPP(90)–FPP(72)	Decomposition 2 Structural	3 Productivity	4 Interaction	Compound annual growth rate of: 5 FPP	6 PROD	7 FPE	8 EMP	Ratio (dir emp)/(ind emp) 9 1972	10 1990	11 Growth
Agriculture, forestry & fishing	1.49	0.25	1.05	0.19	4.66	1.34	-2.57	-1.11	4.02	3.09	-1.30
Mining & quarrying	0.70	0.08	0.85	-0.23	2.69	-1.61	-6.65	-1.96	2.18	1.85	-0.80
Food, beverages & tobacco	0.85	0.02	0.77	0.06	3.12	-0.09	1.42	0.61	5.11	3.52	-1.85
Textiles, apparel & leather	1.12	0.02	1.20	-0.09	3.84	0.11	-0.36	-0.38	3.32	2.89	-0.69
Wood products & furniture	0.60	-0.02	0.59	0.03	2.38	0.32	2.43	-0.80	3.89	2.78	-1.67
Paper, paper products & printing	1.32	0.01	1.30	0.01	4.30	3.00	-0.08	0.54	4.08	3.36	-0.96
Industrial chemicals	1.98	0.02	1.84	0.12	5.61	6.13	-0.21	-0.29	2.25	2.88	1.24
Non-metallic mineral products	0.90	0.03	0.83	0.03	3.25	2.82	-0.69	-0.77	2.04	2.15	0.27
Basic metal industries	20.38	-1.96	21.13	1.21	16.55	1.85	-1.94	0.37	1.94	3.55	3.07
Metal products machinery equipment	2.15	-0.07	2.44	-0.22	5.90	2.56	0.34	0.83	2.08	2.97	1.79
Other manufacturing	1.67	0.11	1.28	0.28	5.04	0.94	-0.10	0.25	1.76	1.71	-0.15
Electricity, gas & water	0.49	-0.05	0.63	-0.09	2.00	1.87	2.86	0.40	3.33	2.39	-1.65
Construction	0.89	0.09	0.75	0.04	3.22	1.66	1.59	0.09	1.69	1.70	0.03
Wholesale & retail trade	0.51	-0.05	0.60	-0.04	2.09	1.72	1.73	0.58	2.96	3.55	0.91
Restaurants & hotels	0.17	-0.01	0.17	0.00	0.79	-0.04	2.81	0.55	2.60	2.03	-1.24
Transport & storage & communication	0.83	0.06	0.86	-0.08	3.08	2.25	0.15	0.51	3.18	2.63	-0.95
Finance & insurance	0.14	-0.02	0.15	0.01	0.66	0.57	1.14	2.46	6.43	4.63	-1.62
Real estate & business services	0.26	0.03	0.23	0.00	1.18	-0.11	3.85	1.45	4.80	4.17	-0.70
Community, social & personal services	1.20	0.41	0.47	0.32	4.02	-0.08	4.38	1.94	3.86	2.60	-1.95
Producers of government services	0.76	0.00	0.76	0.00	2.87	1.11	1.13	0.04	—	—	—
Other producers	3.08	0.82	1.33	0.93	7.29	-0.34	4.03	1.25	8.15	2.45	-5.83

Notes:
FPP(90)–FPP(70): ((final product productivity 1990) – (final product productivity 1970))/(final product productivity 1970); FPP: final product productivity (millions of 1985 yen); PROD: productivity in the industry (millions of 1985 yen); FPE: final product employment activated from the industry (number of persons); EMP: employment in industry (number of persons); Differences: differences in annual compound growth rates of the variables indicated; Growth: annual compound growth rate of the ratio direct employment to indirect employment in the period 1972–90.
Column 1 = Column 2 + Column 3 + Column 4.

Table 4.7 Productivity and employment changes (both final product and industry concepts), France, 1972–90

Industry	Overall change 1 FPP(90)–FPP(72)	Decomposition 2 Structural	3 Productivity	4 Interaction	Compound annual growth rate of: 5 FPP	6 PROD	7 FPE	8 EMP	Ratio (dir emp)/(ind emp) 9 1972	10 1990	11 Growth
Agriculture, forestry & fishing	0.73	-0.01	0.70	0.03	3.09	5.51	-2.60	-3.73	3.30	2.97	-0.59
Mining & quarrying	1.44	-0.10	1.50	0.04	5.08	7.20	-3.41	-3.23	4.35	3.13	-1.80
Food, beverages & tobacco	0.25	-0.13	0.23	0.15	1.24	1.12	0.78	0.04	1.75	1.64	-0.33
Textiles, apparel & leather	0.29	-0.18	0.23	0.23	1.41	2.67	-1.68	-3.78	3.42	3.07	-0.60
Wood products & furniture	0.46	-0.12	0.46	0.12	2.11	3.20	-0.02	-1.04	2.23	2.43	0.49
Paper, paper products & printing	0.25	-0.14	0.21	0.17	1.24	1.91	0.73	0.03	4.14	3.14	-1.52
Industrial chemicals	0.66	-0.07	0.69	0.04	2.87	2.83	0.90	-0.33	1.86	2.35	1.31
Non-metallic mineral products	-0.03	-0.08	-0.12	0.17	-0.20	-1.73	3.32	-2.29	2.52	1.94	-1.44
Basic metal industries	1.01	0.01	0.99	0.01	3.95	4.42	-0.12	-2.39	3.33	2.79	-0.97
Metal products machinery equipment	0.30	-0.23	0.24	0.28	1.47	2.97	0.59	-0.93	2.72	2.55	-0.37
Other manufacturing	0.25	-0.22	0.19	0.28	1.25	2.45	1.14	-0.17	1.70	1.69	-0.03
Electricity, gas & water	0.75	-0.07	0.69	0.13	3.16	3.71	1.67	1.32	2.43	3.38	1.86
Construction	0.06	-0.34	-0.07	0.46	0.31	2.03	-1.10	-1.11	1.92	1.82	-0.28
Wholesale & retail trade	0.08	-0.15	0.02	0.21	0.41	1.63	0.70	0.57	4.35	3.62	-1.01
Restaurants & hotels	-0.02	-0.07	-0.08	0.14	-0.10	-0.16	1.98	1.68	2.78	2.45	-0.71
Transport & storage & communication	0.19	-0.13	0.18	0.14	0.98	2.13	0.94	0.87	3.44	3.35	-0.14
Finance & insurance	-0.66	-0.06	-0.68	0.08	-5.82	-6.43	-3.30	2.04	8.09	5.20	-2.42
Real estate & business services	0.00	0.00	0.00	0.00	0.00	0.00	0.00	4.37	0.00	4.81	—
Community, social & personal services	0.09	-0.06	0.07	0.08	0.46	1.09	3.39	3.43	4.45	4.48	0.03
Producers of government services	0.15	0.00	0.15	0.00	0.76	0.76	1.84	1.84	0.00	0.00	—
Other producers	0.00	0.00	0.00	0.00	0.00	0.00	0.00	0.00	0.00	0.00	—

Notes:

FPP(90)–FPP(72): ((final product productivity 1990) – (final product productivity 1972))/(final product productivity 1972); FPP: final product productivity (millions of 1985 FF); PROD: productivity in the industry (millions of 1985 FF); FPE: final product employment activated from the industry (number of persons); EMP: employment in industry (number of persons); Differences: differences in annual compound growth rates of the variables indicated; Growth: annual compound growth rate of the ratio direct employment to indirect employment in the period 1972–90.
Column 1 = Column 2 + Column 3 + Column 4.

found that, while productivity growth in manufacturing industries is mainly pushed by technological progress as opposed to more efficient organization, the contrary holds in service industries. Meijers (1998) finds that most of the decrease in manufacturing employment in France during the 1977–92 period can be attributed to changes in final demand and changes in technology (productivity). Morrison and Siegel (1997) find that in US manufacturing in the 1959–89 period employment composition (in terms of skills) was mainly affected by technological progress, while the impact of outsourcing tended to be small. Ten Raa and Wolff (1996) analyse total factor productivity growth in 87 US industries for the 1958–87 period and they find that, while outsourcing is present, industry-specific effects (technological progress) mainly lead total factor productivity growth. Feenstra and Hanson (1997) investigate the causes behind wage increases for non-production workers relative to production workers in US manufacturing and find that expenditure on computers has the largest impact, followed by outsourcing in the same two digits industry (either to domestic firms or abroad). Using input–output tables for the period from 1960 to 1984, Gregori (1998) finds that outsourcing explains only a small component of employment trends in the Italian manufacturing and service industries. Uneven demand and productivity growth are the main drive of the observed trends in employment. Finally, Slaughter (1995) finds that USA-based multinational enterprises tended to reduce employment of both production and non-production workers in the 1977–94 period.

Outsourcing also affects the ratio of direct employment (employment in the main industry) to indirect employment (employment in supplying industries, columns 9–11), which is in most cases between 2 and 3. That is, direct employment is still the most important part in final product employment, although it is declining in most manufacturing industries. This may be supported by the high positive correlation between the growth rates of industry-specific employment and of FPE ($\rho_{US} = 0.57^*$, $\rho_{Ger} = 0.66^*$, $\rho_{Fr} = 0.45^*$, $\rho_{Jpn} = 0.78^*$).[15] Overall employment dynamics at the industry level and in the vertically integrated product sector seem to be largely determined by product demand shift in employment.[16] Therefore outsourcing is unlikely to be the major cause of the enormous shift in employment between manufacturing and services industries that occurred over recent decades (see Schettkat and Russo, 1998).

Capital–Labour Substitution and Total Factor Productivity

A specific form of 'outsourcing' is the replacement of labour by capital, which may or may not result in efficiency gains. Although an increase in the capital–labour ratio will usually result in higher labour productivity, the efficiency of the production process will not necessarily increase. In general, we find a positive correlation between capital–labour ratios and labour productivity.[17] The exceptions may be the result of measurement error because increases in capital are ignored in measures of labour productivity. To capture the substitution effect and to gain an impression of efficiency gains (technological progress) we compute simple measures of total factor productivity (TFP). As discussed above, the employment effect of productivity gains will depend heavily on the market context.

In Figure 4.3 we summarize for various industries the employment trend over time and the trends in production, capital stock, capital–labour ratios, labour productivity and total factor productivity. As discussed in Malinvaud (1994), increasing capital–labour ratios may be taken as evidence of labour-saving investment and it is, therefore, interesting to investigate the relationship between capital–labour ratios and employment (see also Rowthorn, 1995). The general picture that emerges from inspection of Figure 4.3 indicates a positive correlation between the capital–labour ratios and employment levels in all areas except manufacturing. Manufacturing may therefore be a good case to illustrate the importance of the market context. Demand conditions for manufactured products in France, Germany and the USA are obviously unfavourable; that is, the labour-saving effect of increasing capital–labour ratios and increasing labour productivity leads to declining employment. Here the labour-saving effect cannot be overcompensated by the market expansion effect. This also holds for the USA, where the employment profile is flat but there is a growing active population, which rose from about 85 million (total population 205 million) in 1970 to about 125 million (total population 250 million) in 1990. Taken in isolation, we would have expected manufacturing employment to increase. Japan is the only country where employment in manufacturing is growing, but this, to a substantial extent, is caused by the export success of the Japanese manufacturing sector.

Capital accumulation may improve total factor productivity (TFP) if new vintages of capital are more efficient (have embodied technological progress). However, causation may be the other way round. If technological progress is not embodied, improving efficiency may attract capital (Dollar and Wolff, 1994). The causality is thus difficult to deduce from empirical observations. In service industries, labour productivity and TFP tend to be positively correlated: productivity gains are caused not only by capital–labour substitution but also by 'real efficiency' gains (Figure 4.4). A negative correlation between TFP and labour productivity (as in manufacturing) hints at embodied technological progress; that is, it is caused by capital accumulation. Once we account for the intervention of capital in the production process, the increase in labour productivity is overcompensated by a decrease in capital productivity and it is thus accompanied by a decrease in TFP.

USA

	Production (Value Added)	Capital Stock	Capital–Labour Ratio	Labour Productivity	Total Factor Productivity

Manufacturing (ISIC 3)

Wholesale and Retail Trade (ISIC 6)

Transport and Communication (ISIC 7)

Finance, Insurance, Real Estate and Business Services (ISIC 8)

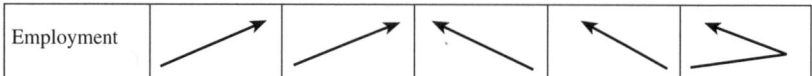

Community, Social and Personal (ISIC 9)

Japan

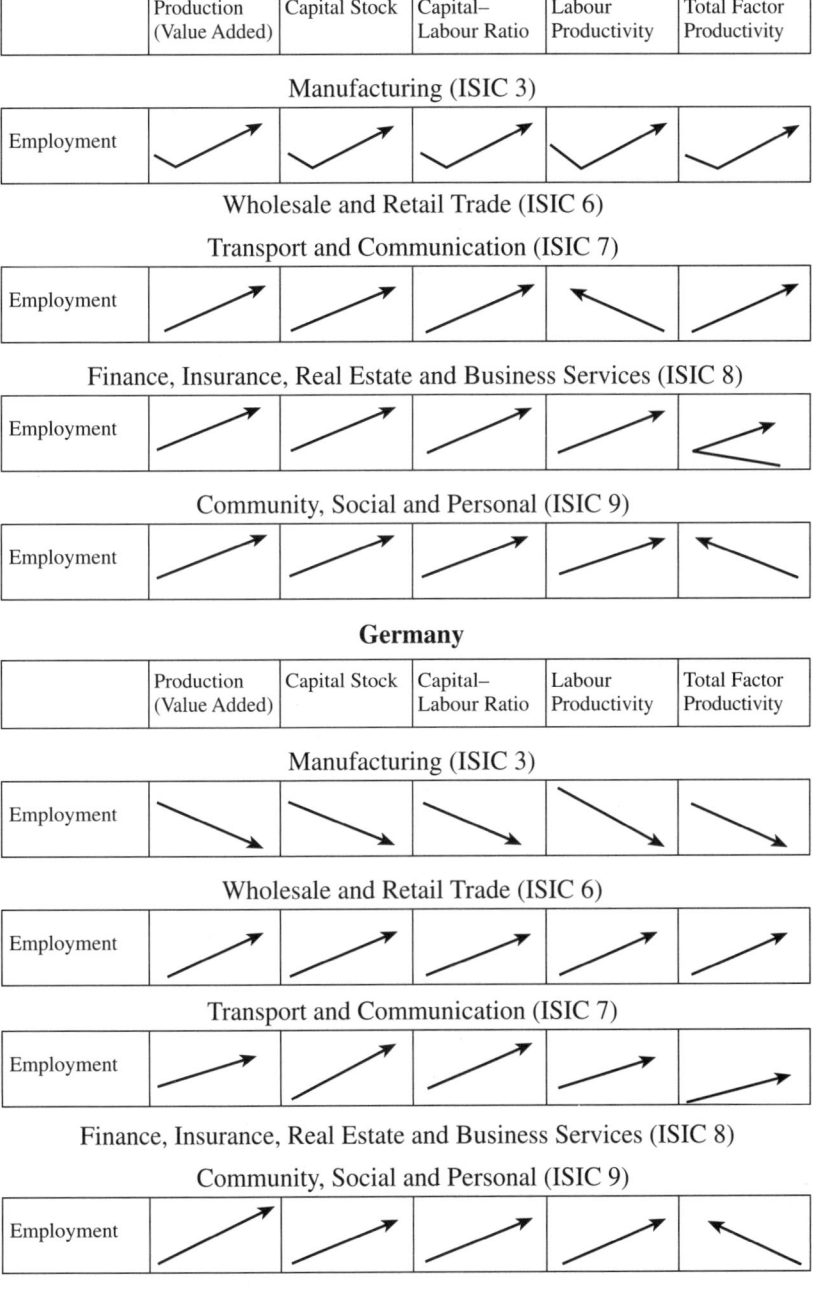

	Production (Value Added)	Capital Stock	Capital–Labour Ratio	Labour Productivity	Total Factor Productivity

Manufacturing (ISIC 3)

Wholesale and Retail Trade (ISIC 6)

Transport and Communication (ISIC 7)

Finance, Insurance, Real Estate and Business Services (ISIC 8)

Community, Social and Personal (ISIC 9)

Germany

	Production (Value Added)	Capital Stock	Capital–Labour Ratio	Labour Productivity	Total Factor Productivity

Manufacturing (ISIC 3)

Wholesale and Retail Trade (ISIC 6)

Transport and Communication (ISIC 7)

Finance, Insurance, Real Estate and Business Services (ISIC 8)

Community, Social and Personal (ISIC 9)

France
Production (Value Added)
Capital Stock
Capital–Labour Ratio
Labour Productivity
Total Factor Productivity
Manufacturing (ISIC 3)
Employment
Wholesale and Retail Trade (ISIC 6)
Employment
Transport and Communication (ISIC 7)
Employment
Finance, Insurance, Real Estate and Business Services (ISIC 8)
Employment
Community, Social and Personal (ISIC 9)
Employment

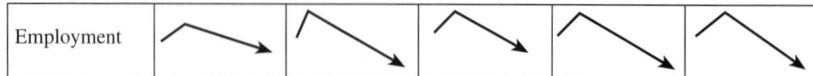

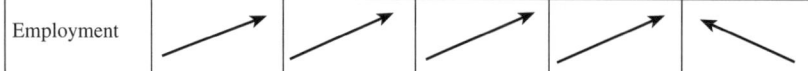

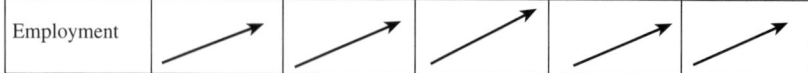

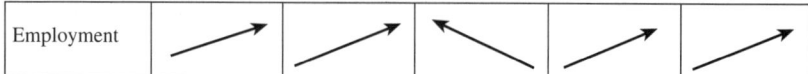

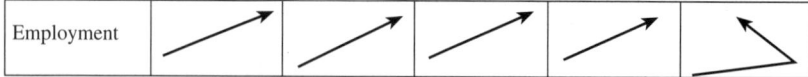

Note: The arrows indicate the direction of time.

Source: OECD: ISDB (period 1970–90).

Figure 4.3 *Employment trend and trends in capital stock, capital–labour ratio, labour productivity and total factor productivity across industries in different countries*

	Labour Productivity France	Labour Productivity Germany	Labour Productivity Japan	Labour Productivity USA

Manufacturing (ISIC 3)

Wholesale and Retail Trade (ISIC 6)

Transport and Communication (ISIC 7)

Finance, Insurance, Real Estate and Business Services (ISIC 8)

Community, Social and Personal (ISIC 9)

Note: Labour productivity is value added per worker; the arrows indicate the direction of time.

Source: OECD: ISDB (period 1970–90).

Figure 4.4   *Total factor productivity and labour productivity across industries in different countries*

4.   CONCLUSIONS

Structural economic dynamics make it very unlikely that an economy will sustain full employment automatically for long periods. The process of adjustment to changing structures involves unemployment resulting from frictions in both the labour and the product market. Moreover, the discussion in section 2 shows that the employment impact of structural economic dynamics is the result of the interaction between demand and supply in product markets.

The industrialized economies have experienced an enormous shift from manufacturing industry to services, as the increasing shares of the latter in nominal GDP and in employment show. The forces underlying these trends are, however, a matter for debate. Explanations range from a constant real economic structure in combination with unbalanced productivity growth, through changes in the inter-industry division of labour (outsourcing), to actual changes in the structure of real final demand. Without compensatory effects, process innovations (or productivity growth) will lead to declining employment.

A changing inter-industry division of labour, as portrayed by national accounts, may be exaggerated because employment decline in an institutionally defined industry may be compensated by employment growth at another stage of the overall production process. In order to evaluate these processes, we have exploited the concept of vertically integrated product sectors. All stages of production contributing to a final product are integrated in this product line approach to the division of the economy, which also allows for the analysis of productivity effects arising from a changing inter-industry division of labour. Again, individual industries may improve their productivity if they outsource fewer productive activities into other industries.

Using an internationally comparable input–output database recently made available by the OECD, together with the International Structural Database (ISDB), we have analysed the importance of the various factors influencing economic structure. Firstly, we investigated structural change with respect to changes in the final demand structure of the economy. Shifts to service industries are most pronounced in final private consumption, partly compensated by reverse trends in exports. With the final product demand (FPD) approach, the shift from manufacturing to service industries is less pronounced than it appears to be in national accounts statistics but it is still present. Therefore structural change is real. It is caused partly by changes in the real demand structure, but other variables must be important as well. We conclude that shifts from manufacturing to services are real, although exaggerated in the institutional division of industry. Structural change is not pure mismeasurement.

Secondly, we investigated whether the institutional division of industry used in conventional national accounts statistics overestimates structural change with respect to changes in the inter-industry division of labour. Outsourcing gained importance over the period we examined and it did contribute to the shift to service industries, but amounted to far less than the expected 'explosion'. There is a slight trend towards increasing outsourcing from manufacturing, but the numbers are certainly not dramatic; outsourcing alone cannot explain the shifts in the inter-industry employment structure. Most surprisingly in the light of the transatlantic explanations of differences in economic structure, which emphasize the high specialization of the US economy as compared to 'backward oriented' Europe, we find the reverse to be true. German manufacturing industries in fact buy relatively more intermediate service inputs (relative to their gross output) than US manufacturing industries.

Thirdly, outsourcing may affect industry-specific productivity. If an industry outsources its less productive activities, productivity in that specific industry may rise even though the productivity in the vertically integrated product sector may not change at all. Taking the final product productivity (FPP) approach (that is, examining the productivity of the vertically integrated product sector), we can analyse the productivity effects of outsourcing. As perhaps one could have expected from the results above, we find intra-industry effects to be the most important. Outsourcing is apparent but affects productivity only slightly. Industry-specific productivity and FPP generally correlate positively, meaning that it is probably technological developments rather than the widely discussed inter-industry division of labour that drive both labour and FPP increases.

Fourthly, the measure of labour productivity may be pushed up if capital–labour ratios increase. Capital–labour substitution is, as it were, a specific form of outsourcing of relatively unproductive activities. To control for capital–labour substitution, we computed simple total factor productivity (TFP) measures. It turns out that the use of labour productivity is not very misleading because the two productivity measures correlate positively over time as well as across industries. Capital–labour substitution itself, that is, abstracting from 'real' efficiency gains, is unlikely to explain declining employment in manufacturing since 'outsourcing' would in this case take place within the manufacturing industries. However, if rising capital–labour ratios improve productivity, this may explain declines in employment under certain demand conditions.

As already pointed out, employment trends are the result of interaction between demand and supply in product markets. Therefore supply indicators such as productivity are insufficient to explain employment trends on their own. If the labour-saving effect of productivity growth dominates the market-expansion effect, it must do so because of product demand trends. Reasons for product demand being sluggish may be linked to market saturation due to lack of product innovation. If this is the case, even if productivity gains translate into price reductions the demand response may be too weak to compensate the labour-saving effect. Market saturation may be particularly relevant for manufacturing products.

To conclude, our study suggests that, although the structure of final demand in the economies has changed in favour of services, and changes in the inter-industry division of labour are present, the latter affect economic structure and productivity only marginally. Thus technology and product demand are most important in relation to the changing economic structure, but their impact on employment is not one-sided. Technological progress and productivity growth may affect employment positively or negatively, depending on demand-side reactions.

NOTES

1. A market consists of a bundle of similar products that can be substituted for one another; that is, which show a high cross-price elasticity.
2. For a discussion of technological progress on employment, see Petit (1995).
3. There may be second-round effects, however. For example, rising income may lead to higher demand, but in that case it is unlikely that demand growth will be concentrated in a specific industry.
4. Price elasticities for individual firms may differ from the overall price elasticity in a specific market (Schettkat, 1997).
5. Information about the data can be found in Appendix 1.
6. The ratio for institutionally defined value added increases by about 15 per cent from 1972 to 1990, whereas the ratio for overall final demand increases by only about 6 per cent.
7. $$\dot{RE}_i = \alpha_1 \dot{RK}_i + \alpha_2 \dot{RFD}_i + \alpha_3 \dot{FDGO}_i + \varepsilon_i \qquad (4.1)$$

   Where the · indicates annual compound growth rates; RE indicates industry i's share in total employment (in the country); RK stands for industry i's capital stock share; RFD is industry i's share in total final demand, $FDGO$ is the final demand to gross output ratio in industry i, and ε is the random disturbance term.
8. This ratio indicates the importance of deliveries received from other industries and/or the importance of supply to other industries.
9. To control for the impact of trade we have also included the relative openness of the market (exports/imports + exports relative to the country average). The addition of this variable to equation (4.1) generally worsens the overall fit of the model. The exception is France, where the estimated parameters (*t*-values in parenthesis) of the augmented model are the following (dependent variable annual compound growth rate of relative employment in industry i):

   RK 0.73 (3.72); RFD 0.51 (3.98); FDGO–0.55 (–3.02); REXP–0.28 (–2.56); $R^2 = 0.67$ F(4, 16) = 8.16.

   REXP is the relative export share (exports/imports + exports) in industry i (relative to the country average). Industries with a higher than average growth in export share have a lower than average growth in employment share.

10. The sign and importance of the parameters estimated did not change when we tried a fixed effect panel model on the logarithm of the levels of the variables (employment, capital stock, final demand and final demand in gross output) included in equation (4.1). We retained the specification in growth rates of the share of the variables (employment, capital stock, final demand and final demand in gross output) because we believe it gives a better idea of the direction of the change in the structure of an economy.

11. The importance of the rationalization and capacity component of investment may differ, however, between individual industries.

12. The aggregate service sector includes wholesale retail and trade, restaurants and hotels, transport and storage, communication, finance and insurance, real estate and business services, community and social and personal services, producers of governmental services, and other producers. The aggregate manufacturing sector includes food, beverages and tobacco, wood products and furniture, paper products and printing, industrial chemical, non-metallic mineral products, basic metal industries, metal product machinery and equipment, and other manufacturing.

13. Of course, the division of labour between firms within an industry may change as well, but this does not change industry structure.

14. A fall in productivity in an industry may be caused by a reduction of average hours worked. Average working hours declined in France from 1962.5 in 1970 to 1668.1 in 1990; in the USA from 1913.0 in 1970 to 1942.6 in 1990; in Germany from 1784.8 in 1978 to 1616 in 1990; data from Japan were not available (source, OECD: ISDB).

15. Where ρ is correlation coefficient, ** is significant at 10 per cent, * is significant at 5 per cent.

16. Siegel and Griliches (1992), in their analysis of the productivity recovery in the US manufacturing sector for the 1972–87 period, find that outsourcing (to services and abroad) does not seem to be correlated with productivity growth (see also Ten Raa and Wolff, 1996).

17. The only exceptions we found are in 'community and personal services' and, in the USA, in 'transport and communication'. In both cases, working-time arrangements may underlie the non-positive correlation between capital–labour ratios and labour productivity.

REFERENCES

Appelbaum, E. and R. Schettkat (1990), 'Determinants of employment developments: A comparison of the United States and the Federal German economies', *Labour and Society*, 15, 13–32.

Appelbaum, E. and R. Schettkat (1995), 'Employment and productivity in industrialized countries', *International Labour Review*, 134, 605–24.

Appelbaum, E. and R. Schettkat (1997), 'Are Prices Unimportant? The Changing Structure of the Industrialized Economies', AWSB Discussion Paper, 97/10.

Archibugi, D. and J. Michie (eds) (1998), *Trade, Growth and Technical Change*, Cambridge: Cambridge University Press.

Ben-David, D. and D.H. Papell (1997), 'International trade and structural change', NBER Working Paper 6069.

Dollar, D. and E.N. Wolff (1994), 'Capital intensity and TFP convergence by industry in manufacturing, 1963–1985', in W.J. Baumol, R.R. Nelson and E.N. Wolff (eds), *Convergence of Productivity*, Oxford: Oxford University Press.

Feenstra, R.C. and G.H. Hanson (1997), 'Productivity measurement and the impact of trade and technology on wages: estimates for the US, 1997–1990', NBER working paper 6052.

Friedman, M. (1968), 'The Role of Monetary Policy', *American Economic Review*, 58, 1–17.

Gouyette, C. and S. Perelman (1997), 'Productivity convergence in OECD service industries', *Structural Change and Economic Dynamics*, 8, 279–96.

Gregori, T. (1998), 'A structural decomposition of service growth in Italy', mimeo, Trieste University.

Leontief, W. (1986), *Input–Output Economics*, Oxford: Oxford University Press.

Malinvaud, E. (1994), *Diagnosing Unemployment*, Cambridge: Cambridge University Press.

Pasinetti, L. (1981), *Structural Change and Economic Growth: an essay in the dynamics of the wealth of nations*, Cambridge: Cambridge University Press.

Petit, P. (1995), 'Employment and technological change', in P. Stoneman (ed.), *Handbook of the Economics of Innovation and Technological Change*, Oxford: Blackwell.

Petit, P., R. Schettkat and B. Verspagen (1997), 'Uneven growth, technology and employment', TSER-MERIT working paper, Maastricht.

Reder, M.W. (1964), 'Wage Structure and Structural Unemployment', *Review of Economic Studies*, 31, 309–22.

Rowthorn, R. (1995), 'Capital formation and unemployment', *Oxford Review of Economic Policy*, 26–39.

Schettkat, R. (1992), 'The labor market dynamics of economics', in *The United States and Germany in Transition*, New York: Praeger.

Schettkat, R. (1997), 'Die Interdependenz von Produkt- und Arbeitsmärkten', *Die Wirtschafts- und Beschäftigungsentwicklung der Industrieländer aus der Produktmarktperspektive*, 35, 720–34.

Schettkat, R. and G. Russo (1998), 'Are structural dynamics a myth?', mimeo, Utrecht University.

Schumpeter, J. (1934), *Theorie der wirtschaftlichen Entwicklung*, Berlin: Duncker & Humbolt.

Siegel, D. and Z. Griliches (1992), 'Purchased services, outsourcing, computers and productivity in manufacturing', in Z. Griliches (ed.), *Output Measurement in the service sectors*, Chicago: NBER, University of Chicago Press.

Singh, A. (1995), 'Institutional requirements for full employment in advanced economies', *International Labour Review*, 134, 471–96.

Slaughter, M.J. (1995), 'Production transfer within multinational enterprises and American wages', NBER Working Paper 5253.

Ten, Raa T. and E.N. Wolff (1996), 'Outsourcing of services and the productivity recovery in U.S. manufacturing in the 1980s', CENTER Discussion Paper 9689, Tilburg University.

APPENDIX 1   DATA DESCRIPTION

The empirical analysis is conducted using sectoral data expressed at *constant prices* (base year 1985) in domestic currencies (so levels are not comparable across countries). Employment is expressed in numbers of workers. We use two input–output tables to construct annual compound growth rates for each country.

Input–output tables summarize the transactions between individual industries within a given economy (Leontief, 1986). The input–output framework divides the economy into industries (or sectors) which at the same time buy products (inputs) from and deliver products (output) to other industries and to satisfy final demand. That is, input–output tables are double-entry tables in which the cells show the transactions between industries. Reading along the rows, figures show the output produced in a specific industry and delivered to other industries (the output of one industry is thus the input for other industries) and to satisfy final demand. Reading down the columns, figures show the inputs received from other industries (input of production) and value added.

Input–output tables decompose final demand into consumption, exports, investments and government spending. Final demand plus (total) sales of intermediate goods (to the rest of the economy) minus imports equals industry's gross output. Gross output can also be obtained as value added produced in a given industry plus the industry's purchases of intermediate goods (from the rest of the economy). The summation of gross output across industries produces the gross domestic product.

The empirical analysis focuses on four countries: Japan, France, Germany and the United States. For each country we used two input–output tables covering the period 1970–90. The input–output tables refer to the following years: USA and France, 1972, 1990; Germany, 1978, 1990; Japan, 1970, 1990.

The data from the OECD input–output database were matched with compatible employment and capital stock figures from the OECD ISBD database. Data on employment in the financial and business service sector in Germany and in the wholesale and retail trade sector in Japan were reported as missing values in the ISDB. The ISDB provides longitudinal information on average hours worked for most countries (except Japan). At the industry level, no data on hours worked were available. For this reason, we used employment measured in persons. Labour productivity is then defined as the value added per worker.

Each economy is divided into 21 industries (listed below), in line with the industry classification used for the OECD International Structural DataBase:

1. Agriculture, forestry and fishing,
2. Mining and quarrying,
3. Food, beverages and tobacco,
4. Textile, apparel and leather,
5. Wood products and furniture,
6. Paper, paper products and printing,
7. Industrial chemical,
8. Non-metallic mineral products,
9. Basic metal industries,
10. Metal products, machinery and equipment,
11. Other manufacturing,
12. Electricity, gas and water,
13. Construction,
14. Wholesale and retail trade,
15. Restaurants and hotels,
16. Transport, storage and communication,
17. Finance and insurance,
18. Real estate and business services,
19. Community and social and personal services,
20. Producers of government services,
21. Other producers.

APPENDIX 2   THE FINAL PRODUCT PRODUCTIVITY CONCEPT

Labour productivity may rise because of outsourcing of less productive activities to other industries. This phenomenon may be facilitated by information and communication technologies, and is the principal explanation for declining employment in manufacturing. However, what we measure as more efficient production in manufacturing may be to a certain extent a misperception as regards the efficiency of the overall production process. What we observe as a productivity increase in a certain industry can, in the extreme, be caused entirely by a shift of less productive activities to other (service) industries.

To identify outsourcing effects, we construct a productivity measure which relates the final product to the overall labour input necessary to produce it, including the labour input into intermediate production. This measure we call *final product productivity*, the productivity of the vertically integrated product sector (Pasinetti, 1981).

The final demand for a product which is mainly produced in industry h can be written as:

$$FD_h = VA_{hh} + \sum_{i \neq h} VA_{ih} \qquad (4A2.1)$$

where FD_h represents final demand for product h (the product of industry h), VA_{hh} represents value added created in industry h incorporated into final demand directed at product h, and VA_{ih} denotes value added created in other industries (within the vertically integrated product sector) incorporated into final demand for product h (intermediate goods produced in industries i and delivered to industry h). We define final product productivity (FPP, the productivity of the vertically integrated product sector) as final demand divided by employment in the vertically integrated product sector (final product employment or FPE).

$$FPP_h = \frac{FD_h}{FPE_h} \qquad (4A2.2)$$

To proceed we make use of two assumptions. First, in each industry, the amount of labour needed to produce one unit of output is independent of the destination of the product. Second, the amount of value added incorporated in the product of that industry is independent of the destination of the product (value added incorporated in the product is equal to the ratio of value added to gross output).

From these assumptions it follows that the product of a specific industry is produced with the same productivity (industry's average productivity) independent of its destination (final demand or intermediate good). In other words, products and production processes are assumed to be homogeneous within industries. From these assumptions it follows that

$$VA_{ih} = \pi_i E_{ih}; \forall h \qquad (4A2.3)$$

where π_i is the productivity (value added/employment) in industry i, h is the subscript for the destination of the product, and

$$E_{ih} = \frac{x_{ih}}{GO_i} \frac{FD_h}{GO_h} E_i \qquad (4A2.4)$$

where E_i is employment in industry i, GO_i is the gross output (final demand + intermediate goods – imports) of industry i and x_{ih} (the entry in the input–output table corresponding to row i and column h) is the amount of intermediate goods which industry h purchases from industry i. The ratio FD_h/GO_h represents the proportion of purchased goods incorporated in final demand relating to industry h (ratio of final demand to gross output: see first assumption. Thus E_{ih} is employment in industry i induced by final demand on industry h through the purchase of intermediate goods necessary to the production process in industry h. Moreover, $E_{hh} = E_h (FD_h/GO_h)$ and FPE_h is thus:

$$FPE_h = E_{hh} + \sum_{i \neq h} E_{ih}. \qquad (4A2.5)$$

The final product productivity can, in turn, be written as follows:

$$FPP_h = \pi_h \frac{E_{hh}}{FPE_h} + \sum_{i \neq h} \pi_i \frac{E_{ih}}{FPE_h}. \qquad (4A2.6)$$

Industry h's final product productivity is given by the weighted average of the productivity in all industries contributing to the vertically integrated product sector. The weights are the employment share of individual industry in overall employment in the vertically integrated product sector (final product employment).

Changes in FPP can be decomposed into a pure productivity effect, a structural effect and an interaction term:

$$\Delta FPP_h = \sum_i \Delta \pi_i a_{ih} + \sum_i \pi_i \Delta a_{ih} + \sum_i \Delta \pi_i \Delta a_{ih}$$

(4A2.7)

$$a_{ih} = \frac{E_{ih}}{FPE_h}$$

where a_{ih} is the industry i's employment share in final product employment.

5. Innovation, demand and employment

Mario Pianta*

---

1. INTRODUCTION

In the 1990s there was a fundamental change in the growth mechanisms of advanced countries; economic growth no longer led to an expansion of employment with rising real wages. The problem assumed different forms in different areas, with persistently high unemployment (but stable income levels) in Europe and a serious polarization of incomes (but high job creation) in the United States.

These developments highlight the novelty of the challenges for economic policies and the variety of outcomes which may emerge in different countries. At the root of this is the breakdown of the postwar model of 'Fordist' growth, which combined a technological paradigm based on mechanical and chemical innovations with the organization of mass production and the expansion of mass consumption that was made possible by a significant redistribution of productivity gains to wage earners. These developments required a complex regulation system which included a specific set of institutions, active public policies on both the demand and the supply side, and major social arrangements leading to the emergence of welfare states.

What is most visible in the current 'post-Fordist' transition is the breakdown of the dynamics between technology, production, demand and employment, which ensured a 'virtuous circle' of growth with (almost) full employment in postwar decades, and the lack of new social and institutional arrangements. This combined breakdown is the result of three main economic 'mismatches' which have progressively emerged, related to the nature of technology, the structure of the economy and the importance of demand.

The Nature of Technological Change

A new technological paradigm based on information and communication technologies (ICTs) has emerged, radically changing the nature and trajectories of innovations. This represents a major discontinuity with the traditional forms of production and organization in firms, and with the associated social arrangements. Firms' strategies for dealing with such changes have led to a variety of efforts to enter new fields of activity, to introduce ICT-based products, to adapt and incorporate the new technologies in existing production systems, or to restructure traditional processes. The results, however, have been slow to emerge, as pointed out by the literature on the 'productivity paradox'; in spite of the extensive introduction of new technologies, productivity growth has slowed down substantially in the last decades in all advanced countries. Only after the recession of the early 1990s were there signs of a pick-up in the USA, the country with both the largest use of ICTs and the fastest growth of demand.

Turning new technologies into more efficient production, new products, consumption patterns and employment opportunities is not an easy task, or one which can be left to the decisions of individual firms and the selection process of the market. New technologies need to be matched by organizational changes, learning processes, emergence of new industries and markets, rule setting and expansion of demand. Several studies on the emergence of technological paradigms and key technologies in the past have pointed out the long time required by the combination of all these elements before their impact on economic growth could become evident. The current bleak employment performance of most countries is just the most visible of the signs of such a 'technological' mismatch.[1]

The Structure of the Economy

A second mismatch is found between the evolving structure of demand and the more entrenched supply structure of national economies (Pasinetti, 1981). The previous growth model was built on the parallel development of mass production and mass consumption of standardized manufacturing goods. Industrial production worked as an engine of growth, raising productivity with scale economies as production increased, a path which could be followed by all industrialized countries.

Now demand patterns are marked by expanding services and ICT-based activities, where the sources of productivity growth are more complex, and production is still much more concentrated in a few countries. Economies with a supply structure closer to the new composition of demand have an obvious advantage in international competition, while others have to undertake a deeper process of structural change. The costs of change are higher the greater is the extension of traditional 'Fordist' production in industries facing restructuring or decline. The sectoral structure of economies is therefore an important factor which can help explain differences in national economic performances.

The Importance of Demand

There is now a dramatic mismatch between the high potential of new products and consumption patterns offered by ICTs, with more varied and 'personalized' goods and services, and the lack of emergence of new large markets with strong demand. The slow learning processes in consumption, the need for social innovation (particularly in the use of time) to 'match' the opportunities of technological innovations, the lack of appropriate institutions and public policies managing such problems are all factors which may explain this (see Chapter 6 in the present volume). But more direct economic constraints on demand come from the restrictive macroeconomic policies pursued by most governments, especially in Europe, and from the unequal distribution of incomes, most extreme in the USA. They have reduced the aggregate demand effects and prevented the emergence of a large demand for new goods and services (especially ICT-based) from wage earners.

The above three 'mismatches' suggest that technological change, the composition of the economy and demand factors can offer a powerful explanation for current unemployment, a more convincing one than that provided by traditional views looking at labour markets alone, emphasizing the lack of flexibility of labour regulations, wage levels or the skill composition of the workforce.

In this chapter the relationships between technological change, demand and employment are examined first by looking at the key links emerging in the strategy of firms and in industry patterns. An analytical framework is proposed in the next section, identifying different patterns of innovation, considering the competitive pressure put on firms, and examining the interaction with demand conditions.

In section 3 the analysis of innovation patterns in Europe in the early 1990s is carried out, using data from the European Community Innovation Survey and focusing on five European countries – Denmark, Germany, Italy, the Netherlands and Norway – with a cross-industry analysis of the way employment outcomes are affected by demand, the intensity of technological change and the nature of firms' innovative strategies (mainly their orientation towards new products or new processes).

In section 4 the performance of industries dominated by such different technological strategies is compared in Europe, the USA and Japan, pointing out the variety of innovative behaviours, the importance of the economic structure and the contrasting employment outcomes.

2.   A MODEL OF THE RELATIONSHIPS BETWEEN INNOVATION, DEMAND AND PERFORMANCE

In order to investigate the effects on employment, it is necessary to examine the links between the innovation process, the economic structure, the forms of competition and demand. Transformations on the supply side, brought about by the innovative activities and the investment patterns of firms, cannot be seen simply as technology-driven developments. Rather, a key role is played also by the demand side, with the aggregate dynamics of consumption, investment and exports, by the sector-specific patterns of change and particular market structures. On top of these economic factors, the outcomes of technological change are associated with the institutional setting, social arrangements and with a broad interplay of social relations.

Figure 5.1 shows the main links at play. Technological change leads to a variety of innovative firm strategies, associated with particular competitive strategies in given markets. In parallel, aggregate and industry demand emerges in particular market structures with given forms of competition. The economic and employment performance of firms results from the interaction between such technological and demand factors; the performance of industries is more directly constrained by sectoral demand dynamics. A few issues emerge as critical in affecting the employment outcomes; they include the type of innovative strategy (dominated by either product or process innova-

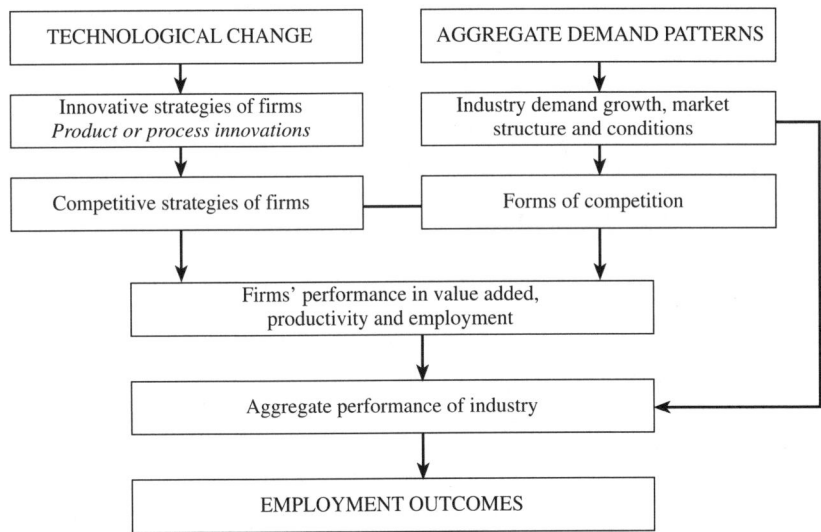

Figure 5.1   Key links between innovation, demand and employment

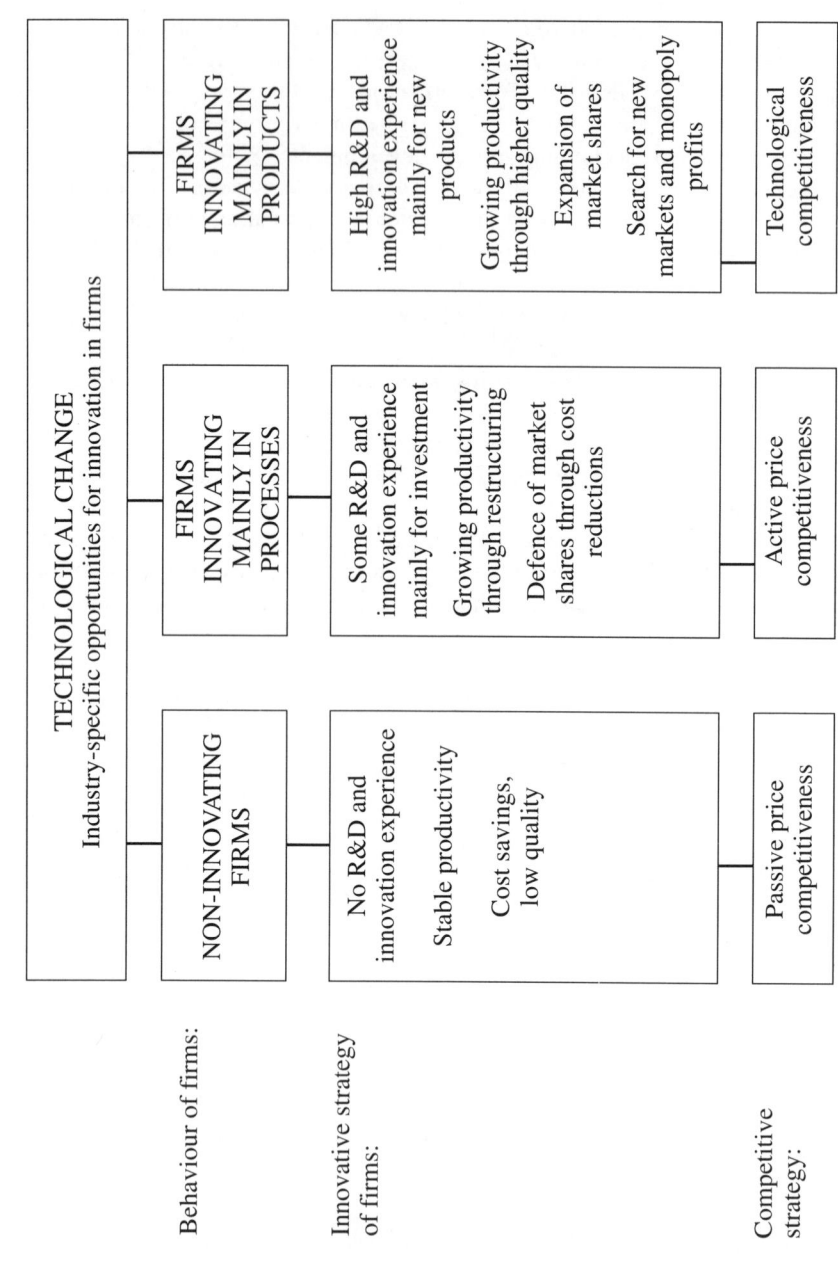

Behaviour of firms:	NON-INNOVATING FIRMS	FIRMS INNOVATING MAINLY IN PROCESSES	FIRMS INNOVATING MAINLY IN PRODUCTS
		TECHNOLOGICAL CHANGE Industry-specific opportunities for innovation in firms	
Innovative strategy of firms:	No R&D and innovation experience Stable productivity Cost savings, low quality	Some R&D and innovation experience mainly for investment Growing productivity through restructuring Defence of market shares through cost reductions	High R&D and innovation experience mainly for new products Growing productivity through higher quality Expansion of market shares Search for new markets and monopoly profits
Competitive strategy:	Passive price competitiveness	Active price competitiveness	Technological competitiveness

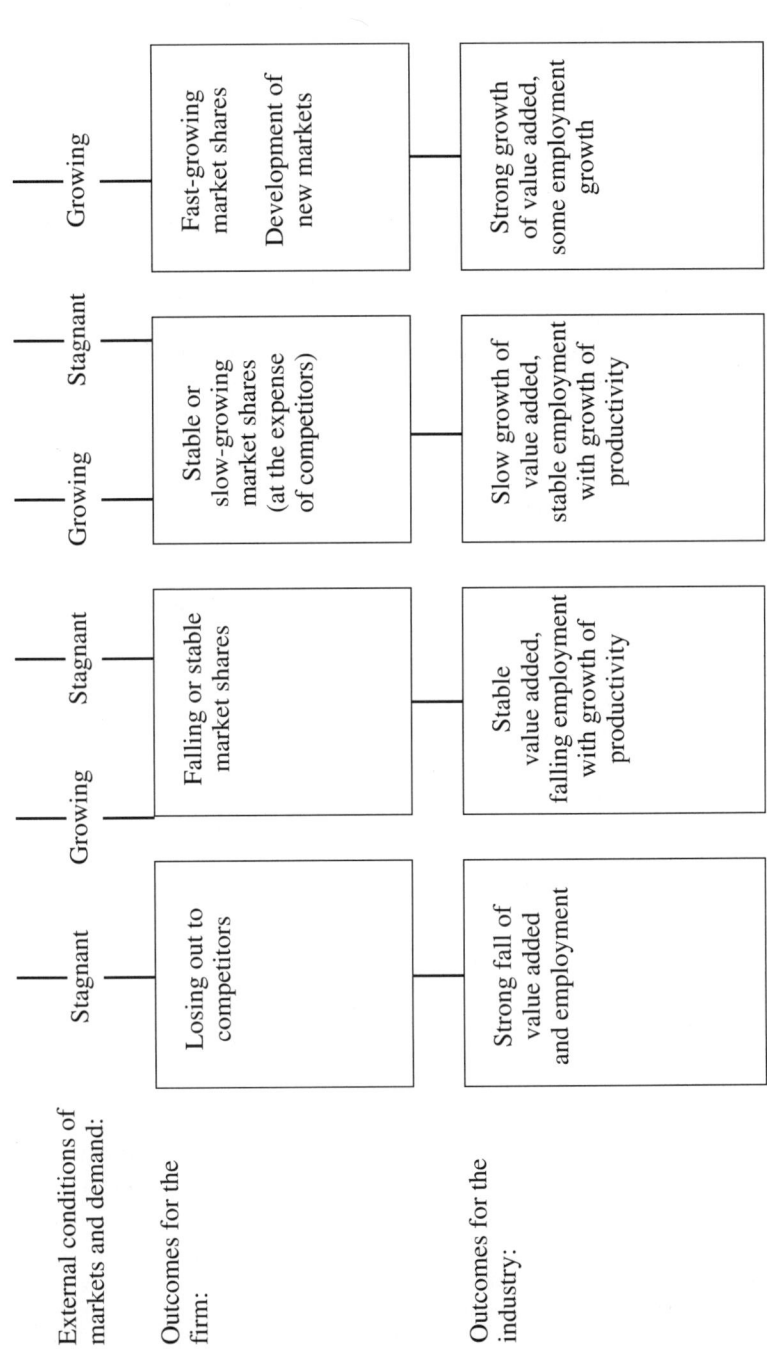

External conditions of markets and demand:

| Stagnant | Growing | Stagnant | Growing | Stagnant | Growing |

Outcomes for the firm:

| Losing out to competitors | Falling or stable market shares | Stable or slow-growing market shares (at the expense of competitors) | Fast-growing market shares. Development of new markets |

Outcomes for the industry:

| Strong fall of value added and employment | Stable value added, falling employment with growth of productivity | Slow growth of value added, stable employment with growth of productivity | Strong growth of value added, some employment growth |

Figure 5.2   A framework for investigating the impact of innovation on growth and employment

147

tions), the dynamics of demand, and the level of the analysis (firm or industry level). They are discussed at length below, while the form of competition – another relevant issue – is not further elaborated here.

Figure 5.2 suggests how these links might be turned into a set of assumptions and expected outcomes which could be used as an operational tool for empirical analysis. Technological change creates industry-specific opportunities for innovation in firms; firms' strategies can neglect them, or turn them in different directions.

A key distinction here is that between a strategy based on process innovations (introduced mainly through new investment) and the search for product innovations (based on internal innovative activities as well as on new intermediate or capital goods). This distinction is essential because they contribute in different ways to the process of technological change. Process innovations lead to improvements in the efficiency of production of particular goods and services, while product (or service) innovations – either incremental or radical – increase the quality and variety of goods and may open up new markets, when the replacement of old products is not the dominant pattern within product innovations. They have, in general terms, contrasting employment effects: increasing productivity and replacing labour in the case of process innovations; creating new markets, production and jobs in the case of product innovations. Obviously, the distinction between process and product innovations should not be exaggerated. In some cases, especially in services, the two are closely interlinked, and, in the case of the introduction of radically new products, innovations also in processes are usually required. So there is a degree of complementarity between the two which should not be ignored, but in most firms and industries it is possible to identify the dominant orientation of innovative efforts.

The need to assess the specific nature of innovation does not stop at identifying the dominance of new products or processes. The specific activities undertaken by firms must be considered, beyond the traditional economic analysis of R&D or patenting indicators. The recent results of the European innovation surveys, discussed in the next section, show that 'disembodied' innovative efforts go far beyond R&D and include a variety of activities, such as design, trial production, exploratory marketing and the acquisition of new knowledge and skills. In addition, innovative activities 'embodied' in new equipment and machinery can now be identified with greater precision. The result is that, in some sectors, in traditional and mass production industries, as well as in services, internal R&D activities may be negligible but, nevertheless, strong innovation efforts may be undertaken. Again the composition of the innovative activities carried out in firms and industries can shed new light on the strategy and the objectives which are pursued.

Summing up these alternative firm behaviours and industry patterns, a strategy focusing on product innovation follows a search for technological competitiveness, based on high productivity rooted in quality advantages and the control of new and dynamic markets. This is typical of firms at the technological frontier, or which are leaders in their market segments or entering new fields of activity.

A focus on process innovations follows from a strategy of active price competitiveness in established markets with productivity growth rooted in innovation-based restructurings. This is typical of mature markets with more intense competition, and of firms adopting a 'follower' strategy. Non-innovators, on the other hand, may survive essentially with cost savings in what can be termed a passive price competitiveness strategy.

As a result of the industrial specificity of technological opportunities and market conditions, it is possible to identify the industries which tend to be dominated by one of these firms' behaviours. Moreover, industries in different countries are subject to a competitive pressure in increasingly global markets similar to the one that firms have to face in a given market. However, the analogy between firm- and industry-level processes stops here, and a fundamental difference is found when the innovation–employment relation is investigated.

Empirical studies have shown that most innovative firms perform better than non-innovative ones in terms of output and employment, regardless of industry, size or other characteristics. This does not clarify, however, whether such performances are obtained by expanding markets and employment or simply by stiffening competition with other firms and taking business and jobs from them. The only way to assess this is to look at the sectoral patterns; the evolution of sectoral value added and employment may show whether the gains of innovative firms have been greater or smaller than the losses of non-innovative ones.

Demand comes back into the picture here. While an individual firm faces a large potential demand, and its performance essentially depends on its competitive success, an industry faces a real demand constraint, even when a strong export orientation exists. When demand grows rapidly, a variety of firms' strategies are possible, the competitive pressure is reduced and it is more likely that a net positive employment outcome may emerge from the processes of innovation and competition among firms. Conversely, in a context of weak demand, net job losses are much more likely to emerge, as competition gets stronger and firms' innovative strategies are mainly aimed at expanding market shares – via cost cutting, process innovation and so on – at the expense of domestic competitors and, in particular, of non-innovative firms.

Technologically competitive firms (and industries in particular countries) concentrating on product innovations tend to expand (or preserve) their market shares regardless of the dynamics of demand. Firms (and industries) relying mainly on cost-reducing process innovations may expand production only in growing markets, while in conditions of stagnant demand they are likely to lose out to competitors with new, higher-quality products. Similarly, non-innovators are likely to survive only in markets with sustained demand and little competition, and to disappear when demand declines and price and non-price competition increases.

This complex combination of innovative strategies, competitive conditions and demand patterns has to be understood in order to investigate the employment outcomes. This is best done (also in the empirical work of the next section) by shifting the analysis at the industry level, so that the evolution of supply can be linked to that of demand within the existing market structures.

There is a further consideration with demand conditions. A vast body of literature has examined the association between changes in technological paradigms and long cycles of economic growth; between the introduction of innovations and business cycles; between demand–pull factors and the shaping of new technologies. While the role of demand can be clearly pointed out in the study of the emergence of particular innovations, it is more difficult to disentangle the interaction of demand and supply factors when the impact of innovation is examined across all industries in shorter periods of growth.

The structure of demand, price elasticity and income elasticity are important for industry-specific employment trends. The higher the rate of demand growth in an industry, the higher employment growth will be. However, the positive demand effects may be compensated or even overcompensated for by productivity growth. Indeed, there is empirical evidence that the positive correlation between industry-specific productivity growth and employment has turned into a negative one in more recent years (Appelbaum and Schettkat, 1995). Demand growth therefore does not necessarily lead to employment growth, as changes in technologies, organizations, skill composition and business structure may extract higher productivity from the same amount of labour.

In the empirical study of the next section the most practical indicator of demand patterns is used: the growth of value added, which represents the *ex post* demand to a country's industry. More refined studies of demand dynamics would need to consider its components (consumption and investment, domestic and foreign), the different market structures, stages of maturity and competition regimes of individual industries.

Returning to Figure 5.2, the interaction between the innovation and competitive strategy of firms and sectors, and the demand patterns and market structure of industries is expected to explain the employment performance of industries. In the empirical investigation of the next section employment changes in manufacturing industries are assumed to result from the dynamics of demand, and structural and technological change, considering a variety of innovative activities and the prevalence of product or process innovations. Better employment outcomes will probably come from a combination of dominance of product innovations and growing demand. When an opposite direction of technological change is dominant, higher innovative efforts could simply increase the pace of labour-displacing process innovations.

3. THE LESSONS FROM INNOVATION SURVEYS

The usual technological indicators – patents, R&D and investment – used as proxies for disembodied and embodied technological activities account for only some aspects of the complex innovation process, and this results in a major limitation of available studies. In particular, activities such as non-formalized research, design and engineering are not covered by such variables, nor is it possible to identify the part of investment related to innovation.

A solution to such a problem is offered by the use of innovation survey data, which provide an enlarged set of quantitative and qualitative indications on firms' innovative activities. Innovation surveys have been carried out in several European countries in the framework of the EU-sponsored Community Innovation Survey (CIS). They identify the firms which have introduced innovations (at least one product or process innovation) in the period 1990–92. The appendix to this chapter provides all the information needed on the data used in this empirical analysis.

In this section the relationship between technological change and employment is examined for five European countries (Denmark, Germany, Italy, the Netherlands and Norway) in the early 1990s at the level of 21 manufacturing sectors. A database has been built, based on the work of Evangelista *et al.* (1998), including the firms with non-missing values for the key innovation variables to be considered.

Table 5.1 lists the main variables taken from the innovation surveys and the number of firms included, showing the values for total manufacturing for each of the five countries considered. First, the composition of total innovation expenditure is indicated, showing the shares of R&D, design and trial production, on the one hand, and of innovative investment, on the other. The importance of the latter is remarkable, accounting on average for about half of total innovation expenditure, although with high variations in the countries with fewer observations.

The detailed study by Evangelista *et al.* (1998) on innovative expenditure in all European firms (not limited to the five countries here considered and covering 8729 firms in all) has shown that 50 per cent of expenditure is due to innovative investment, and the other half includes 20 per cent only of R&D,

Table 5.1   Main variables from the European Community Innovation Survey, country averages

	Share of research, design, trial production, etc. in total innov. expend. (%)	Share of innov. investments in total innov. expend. (%)	Share of R&D related to product innovations (%)	Share of new products in sales (%)	Share of new products in exports (%)	Number of firms in the database
Denmark	61.5	38.5	71.5	44.6	43.1	215
Germany	43.1	57.0	76.1	55.3	52.3	1 042
Italy	45.2	54.8	72.5	30.3	38.3	4 193
Netherlands	23.4	76.6	61.4	37.2	n.a.	782
Norway	74.2	25.8	71.8	33.7	31.7	224

while design accounts for 10 per cent, licences and patents for 2 per cent, trial production for 11 per cent, marketing for 3 per cent and other expenditures for 4 per cent. This comprehensive view of all innovative activities carried out by firms highlights the rather limited role of R&D (which is nevertheless crucial in some industries and in the search for radical innovations), the variety of different innovative efforts and the large role of technological change embodied in new machinery and equipment (typical in the case of process innovations or incremental improvements in products).

Differences in the composition of innovative expenditure from country to country are substantial, and might be due to the specificity of national surveys. Therefore in our analysis innovation expenditure will be mainly examined in total.

The next column in Table 5.1 shows the importance of product innovations, calculated on the share of R&D devoted to product innovations. Here there is strong consistency between the five countries, with product innovations accounting for about 70 per cent of R&D activities, according to firms' answers to the surveys. It is reasonable that such a high share of R&D is devoted to the search for new products; conversely (although no specific information on this is available from the surveys) we can assume that an even higher share of innovative investments is related to process innovations; from these data on the type and objective of firms' expenditure, the distinction in the orientation and content of innovative strategies emerges even more clearly.

Two more variables on the economic outcome of innovations are listed in Table 5.1: the share of firms' sales and exports due to new products. They are key indicators of how important product innovations (including both incremental and radical ones) are in the performance of firms. The evidence is that their economic impact is high, ranging from one-third to half of the sales and exports of the innovative firms considered. Therefore, as expected, product innovations do have a crucial impact on the competitive performance of firms, both in their growth record and in their success in international markets.

While these summary data are averages for all manufacturing industry, the rest of the empirical analysis in this section is carried out at the industry level, as the distribution of these variables across industries provides crucial insights into the link between technological change, properly qualified, demand and employment changes.[2]

An overview of the information offered by the innovation variables at the industry level is shown in Figure 5.3, illustrating the link between the importance of product innovation in R&D efforts and the share of sales due to new products, pooling data for the five countries considered in the database. It shows a close positive relation across sectors, suggesting a strong coherence between the orientation of industries' R&D strategies towards product inno-

Sectoral changes and demand

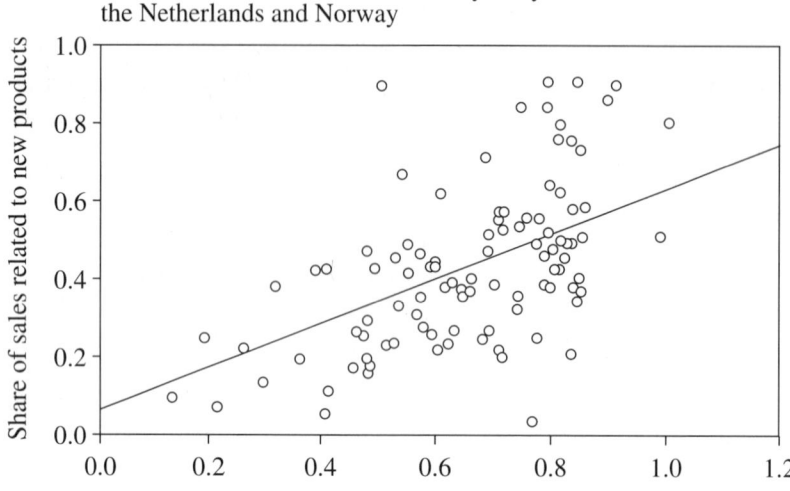

Figure 5.3   Product innovations in R&D and in sales

vations and the outcome in terms of sales of innovated products.[3] In markets where the innovators can rely on a large output of innovated products, the nature of competition is more removed from price factors, reflecting the technological competitiveness strategy described in Figure 5.2, a strategy that is indeed rooted, as argued above, in innovative efforts mainly oriented towards product innovations.

The rich database available for these five European countries makes it possible to investigate the role of the different factors influencing employment change, including the relevance and nature of innovative activities and demand dynamics. The most effective way of examining the joint influence of these factors on employment is to carry out a regression analysis across the 21 industries considered. We try to explain employment changes in 1989–93 with variables accounting for changes in demand and industrial structure (value added and gross investment), for innovation intensity (total innovation expenditure per employee and R&D and design per employee) and for the prevalence of product innovations (share of R&D related to product innovations). Table 5.2 shows the results of the regressions carried out across 21 manufacturing sectors, for the pool of the five European countries and for each of them (see the appendix for details on data and methods).

When all countries are included, value added, gross investment and product innovations have a positive impact on employment changes; innovative

Table 5.2 Regression estimates (dependent variable: rate of change of employment, 21 manufacturing industries)

	Pool of 5 countries		Denmark	Germany	Italy		Netherlands	Norway
	a	b	b	b	a	b	b	b
Change of value added	0.38 (4.06) ***	0.41 (4.35) ***	0.35 (1.94) *	0.58 (4.51) ***	0.22 (2.86) ***	0.18 (1.94) *	0.96 (2.79) **	0.99 (3.23) ***
Change of gross fixed investment	0.04 (2.45) **						0.11 (2.31) **	
Total innovation expenditure per employee	-2.47E-07 (-1.84) *				-9.13E-07 (-3.01) ***		-3.41E-07 (-2.07) *	-2.24E-07 (-0.94)
Exp. for R&D and design per employee		-4.76E-07 (-2.17) **	-6.66E-08 (-0.04)	-4.42E-06 (-1.95) *		-9.46E-07 (-2.06) *		
Share of R&D related to product innovation	0.03 (2.24) **	0.03 (2.48) **						
Constant	-0.03 (-3.74) ***	-0.04 (-4.16) ***	-0.01 (-1.75)	0.02 (1.89) *	-0.01 (-3.29) ***	-0.02 (-5.83) ***	0.002 (0.27)	-0.53 (-8.08) ***
Adjusted R-squared	0.21	0.21	0.09	0.55	0.50	0.39	0.38	0.35
F	7.56 ***	8.65 ***	1.87	13.2 ***	11.1 ***	7.55 ***	5.11 ***	5.90 **
Number of cases	101	102	19	21	21	21	21	19

Notes: t-statistics in parentheses; significance levels: *90%, **95%, ***99%.

155

expenditure per employee (or R&D and design expenditure per employee) turn out with negative signs. All coefficients are significant and their signs are stable in several variations of the regression. The results show that employment change positively follows the dynamics of demand (as proxied by value added), of structural change (as indicated by gross investment) and the orientation of innovation towards new products, while a higher intensity of innovative expenditure *per se* has a negative impact on jobs.

The results for individual countries (which have a rather limited number of observations) are generally consistent with those of the pooled regression. In all countries the growth of value added has a positive and significant impact on employment change. For Denmark and Germany, this variable is combined with the negative impact of the R&D and design expenditure per employee (significant only in the latter country). In the case of Italy, the intensity of both total innovative expenditure and R&D and design expenditure are negatively and significantly associated with employment change. For the Netherlands, the positive impact of change in gross investment is added to the negative role of total innovative expenditure per employee (both with significant coefficients). In the case of Norway, a negative (non-significant) impact of total innovative expenditure is found.

These findings shed new light on the employment impact of innovative activities in Europe. In explaining the changes (in most cases the decrease) of employment in European industries, as expected, demand and structural change factors, and the relevance of product innovations, have a positive impact. Once structural change and the nature of technological change are considered, a greater innovation (or R&D and design) expenditure is associated with worse employment outcomes, suggesting a prevailing pattern of labour-replacing technological change.

The evidence emerging from these relationships is that, in the five European countries under study, in the cycle of growth which includes the recession of the early 1990s, demand factors have been crucial in sustaining employment, while technological factors show contrasting effects of product innovations (positive) and general technological efforts (negative).[5] These findings point out the main mechanisms which can explain the large losses in manufacturing jobs in Europe in the early 1990s. Total manufacturing employment in the aggregate of the five countries fell from 15.6 million in 1989 to 13.7 million in 1995, a loss of one-eighth of total employment in six years, with annual rates of decrease ranging from -1.3 per cent in the Netherlands to -2.3 per cent in Germany.

This empirical evidence may be interpreted to conclude, within the framework proposed in Figure 5.2, that the heavy job losses in Europe in the early 1990s were largely the result of a pattern of technological change dominated by labour-saving process innovations. The positive effects of demand growth, structural change and product innovations were too weak to counterbalance the large labour-saving effects. This outcome closely resembles what we have defined above as a strategy of active price competitiveness in a context of weak demand growth and increased international competition in most industries.

A major limitation of this evidence is the small number of countries for which innovation survey data are available. In the next section an attempt to generalize these results to the more advanced countries is proposed, comparing Europe, the USA and Japan.

4.   A COMPARISON OF EUROPE, THE USA AND JAPAN

The analysis so far has shown that the question of the employment impact of innovation has to be asked in more specific contexts, identifying particular patterns of technological change and demand conditions. The empirical work of the previous section has tested the ability of this framework to identify the employment impact of the nature and dynamics of innovation and demand, using detailed information on the intensity and orientation of technological change. Now the framework of Figure 5.2 can be used in a more synthetic way, looking at the major characteristics and outcomes of this process, extending the investigation to the more advanced countries and contrasting, in particular, the performance of Europe with that of the USA and Japan.

The lack of detailed innovation survey data for other countries makes it impossible to extend the previous methodology to broader international comparisons. What is proposed here, therefore, is to concentrate the attention on a few major aspects which can be investigated with existing economic indicators. They include the product or process innovation orientation of industry, the dynamics of productivity, which is expected to reflect the success of innovative strategies, the structural composition of industry and the aggregate pattern of economic and employment growth.

The analysis focuses here on the four largest European countries (Germany, France, the UK and Italy), this European group as a whole, the USA and Japan, and the aggregate of these six major economies: the G6 group.

The need to discriminate the nature of the innovative efforts suggests that industries could be divided into those with a prevalence of product innovations and those dominated by process innovations. It is expected that significant differences can be found in the performance of such groups, reflecting the nature and operation of technological and structural change in different countries.

The partition of industries (the full list is included in the appendix to this chapter) is based on the shares of R&D related to product innovation in the five European countries examined in the previous section (this variable is strongly associated with other indicators of the relevance of product innovations, as shown in Figure 5.3). The industries dominated by product innovations mainly include the chemical, electronic, transport and machinery sectors, in addition to furniture and other industries and leather and footwear. They are characterized by generally oligopolistic markets, greater innovative intensities (not just in R&D, but also in design and innovative investment) and by higher shares of new products in sales and exports; in this field higher rates of demand growth and job creation may be expected.

The process innovation-oriented group includes industries in competitive (but sometimes oligopolistic and segmented) markets, as well as traditional industries facing structural decline or strong cost-based foreign competition. Here the innovation strategies of firms are characterized by labour-saving innovations leading to higher productivity and pressure to lower costs. The net impact on employment will be the result of the direct labour-saving effects and of the evolution of market shares.

The strong sector-specificity of the patterns of technological change suggests that the breakdown between product and process-oriented industries can be generalized to other advanced countries, making it possible to compare the performances of Europe, the USA and Japan. The key indicator used here to this purpose is the growth of productivity (value added per employee), contrasting the long period of sustained growth from 1975 to 1989 with the period 1989–94, as shown in Table 5.3. Productivity growth can result from the ability to expand markets and take advantage of economies of scale and product innovations, or from efforts at labour-saving improvements in processes and organizations. Splitting the industries dominated by product or process innovations helps identify the relative importance of the two paths to productivity growth.

Table 5.3 shows that productivity growth in the whole of the G6 countries was much higher in the 1970s and 1980s than in the 1990s, and in both periods higher in product innovation-based industries than in those dominated by process innovations. This reflects the faster growth of markets associated with the emergence of new products and the positive impact of expanding demand on productivity, while the industries dominated by strategies based on restructuring and process innovations have shown only modest productivity gains.

This overall picture for advanced countries conceals different national patterns. The USA is broadly in line with the G6 picture; Japan has more extreme performances, with the highest productivity growth in product industry until 1989, and the lowest after 1989 (even a falling one in process industry) owing to the employment protection in a period of slower growth.

Europe is close to the G6 pattern in the first period, but after 1989 productivity growth in process industries outpaces that in product-based ones,

Table 5.3   Performance of product and process innovation-based industries

Countries	Productivity growth avg. annual rates of change				Share of product innovation industries in manuf. value added		Employment growth in total manufacturing avg. annual change	
	1975–89		1989–94		1975	1994	1975–89	1989–94
	Product	Process	Product	Process				
USA	3.78	2.37	3.48	0.59	45.9	56.5	0.51	−1.01
Japan	6.33	2.50	1.39	−0.36	40.8	60.3	0.47	0.33
Germany	2.30	1.84	0.99	2.56	48.1	54.3	−0.37	−1.63
France	3.87	2.79	1.88	1.89	43.5	49.7	−1.60	−1.59
UK	3.76	3.43	1.69	1.32	49.0	54.5	−2.32	−1.62
Italy	5.44	4.65	2.12	2.87	40.8	43.7	−0.70	−2.07
Europe 4	3.59	2.94	1.41	2.17	46.0	51.0	−1.18	−1.72
G6	4.35	2.66	2.20	0.96	44.9	55.5	−0.24	−0.93

reversing the general pattern found so far (the UK being an exception). This outcome is consistent with our findings of a dominant strategy of labour-saving process innovations which in the 1990s was combined with a slow growth of demand, producing in Europe massive job losses, in contrast to the modest employment decline of the USA and the modest growth of Japan in 1989–94, pointed out in the last column of Table 5.3.

The US and Japanese product innovation industries performed better in productivity terms than process-based industries, and this suggests a prevailing strategy of technological competitiveness based on the introduction of new products and the creation of new markets, while the expansion of demand made room for strong increases in productivity. The model of productivity growth typical of the 1970s and 1980s in all advanced countries appears to persist in the USA and Japan in the 1990s, while Europe seems to be withdrawing to a strategy of active price competitiveness dominated by labour-saving process innovations.

Besides this combination of innovative strategies and productivity performances, a second key aspect to be considered has to do with the structural composition of manufacturing industry in Europe, the USA and Japan, indicated in Table 5.3 by the share of value added in product innovation industries in 1975 and 1994. Product innovation-based industries have expanded everywhere, but at markedly different rates. In 1975, they accounted for 45 per cent of manufacturing value added of G6 countries; 20 years later the share had reached 55 per cent. The largest increase has been that of Japan (from 41 per cent to 60 per cent), followed by a ten-point growth of the USA and by an increase of only five percentage points in Europe.

The extent of industries characterized by process innovations has remained larger in Europe than in its major competitors, and therefore it is no surprise that the negative employment effects of technological change in such sectors are more strongly felt in Europe. The composition effect of manufacturing industry adds to the effects of the dominance of labour-saving strategies in leading to worse employment outcomes (see Pianta and Vivarelli, 1999).

A third aspect to be considered is the diverging overall economic and employment performances of Europe, the USA and Japan, rooted in the different dynamics of demand. In 1975–89, Europe was already losing jobs at the rate of more than 1 per cent a year, while the USA and Japan expanded employment by about 0.5 per cent a year. After 1989, Europe's decline deepened at an annual rate of –1.7 per cent, and the USA started to lose jobs at –1 per cent a year, while Japan continued to show a modest employment growth (which, however, turned into a decline in the economic stagnation of more recent years).

Aggregate demand growth and different macroeconomic policies – more restrictive in Europe, more expansionary in the USA and Japan – have played a key role in shaping the diverging employment performances of the three areas. Again the set of factors and of their combinations pointed out in our analytical framework of Figure 5.2 appears consistent with the different performances summarized above for Europe, the USA and Japan, suggesting that the better employment patterns of the latter countries result from the combination of technological strategies more focused on product than on process innovations; a structural composition of the industry leading to a stronger presence in industries with expanding markets; and a stronger aggregate demand dynamics and more expansionary macroeconomic policies.

Testing these relationships in empirical terms is difficult owing to the lack of adequate technological data, and would be beyond the scope of this chapter anyway. It is equally beyond its aims to contrast this explanation of employment (and unemployment) based on technological and demand factors with the interpretation, dominant in the literature and in policy (see OECD, 1994), concentrating on the dynamics of labour markets. While wages, skills, rigidities and labour market regulations may play a role in current (un)employment patterns, a full explanation can hardly ignore the relevance of technological change and demand dynamics which emerged in this analysis.

What the evidence in this section does show is the contrasting patterns shown by Europe, the USA and Japan, which shed some light on the different nature of technological change and role of demand patterns in the evolution of manufacturing. For advanced economies, product innovation-based sectors represent a source of strength in the economic structure, a presence in faster-growing industries at the global level, and activities where innovation can have fewer labour-saving consequences. Conversely, industries dominated by process innovation are more exposed to international competition, tend to grow less and lose jobs faster than product-oriented industries, resulting in worse employment performances of the countries where such sectors dominate the industry structure.

The position of Europe offers few signs of optimism. The four European countries considered here have expanded value added more slowly and reduced employment faster than the USA and Japan. They show no positive association between growth and employment performances. The presence of industries characterized by product innovations is smaller, leading to slower demand growth and greater job losses. The prevalence of process innovations is associated with strategies aiming at preserving cost competitiveness in the face of increasing international competition, at the cost of reducing the existing domestic industrial base, without expanding it in new, faster growing and product-oriented sectors. In a macroeconomic context marked by continuing restraint on demand, these dismal performances are likely to continue in the future, further weakening Europe's industrial base and employment performance.

NOTES

* I thank Pascal Petit and Rinaldo Evangelista for their comments on earlier drafts of this chapter.
1. Recent studies on technology, growth and employment addressing some of the aspects covered in this chapter include Freeman *et al.* (1982), Freeman and Soete (1987, 1994), Boyer (1988), OECD (1996), Petit (1995a, 1995b), Pianta *et al.* (1996), Pini (1995) and Vivarelli (1995).
2. Early uses of the data for Italian industry at the sectoral level include Evangelista (1995, 1999), Vivarelli *et al.* (1996) and Pianta (1996a), who have found a generally negative employment impact of technological change. An update and an extension to service sectors is found in Pianta (1998) and Evangelista and Perani (1998). At the firm level, the study by Cesaratto and Stirati (1996) compares 6000 innovating and 9000 non-innovating Italian firms with the employment and economic performance of the 26 000 firms included in the survey on economic activity. The number of employees appears an unreliable measure as it is gross of temporary layoffs, and the number of hours worked by blue-collar workers is also investigated. Overall, innovation shows a negative relation to employment, although small innovating firms perform better than any other group in all indicators. The expected contrasting consequences of product and process innovations are also found.
3. When the importance of product innovations is plotted against changes in employment (which are negative in the majority of sectors) no clear relationship emerges, although most sectors with above-average employment growth show a strong prevalence of product innovations. When employment change is plotted against total innovative expenditure per employee, countries show different patterns. In Germany, a weak positive relation is found, while for Italy a negative association emerges, as shown in Pianta (1996b). This finding for Italy is affected by the specific industrial structure of the country, by the prevalence of labour-saving process innovations within firms' innovative efforts, and by the strong productivity growth obtained at the expense of employment also in high-tech industries.
4. It could be argued that the consequences of the recession of the early 1990s are making the picture offered by a data set ending in 1993 gloomier than may be the case, especially on the negative employment impact of innovation intensities. Additional evidence on this comes from the availability of sectoral employment data for Italy up to 1996. Several years after the end of the recession, the negative relationship between sectoral innovation intensities and job performances is confirmed. When employment changes from 1992 to 1996 (an appropriate period to search for the impact of the innovative efforts of 1992) are investigated, their relationship to the innovation intensity of manufacturing sectors remains the same as shown in the regressions for Italy of Table 5.2.

REFERENCES

Appelbaum, E. and R. Schettkat (1995), 'Employment and productivity in industrialized economies', *International Labour Review*, 134 (4–5), 605–23.

Archibugi, D. and M. Pianta (1996), 'Innovation surveys and patents as technology indicators: the state of the art', in OECD, *Innovation, Patents and Technological Strategies*, Paris: OECD.

Archibugi, D., R. Evangelista, G. Perani and F. Rapiti (1996), 'Nature and impact of innovation in manufacturing industry: some evidence from the Italian innovation survey', *Research Policy*, 26, 521–36.

Boyer, R. (1988), 'Formalising growth regimes', in G. Dosi, C. Freeman, R. Nelson, G. Silverberg and L. Soete (eds), *Technical Change and Economic Theory*, London: Pinter.

Cesaratto, S. and A. Stirati (1996), 'The impact of innovation on employment in the manufacturing sector in Italy. Results from CIS', paper for the Eurostat-DG XIII conference on 'Innovation measurement and policies', 20–21 May, Luxembourg.

Evangelista, R. (1995), 'Innovazione e occupazione nell'industria italiana: un'analisi per imprese e settori', *L'Industria*, 1, 107–26.

Evangelista, R. (1999), *Knowledge and Investment. The sources of innovation in industry*, Cheltenham, UK and Northampton, MA, USA: Edward Elgar.

Evangelista, R. and G. Perani (1998), 'Innovation and employment in services. Results from the Italian innovation survey', paper for the EAEPE Conference, 5–8 November, Lisbon.

Evangelista, R., T. Sandven, G. Sirilli and K. Smith (1998), 'Measuring innovation in European industry', *International Journal of the Economics of Business*, 5(3), 311–33.

Freeman, C. and L. Soete (eds) (1987), *Technical Change and Full Employment*, Oxford: Blackwell.

Freeman, C. and L. Soete (1994), *Work For All or Mass Unemployment?*, London: Pinter.

Freeman, C., J. Clark and L. Soete (1982), *Unemployment and Technical Innovation*, London: Pinter.

OECD (1994), *The OECD Jobs Study. Part I: Labour market trends and underlying forces of change*, Paris: OECD.

OECD (1996), *Technology, Productivity and Job Creation*, 2 vols, Paris: OECD.

Pasinetti, L. (1981), *Structural Change and Economic Growth*, Cambridge: Cambridge University Press.

Petit, P. (1995a), 'Employment and Technical Change', in P. Stoneman (ed.), *Handbook of the Economics of Innovation and Technological Change*, Amsterdam: North-Holland, 366–408.

Petit, P. (1995b), 'Technology and employment: key questions in a context of high unemployment', *Science Technology Industry Review*, 15.

Pianta, M. (1996a), 'L'innovazione nell'industria italiana e gli effetti economici e occupazionali', *Economia e Politica Industriale*, 89, 261–80.

Pianta, M. (1996b), 'Technology and jobs in the 1990s. The results of the German and Italian innovation surveys', paper for the TSER project meeting, 22–23 November, Paris, CEPREMAP.

Pianta, M. (1998), 'Innovazione e occupazione in Italia e in Europa', in P. Guerrieri and M. Pianta (eds), *Tecnologia, crescita e occupazione*, Naples: Cuen.

Pianta, M. and M. Vivarelli (1999), 'Employment dynamics and structural change in Europe', in J. Fagerberg, P. Guerrieri and B. Verspagen (eds), *The Economic Challenge for Europe: Adapting to Innovation-based Growth*, Cheltenham, UK and Northampton, MA, USA: Edward Elgar.

Pianta, M., R. Evangelista and G. Perani (1996), 'The dynamics of innovation and employment: an international comparison', *Science Technology Industry Review*, 18.

Pini, P. (1995), 'Economic growth, technological change and employment: empirical evidence for a cumulative growth model with external causation for nine OECD countries, 1960–1990', *Structural Change and Economic Dynamics*, 185–213.

Vivarelli, M. (1995), *The Economics of Technology and Employment*, Aldershot, UK and Brookfield, USA: Edward Elgar.

Vivarelli, M., R. Evangelista and M. Pianta (1996), 'Innovation and employment in Italian manufacturing industry', *Research Policy*, 25, 1013–26.

APPENDIX

The first results of the Community Innovation Survey (CIS) were presented at the conference on 'Innovation measurement and policies', organized by Eurostat and the European Commission DG XIII, and held in Luxembourg on 20–21 May 1996.

A major advance in technology indicators made possible by CIS is the definition of the total expenditure devoted to innovation by firms, including expenditure for R&D, design, trial production, innovative investment, acquisition of patents and licences and exploratory marketing. The nature and structure of this expenditure is described in Evangelista *et al.* (1998), on the basis of a larger research report by ISRDS–CNR and STEP from which the data used here are taken. An overview of early studies based on innovation surveys is in Archibugi and Pianta (1996). The results of the Italian survey are investigated in Archibugi *et al.* (1996).

The analysis of section 3 was carried out on Denmark, Germany, Italy, the Netherlands and Norway because these are the countries with the more solid statistical results of the innovation surveys. In Italy, 22 787 firms responded to the innovation survey, 7553 of which were innovative. In Germany, 3879 firms responded and two-thirds introduced an innovation in the 1990–92 period. The number of responses were 4094 in the Netherlands, 674 in Denmark and 982 in Norway. Other countries had lower numbers of replies and worse response rates, or did not include the question on innovation expenditure (France). In the database used in this chapter the number of firms included is lower, as shown in Table 5.1, because those with missing values for the innovation variables here considered have been excluded.

The key innovation variables used in the analysis of section 3 are the following: total innovation expenditure per employee in 1992, expenditure for R&D and design per employee in 1992, and share of R&D related to product innovations in 1992. The first two variables are calculated with reference to the firms which have introduced an innovation in 1990–92. The innovative intensities have been calculated by dividing the expenditure of the firms of a sector by the number of employees of the innovating firms of that sector (an alternative method, producing very similar results, is the use of the sales of innovating firms). These sectoral innovative intensities can be considered as an indicator of the innovative effort of individual industries, and are related to the performances of the whole of the sectors.

The third variable was the product innovation indicator available for all five countries. In some cases, such as Italy, information on the number of firms introducing product innovations only and on the share of sales related to product innovation was also available; across industries, the results obtained are broadly consistent with those of the variable shown here.

These innovation variables have been combined with economic data for 21 manufacturing sectors (3-digit ISIC classes) drawn from the OECD STAN 1995 database. These variables include: average annual rate of change of real value added, 1989–94 (for Italy, 1989–93), average annual rate of change of gross fixed capital formation, 1989–92 (for Denmark and Italy, 1989–91; for the Netherlands, 1989–93), and average annual rate of change of employment, 1989–93 (dependent variable) (for Denmark, 1989–91; for Germany, 1989–92).

It is difficult to estimate the lag between the innovative effort and the emergence of the economic impact; employment and output performances have been calculated using the average annual rates of change from 1989 (the start of the downturn) to the latest available year, with some differences due to missing data. We expect that the inter-industry differences investigated here are likely to emerge even with an imperfect structure of lags. The regressions reported in Table 5.2 are all carried out with the ordinary least squares method.

In the analysis of section 4, economic variables at the industry level have been aggregated into two groups, defined on the basis of the share of R&D devoted to product innovation as reported by the firms surveyed in Europe.

The list of industries in product innovation-based industries includes the following: Leather and footwear, Furniture and other industries, Chemical products, Electrical apparatus, Radio, TV and communication equipment, Machinery and equipment, Office and computing machinery, Motor vehicles, Other transport and Professional goods.

The list of industries in process innovation-based industries includes Food, beverages and tobacco, Textiles, Wearing apparel, Wood products, Paper and products, Printing and publishing, Rubber and plastics, Non-metallic mineral products, Basic metal industries and Metal products. The class of Petroleum refineries has been excluded because of its specificity and the low number of firms included in this sector. Therefore each group includes ten 3-digit ISIC classes.

Data for Table 5.3 are drawn from the OECD STAN 1996 database. Productivity is calculated as constant prices value added divided by employment.

6. Technical change and employment growth in services: analytical and policy challenges

Pascal Petit and Luc Soete[1]

INTRODUCTION

It is generally acknowledged that employment in our economies is increasingly dependent on services. As in other highly developed economies, the European countries are continuing their gradual move towards a service-based economy with today nearly 70 per cent of the total labour force being employed in service activities. It is also generally acknowledged that services provide the key to future employment growth. Neither agriculture nor manufacturing has been able to generate sufficient output growth to offset, in the last two decades, the productivity growth following the diffusion of labour-saving machinery and the reorganization of work and trades, impelled by increasing international competition. And while some high-tech manufacturing sectors have succeeded through the introduction of new and improved high-income-elastic consumer goods to generate new employment opportunities, their number has been falling steadily over time. Particularly in Europe, high-tech manufacturing sectors no longer witnessed any employment growth over the 1990s. Still, given the generally acknowledged importance of services for future output and employment growth, relatively little attention has been paid to technical change in services.

But technical change in services is a key issue to understand to what extent it will help to develop new markets and welfare or will be furthering the trends of automation. The future of work is at stake in these processes and the answers are not straightforward. Technical change in services has its specificities.

In the first place the development of goods and service markets is not submitted to the same type of constraints. The localization of services and the interaction between customers and producers that occurs in these trades impose specific constraints on the development of new products and of new processes. Both process and product innovations in services will thus be more severely constrained by the willingness, abilities and original tastes and habits of the customers than they are for goods. Process innovations in manufacturing of goods are neutral for the product market and product innovations can be channelled by widespread advertising, marketing techniques or straightforward and rapidly diffusing demonstration effects.

Information and communication technologies (ICTs) have also a specific impact in that respect as these technologies transform the basic context in which services can be perceived and delivered. We say they change the tradeability of services and expand the potential of fields and forms of new markets. Conversely, ICTs transform the markets of goods, with more customerization and lasting relations, thus bringing characteristics of services to these markets.

These transformations clearly depend on the initial types of arrangements under which service markets are organized. These arrangements are very country-specific, implying local cultures and customs. They also depend on the skill structure of the country. Still one guesses that there is no determinism and that changes in regulations (which is more realistic than to speak of deregulations) and in policies may be quite important for the type and magnitude of new activities and therefore for the future of work.

This chapter does not attempt to answer all these broad issues. It aims to draw up some of the analytical arguments that we need to think in more appropriate terms about what is the dynamics of technical change in services, how it relates to the dynamics of employment in these activities and in which directions we should look for design policies at all levels.

We summarize in section 1 the traditional technology and employment debate and point to some of the contemporary challenges. The spread of service activities and the extent of internationalization of markets and production processes make it difficult to assess any 'compensation scheme', whereby the jobs destroyed in some trades by the emergence of new techniques and products would be more than offset by the gains in some other national trades. Opportunities for other countries or other activities to reap the benefits from the changes are too numerous. The second section traces the initial differences between innovations in goods and in services (generically speaking) and how ICTs, somehow, bridge some of these gaps.

Section 3 looks for the conditions under which service activities could presently act as an engine of growth for the whole economy. It goes on to stress the importance for the expansion of service markets and for employment of (a) user–producer relations, (b) the skill structure and (c) the time budget constraint.

The fourth section expands on the employment issue, taking into consideration, first, the high diversity of activities under consideration and, second, how the form of changes brought by ICTs depend upon skills and cultures.

The fifth, and concluding, section explores, on this basis, which kind of policies could be pursued to develop markets as well as the number and quality of jobs contributing to such development.

1.   TECHNOLOGY, GROWTH AND EMPLOYMENT: THE END OF A VIRTUOUS CIRCLE?

The relationship between technology, growth and employment has traditionally been the subject of many contributions in economics.[2] While controversial, and the subject of intense debate over the last two centuries, the issue appears straightforward at least from the macroeconomic perspective. Either the introduction of new technologies leads to more efficient production processes, reducing costs by saving on labour, capital, materials, energy or any other factor of production, or it leads more directly to the development of new products that generate new demand. In either case, more welfare is created: in the first, through more efficient production combinations that liberate scarce input resources; in the second, by satisfying new wants.

The extent to which this higher welfare or increased productivity feeds back into employment growth depends on the extent to which firms translate productivity gains into lower prices and new investment and consumers respond to lower prices in terms of greater demand. The job losses that often follow the introduction of a new labour-saving process, for example, are compensated by the job creation associated with the output growth following the decline in prices, by additional employment creation in other sectors, particularly the new technology-supplying sector, and by the possible substitution of labour for capital following the downward wage adjustment that clears the labour market.

However, the extent to which new or improved products generate new employment growth depends on whether old products are replaced by new ones and on the responsiveness of consumers to the new or improved goods or services (reflected in the size of the income elasticity of demand). As long as there are unsatisfied needs in the economy and as long as labour and product markets are sufficiently flexible, technological change, even in the form of new labour-saving production processes, does not reduce aggregate employment but generates more growth and jobs.

Most of the controversies that have dominated the economics literature on this issue over the last decades have centred on the automatic nature of the various compensation effects described above. Many contributions have questioned the way in which cost reductions following the introduction of new technologies are effectively translated into lower prices and are likely to lead to more output growth: the functioning and flexibility of product markets depend in part on the firm's monopoly power, the degree of economies of scale, and various other factors influencing 'price stickiness'. Similar issues can be raised with respect to employment growth and the functioning of labour markets; they range from downward wage flexibility to the many mismatches typical of relatively heterogeneous labour markets. In either case, it is less technology that is at the centre of the debate than the speed and clearing function of the product and labour markets.[3] The relevant policy issues fall primarily under the heading of improving the functioning of labour and/or product markets.[4]

Other contributions in the classical economics tradition have questioned the possibility of *ex post* substitutions between labour and other factors of production. At least in the short term, the implications of a more rigid fixed set of production coefficients for analysing technical change and employment are relatively straightforward. Labour-saving technological change embodied in new investment could, if wages adjust slowly, lead to unemployment because of insufficient investment to maintain the full-employment capital stock;[5] this is the so-called 'capital-shortage' unemployment.[6] There was a lively debate during the 1980s on the extent to which the increase in unemployment in European countries in the 1970s could be due to this phenomenon.

Yet other contributions question the automatic nature of the link between input-saving new technologies and productivity gains. Most of these studies (which often attempt to explain the 'productivity paradox') are empirical in focus and attempt to find reasonable explanations for the disappointing performance of productivity growth in most OECD countries over the last two decades, despite rapid growth in knowledge investment, in particular in private sector R&D, and the emergence of the new cluster of information and communication technologies. The OECD summarized much of this debate in the so-called 'Sundqvist Report' (OECD, 1986) and the subsequent 'Technology and the Economy' programme (OECD, 1992). However, the discussion is far from over. In particular, there have recently been a large number of empirical and theoretical contributions from growth economists (for example, Young, 1995; Mankiw, 1995).

Finally, some recent contributions have focused explicitly on the international 'open economy' framework within which most compensation mechanisms are likely to operate. As a result, the relatively straightforward linkages between technology, productivity growth and job creation mentioned above appear much more complex. A relatively simple elaboration in terms of employment compensation due to foreign demand, for example through export and import elasticities, complicates the matter greatly (Stoneman, 1984). More complete pictures including not only trade but also the effects of international spillovers of technology on productivity growth or international capital mobility make it much more difficult to identify the key links between the introduction of a new technology and the ensuing domestic employment impact.

Many of the recent concerns about the implications of technological change for employment appear to relate to these international compensation mechanisms and to the way that gains from technological change are distributed internationally. In the gloomy vision of some popular authors,[7] 'wages in the most advanced economies are being eroded owing to the emergence of a global market-place where low-paid workers compete for the few jobs created by footloose global corporations' (Rifkin, 1995). Others (such as Freeman, 1995) stress that the wages of developed economies are not set in Beijing, because a lot of jobs are in trades which do not face so directly the competition of very low wages countries, because either the products are more sophisticated or differentiated and submitted to non-price competitiveness or the trades are local and sheltered from outside competition, as in some service activities. Thus, even when the internationalization of manufacturing and service industries is expanding, spurred by low costs of transport and communication, the balance of the interactions between technology and employment much depends on the type of competition prevailing on product markets and on service markets, which concentrate two-thirds of employment.

While it is still generally agreed that in a 'world' economy framework, input-saving technical change leads, through increases in productivity, to higher welfare, wages and growth, and thus generates new employment, the impact on individual countries is now much more complex and is based on a broad range of macroeconomic and microeconomic adjustment mechanisms. At the same time, the premium placed on the role of knowledge and on the acquisition of skills in this global environment implies that international differences in the pattern of employment and unemployment in industrialized economies may be coming to depend increasingly on the capacity of national economies to innovate, enter new 'service' areas and/or absorb new technology more rapidly.

2. NEW INFORMATION AND COMMUNICATION TECHNOLOGIES: BRIDGING TIME AND DISTANCE

The dramatically increased capacity to store, process and disseminate information at minimal cost has been described most extensively in the context of industrial (or agricultural) production processes. Predating even the early 'Information Technology' literature, the so-called 'automation debate', popular in the USA in the mid-1960s, described how labour-saving 'robotics' would raise industrial productivity and bring about major organizational changes. In line with this literature, many IT analyses have always wondered how, confronted by such pervasive cost-reducing technologies, economies would be able to generate sufficient new employment (the various price and substitution elasticities being too low to bring about sufficient employment compensation[8]). More recently, the specific impact of new information technology on services has re-entered this debate. It could be argued that the impact on services will be more of an opposite nature compared to the impact on manufacturing.

In many ways services can be defined[9] as those activities (sectors) where *output is essentially consumed when produced*. While this might well be considered a rather narrow definition and one which covers only a limited number of sectors currently falling under the statistical definition of service sectors, it is an analytically useful definition because it highlights the intrinsic immaterial, intangible nature of many service activities, whether they are personal services, such as hair-cutting, entertainment, such as an opera performance, education, such as teaching, health, such as a doctor's visit, or public services, such as applying for welfare services. With intermediary services such as transport, communication, finance and trade, this simultaneity still holds, but is partially altered. Intermediary services are effectively delivered more or less on a permanent or fixed basis, whether they are used or not. The frequency of provision may vary, but in a scheduled way (timetables of services need to be available). The logistic support it gives is independant, in the short run, of demand. Good management should certainly adapt the level of production to the needs, while sizeable short-term productivity cycles remain a characteristic of these service industries. The link between production and consumption is somehow even more altered when considering such business services as marketing, R&D, consultancy, accounting and advertising.

It follows that services range from activities where production and consumption cannot be dissociated to all kinds of loose linkages between production and consumption. Still it is this similarity feature of production and consumption which has generally limited productivity improvements in such activities.

Information and communication technologies, almost by definition, allow for the increased tradeability of service activities, particularly those which have been most constrained by the geographical or time proximity of production and consumption. By releasing somehow these constraints, information technology will make possible the separation of production from consumption in a large number of such activities, thereby increasing the possible trade of such activities.[10]

On Tradeability

The notion of tradeability when applied to a commodity seems to refer to the propensity to be more or less readily accessible in time and space. The notion is vague and has been mainly used in international economics to distinguish between commodities which were traded internationally and others. Tradeability obviously depends in the first place on the context in terms of logistics organizing the market, such as the distribution system, the communication system or the transport system. We shall refer for this set of conditions to the standard means of market provisions. On top of that, the quality and characteristics of one product may be more or less easy to identify and require some specific knowledge or abilities.

Tradeability of any given product finally has two dimensions: how easy and costly its provision is and how clear and straightforward its use is. Provisionability clearly depends on transport and communications costs as well as on after-sale services. User friendliness depends on information, regulations, insurances and knowledge. This dichotomy somehow decomposes the 'transaction costs' which remain even on organized product markets.

An (incremental) innovation can enhance the tradeability of a product by improving either the context of provision or the straightforward content of a product.

Product and Process Cycles in Innovation Schemes

ICTs play an essential role in the transformation of information into knowledge as well as in the 'codification' of knowledge. The latter implies that knowledge is transformed into 'information' which can either be embodied in new material goods (machines or new consumer goods) or be easily transmitted through information infrastructures. It is a process of reduction and conversion which makes the embodiment or transmission, verification, storage and reproduction of knowledge especially easy.[11] In contrast with codified knowledge, tacit knowledge refers to knowledge which cannot be easily transferred because it has not been stated in an explicit form. One important kind of tacit knowledge is skills. The skilled person follows rules which are not all known as such by the person following them. They are linked to activities acquired through learning but often of a non-routine kind.[12] The most important impact of new ICTs is that they move the border between tacit and codified knowledge. They make it technically possible and economically attractive to codify kinds of knowledge which so far have remained in a tacit form.

The embodiment of codified knowledge in material goods has been typical of the dramatically increased performance of many new capital and consumer goods, incorporating many new electronic information and communication devices. The latter in turn have been at the core of the continuous productivity, investment and consumer demand growth in Western societies. As emphasized by authors criticizing the early 'post-industrial society' literature,[13] this process could also be described as a process of 'industrialization' of services: the continuous replacement of particular service activities by household material goods, embodying at least the 'codified' knowledge part (washing machines, televisions, dryers and so on). The more recent electronic improvement in these products has further increased the 'household' performance of these products, freeing further household time. While the quality of these new material goods will not always substitute for the service activity they replace (a dishwasher is a good example), the codification process will be to some extent complete. The product might lack user friendliness (the typical example being the video player), but the user is not required to possess, or to understand, the knowledge embodied in the machine.

In services, by contrast, while the codification of knowledge will have made such knowledge more accessible than before to all sectors and agents in the economy linked to information networks or with the knowledge of how to access such networks, its immaterial nature will imply that the codification will never be complete. The codification process will rarely even reduce the relative importance of tacit knowledge in the form of skills, competencies and other elements of tacit knowledge – rather the contrary. These latter activities will become the main value of the service activity: the 'content'. While part of the latter might be based on pure tacitness, such as talent or creativity, the largest part will be greatly dependent on continuous new knowledge accumulation – learning – which will typically be based on the spiral movement whereby tacit knowledge is transformed into codified knowledge, followed by a movement back where new kinds of tacit knowledge are developed in close interaction with the new piece of codified knowledge. Such a spiral movement is at the very core of individual as well as organizational learning.

On Goods and Services

If we come back to the characterization we gave above on services, namely that they are produced and consumed *at the same time* and *on the same spot*, we see that it implies a low tradeability with respect to the two dimensions. In the first place, services cannot be stored, otherwise production and consumption could easily be separated. The market provision of services is therefore severely constrained. For the classics (and Smith in particular) this non-storability drastically limited the ability of service activities to take part in the accumulation process and therefore services were considered as not creating value. Secondly, the fact that the service is consumed while produced leaves some uncertainty on its very content; the transactions are thus more open to asymmetries of information and hazards of different kinds.

On the whole this simultaneity of production and consumption clearly makes services less tradeable than goods. Still, the border between the two types of production is not so clear-cut. We mentioned in the first place that the simultaneity of production and consumption is more or less strict depending on which service we consider. Secondly, some goods such as some equipment goods, highly customized or produced upon specific order, have a low tradeability according to our definition. In that case a clear but specific content goes altogether most of the time with a more difficult type of provision (implying delays and special requirements for transport).

At a given time and space the set of products available thus displays a wide spectrum of tradeability indexes, where goods depending to some extent on their degree of 'standardization' have on average a higher rating than services. Moreover, these tradeability characteristics change over time. Even in the absence of any technological change and of any change in the logistics of service provision, simple learning processes would lead to some steady increases in tradeability. The crucial role of regulations in determining product tradeability should not be forgotten either.

Changes in the context of provision and in the content of products, which are produced by the diffusion of ICTs, can therefore radically modify the tradeability pattern of the products.

ICTs Can Alter this Simultaneity of Production and Consumption

Considering the new facilities brought by ICTs, services can be delivered in various places simultaneously with their production. The concept of production itself is spread over time when deliveries are automated (pushing a button in various automated tellers or similar). If services are also something you get in indefinite amounts, providing you show up at some counter (to get information, training and so on) the space simultaneity of production and consumption of services is also altered. The provisionability of services is thus greatly improved by releasing the constraint that services were consumed when and where produced. Besides, the problems raised in appraising the content of services may also be reduced, as ICTs can help to standardize and diffuse information on the products. On both dimensions, provision and content, the tradeability of services is thus improved, with a greater emphasis on provision improvement.

Conversely, ICTs seem to improve greatly the information and the control over the quality and the use of the goods we buy. This rise in the ability to certify the quality of goods and to control their use (the above content dimension) is even more crucial than the improvement in the provisionability of goods, brought by intermediary services regenerated by ICTs (such as transport, distribution, finance and telecommunication).

Figure 6.1 illustrates simply these asymmetric improvements in the two components of tradeability for both industry and services. By and large, it suggests that the spectrums of tradeability of the sets of goods and of services are somehow converging. In effect, the tradeability conditions between goods and services are getting more similar through the effects of ICTs. This convergence results also from the fact that ICTs could well be characterized as reducing the time/storage dimension for goods and as bringing a time/storage dimension between production and consumption in services. Many of the most distinctive characteristics of the new information and communication technologies are effectively related directly to the potential of the new technology to link up networks of component and material suppliers, thus allowing for reductions in storage and production time costs – typified in the so-called 'Just-in-Time' production system. At the same time, the increased flexibility associated with the new technology allows for a closer integration of production with demand, thus reducing the firm's own storage and inventory costs, which could be typified as 'Just-in-Time' selling. Both features aim at reducing the *time/storage* dimension between production and consumption, but the 'tradeability' of products is not hampered because the product is more customerized (the buyer is made more confident that the product will meet his specific needs). In fact more customerized products, delivered just in time, transform the tradeability pattern of goods. Thus the paradox of opposed effects of ICTs on goods and services production disappears if one admits that ICTs eventually have an impact on two different things: the nature of the product itself and its provision. Figure 6.1 schematizes the relative convergence between the pattern of tradeability of goods and services.

Moreover, tradeability appears to be a notion highly dependent on the general context of provision (for the provisionability dimension) and of regulations and customs (for the conditions of use of the product). ICTs, in improving the logistics of intermediaries' activities which are organizing markets,[14] have shifted upwards the general level of tradeability. But, much like the notion of competitiveness, it is the relative level of tradeability which matters to the assessment of the new potential of product markets spurred by ICTs.

Finally, the fact that ICTs are making services more tradeable and more like manufactured goods on the one hand, and that ICTs favour the differentiation of products on the other, all lead to modifying the conditions of consumption.

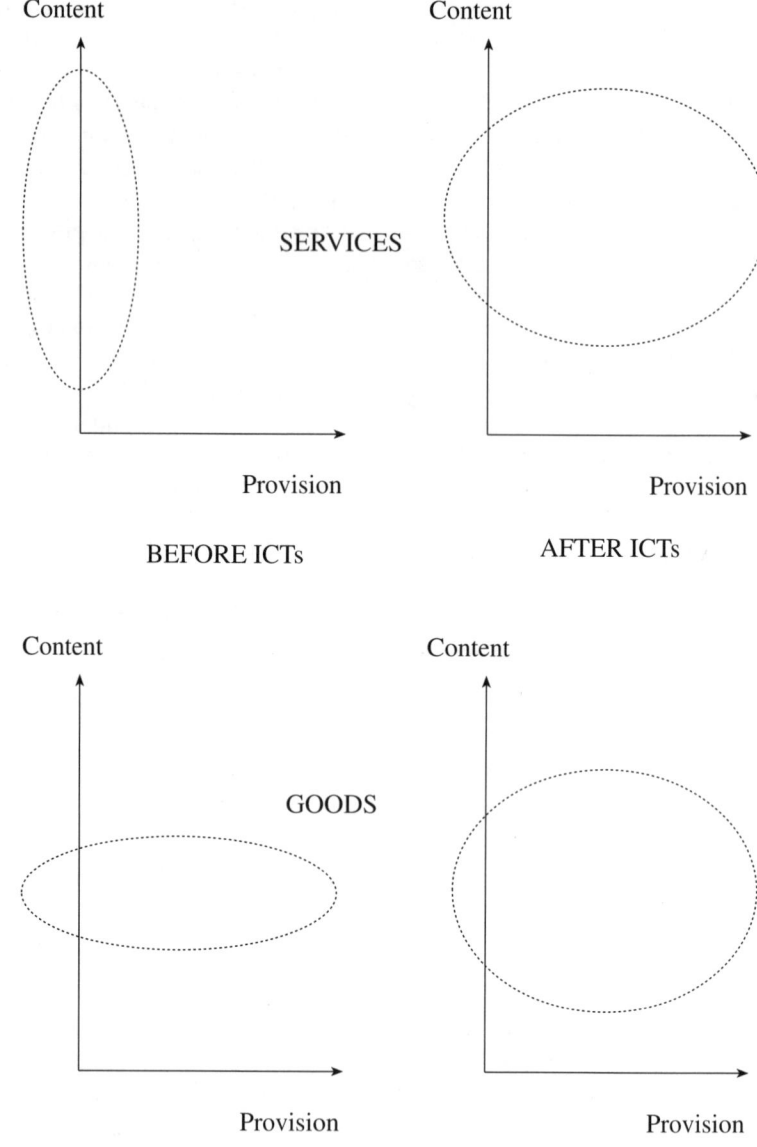

Figure 6.1   ICTs' effects on the tradeability of goods and services

A More Binding Time Constraint for the Consumers in a World of Enlarged Choices

We are used to thinking that new products lead to the discarding of old ones. Not only does innovation bring scrapping of old equipment and accelerate replacements but, according to the standard view, old varieties, often after a last fight (see, for example, Rosenberg's 'sailing ship effect') disappear. Historically, this stylized fact applies more accurately to innovation in goods than to innovation in services. In services, and especially in personal and social services as well as in intermediate services, it is quite characteristic that the new does not chase out the old to the extent that it does with goods. Services, modernized by some automatization process, are often seen as different services and the old form tends to become a service 'à l'ancienne' with an improved standard. Such has been the case with restaurants and hotels, but also with arts performances, distribution and personal care. Finally, innovation in services seemed to end up with a whole range of differentiated products.[15] A similar process happened with goods and 'antiques', but this remained a fairly marginal activity, while the decoupling of services by the modernization, partial or total, of their process of production contributes in the case of final services in distribution, catering and leisure activities) to enlarging the basket of activities at the consumer's disposal. Another way to look at this phenomenon is once again to note that, to the extent that in services production and consumption are tied, any process innovation is perceived by the consumer and therefore is also by nature a product innovation.

Insofar as manufacturing activities have gained similar service characteristics in using ICTs, while services themselves have become more tradeable, the basket of commodities available for the consumer has been greatly enlarged and transformed in nature, as their increased service dimension implies a more time-consuming consumption. As a result, more numerous and more time-consuming products have given de facto more importance than in the past to the time budget constraint, at least for the middle- and high-income groups. In effect, according to most recent surveys, while patterns of consumption have not been changing radically with the arrival of the new products, every choice seems increasingly to face competitive pressure from alternative time uses (Haddon and Silverstone, 1996).

To summarize our argument, we would claim that, for a long period of time, new industrial goods have been substitutes for old services (the industrialization of services hypothesis) therefore saving time for households which could be used to work and increase the capital stock of equipment. The relative convergence between some manufactured productions and services that is part of the diffusion of ICTs as we have underlined, leads us to insist much more on the time constraint. Services take time and modernization both accelerates some brands of services (sometimes relying on self-servicing) and also expands their range. More sophisticated goods have to be serviced and, although they save time in most cases, they expand the range of choices in such ways that one feels more strongly the time budget constraint when facing these enlarged choices.

The time constraint is sometimes greater than the budget constraint. This was a typical pattern for rich people but it seems now to concern a much larger set of people. One might think of youngsters having an increasingly difficult time managing the time constraint between school education, home education, TV, multi-media entertainment, physical entertainment and contributions to housework.

Whether or not these changes contribute to giving a new role to services in the process of economic growth becomes a key question to address for the future of growth and employment channelled by the ICTs.

3.   SERVICES: THE NEW ECONOMIC DRIVING FACTOR?

Since the emergence of ICTs and their impact on the tradeability of many service activities, which among other things partly blurred the frontier between goods and services, and since service activities correspond on average to two-thirds of economic activity in the EU countries, it is time to examine the role of these services in the process of economic growth.

Services as an Engine of Growth

Manufacturing has long been considered as an engine of growth for its capacity to organize and restructure production in ways allowing steady productivity gains. Economies of scale, such as replication on a larger scale of production processes, have been a favourite means to sustain this dynamics. It went together with the old Smithian principle that large markets allowed bigger scales of production, which in turn permitted a broader division of labour. Allyn Young (1928) emphasized the fact that such division occurred both within firms and between firms and that in all cases it stimulated technological change, which in turn fostered demand, so that economic growth propagated itself in cumulative ways. This was basically the mechanism referred to by Kaldor when speaking of manufacturing as an engine of growth.

Can one identify a similar cumulative dynamics in service activities? Certainly not in the pre-ICT period. Service activities were then seen in the cumulative causation model as necessary conditions, complementary to the manufacturing engine of growth in order to organize markets (the provision of market access was a function of intermediary service activities). Meanwhile, personal services were looked upon basically in relation to the prevailing conditions on the labour market (see the sponge effect in the presentation by Kaldor of the determinants of employment in personal services).

The question is thus whether innovation, driven by ICTs, can launch a cumulative mechanism based on services somewhat similar to the one experienced in the past in manufacturing. Another way to rephrase this, following our previous definition, is to appreciate how innovation processes in services enhance their tradeability and help to expand their markets in ways which in turn cumulatively improve their efficiency and tradeability. The conditions for such a growth principle to be effective depend on the organizational issues raised by the diffusion of ICTs in services, in particular on the relation between processed information, knowledge accumulation and elaborated routines.

There is thus a need to compare the schemes of innovation in manufacturing and in services. The learning processes implied in cases of innovation in goods and in services are, as we argued above with respect to the different impacts of 'codification', rather different. They characterize to some extent the various patterns of cumulative growth that can occur. In the case of goods, the learning process is centred upon the product itself. Producers are learning how to adapt the new product to tastes and how to take advantage of expanding markets to make productivity gains which in turn will help to increase the market and improve the product. It corresponds to the first phase of a Vernon product cycle. Users have of course their say in the process but that say is by and large limited to a process of adjusting to the quality of the product. We would even go a step further: the main driving factor for innovation is performance or quality improvement, with the aim of convincing the average consumer that what he needs is the best, professional quality. In doing so the innovating firms can avoid, at least for some time, price competition. When the second phase of the maturing product is reached and standardization and imitation are taking place (for example, when competitors with low wage costs take over a stabilized production process), productivity growth is the only answer, but will depend heavily on the extent to which economies of scale can still be achieved. Conditions for sustained innovation and market expansion may thus depend on adequate demand policies.

By contrast, the similar dynamics in services tends to start from the opposite process innovation side, as suggested by Barras (1986). ICTs help to transform parts of the production process of services, mainly by codifying knowledge and processing accordingly information in one part of the old process.[16] The drive behind this substitution is in the first instance an increase in the tradeability and market for an existing product. While it is not meant to modify the product, it will do so, in two stages. In our view though, and contrary to manufacturing, this will often imply in the first stage a product with lower-quality characteristics, compensated for by faster delivery. The driving force behind service innovation is thus not just process innovation; it is also cheap mass provision of a possibly lower-quality product. However, parallel to what was said above about manufacturing, the second stage will involve an explosion of new product innovation, involving high-quality, often personalized, services, using the new process technology for the specific aims and needs of particular users. It is through the combined effects of learning by doing and learning by using that the innovative content of the 'old' service product, produced with the new automated process, is progressively enlarged. Electronic networks have often evolved this way, as have a lot of new telecommunication products.

In other words, in this reverse product cycle, productivity gains are conditioned by improvement in the quality of the service products and process innovations alone are not sufficient (as they risk being associated with lower-quality products). This is a much more hazardous way to fuel a process of cumulative growth than is the case with manufacturing goods. It requires skill from the producers to enlarge the process into a meaningful product innovation but it requires also some learning from the consumer to direct and legitimize the quality improvement of the services. In this dual process of product innovation in services (mass products on one side, highly customerized services on the other) the challenges are different according to whether households or businesses are concerned. In particular, the markets of tailored new services are more difficult to develop with households as they require rather intense and long-term user–producer interfaces which only medium and large businesses usually enjoy.

Besides, the implications of this continuous shift in value from manufactured goods embodying increasing amounts of 'codifiable' knowledge towards service-based 'tacit' knowledge activities are typical of the new emerging Information Society. It explains the attempts of electronic and computing manufacturing firms to enter information content activities. Within services, it explains the move of 'carrier' operating firms, being most directly confronted with the codification of knowledge and its distribution, to enter content sectors (media, education, culture). This difference between innovation schemes in manufacturing and in services is much enhanced by the upstream dynamics of ICTs driven by the current miniaturization of microprocessors. It reinforces in all activities the process-driven dimension of technological change.

Figure 6.2 seeks to summarize these two schemes of innovation predominant, respectively, in goods and in services and to relate them to the specific

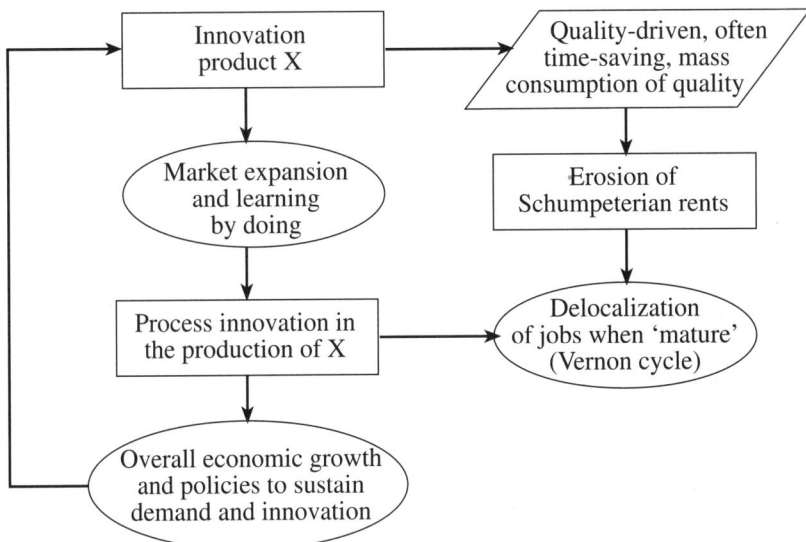

Figure 6.2(a)   Innovation scheme in manufacturing

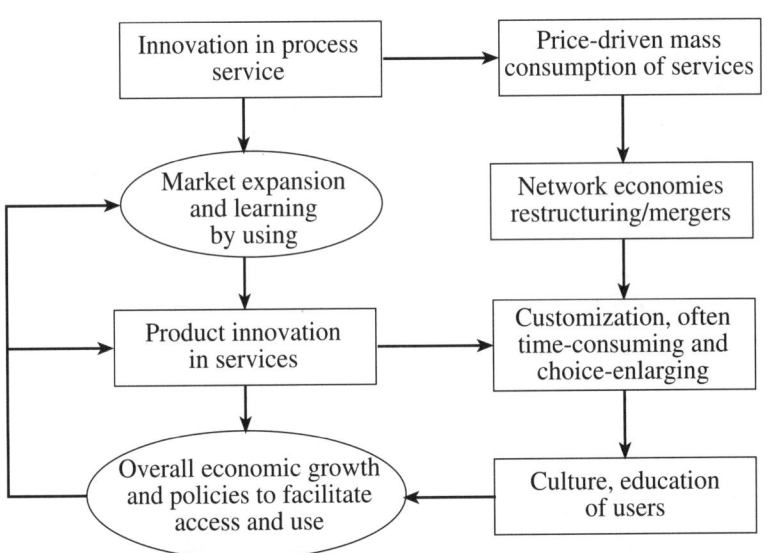

Figure 6.2(b)   Innovation scheme in services

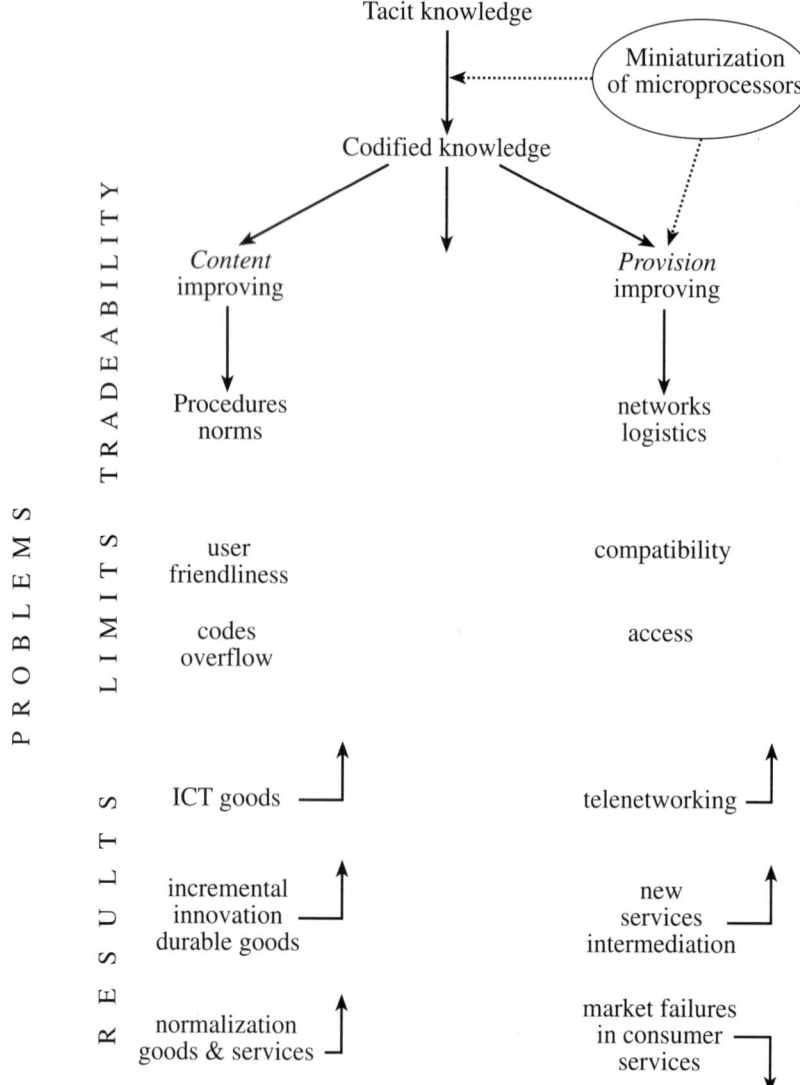

Figure 6.2(c)   ICTs on tradeability

effect of ICTs. It stresses that the outcome of innovation processes depends more extensively on learning processes on the part of the users in the case of services than on the ready-made quality improvements of the suppliers in the case of manufacturing. It also brings to the forefront the need for a shift in policies to sustain the innovation process. In the first case, policies to reflate demand, whereby the gains in productivity are reflected in gains in wages and domestic consumption, are essential to maintain the virtuous circle between productivity gains and new demand. The employment compensation mechanisms operate primarily through income-elastic demand for new and improved goods. The only changes brought by ICTs will be to shift the focus of innovation policy away from supply-dominated science and technology support policies to policies aimed at the translation of new scientific and technological breakthroughs into new innovations. Typically, most of the current EU innovation policies, as their name (VALUE) indicates, correspond to this aim.

In the second case, where the service-oriented innovation scheme prevails, policies will be much more diversified, helping users, in the first instance, to coordinate themselves, deregulating particular service markets and breaking up where necessary cartel agreements and providing incentives to new firms to develop services using the new provision channels of old services.

A Twin-motor Engine of Economic Growth

In fact both innovation schemes are obviously interrelated, in a world where manufacturing and service activities tend to be more and more connected. The debate over the 'engine of growth' has changed with the development of inter-firm relationships, and especially with the growth of business services. While intermediation services (such as transport, banking, distribution and communication) have always been steadily growing, along with the overall development of services, the key feature of the last phases of tertiarization (over the last two decades) has been the growth of business services (as shown in Table 6.1 and in the employment figures shown in section 4 below). This change is fully in line with the development of network and external relations that ICTs favour.

Therefore the growth dynamics clearly depends as much on the in-house know-how of the firms as on the facilities brought by logistics of business (and intermediation) services. This is the common thrust of all the arguments to be found in the highly diversified contributions to the literature on endogenous growth. It places the debate over the engine of growth sector in a new perspective, that our reference to a twin-engine of growth attempts to recall.

However, it is difficult to assess which are the main linkages and how they affect the overall economic dynamics. Complex links have already been mentioned, regarding both innovation schemes. Three of them are worth repeating. One has to do with the change in tradeability (regarded as an innovation) generated by changes either in provisionability or in the product content. The second is tied to the time-consuming or time-saving bias of

Table 6.1   *Tertiarization: the rise of business services in recent decades*
            (service employment by sub-sector as a percentage of total
            employment)

	Producer services	Distributive services	Personal services	Social services	Total services
France					
1960	3.5	16.8	7.9	16.0	44.1
1973	6.0	18.6	7.5	19.2	51.3
1987	9.0	20.1	7.9	26.4	63.4
Germany					
1960	3.4	17.5	7.4	10.3	38.6
1973	5.2	18.1	6.5	16.3	46.1
1987	7.7	18.1	8.1	21.6	55.5
Japan					
1960	3.3	18.5	7.5	8.2	37.5
1973	6.5	23.3	8.0	10.5	49.1
1987	10.2	25.1	10.2	13.0	58.5
Netherlands					
1960	4.2	20.4	8.5	14.7	47.8
1973	6.8	20.5	7.6	22.8	57.7
1987	10.8	21.3	8.6	28.4	69.1*
Sweden					
1960	3.5	19.4	8.4	16.3	47.6
1973	5.1	19.8	6.6	26.2	57.7
1987	7.2	19.2	5.9	35.1	67.4
United Kingdom					
1960	4.4	20.6	8.0	15.8	48.8
1973	6.5	20.1	7.9	20.8	55.3
1987	10.4	21.0	10.1	25.3	67.0
United States					
1960	6.4	22.2	11.3	21.2	61.1
1973	8.7	21.5	10.9	25.1	66.4
1987	10.4	21.3	10.1	25.3	67.0
Average					
1960	4.1	19.3	8.4	14.6	46.5
1973	6.5	20.3	7.8	20.2	54.8
1987	9.8	20.9	8.7	25.1	64.5

Notes: *Includes 2.1 per cent temporary workers employed at employment agencies who cannot be allocated to one of the four sub-sectors.

Source: Derived and updated from Appendix D in Elfring (1991).

contemporary innovation depending on the kind of product we consider. The third stems from the difference made in the learning processes at work in the different innovation schemes. We shall come back to these dimensions when looking at the dynamics of employment as in services. Effectively, as the way in which services are produced is part of their content, the criteria of employment (shares of qualified labour, of part-time or temporary jobs and so on) are also characteristics of the patterns of innovation. Meanwhile a two-sector growth model is given in the appendix to this chapter to specify the interplay between these three linkages and the macroeconomic dynamics.

4. STRUCTURE OF EMPLOYMENT AND CHANGES IN SERVICE ACTIVITIES

So far we have been mainly concerned with the dynamics of service markets and implicitly considered that employment could be fully determined by the levels of activity. The causality is not so one-sided, especially in the case of services.

Traditionally, in personal services the level and structure of employment have always been largely influenced by the conditions prevailing on the local labour market as the production in services can be more easily divided into tasks in accordance with local labour supplies. But this 'sponge effect' (as Kaldor named it) obviously acts on the quality and therefore the nature of the service products under consideration. Culture and traditions strongly conditioned how these personal service products were perceived by the users. The expansion of business services has been faced by similar challenges. One can effectively find, in this category of business services, highly-qualified jobs as well as poor jobs, which differentiate strongly between these activities.

In the present context of developed economies, largely engaged in tertiary activities and willing to take 'their' advantages of the new ICTs, the interdependence between the stock of human capital, in broad multidimensional terms, and the growth path has been reinforced. The issue is at the core of recent works on endogenous economic growth (see Lucas and Romer's various contributions over the past decade). However, the issue is not as linear as is often assumed in the sense that 'more human capital' is not always positively correlated with more economic growth. Clearly, some matchings are required between the education of workers, the forms of on-the-job training, the availability of efficient producer and intermediary services, along with the capabilities of users. We want hereafter to explore some dimensions of this complex nexus, directly linked with the use of ICTs in service activities.

The assessment, made in the previous section, of what would be the main characteristics of a growth process more centred on service activities leads us to stress the new role bestowed on users in the learning processes. On one side the choices of products and activities for all users (final or intermediate) have been enlarged and consist of more time-consuming products. Therefore choices between alternatives within the time budget constraint are more compelling and lead the consumers/producers, through some new learning processes, to modify slowly their way of life. On the other side, the dynamics of innovation, which starts most often from process innovations directly driven by the diffusion of regularly improved ICTs, relies more largely for its expansion on positive feedbacks from potential users. The development of these learning effects modifies the content and the provision of new services. A networking effect with positive externalities sustains the diffusion of radically new services.

However, we know little of these learning processes. The productivity slowdown, much more marked in services than in manufacturing activities (see Roach, 1991), suggests that organizational mismatches and market failures may be more important in service activities than elsewhere. Moreover, not all service activities are in the same position, if only because, as we noted, their tradeability differs. In all cases the above implies that the dynamics of cumulative growth may depend to a larger extent than previously experienced on the quality of the labour force. We shall consequently look hereafter at employment in services with two considerations in mind: that the growth potential may specifically depend on the quality and size of the labour force and that situations may differ widely from one service activity to the other.

In presenting statistical trends, use will be made of the official statistical classification of service industries and the classification of services in four one-digit ISIC (International Standard Industrial Classification) sectors: wholesale and retail trade, hotels and restaurants; transport, storage and communications; finance, insurance, real estate and business services (FIRB); and community, social and personal services (CSPS).

Trends in Service Industries Employment

The crucial importance of services for overall employment growth in the EU, but also in the USA, is illustrated in Figure 6.3, representing employment trends for the period 1980–96 for the EU, the USA, Japan and OECD total for services, manufacturing and total employment. Even in Japan, services employment has now become essential for overall employment growth, manufacturing employment having fallen substantially since the early 1990s.

As Figure 6.4 illustrates, amongst the sectors with the most substantial employment growth in the EU for the period 1970–93, service sectors (real estate and business services; social services; restaurants and hotels; and fi-

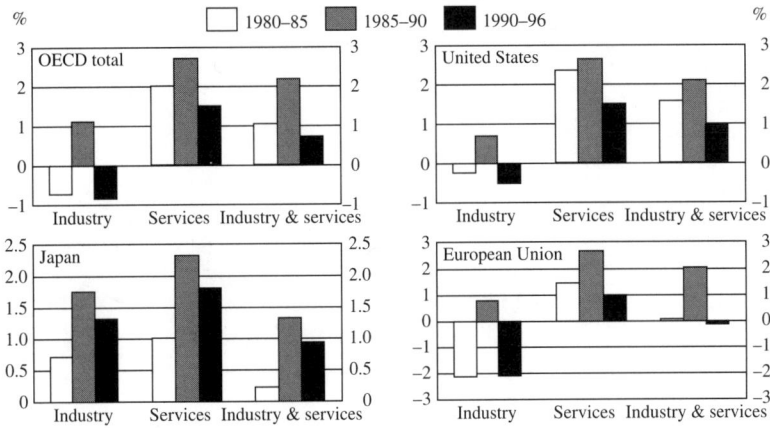

Note: Industry covers mining, manufacturing, utilities and construction; services covers all services, including government.

Source: OECD, *Labour Force Statistics*, December 1997.

Figure 6.3   *Employment growth in industry and services (compound annual growth rates)*

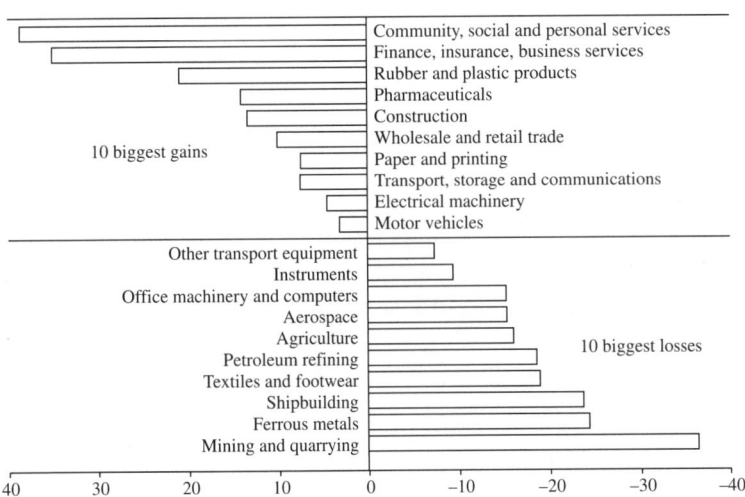

Source: OECD, STAN and ISDB databases, December 1997.

Figure 6.4   *Job gains and losses, by industry, OECD total (percentage change from 1985 to 1995)*

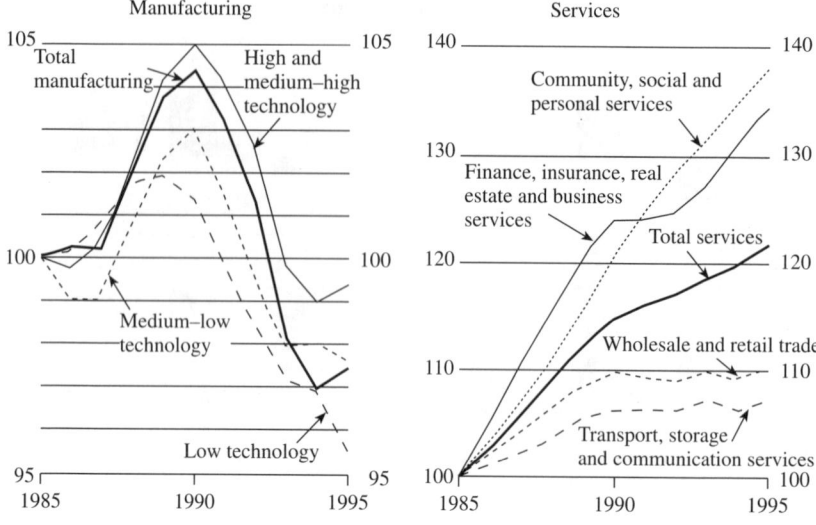

Source: OECD calculations from STAN and ISDB databases, December 1997.

*Figure 6.5   OECD employment trends in manufacturing and services
industries (1985 = 100)*

nance and insurance in the first four places) dominate with only a couple of
high-tech manufacturing sectors (computers, pharmaceuticals, communications, aerospace) witnessing above average employment growth. This pattern
is more or less similar for the USA.

Figure 6.5 represents the trends in employment for four broad service
sectors: financing, insurance, real estate and business services (FIRB, ISIC
8); community, social and personal services (CSPS, ISIC 9); wholesale and
retail trade, restaurants and hotels (ISIC 6); and transport, storage and communications (ISIC 7); as well as manufacturing (ISIC 3) for the OECD
countries. Interestingly, the FIRB and CSPS service sectors witnessed the
most rapid employment growth.

These two service sectors, FIRB and CSPS, illustrate well the totally
different impact of ICTs on employment. In the case of FIRB, ICTs have
led over this period to a substantial increase in the tradeability of such
services; in the case of CSPS, ICTs have had practically no impact on
tradeability, most of these services depending crucially on physical contact
and presence in delivering such services (even if ICTs are more and more
put to use in the 'back office' of these CSPS services, they do not change
their content). Basically, the magnitude of these services depends on the
relations between the domestic and the formal economy (public and pri-

Table 6.2 Employment ratios by NACE sectors (percentage of population between 15 and 64 years of age)

	BEL	DNK	GER	GRC	SP	FRA	IRL	ITA	LUX	NLD	AUS	PORT	FIN	SWE	UK	EU 15
Trade	8.3	10.2	8.9	9.2	7.9	8.2	7.8	8.9	8.6	10.4	11.1	10.1	7.1	8.6	10.8	9.1
Hotels	1.8	2.0	2.0	3.4	2.8	2.0	3.1	2.2	2.7	1.9	3.7	3.2	2.0	1.8	3.1	2.4
Transport and telecom	4.3	5.3	3.4	3.7	2.8	3.8	2.6	2.8	4.0	3.9	4.6	2.7	4.4	4.7	4.3	3.6
Finance	2.4	2.4	2.3	1.4	1.3	1.9	2.2	1.7	5.9	2.1	2.4	1.9	1.4	1.3	3.0	2.1
Business services	3.6	5.4	4.0	2.3	2.8	5.1	3.6	2.7	3.6	6.3	4.5	3.1	5.2	4.7	6.6	4.3
Administration	5.4	4.7	5.5	4.0	3.1	5.7	3.2	3.8	6.4	4.8	4.6	4.5	3.0	3.6	4.2	4.6
Education	5.1	5.7	3.3	3.3	2.8	4.5	4.0	3.8	3.4	4.1	4.0	4.5	4.5	5.2	5.4	4.1
Health	6.0	12.7	5.8	2.5	2.6	6.2	4.7	2.9	4.3	8.6	5.3	2.0	8.6	14.1	7.5	5.6
Other services	2.1	3.8	3.2	1.9	1.7	2.5	3.1	2.1	1.9	2.4	2.9	2.9	3.2	5.8	3.6	2.7
Domestic services	0.1	0.2	0.2	0.6	1.4	1.3	0.0	0.5	0.6	0.2	0.3	1.4	0.1	0.0	0.4	0.6
International organizations	0.3	0.0	0.1	0.0	0.0	0.1	0.0	0.0	2.8	0.0	0.1	1.0	0.2	0.0	0.1	0.1
Total services	39.4	52.6	38.7	32.3	29.8	41.4	34.5	31.4	44.3	44.9	43.5	37.3	39.6	49.8	49.0	39.1
Agriculture and industry	17.2	22.9	23.9	24.6	18.0	18.9	21.8	20.0	15.3	20.3	26.3	28.7	22.0	20.5	20.7	21.2
Total employment	56.6	75.5	62.6	56.9	47.2	60.3	56.3	51.4	59.6	65.1	69.8	66.0	61.7	70.3	69.8	60.3

vate), all of which are very country-specific, as can be shown simply through the wide range of employment ratios to be met in the European countries (see Table 6.2).

It is also worth noticing that the productivity gains have on average been really low in these two kinds of activities. We have here a noticeable manifestation of a productivity paradox, considering that investments in ICTs have been relatively important. By contrast, intermediate services, where investments in ICTs have been especially important, displayed relatively enhanced productivity gains, as expected, but with little expansion of markets and therefore little employment growth.

Looking precisely at industries by country could bring more insight into this question, but data on real growth are problematic, on two grounds. In the first place, activities are not organized in a similar way and sectors may not correspond from one country to the other; secondly, there is a severe measurement problem in most cases where, precisely, ICTs seem largely to have transformed the content and provision of activities. The quality improvement of those services may well have been underassessed in the national accounts.

Over the last decade, the mismeasurement of consumer surplus may have become quite sizeable (see Nakamura, 1995; Boskin, 1996; Moulton, 1996). Such mismatch though is not a simple statistical flaw. It points to an important 'under appreciation' of the quality improvement of some products. Careful studies, using hedonic price indexes, could tell how much of this 'evaporation' of the consumer surplus can be blamed on statistical methods and how much is due to some 'deficit' on the part of users. This issue is rather important for the dynamics of the CSPS sector in times of ICTs and for its employment potential: all the more so now that the CSPS sector represents more than one-third of total employment in the EU and the USA, substantially more than the whole of manufacturing.

A similar issue of mismeasurement of real term values is raised for the FIRB sector, where such measures have always been problematic. The question is all the more important in that technological change not only modifies the content of these services but also strongly blurs the frontiers between activities.

Firstly, services, partly as a result of the increased tradeability of service activities in financial, communication and other business services, and partly as the result of the increased 'outsourcing' of intermediate inputs, have in other words become much more dependent on cyclical swings, causing similar upturns and downturns in service activities. As an increasing number of such service activities are becoming deregulated or opened up to international competition, these sectors are likely to become much more vulnerable to economic contractions and their traditional role as sheltered sectors of employment reservoir is becoming significantly reduced.

Secondly, major structural changes develop as a response to the challenge of internationalization and technological change. These include the following:

- potentially major shifts between sectors and services (for example, as retail banks restructure away from physical branches and offer a wide range of services electronically while expanding from finance into other sectors, such as travel, entertainment and shopping);
- new alliances and industrial groupings between different sectors (for example, between media and communications, leisure and education, finance and computing);
- accentuation of the trend to globalize and deliver services internationally;
- increasingly close links between suppliers and providers, supported by electronic data information and vastly more accessible and improved inter-enterprise networking and connectivity;
- greater opportunities for small and medium enterprises (SMEs) through universal access;
- greater openness and participation of customers, consumers and businesses.

With such restructuring of service activities, straightforward extrapolations of past trends are of little help. Insights on the development paths to be followed have to come from analytical arguments.

Global Competition and Changes in Work and Occupations in Service Industries

The structural changes mentioned above are accompanied by considerable changes in the way people work and are employed in service industries. Let us start with the essence of the new global competition which concerns some of these services.

As a consequence of the increased potential for international codification and transferability, the new information and communication technologies can to some extent be considered as the first truly 'global' technology. The possibility of ICTs codifying information and knowledge over both distance and time brings about more global access. Knowledge, including economic knowledge, becomes available worldwide. While the local capacities to use or even have access to such knowledge will vary widely, the potential is there. ICTs, in other words, bring to the forefront the enormous potential for catching up, based upon the transparency of economic advantages, while stressing at the same time the crucial importance of tacit knowledge to access international codified knowledge. For technologically leading countries or firms, this implies increasing erosion of monopoly rents associated with innovation and shortening of product life cycles.

At the same time, the ability to codify relevant knowledge in creative ways acquires more and more strategic value and will affect competitiveness at all levels. Network access as well as the competence to sort out the relevant information and to use it for economic purposes become of critical importance for performance and income distribution. Specific skills relevant to the use of information become of strategic importance. More routine skills, by contrast, might become largely codifiable and their importance dramatically reduced.

For services this might imply significant relocalization possibilities for many routine functions. The increased potential for teleworking does not stop at the border. The rapid growth in teleservices in less favoured regions, such as Ireland, .is illustrative of this potential for relocalization of hitherto untradeable service functions. In essence, this is a process of international division of labour whereby service sectors are discovering advantages of international relocation. The impact of the decline in communication costs following the widespread use of global ICTs on the international trade of services can be compared with the impact of the decline over the last 30 years in transport costs on the international trade of commodities and manufactured goods. This threat does not mean that a huge proportion of service activities will be delocalized, if only because delocalized activities which deal with codified information can in turn be fully computerized, while activities 'back home' can develop in expanding their capital of tacit knowledge.

Still the increased competition and market orientation that are the consequence of the above changes affect, to a greater or lesser degree, the work organization in all activities. By and large, such increased tradeability of services implies substantial shifts in the occupational and skill structure of service industries employment. The following broad trends can be expected:

- less security of employment and of careers in traditional areas (such as front and back office clerical work in banks, post offices and so on);
- a much more explicit need for staff to be responsive to customers, able to adapt to and offer new services;
- an increased ability to adjust to internal changes in company structure, such as reduction of the traditional hierarchy;
- a need for quickly acquiring new skills – particularly for more worldwide communication;
- the need for public sector workers to adapt to the market rigours typical of private enterprises;
- lifelong learning or readiness for a continuous acquisition of skills and knowledge;

● expansion of opportunities in services where human interaction remains the essential element: teaching, health, entertainment, leisure, social services.

These changes will represent a considerable change in the nature and content of work as well as in the organization of the workplace. Making the best of these trends in terms of growth and employment, when their outcomes seem far-ranging from the worst to the best, has become a major challenge, all the more so since countries are starting from rather different positions regarding occupational structures. Services industries have typically been characterized by their extensive use of white-collar workers and the current growth of employment in services has reinforced this characteristic. In both the manufacturing and service sectors the number of blue-collar workers employed is decreasing.

Still large differences appear in the occupational composition by sector across countries when looking at the skill level of the new jobs created (see Figure 6.6). The distribution of the changes in employment by occupation between 'professional and technical workers', the most highly-skilled occupational category, and the relatively low-skill categories, such as 'sales workers', 'clerical and related workers' and 'service workers' seems important and somehow displays the differences in national trajectories. The question then is whether these changes in occupational skills are improving or worsening the position of the various countries.

Interdependence of Technological Change, Skill and Culture

The ability to codify knowledge in creative ways, mentioned above as a factor of competitiveness in a world where service activities are internationalizing, seems to imply that technological change will lead to upgrading the average skill level in the activities concerned. Therefore one would find in services that the technological change that ICTs represent is skill bias, as has been suggested concerning manufacturing industries (see Berman *et al.*, 1994). However, the argument might run somewhat differently when considering the whole range of services.

If we follow the argument presented when schematizing innovation in services (see Figure 6.2b), technological change in services is to be assumed to depend on both the skill structure of the producers of the services and on the 'skill structure' or cultural background of the consumers of these services. The assumption of such 'reverse causality' stems directly from the reverse product cycle that we identified as an important pattern of innovation in service activities.

To clarify how this interaction between consumers' capabilities and products innovation functions, it is useful to distinguish service industries according

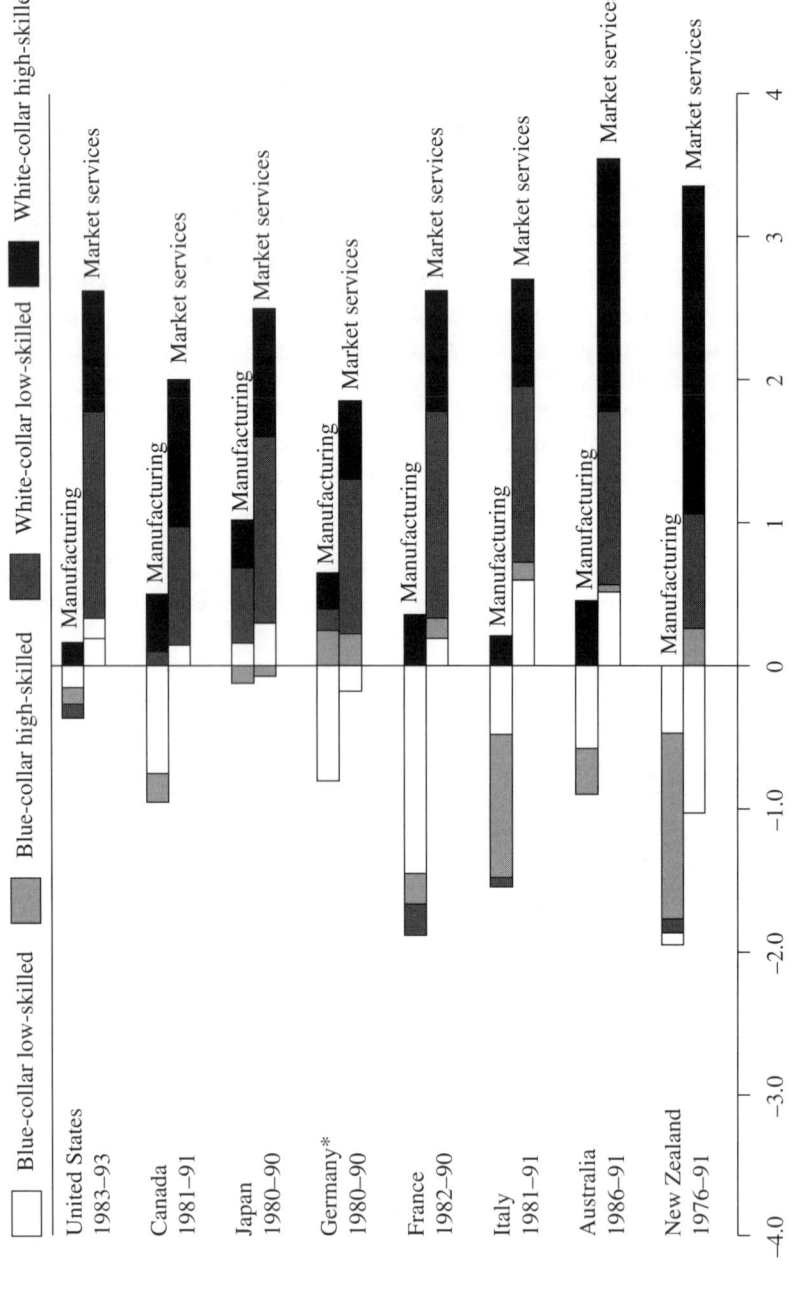

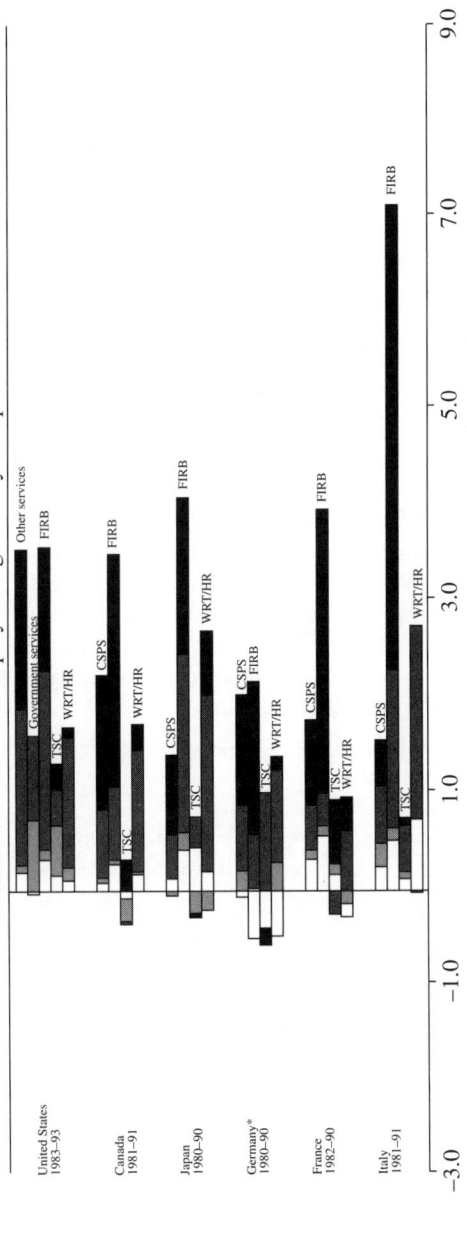

Breakdown of services employment growth by occupation

Notes:
See text for definitions: growth rate are annual average growth rates.
*The white-collar high-skilled group in Germany excludes some occupations and is thus underestimated.
Abbreviations: WRT/HR, 'Wholesale and retail trade, hotels and restaurants' (ISIC 6); TSC, 'Transport, storage and communications' (ISIC 7); FIRB, 'Finance, insurance, real estate and business services' (ISIC 8); CSPS, 'Community, social and personal services' (ISIC 9).

Source: OCED Secretariat calculations from national data, STI/EAS Division.

Figure 6.6 Employment growth, by skill level, in manufacturing and services

to the extent to which the reverse cycle hypothesis applies. Basically, this innovation scheme fits rather well the case of industries with a large number of registered customers, engaged in recurrent transactions, as in banks and insurance, but also in large systems of distribution, transport and communication. These intermediary activities are organized in networks addressing diversified communities of customers. The need to cut down the running costs of these networks is the main drive for process innovations. How these process innovations may lead to product innovations is at the core of the thesis of the reverse product cycle. The question is thus to see how changes in the way services are delivered on these networks create conditions for product innovations to emerge from active user–producer interactions.

The answer lies at two levels. At the first level some intermediary activities may attract sub-sets of customers, but the exact forms of new services suitable for consumers and provisionable by services producers have to be elaborated in a joint learning process. The number of services that can be extensively developed straightforwardly without this cross-learning process is limited.[17] Producers have to engage in processes of trial and error through which they may develop specific services for segments of their customers. Customers' reactions help to define the content of the products. Basically, this clustering of new products much depends on the cultural and educational backgrounds of the groups of consumers concerned, whether they are final users or small and medium enterprises. We have then a process of product differentiation which can lead to product innovations according to the creativeness of the interplay between the two learning processes involved. Price differentiation to adjust to the various needs and possibilities of the customers is a common practice of the intermediary services. Product differentiation is more difficult to achieve and requires adaptive flexibility which large systems of intermediation may have lost.

The second level of answer precisely takes this rigidity into account in stressing that often SMEs of services take part in this process of innovation, either by taking over developments from large networked industries or by selecting directly such niches in making use of the improvement of the intermediary services. We then have an extended version of the reverse product cycle where firms innovate to fulfil specific needs using the technological innovations made in the production process of intermediary services.

At both levels (of the intermediary services or of their users) the potential for product innovation depends on the abilities of the users and on the quality of the feedback they give. It conditions the extent to which the codification of information in the new services can leave room for the development of personal involvement and accumulation of new tacit knowledge. Conversely, if one aims to develop products in accordance with the capabilities and likings of the consumers, the qualifications required of the labour force delivering the service may also mix various skills. In particular, semi-skilled workers might be more productive intermediaries because they are closer to the consumers than high-skilled or low-skilled attendants.

All of which suggests that successful learning processes (able to expand markets) could require more balanced (more realistic in a way) approaches to the skill requirements in a lot of service trades.

5. POLICIES AND PERSPECTIVES

We have so far only assessed some of the key dimensions and questions raised by a growth pattern where service activities are preponderant and technological change is intensively fuelled by the steady diffusion of ICTs. The bases for new virtuous circles are still unclear and highly conditional. Even if we have stressed a convergence between manufacturing and service activities, it does not follow that the forms of industrial organization have converged towards some best-practice form of organization. Services have always been very country-specific owing to the set of institutional arrangements that their market provision requires; the fact that ICTs have transformed their content and their 'provisionability', which partially shows in the fact that they are more sensitive to the business cycle, does not radically alter this feature. National institutional contexts matter. Regionalization processes, as experienced in Europe, or the liberalization of trade and foreign investments, have also contributed to some convergence among the production processes; however, we are far from having industrial fabrics in service activities which can be easily compared internationally. This implies that, if any new growth model is emerging, there is a strong likelihood that these models will differ between countries. We have not reached that stage where the identification of growth patterns can take into account national differences, even if the overall coherence of such patterns may differ markedly from one case to the other.

The fact that the cultural backgrounds of consumers were important conditions featuring the growth path of national economies is also an important reason to affirm that the 'new' growth model will be plural.

Even if our knowledge and assessment of these new schemes remain patchy, because changes are still going on and may well extend over long periods of time, we should be able to conclude with some policy orientations as well as indicating some crucial lines of research.

We can finally list three policy orientations regarding, respectively, investment and related industrial policies, labour market policies and, finally education and training policies.

We have stressed in this chapter all the organizational difficulties raised by the market provision of services, using the facilities of ICTs. Asymmetries of information and externalities remain numerous in these activities in these new networks and hinder some of the growth potential of this new fabric. Therefore public interventions to coordinate actions, certificate intangible productions and internalize positive externalities are very worthwhile. These public investments can take many forms, still this public intervention is hindered by trends towards deregulation, precisely in the intermediate services such as banking, telecommunications and transport where appropriate industrial policies should help to develop the logistics and normative framework required to coordinate the actions of private agents and internalize their effects. While the deregulations have been largely provoked by the obsolescence of the old regulatory frameworks, often inherited from the 1930s and 1940s, they have been too often viewed as a necessary withdrawal of public intervention. Though new regulatory frameworks need to be elaborated (and ICTs need to open new possibilities in that respect) in order to take full benefits of the current structural changes, this can be done all the more easily when it is accompanied by appropriate infrastructure investments, which can be tangible (especially in telecommunications networks, for instance) or intangible (in the form of specialized training schemes, broader intermediation institutions or social networks to facilitate the access of the populace with various cultural levels to new services). A wide range of measures can be envisaged at the European, national and (mainly) regional level that we can group under the heading of new industrial policies. The current problem is the total lack of commitment to such policies, given the absolute priority given to monetary rigour and fiscal consolidation.

A second set of policy issues concerns the labour market. Large amounts of public money are devoted, at least in European countries, to coping with the rise and persistence of unemployment. These measures are costly and their efficiency is questioned; their main flaw is that they are subject to important deadweight loss effects (meaning that most measures have little net incentive effect but possibly big distorting price effects, largely uncontrolled). Keeping in mind the change towards a growth regime where services and time budgets play a different role may help to reconstruct these labour market policies in ways which provide the safety net (guaranteed income), while inserting people who are out of work in any of the training schemes that are part of some learning processes, either on the consumer side (networks accessing any of the big social systems) or on the production side (networks retraining the labour force according to needs and wants) or both. Such a comprehensive approach to labour market policies should therefore be developed in setting up any scheme to reduce working time, where what the people out of work will do is as important as where and how the reduction will be implemented.

The third and last set of policies is familiar as it concerns education and training policies. The fact that it should concern, not so much initial formation, but lifelong training schemes is also widely accepted. How it can be implemented or which principles are implied is less straightforward. Our analysis suggests that we should look at the various types of learning required not only at all stages of a working life, but at all stages of consumer life. This includes also looking at the rate of obsolescence of knowledge, on one side, and at the cultural backgrounds and their evolution over time, on the other. Returns to illiteracy and rigidity of cultural patterns regarding the use of ICTs are also important features revealing the difficulties of maintaining certain levels of training. For example, people engaged in jobs requiring lower qualification than they obtained, is another source of deskilling. However, such deskilling at work could be compensated for by 'reskilling' in consumption or other non-work activities. Policies have to be realistic and take into account the poor performance of some groups and the barriers to access of some ICT uses. If training policies only aim at some general improvement they may well strengthen a tendency to discriminate that one finds at work (with skill-biased modernization) and at home (with cultural barriers to access to some service provisions).

Finally, it is also worthwhile to recall some of the lines for future research suggested in our analysis. A first question concerns the dynamics of codification. How can we consider that codified knowledge is accumulating, what is its economic rate of depreciation, which includes 'physical obsolescence' and being overtaken by competition? These questions are crucial to clarifying how codification of knowledge leaves room for increases in tacit knowledge and efficiency. There are also questions on the rate of return to innovations. Too fast depreciation with a lack of similar speedy organizational change will reduce too rapidly or strongly the returns from innovation of the intangible investments under consideration (social as well as private benefits of innovation in a world where network externalities are omnipresent may also be concerned).

A second line of research, following from the above, has to do with our approach to educational and cultural issues. How far are skills specific or general; how do they combine during life cycles? Are training periods specific out-of-work periods or can they be mixed with on-the-job or in consumption learning? In which case, could training be dealt with in the organization of work? How does this apply respectively to large, small and medium firms? Conversely, what is the nature of the cultural barriers to access? By which practices could they be overcome or adapted? Does this training need to be channelled by formal social networking? Could it be conveyed by adapted forms of products and uses? We know relatively little on how ways of living and consumption patterns react to differentiated ranges of new products and services.

A third line of research would go back towards the identification of what would be the growth model in a fully developed 'information economy'. Of special interest would be the way such economies evolve in a context of growing internationalization. A large proportion of service activities take part in that process of internationalization. Does this lead to specific forms of competitiveness, or do these service organizations evolve towards some best-practice universal pattern of provision? By contrast, it would be interesting to see how country-specific the organization of social and personal services can remain. These issues are directly linked to research trying to characterize the nature of the globalization and the extent of the convergence of production processes and consumption patterns that internationalization brings in a world market with a rapid diffusion of steadily improving ICTs.

NOTES

1. This chapter is a revised and extended version of a paper presented by the authors at the conference on 'Technology, Employment and Labour Markets' held in Athens University of Economics and Business, 16–18 May 1996, published as a 1997 working paper by the Enrico Mattei Fondation, Milan.
2. There is now a voluminous empirical and theoretical literature on this subject (see the surveys by Freeman and Soete, 1987; Petit, 1995). In the 20th century alone, we may distinguish four sets of economic debate on the relationship between technology and employment. The first, probably the most 'classical' in its origins, took place during the economic depression of the 1930s. Contributors included Hansen (1931, 1932), Kaldor (1933), Weintraub (1937) and Neisser (1942). Many of the issues and concerns raised by these authors sound quite familiar today, particularly in the context of the notion of increasing returns in current 'new' growth models (for example, Aghion and Howitt, 1991). The second debate focused mainly on postwar United States and the fear of 'automation'. In the 1960s, levels of unemployment were higher in the United States than in Europe, and many blamed technological change. As a result, a National Commission on Automation was appointed and produced a massive six-volume report (US National Commission, 1966). This debate had little influence in Japan and in the European countries that were rapidly catching up. The third debate, which began in the late 1970s, was particularly active in Europe. It focused on the emergence of the cluster of computer-based communication, information and automation techniques associated with microelectronics, which appeared at first glance to have great labour-displacing implications (for example, Freeman *et al.*, 1982; OECD/ICCP, 1984; Katsoulacos, 1984). The fear that these displacement effects might dominate the compensating job creation effects for quite some time recalled in many ways the classical debate. As there, it appeared to be a reflection of the times: there was a set of 'revolutionary' new technologies and persisting high unemployment. The most recent debate focuses much more on the global aspects of the new information and communication technologies and the possible erosion of employment and high living standards in the advanced countries. Originating to some extent in the United States, and linked to the political debate surrounding NAFTA, it quickly spread all over the world.
3. As von Mises (1936, p. 485) put it: 'Lack of wages would be a better term than lack of employment, for what the unemployed person misses is not work but the remuneration of work. The point [is] not that the "unemployed" [cannot] find work, but that they [are] not willing to work at the wages they [can] get in the labour market for the particular work they [are] able and willing to perform'.
4. Both the OECD *Jobs Study* (1994a,1994b) and McKinsey Global Institute (1994) can be said to have focused primarily on these market issues. The former emphasized the func-

tioning of labour markets, the latter the functioning of product markets, particularly in services.

5. This result is obtained by Venables (1985), for instance, with the use of general equilibrium setting with fixed coefficients.

6. 'Capital shortage' unemployment reflects a seeming lack of 'productive' capital to employ part of an 'adequately' skilled and suitably located labour force. Capital shortage unemployment can occur as a result of both lack of physical capacity and economic obsolescence; if variable costs exceed price, capital will not be used (and the corresponding jobs will disappear) even though such capacity could be operated in a physical sense (OECD, 1984).

7. See, for example, Aranowitz and DiFazio (1994) and Rifkin (1995). In many ways, and as noted by some trade economists (Krugman and Venables, 1994), such views are to some extent reminiscent of the old Prebisch–Singer *dependencia* arguments, but applied to the advanced countries. In the old core–periphery models, 'immiserising' growth in the developing countries would take place because all the benefits of increased efficiency gains in raw materials, agricultural and labour-intensive manufacturing production were passed on to the advanced economies, fo example through lower prices or higher repatriated profits. In the current view, the pattern is the opposite: most of the benefits of technological change are passed on to some of the rapidly industrializing countries through more rapid international diffusion of technology from the advanced countries, the reinvestment of profits and relocation of production to those industrializing countries, and the erosion of various monopoly rents in the advanced countries, including wages. In principle, though, and in contrast to the Prebisch–Singer model, such a redistribution process should lead, as trade theory would predict, to the convergence of growth and income.

8. Interestingly, this is still the main argument of those studies in this area which limit their focus to manufacturing (Pianta, 1996; Reati, 1995; see also comments from the Commission on the OECD G7 study).

9. For early analyses along these lines, see Quinn (1986) and Soete (1987).

10. This was certainly the case with regard to the invention of printing in the Middle Ages and the impact this first new information technology had on the limited tradeable 'service' activity of monks copying manuscripts by hand. It was the *time/storage* dimension of the new printing technology which opened up access to information in the most dramatic and pervasive way and led, to use Marx's words, to the 'renaissance of science', the growth of universities, education, libraries, the spreading of culture and so on. This opening-up, 'tradeability' effect would become of far more importance to the future growth and development of Western society than the emergence of a new, in this case purely manufacture-based, printing industry.

11. See in particular David and Foray (1995) and Ergas (1994).

12. One might think of such activities as gardening, cycling or housekeeping.

13. See, amongst others, Gershuny and Miles (1983).

14. Though the transformation of intermediary services is not homothetic and therefore the effects on the organization of markets and production are biased in favour of some means of intermediation. Thus transport costs may have fallen systematically over the postwar period, but they have risen significantly in relation to communication costs over the last decade.

15. With, at the extreme, the self-service where innovation is turned into new goods for personal use. (TV sets and cars can be seen in such perspectives as following on from long lines of innovation in entertainment and transport industries.)

16. This modernization of the production process can be done by direct use of ICTs in the process or indirectly by using modernized intermediary services (banks, transport, communications), as we shall stress in section 4. Such extension gives a much wider scope to the thesis of the reverse product cycle that we have referred to.

17. Banks and distribution services experienced some difficulty in the early 1980s in entering the market of tourist services straight away.

BIBLIOGRAPHY

Aghion, P. and P. Howitt (1991), 'Unemployment: A Symptom of Stagnation or a Side-effect of Growth?', *European Economic Review*, 35, 535–41.

Aranowitz, S. and W. DiFazio (1994), *The Jobless Future: Sci-tech and the Dogma of Work*, Minneapolis: University of Minnesota Press.

Barras, R. (1986), 'Towards a Theory of Innovation in Services', *Research Policy*, 15, 161–73.

Berman, E., J. Bound and Z. Griliches (1994), 'Changes in the Demand for Skilled Labor within U.S. Manufacturing: Evidence from the Annual Survey of Manufactures', *Quarterly Journal of Economics*, CIX, 367–98.

Boskin, M. (ed.) (1996), 'Towards a more Accurate Measure of the Cost of Living', Final Report to the Senate Finance Committee, December.

David P. and D. Foray (1995), 'Accessing and Expanding the Science and Technology Knowledge-base', *STI Review*, 16, OECD, Paris.

Elfring (1991), 'An International Companion of Service Sector Employment Growth in ECE', Economic Discussion Paper, UN: Geneva.

Ergas, H. (1994), 'The New Faces of Technological Change and Some of its Consequences', mimeo.

Freeman, C. and L. Soete (1987), *Technical Change and Full Employment*, Oxford: Basil Blackwell.

Freeman, C., J.A. Clark and L. Soete (1982), *Unemployment and Technical Innovation: A Study of Long Waves and Economic Development*, London: Frances Pinter.

Freeman, R.B. (1995), 'Are your Wages Set in Beijing?', *Journal of Economic Perspectives*, (3), Summer.

Gallouj, C. and F. Gallouj (1997), 'L'innovation dans les services', *Economica*, Paris.

Gallouj, F. and O. Weinstein (1997), 'Innovation in Services', *Research Policy*, 26, 537–56.

Gershuny, J. (1978), *After Industrial Society?*, London: Macmillan.

Gershuny, J. and I. Miles (1983), *The New Service Economy: The Transformation of Employment in Industrial Societies*, London: Frances Pinter.

Haddon, L. and R. Silverstone (1996), 'Home Information and Communication Technologies and the Information Society', mimeo, Centre of Culture and Communication, University of Sussex, Brighton.

Hansens, A. (1931), 'Institutional Features and Technological Unemployment', *Quarterly Journal of Economics*.

Hansens, A. (1932), 'The Theory of Technological Progress and Dislocation of Employment', *American Economic Review*.

Kaldor, N. (1933), 'A Case Against Technological Progress', *Economica*.

Katsoulacos, Y. (1984), 'Product Innovation and Employment', *European Economic Review*, 28, 83–108.

Krugman P. and A. Venables (1994), 'Globalization and the Inequality of Nations', CEPR Discussion Paper Series, no. 1015, September.

Mankiw, G. (1995), 'The Growth of Nations', *Brookings Papers on Economic Activity*, 1, 275–310.

McKinsey Global Institute (1994), 'Employment Performance', Washington, DC, November.

Mises, Ludwig von (1936), *Socialism*, London: Jonathan Cape.

Moulton, B. (1996), 'Bias in the Consumer Price Index: what is the evidence?', *Journal of Economic Perspectives*, 10 (Fall), 159–78.

Nakamura, L. (1995), 'Is US Economic Performance Really that Bad?', Working Paper 95–21, Federal Reserve Bank of Philadelphia, October.

Neisser, H. (1942), 'Permanent Technological Unemployment', *American Economic Review*, 32(1), 50–71.

OECD (1986), *Flexibility in the Labour Market: the Current Debate*, Paris.

OECD (1992), *Technology and the Economy: The Key Relationships*, Paris.

OECD (1994a), *The OECD Jobs Study: Evidence and Explanations; Part I: Labour Market Trends and Underlying Forces of Change*, Paris.

OECD (1994b), *The OECD Jobs Study: Facts, Analysis, Strategies*, Paris.

OECD/ICCP (1984), *Information Technology, Employment and Economic Growth*, ICCP 17, Paris: OECD.

Oliner, S.D. and D.E. Sichel (1994), 'Computers and Output Growth Revisited: How Big Is the Puzzle?', *Brookings Papers on Economic Activity*, 2.

Petit, P. (1991), 'Diffusion of Information Technologies and the Productivity Black Hole: with an application to the case of France', Working Paper, CEPREMAP, Paris.

Petit, P. (1995), 'Employment and Technological Change', in P. Stoneman (ed.), *Handbook of the Economics of Innovation and Technological Change*, Oxford: Blackwell, pp. 366–408.

Pianta, M. (1996), 'S&T Specialisation and Employment Patterns', paper presented at the OECD/KUF conference on 'Creativity, Innovation and Job Creation', 11–12 January, Oslo.

Quinn, L. (1986), 'Technology Adoption: The Service Industries', in R. Landau and N. Rosenberg (eds), *The Positive Sum Strategy*, Washington, DC: National Academy Press.

Reati, A. (1995), 'Radical Innovations and Long Waves in Pasinetti's Model of Structural Change: Output and Employment', Economic Paper Series no. 109, Directorate-General for Economic and Financial Affairs, European Commission.

Rifkin, J. (1995), *The End of Work : The Decline of the Global Labor Force and the Dawn of the Post-Market Era*, New York: G.P. Putnam's Sons.

Roach, S.S. (1991), *Pitfalls on the 'New' Assembly Line: can services learn from manufacturing?*, Paris: OECD, pp. 119–29.

Soete, L. (1987), 'The Emerging Information Technology Sector', in C. Freeman and L. Soete (eds), *Technical Change and Full Employment*, Oxford: Basil Blackwell.

Stoneman, P. (1984), 'An Analytical Framework for an Economic Perspective on the Impact of New Information Technologies', OECD/ICCP ITEP project, Paris.

US National Commission on Technology, Automation and Economic Progress (1966), *Report and Appendices*, Vols 1–6, Washington, DC.

Venables, A. (1985), 'The Economic Implications of a Discrete Technical Change', *Oxford Economic Papers*, 37, 230–48.

Weintraub, D. (1937), 'Unemployment and Increasing Productivity', in National Resources Committee (ed.), *Technological Trends and National Policy*, Washington, DC.

Young, A. (1928), 'Increasing Returns and Technical Progress', *Economic Journal*, December.

Young, A. (1995), 'Growth Without Scale Effects', NBER Working Paper no. 5211, Cambridge, MA.

7. The European unemployment problem: a structural approach

Michael Landesmann and Robert Stehrer[1]

1. INTRODUCTION

In this chapter we examine an important dimension of structural change which has so far been insufficiently accounted for in the discussion of high levels of European unemployment, especially its development over the 1980s and 1990s. A detailed examination of sectoral employment patterns shows that a significant group of (continental) European economies experienced continued high (and even growing) rates of labour shedding out of manufacturing over the late 1980s and the 1990s ('deindustrialization') and, at the same time, a significant break in rates of employment absorption in the social services sector (ISIC 9: community, social and personal services), a sector which now accounts for the highest share in total employment of all sectors. Neither of these features can be found for the USA (or the UK) and hence we argue that they contribute towards an explanation of the additional hikes in unemployment rates in (continental) European economies as against the USA and the UK from the mid-1980s onwards. The analysis proceeds through a careful examination of changing sectoral employment patterns across different OECD economies and discusses reasons for inter-country differences in the time patterns of deindustrialization and employment absorption in the different service sectors. We also discuss theoretical reasons why the overall unemployment situation should be affected by the characteristics and the speed of sectoral structural change. Further, we suggest a theoretical model showing why the structural break in the relative employment absorption capacity of the welfare services sector might have taken place in the European economies owing to changes in redistributional policies.

2. THEORIES OF STRUCTURAL CHANGE AND THE UNEMPLOYMENT PROBLEM

In this section we give a short overview of some theories of structural change which also imply dynamics in the structure of employment. Theories of structural change and economic growth have a long pedigree (going back to the classical contributions by Quesnay, Smith and Ricardo); however, they have scarcely been brought to bear more recently on the subject of macroeconomic unemployment.[2] On the other hand there is a more recent literature and applied research on 'structural unemployment' (see, for example, the well-known study by Davis *et al.*, 1997). These investigations explain unemployment mainly by rigidities in the labour market (wage inflexibility, skill mismatches, low regional and/or sectoral mobility of workers, and so on). From our point of view this literature insufficiently specifies the structural change issues which cause the requirements for adjustment. The adjustment processes are investigated at the microeconomic level (across individual workers) whereas issues related to sectoral adjustments in the economy as a whole are mostly ignored. Further, the empirical studies are not at all conclusive as to the significance of 'structural unemployment' and do not contribute substantially towards an explanation of the recent high levels of unemployment in Europe.

Thus we face a situation of 'double ignorance': the existing theories of longer-term structural change do not focus on the unemployment problem, and in the microeconomic literature on 'structural unemployment' the actual theme of structural change is not explicitly formulated.[3] In the following we shall address the issue of structural unemployment in the context of an analysis of sectoral uneven employment growth. This is in line with traditional theories of structural economic dynamics (see, for example, Pasinetti, 1981, 1993). Structural shifts in sectoral employment patterns are – in our view – an important contributing factor to explain the recent unemployment experience in Europe and also contribute to an explanation of some of the differences in the employment experience of the USA and Europe. This should, of course, not be seen as belittling the other problem of adjustments at the microeconomic level and, in fact, the two types of analysis should be seen as complementary.[4]

In Pasinetti's work some important issues are highlighted which we also see as crucial for modelling employment problems of structural change. This is, firstly, the introduction of structural change from the demand side by specifying non-linear Engel curves; secondly, uneven productivity advances in different sectors of the economy (a theme also elaborated in a number of contributions by Baumol and his associates: Baumol *et al.*, 1985; Baumol, 1987) which in turn affect the composition of output and particularly of employment in a dynamic economy; and, thirdly, the (potentially) arising effective demand problem in the course of structural adjustment to these two types of forces. We shall refer to each of these elements in the context of the more specific historical developments which confronted OECD economies in the 1980s and 1990s in the following sections of the chapter.

The Mechanics of Sectoral Labour Shedding and Labour Absorption

In this section we present a simple accounting framework in which the European unemployment problem in the 1980s and 1990s can be addressed from a structural point of view. Seen algebraically, the potential for structural unemployment can be stated in a very simple manner: structural change (in the form of technological change, changes in the structure of demand in closed or open economies, and environmental and policy constraints imposed on the structure of an economy) requires continued adjustment in the allocation of labour across sectors. Under the assumption of a stationary workforce, workers which are shed in some sectors must be absorbed by other sectors which would guarantee full employment (if we start at the full employment level). With a growing workforce (growing population or rising labour force participation) additional workers (or hours worked) must find employment in all sectors, but in effect they are mainly absorbed by a sub-set of (labour-absorbing) sectors.

'Structural unemployment' cannot arise if we assume frictionless shifting and absorption of workers (and hours supplied). Hence an analysis of these frictions has to be part of any theory of 'structural unemployment'; in our approach we emphasize that the analysis of such frictions has to focus specifically upon stock adjustment problems which arise in the (historical) context of sectoral growth (including technological and productivity) dynamics. In a nutshell, there are two issues involved in 'matching' an available workforce to the existing structure of employment opportunities: (a) the two have to 'match' in terms of the structural (stock) characteristics of both the labour force (skill, demographic and geographical/mobility characteristics) and the potentially available jobs (their skill, motivational and geographic requirements); (b) they also have to match in terms of the more traditional economic concept of the intersection of labour supply and labour demand curves; that is, given the structural characteristics of workers and jobs there has to be a matching in terms of the wage/price regulating mechanisms in the labour (and product) markets. The two issues are obviously not independent as (b) is defined in terms of (exogenously) given structural characteristics, while (a) focuses upon these structural characteristics, including their evolution over time. A 'structural' analysis of 'matching problems' in fact analyses how the structural characteristics of both sides of the labour market (labour force and employment opportunities) change over time. Matching problems may become more or less severe as the structural characteristics of the stocks of workers and of jobs change over time. There is also the other type of relationship between (a) and (b): the same type of structural features of labour supply and employment opportunities can lead to greater or smaller matching problems of the (b) type, depending upon the way the more traditional mechanisms of the labour (and product) markets operate (degree of flexibility of wages and the shapes of the labour supply and labour demand schedules defined as a function of the real and product wage rates).

Changes in the structural characteristics of workers and jobs and evolving stock adjustment processes are thus a central part of the analysis of 'structural unemployment'. In general, one can argue that the type (a) matching problem is a function of the degree of structural change itself: if there is a fundamental change in the characteristics of the (potentially) available job opportunities, due for example to technological change, shifts in the structure of sectoral demand, and so on, then the 'distance' in the characteristics of the available labour force and the structure of the demand from a new (compositionally different) set of employment opportunities is wide and hence the stock adjustment problem is large; big strains are going to be put on the type (b) market mechanism to induce a matching process to avoid high 'structural unemployment'. Over time the type (b) matching mechanism will induce stock adjustment processes so that the 'structural mismatch' might decline (for example, a scarcity premium on skills might induce more skilling); however, the structural mismatch might also grow over time (as with unemployment effects on skill erosion and motivation loss). There is hence a clear time dimension to the structural mismatch problem induced by structural change (which, by the way, can also be initiated by compositional shifts in the labour supply, such as a sudden inflow of immigrants). As a working hypothesis, we shall maintain that the severity of shifts in the structural (compositional) characteristics in labour demand (or labour supply) increases the potential of type (a) mismatch and thus might go along with higher 'structural unemployment'. The degree to which such a 'structural mismatch' generates unemployment and sustains it over time depends, among other things, upon type (b) matching mechanisms in labour and product markets.

In the course of structural changes and dynamics of employment patterns, a very important part is played by the entries and exits of workers. The entrants are mainly younger workers coming from the educational system, whereas exits are mainly from employment to retirement and withdrawals (temporarily or permanently) from the active labour force. The entry–exit dynamics is of course shaped by institutional (country-specific) settings (retirement systems, social security systems, and so on). At this point it is sufficient to mention that a great part of the mobility and flexibility at the

micro level which is required by structural change is through the replacement
of workers through entry and exit of workers.

As will be seen in the following sections of the chapter, we shall focus on
two particular features of structural change in employment structures in
OECD economies over the 1980s and 1990s: the speed and severity of the
process of labour shedding in the industrial sector ('deindustrialization') and
employment absorption in particular segments of the tertiary (service) sectors. As we shall see, there are significant differences in the trajectories of the
different economies (and particularly between the USA and a sizeable group
of European economies) in these two respects. Within the service sector, a
particularly important role is played in these US–European comparisons by
what can be called the 'welfare services' sector (ISIC 9). This sector accounts
for a sizeable share of total employment in OECD economies and we shall
see that the dynamics of employment absorption in this sector differed substantially between various groups of economies.

Let us shortly proceed to reviewing the mechanics of the dynamics of labour
absorption in different sectors which will be of use when we discuss the
empirics of employment growth in tertiary activities in OECD economies in
section 3. In principle, the ability of a sector i to absorb labour Δe_i is a function
of two variables: the growth of output (dependent on demand) and the level and
change of the employment coefficient (the inverse of labour productivity):

$$\Delta e_i = \left(\frac{e_i}{q_i}\right)\Delta q_i + \Delta\left(\frac{e_i}{q_i}\right)q_i$$

where e_i denotes employment and q_i denotes output in sector i. Hence a decomposition of sectoral employment growth into its various components shows that
a particular sector's employment absorption capacity (that is, the number of
jobs or hours of work it generates over time) is a function of, first, output
growth exceeding (labour) productivity growth; second, a high labour intensity;
and third, its size (measured by its initial share in output or employment).

Historically, most major shifts in employment structures were characterized by a shift towards a sector which had – at least initially – high labour
intensity (which was true for early manufacturing and later for the various
tertiary activities) and also high growth of demand. As is well known from
elementary economics, the growth of demand depends upon two types of
factors: the shift of demand towards this sector at constant relative prices in
response to income or wealth changes which is measured by the income
elasticity, and the shift due to the reaction of demand to changing relative
prices; this reaction is measured by the elasticity of substitution. Relative
price movements are, in turn, over the longer term, substantially affected by
relative (total) factor productivity growth.

This gives us a number of categories of (potentially) labour-absorbing sectors. Sectors with high income elasticities are potentially labour-absorbing if there are (a) high productivity growth and high substitution elasticities (larger than one in absolute terms) or (b) slow productivity growth and low substitution elasticities (less than one). In exceptional circumstances, sectors with only mediocre income elasticities (around one) can be substantial labour-absorbing sectors at a point of time when they account for a large share in overall employment and either (c) relative productivity growth and substitution elasticities are very high or (d) relative (labour) productivity growth and substitution elasticities are rather low.

Constraints on Labour Absorption: the Role of the Welfare Services Sector

In this section we present a theoretical discussion concerning the employment absorption capacity of the welfare services sector and its potential constraints (see also Landesmann and Pichelmann, 1998). With regard to the classification introduced at the end of the previous sub-section, we shall make the argument that the welfare services sector falls at different times into categories (b) or (d). It is a sector which now accounts in all OECD countries for a share in total employment of about 30–40 per cent (more than any other service sector; for evidence, see section 3); estimates as to its income elasticity differ, but it includes sectors with very high income elasticities (education, health); it is a sector with low labour productivity growth and high employment intensity. As to substitution elasticities, we shall argue that there can be changes in regimes. In most OECD countries, but in Europe in particular, welfare services are to a substantial degree publicly funded and the measurement of substitution elasticities involves the way public spending programmes respond to the workings of Baumol's 'cost disease' (the fact that its productivity growth potential is very low). Our argument will substantially rely on emphasizing the possibility of a regime switch in the implicit substitution elasticities driving the expenditure patterns in different OECD economies, given the other parameters, and thus driving employment absorption in that sector (that is, slow rate of productivity growth, high employment intensity, large size).

The data which we shall present in section 3 show the following stylized facts.

1. The employment absorption in the social services sector was particularly high in Europe in the 1970s and early 1980s (much higher than in the USA). As compared to this, the USA had much more balanced growth of employment across the whole range of service activities.

2. By the mid-to-late 1980s, a significant number of European economies had experienced a break in the rates of employment absorption of the welfare services sector; such a break is not visible for the USA.
3. Furthermore, the break occurred at a time when a significant number of European economies continued to experience a persistent process of 'deindustrialization' (labour shedding in the industrial sector).
4. By this time, deindustrialization had flattened considerably in the USA.

We shall discuss below a stylized model which explains the patterns of employment absorption in the welfare services sector in Europe described above. The following are the ingredients of such a model. First, welfare services suffer from Baumol's 'cost disease' problem: the scope for productivity growth is very low, especially in that there is consumer pressure for the 'quality' of services not to decline (this requires, for example, a particular ratio of nurses to patients, teachers to students, and so on).

Second, an argument which applies to welfare services, but less so to other service activities which are also employment-intensive (such as distributive trades, restaurants and hotels) is that a significant fall in the relative wages in that sector to counter the low productivity growth problem is not an option: the reason is that the sector requires a significant proportion of skilled and motivated employees to ensure the (politically and socially) required 'quality standards'; a sharp drop in the relative wage rate will lead to a negative selection of job applicants and motivational problems in jobs in which job performance is not easy to monitor; employment in some other service areas (such as in distribution) rely much less on skilled labour and monitoring is easier.

Third, given that Baumol's cost disease works strongly in this sector as a result of both low labour productivity growth and a relative wage (efficiency) constraint, purchasers (and subsidizers) of that sector's output will have to bear the growing costs in terms of relative price (and subsidy) increases.

Fourth, the above is true for both the USA and Europe; let us now come to the differences: our stylized European model suggests a model in which a constraint was imposed on 'quality differentiation' in the provision of welfare services: roughly the same type of health service, schooling, old age provisions and so on was provided for all households; in the USA there was much more scope for 'quality differentiation'. 'Quality differentiation' induces producers to absorb consumer rents which can be particularly exploited when income inequality is high. The US system is one with high income inequality and high 'quality differentiation' in the provision of welfare services and this provides a model of continuous expansion of (differentiated) output over time. In the European model, given that there is a constraint in terms of relative uniformity in quality in the provision of social services, growth of social services provision will have to rely on a different mechanism than in the US model. Since quality differentiation is constrained, it cannot rely on the exploitation of 'quality rents' through the provision of higher-quality services to high-income households. In fact, given the uniformity of quality provision, the suppliers will be faced by falling income elasticities for their (uniformly supplied) low/medium-quality service at high levels of income. Instead, the European model will have to rely for its growth in the social services sector on some form of redistribution (this can take the form of direct income transfers to low-income households or subsidization of a certain proportion of the purchase price of social services, or direct cost subsidization of the social service industries). This allows the group of low-income households to move into a region of their demand schedules characterized by relatively high income elasticities and also allow a general increase in the (uniformly supplied) quality of the service product.

Finally, as the European model of welfare services provision on the redistributive mechanism outlined above (modelled for example as public subsidy to poorer households per unit of service provision), this leads also to a growing subsidy to GDP burden as a result of Baumol's 'cost disease' dynamics (low labour productivity growth combined with an efficiency wage rigidity for welfare services employees). Once the increase in the subsidy to GDP ratio is stopped owing to fiscal/political constraints, this growth mechanism in the European model collapses (for details, see the stylized model in Appendix A of this chapter).

The above is a story which can provide an explanation for structural breaks in the employment absorption pattern of the welfare service sectors in continental Europe in the mid-to-late 1980s; these breaks are clearly evident in an examination of the sectoral employment time series for a significant number of continental European economies (see below). This, together with the continued process of deindustrialization in a large number of continental European economies (lagged in comparison with the USA and the UK), supplies the elements for interpreting some of the hikes in the European unemployment figures in the late 1980s and the 1990s as 'structural unemployment'.

3. EMPIRICAL EVIDENCE

In this section we examine the different patterns of development and the courses of structural change in various European and non-European countries. We use the OECD Labour Force Statistics (LFS) to describe the employment patterns mainly for the period 1975 to 1994 (with exceptions for some countries owing to data problems). Further, we wish to indicate that, first, we use without exception employment time series and no other data on variables such as output, price or productivity and, second, we also refrain in this section from using at this stage unemployment or demographic data, which means that we do not refer to labour force participation rates across OECD economies or demographic developments. As mentioned in the introductory sections, our analysis stresses the importance of structural change in sectoral employment structures in OECD economies and their inter-country differences over the 1980s and 1990s; we introduce this factor into the current debate on European unemployment.

The labour force statistics are differentiated into nine ISIC sectors (see Table 7.1). For some of the analysis below, we have summed up the sectors at a more aggregate level, which we denote by I (agriculture), II (industry), and III (services), corresponding to the 'classical' sector scheme employed by Fourastié (1949) and Clark (1957). However, we shall quickly move on to argue that for explaining structural changes, and especially unemployment issues, in Europe one has to take a more differentiated look at the service sectors.

Description of Employment Patterns

The issue of structural employment change and its potential for explaining European unemployment will be discussed in five sections. First, we give a short description of the employment patterns along the three broadly defined 'classical' sectors I, II and III. We argue that, in order to contribute towards a 'structural' explanation of Europe's recent unemployment experience, one has to take into account the potential for labour absorption by the service sectors (III) in relation to the labour-shedding trends of the industrial and agricultural sectors (II and I). Second, we differentiate sector III; sectors 6–9

Table 7.1   ISIC classification

ISIC	Description	
1	Agriculture, hunting, foresting, and fishing	I
2	Mining and quarrying	
3	Manufacturing	
4	Electricity, gas and water	II
5	Construction	
6	Wholesale and retail trade, restaurants and hotels	
7	Transport, storage and communication	
8	Financing, insurance, real estate and business services	III
9	Community, social and personal services	

defined above have different labour-absorbing potentials, which depend on their shares in overall employment and their (relative) growth rates. Here we show that sector 9 (social services) especially is crucial, particularly in the European context. Third, we present an overview of development patterns of various countries. The various countries experienced similar patterns of development but at various times (that is, the phasing of labour shedding and labour absorption across sectors differed). Fourth, we show by econometric means that in most European countries – as opposed to the non-European countries – there was a break in labour absorption in sector 9 some time in the period 1980–90. We also test for evidence of a prolonged and/or delayed process of deindustrialization in continental Europe (relative to the UK and the USA). These two factors, we shall argue, contribute towards an explanation of the high levels of unemployment in European countries. Fifth, we use this argumentation to explain cross-country differences in changes in unemployment rates (see section 4).

Broad sectoral employment shifts
In this section we want to discuss briefly the development of employment patterns in sectors I, II and III across the OECD economies. Of course, the general pattern and evolution of employment shares in these broadly defined sectors is well known and we shall only highlight the general development and some differences between groups of countries in a brief manner.

Figure 7.1 shows the shares of employment in these three broadly defined sectors for certain groups of countries in 1975, 1985 and 1994.[5] The groups of countries identified are the northern and southern EU countries, the Scandinavian countries,[6] the USA, the UK and Japan. Panel A of Figure 7.1 shows the employment shares in agriculture (I) which in most countries are now below 5 per cent. Further, countries with relatively high shares in agriculture employment in 1975 (mainly the countries of EU-South) experienced a tremendous decline in these shares over the last 20 years and are converging rapidly on the average level of about 4–5 per cent. This is especially true for the southern European countries (especially Portugal, Spain and Italy and – to a much lesser extent – Greece) and also Iceland (sector I includes also fisheries) but at a less rapid pace. The largest decline in the share of agricultural employment was in Turkey (from 20 per cent in 1975 to 5 per cent in 1994).

The industrial sector (II) (see Panel B) has more or less steadily (see below) declined in all countries from a share of about 45–50 per cent to less than 30 per cent in 1994. But there are two remarkable exceptions. In Japan, the share of employment in industry is declining very slowly and had reached about 38 per cent in 1994. The other exception is again Turkey. In this country the employment share in industry *rose* from 30 per cent to about 45

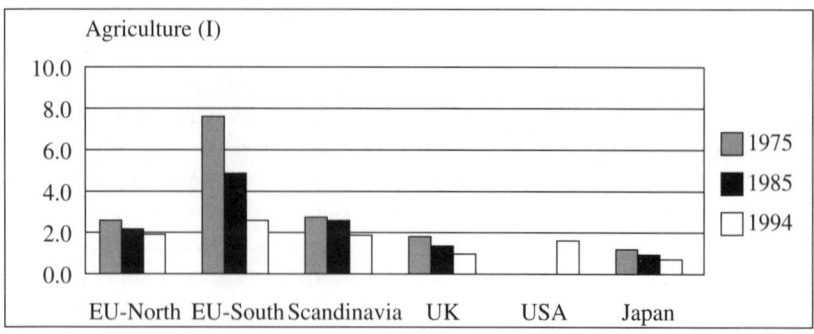

Panel A

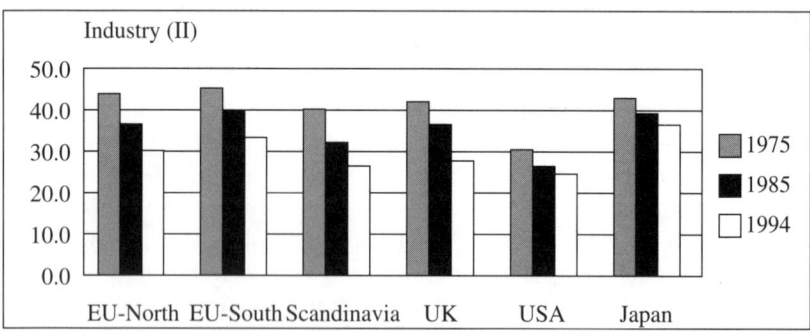

Panel B

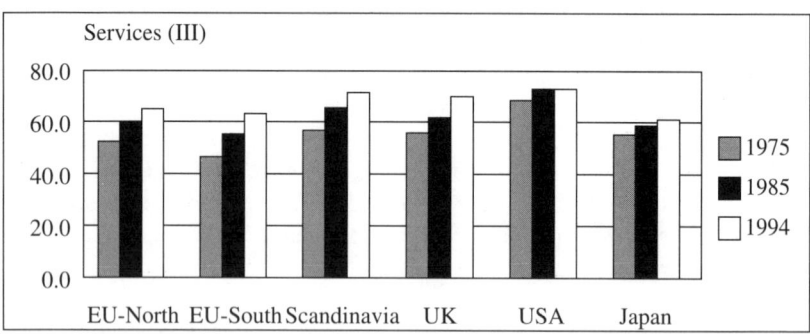

Panel C

Figure 7.1   Shares in sectors I, II and III

per cent. With respect to sector II the southern European countries did not look that different from the other countries as in sector I, either in shares in 1975 (which were only slightly higher than that for other countries) or in its general pattern of development. Hence these countries faced, in addition to shrinking agricultural employment, a strong process of deindustrialization, more or less in line with the more advanced EU economies. Below, we shall look more closely at the time patterns of the process of deindustrialization for different countries. Generally, most countries experienced huge changes in their employment patterns over the period 1975 to 1994, with at times dramatic declines in sectors which formerly accounted for a sizeable proportion of total employment.

Panel C of Figure 7.1 shows the great (and growing) importance of the service sectors for employment. In most countries nowadays more than 50 per cent and up to more than 70 per cent of the employed population is engaged in the service sector. In this sense, the service sector was, over this period of time and at this broad level of aggregation, the employment-absorbing sector, as the industrial sector was in the period of the first Industrial Revolution.[7] Seemingly, the non-European countries (with the exception of Japan) started from a higher share of this sector in 1975, but experienced a slower growth after that. In particular, the USA already had a share of 70 per cent in 1975, which rose only slightly to a share of about 75 per cent. The southern European countries are still lagging somewhat behind the average share, as they started from a lower level, but also experienced large shifts of employment towards services.

In fact, a general trend towards convergence in the sectoral employment shares can be discerned in these figures: all countries converged to a share of less than 5 per cent in agriculture, about 60–70 per cent in services and about 30 per cent in the industrial sector, although some countries are still lagging behind. The above supports the contention that a serious discussion of employment and unemployment performances of different economies would have to include an analysis of the pattern of employment-absorption capacities of the service sectors, together with an analysis of the phasing and severity of labour shedding in industry and, for some countries, in agriculture.

Deindustrialization and tertiarization

Making use of the sectoral disaggregation introduced above (see Table 7.1) we can take a closer look at sectors II (industry) and III (services). Within industry, sector 3 (manufacturing) is the most important, accounting for about 30 per cent of total employment in 1975, and experienced in most cases the highest (negative) shifts in employment shares over the period 1975 to 1994 (see Figure 7.2). Given these two facts, the labour-shedding processes

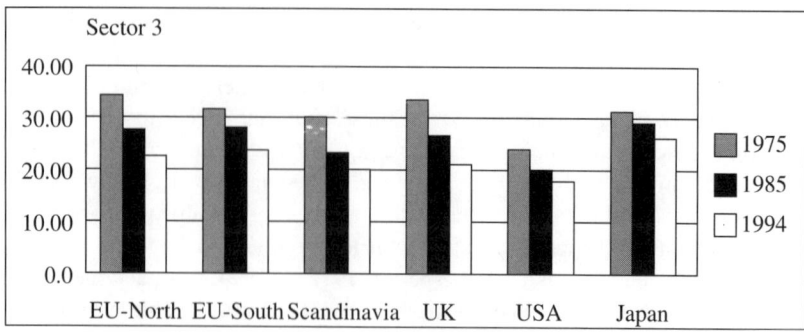

Figure 7.2 Employment shares in manufacturing (sector 3)

in absolute values were mainly due to developments in this sector. This may also explain why this sector was very much at the centre of discussion about the high unemployment rates in Europe, especially in the 1980s (see, for example, Rowthorn and Wells, 1987). Starting from a share of more than 30 per cent (of total employment) in almost all countries in 1975, it declined to about 20 per cent with some country differences; the US declined from about 25 per cent in 1975 to 18 per cent in 1994. The exception is Turkey, which is the only country experiencing a positive shift towards this sector. Amongst the non-European countries – which generally have a lower share of employment in this sector than the European countries – only Japan had an employment share of more than 25 per cent in 1994, as opposed to about 15–20 per cent in the other non-European countries.

Labour-shedding trends in manufacturing is only one side of the analysis of the impact of structural change on unemployment. The other side has to do with the labour absorption capacity of other sectors. As we have seen above, the labour-absorbing sectors were the service sectors included in III. Here we want to differentiate between the various service sectors 6–9 (see Table 7.1). Figure 7.3 shows the shares in 1975, 1985 and 1994. The general pattern is that employment shares are highest in sectors 6 and 9 but are growing very rapidly, mainly in sector 8, although from a very low level. With regard to sector 6, the striking difference between the European and the non-European countries is that the shares are about 15 to 20 per cent in 1994 in the European economies (with even lower levels in Belgium, Italy, Sweden and Turkey) as against shares of about 25 per cent in the non-European countries. Sector 7 shows more or less constant shares in total employment, with declining trends in those countries in which the share of employment was already relatively high compared to other countries. Thus sector 7 seems to experience a convergence towards shares of about 6 to 8 per cent in almost all countries. Employment shares in sector 8

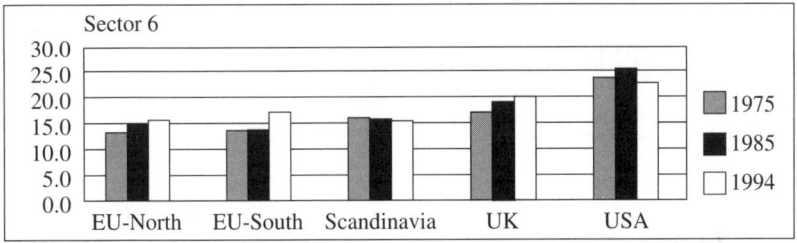

Panel A

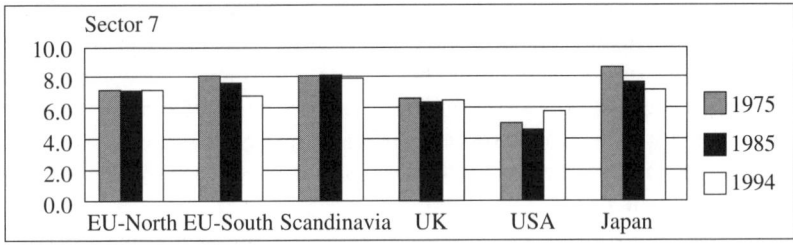

Panel B

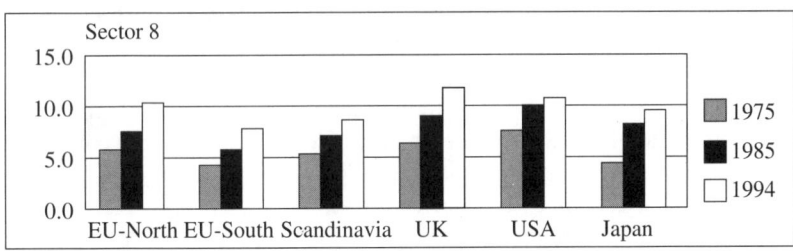

Panel C

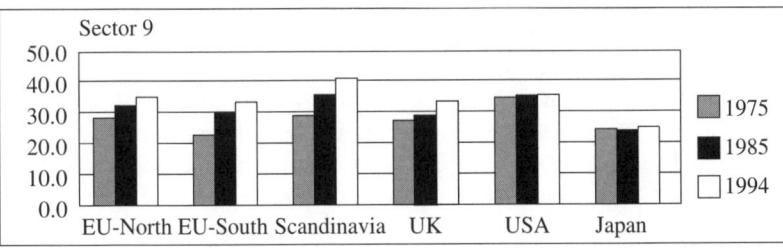

Panel D

Figure 7.3   Employment shares in services (sectors 6–9)

are growing rapidly (but starting from a low absolute level) and there is a more differentiated development across countries. From an aggregate employment perspective, sector 9 is the most important, as this sector has the relatively highest employment share within services and has also experienced rapid growth in employment levels.

Dynamic Patterns of Structural Change

General discussion

We turn now to examine the time pattern of structural change in different countries, which we started to discuss above. We do not give an exhaustive description, but only wish to distinguish some 'typical' patterns in European as opposed to some non-European countries. Different patterns of structural change can be distinguished at three levels. First, countries can experience equal or at least similar patterns, but at different times (see, for example, Chenery and Syrquin, 1975); second, countries also experience different patterns of structural change (or development), but the countries converge on a similar structural pattern; and third, countries experience totally different development patterns.

The first variety means that countries experience the same pattern of structural change, but at different times (that is, the time trajectories are equal, but shifted on the time dimension). The second variety is best illustrated by an example. Some countries are experiencing a process of deagrarianization and tertiarization without ever having been industrialized to the extent of other countries (as a consequence of different trade structures, technological paradigms and so on); but the actual structural employment patterns are converging towards those of the other (more advanced) countries. In terms of trajectories, this means that the time trajectories are different, but 'end' in a similar structural pattern. Thirdly, some countries may remain agrarian economies for a long time, thus showing a different economic structure initially, and are also unable to develop a structure converging towards that of more advanced economies.[8] Reasons for this could be that they are naturally resource-endowed (for example, oil economies, fisheries).

Although this third possibility can be true for developing (or transition) economies, we think that the countries in our sample are best described by the first and second types of pattern. That is, the countries show a (more or less) similar pattern of development and structural change, as can be seen, for example, from the converging patterns of employment structures discussed above, but the trajectories might occur at different times. Of course, the patterns are not exactly replicated, as countries have different institutions, different endowments/and or other reasons for distinct specialization and trade structures. Furthermore, as we are thinking of employment rather than output shares, we also have to take into account that countries may have different sectoral relative productivity levels.[9]

Description of actual patterns

In this section we examine the dynamics of structural change in employment in the OECD economies over the last 20 years. In particular, we take a closer look at *changes in the dynamics of structural change* in sector 3 and the service sectors 6–9. For some countries (the southern European countries) we also consider the decline of employment in sector 1 (agriculture) as for these countries there was a huge decline in employment in this sector, which increases the pressure on the overall employment situation. As we have seen above, in most countries sector 3, and for some countries also sector 1, are the most important labour-shedding sectors and sectors 6–9 (especially 6 and 9) are in general the most important labour-absorbing sectors.

In the following we give a short description of the changing dynamics of structural change, as these dynamics are obviously not smooth processes. We show in particular that the labour-shedding and labour-absorption pattern of the sectors is varying in time and across countries. In this section we rely on some descriptive statistics. In the next section we shall present the result of an econometric investigation of structural breaks in employment shedding and employment absorption in sectors 3 and 9, respectively. We consider this to be a link in the understanding of the different unemployment performances between European countries and the USA which has not received sufficient attention in the discussion so far.

In order to provide a descriptive overview, we divide the period 1975 to 1994 into four sub-periods, 1975–79, 1979–84, 1984–89 and 1989–94. These sub-periods correspond more or less to the peaks in the growth cycles. Of course, the peaks differ from country to country, but this general time structure should suffice for the preliminary overview given here.[10]

To highlight the differences between sectors, countries and periods, we constructed the following indicator. First, we calculated the (exponential) growth rate g_j^c of total employment e_t^c for each country c and each of the four sub-periods $j = 1...4$:

$$g_j^c = \frac{\ln\left(\dfrac{e_{T_j}^c}{e_{0_j}^c}\right)}{T_j - 0_j}$$

where 0_j and T_j denote the first and last year of the sub-period under consideration, respectively. We then worked out a hypothetical sectoral development under the assumption that employment in each sector had grown with the rate of growth of total employment (which amounts to maintaining a constant share in total employment). The hypothetical employment path is thus based on the assumption of even sectoral development: the rate of (labour-saving) technological progress and the rate of demand growth is equal in each sector, leading to balanced sectoral employment growth. Thus the hypothetical trajectory of sectoral employment is calculated as:

$$\tilde{e}^c_{i,t_j} = e^c_{i,0_j} \exp^{g^c_j t_j}.$$

Third, we subtract the employment levels on the hypothetical path from the actual ones to get the deviation from the hypothetical path and express it as a percentage of the (initial) employment level of sector i at time $t_j = 0$:

$$\Delta^c_{i,t_j} = \frac{\left(e^c_{i,t_j} - \tilde{e}^c_{i,t_j} \right) 100}{e^c_{i,0_j}}.$$

This exercise was done for each of the 20 countries studied.[11] In Appendix B (Figures 7B.1 and 7B.2 and Table 7B.1) the percentages are listed for sectors 1, 3, and 6–9, always for the last year of each of the four sub-periods, T_j, and for each country. If this indicator equals 0, then the employment growth of sector i will have been equal to the total employment growth; values lower (higher) than 0 thus mean a lower (higher) growth than overall employment growth. Further, the height of the bars gives a hint of the importance of the sectors in its labour-shedding and labour-absorbing performance in relation to total employment growth.

Next we discuss some of the striking features of the changes in the dynamics of structural change for some sub-groups of countries. Figure 7.4 shows the pattern for the labour-shedding sectors 1 (Panel A) and 3 (Panel B). As we can see, sector 1 (agriculture) is for all sub-groups of countries a relatively labour-shedding sector.[12] The values observable are, however, very small. The exceptions are of course the southern European countries, where there was a dramatic decline in the employment share of this sector (see above). In comparison to sector 3, the values are smaller in almost all country groups (exceptions are the EU-South countries). This indicates that the relative labour shedding in sector 3 (manufacturing) was much more important overall.

Turning to the manufacturing sector (sector 3), we want to stress two facts: first, there is a difference between the European countries and the USA, with the USA having in general lower relative labour-shedding rates in this sector; second, the EU-North countries show a rather constant higher rate of deindustrialization with a temporary dip in the third sub-period followed by a speeding up of the process of deindustrialization again in the fourth period. In the Scandinavian countries, there were very high rates of deindustrialization over the first two periods, followed by some decline in the two last sub-

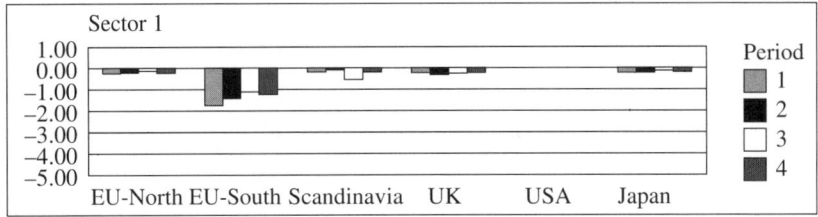

Panel A

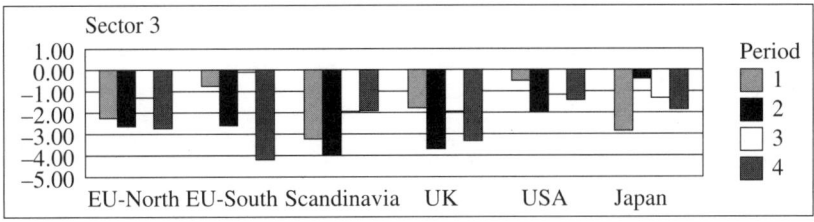

Panel B

Figure 7.4   Dynamics of change, sectors 1 and 3

periods. The southern European countries experienced a very high rate of deindustrialization in the last period (in addition to the labour shedding in agriculture). In Japan, there is also some evidence of a speeding up of deindustrialization over the last three periods (but at lower relative rates than in EU-North). This points towards a prolonged (and even speeding up) experience of deindustrialization in a range of European economies (and Japan) which is not found in the USA.

Next we turn to the labour-absorbing sectors. Figure 7.5 depicts the pattern of structural change in the service sectors. To highlight also the differences in the importance of the sectors we graded the vertical axes identically. In comparing the sectors it can be seen that the values for sector 7 (transport, storage and communication) are very low, so that this sector is not very important for the discussion of relative labour absorption and labour shedding.

Although sector 8 (financing, insurance, real estate and business services) is an expanding sector, with high rates of employment growth, its contribution to overall labour absorption is affected by its relatively small initial share. It performs rather differently across the country groups, with an earlier speeding up of relative employment growth in that sector in the USA and the UK as compared to EU-North and EU-South economies; the sector became relatively more expansive in EU-North and EU-South over the last 10 years, while it was declining (although it remained labour-absorbing) in Scandina-

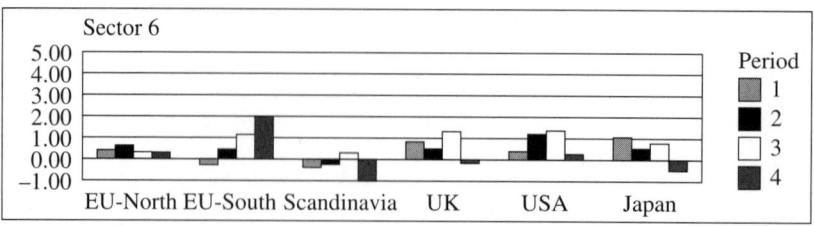

Panel A

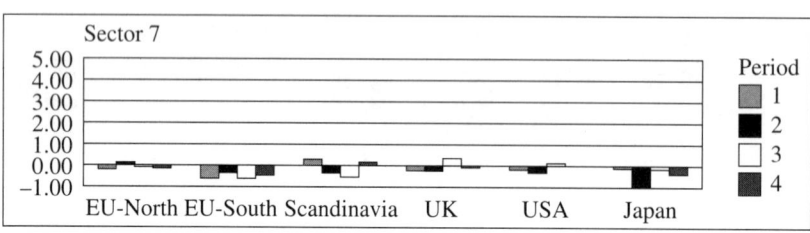

Panel B

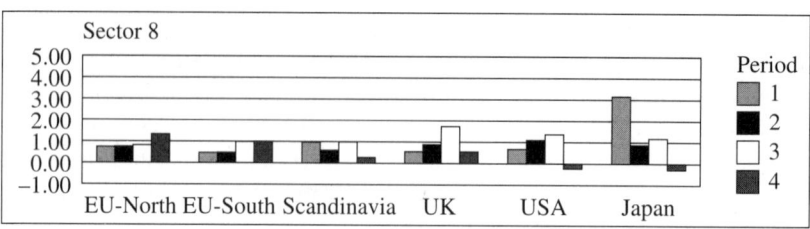

Panel C

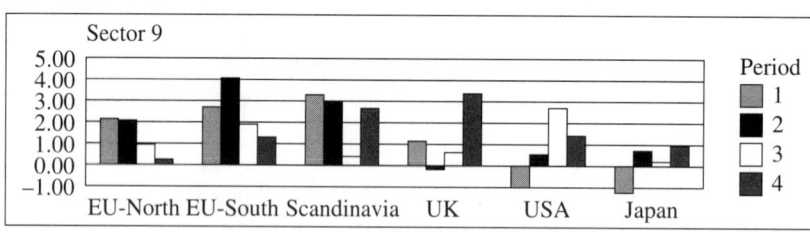

Panel D

Figure 7.5 Dynamics of change, sectors 6–9

via and especially in Japan (slightly negative over the last period). Given the relatively greater weight of this sector in the US and UK economies (see above), the sector was most employment absorbing in the UK and the USA over the period as a whole, but experienced lower rates of relative employment growth in the last sub-period, with even a negative rate differential for the USA.

There remain two sectors with a potential for labour absorption on a scale which could counteract the labour shedding from the manufacturing sector. The first of them, sector 6 (wholesale and retail trade, restaurants and hotels) was very much discussed in the 'job miracle' debate in the USA, and, indeed, one sees a much higher (and even growing) rate of relative labour absorption in the USA and the UK than in Scandinavia and EU-North. But there was again a decline in the last sub-period in the UK and the USA. Japan also experienced a high rate of relative labour absorption over the earlier periods, but it was declining and became even negative (relatively labour shedding) over the period 1989–94. Only in the EU-South countries did labour absorption in sector 7 proceed at very high rates. This is most certainly the result of the convergence process discussed earlier on, as these countries started with an underrepresentation of this sector (see Figures 7.1 and 7.3) and hence relative employment growth in this sector represents a catching-up process.

There remains sector 9 (community, social and personal services) as a vital sector for employment absorption. As mentioned above, this sector is the most important potential labour-absorbing sector as it has the highest share in total employment of all the service sectors (see above) and the potential for productivity growth is low. From Figures 7.4 and 7.5, we can see that the bars of the diagrams for sectors 3 and 9 have more or less the same height in absolute terms (2–3 per cent on average), with opposite signs. As to intercountry variations, we can see here a crucial difference: in the EU-North and EU-South countries, the rates of relative employment absorption in sector 9 are initially much greater than in the USA, but are declining, whereas the rates of labour shedding in sector 3 remain constant or are speeding up (see discussion above). In Scandinavia, the situation is better, as the rate of relative labour shedding declines in sector 3 but from a very high level. In the UK and the USA, on the other hand, we observe high and/or growing relative labour absorption rates in sector 9 over the last two sub-periods. Japan faces a situation like that of Europe with growing rates of relative labour shedding in manufacturing combined with very low rates of relative labour absorption in sector 9.

Empirical Evidence of Structural Breaks

In this section we present econometric evidence of the occurrence of structural breaks in relative employment growth of sectors 3 and 9. We looked for statistically significant breaks in the growth rates of the shares of these two important sectors (manufacturing and welfare services) in total employment.[13] The main problem was first to find the year(s) in which a structural break has occurred, if at all. For this purpose we applied some tests of structural breaks. In fact, we used the CUSUM, the CUSUM-SQUARE and the Chow tests. These tests give some hint if a break has occurred at all and the first and third tests indicate also the year in which this has probably occurred. The problem is that for a number of reasons these tests (sometimes) produce different results.[14] By inspection of the log-linearized time series one can also detect the years of the suspected breaks, although these have to be tested econometrically afterwards.[15]

In a second step we estimated piecewise linear regression functions (spline regression functions) for the diverse time periods for each country. These regressions yield values for the different growth rates over the time periods and can further be used to test for significance of differences in growth rates. In the data sample we included 18 countries[16] (15 European and three non-European countries) and the employment data over the period 1975–94. In the following we only present the results of the spline regressions and some test statistics.

Structural breaks in relative employment growth/contraction in manufacturing

Table 7.2 shows the results for changes in the growth rates of the employment shares of the manufacturing sector for the 18 countries. The values labelled d1, d2 and d3 are the estimation coefficients for the dummy variables.[17] The values below these coefficients give the *t*-statistics. If there is no value inserted in d2 or d3 this means that we did not find a structural break. Below these values, the F statistic for the first and (eventually) second structural break are listed. Rows GR1, GR2 and GR3 then give the growth rates of the shares in the sub-periods. The years separating these sub-periods are listed in rows 'Break 1' or 'Break 2'. The last three rows give the overall test statistics of the regression.

The results are in line with the hypothesis stated above. We find significantly lower growth rates (higher negative growth rates) in the second sub-period (the period after the first break) for Austria, Finland, France, the Netherlands, Sweden, Spain, Iceland and Portugal. In Austria, France, the Netherlands, Sweden and Portugal these breaks occurred at the end of the 1980s or beginning of the 1990s, which is in line with our thesis of delayed deindustrialization. In the Netherlands and Sweden we see two phases of deindustrialization. For Belgium, Norway, Denmark, Ireland and Italy there is evidence of a slowing down of deindustrialization. Furthermore, for Great Britain, the USA, Australia and Japan we did not find evidence for a structural break over this period.

Structural breaks in the welfare services sector

The same procedures were used to test for structural breaks in relative employment absorption in the welfare services sector (sector 9). Again, we find support for our hypothesis above for most of the countries. In most European countries – exceptions are Finland, Great Britain and Iceland – we found a significant decrease in the relative growth rates in the 1980s. Only Great Britain shows a significantly higher growth rate from 1989 onwards. Norway and Sweden experienced a lower relative growth rate from 1984 to 1989. For the non-European countries we did not find empirical evidence for structural breaks. Table 7.3 summarizes the results of the spline regressions for sector 9.[18]

Summary

These results can be summarized by a matrix indicating whether there was a significant decrease ($\downarrow$) or increase ($\uparrow$) in the relative growth rates of employment shares of sectors 3 and 9 (see Table 7.4).

Thus most European countries can be found in the group of countries where the relative growth rates in both sectors declined: rising labour shedding in the manufacturing sector and a weakening of labour absorption in the welfare services sector. In three countries (Denmark, Belgium and Italy) labour absorption of sector 9 decreased but labour shedding in manufacturing also decreased. Finally, only in Great Britain did relative labour absorption in sector 9 increase; however, this followed a period of negative relative labour absorption in that sector over the period 1979–84 (see above), whereas the relative rate of employment decline in the manufacturing sector remained constant. Finland and Iceland experienced a higher rate of deindustrialization, but no significant decrease in the relative labour absorption in sector 9.

Sweden and Norway show another pattern. There was a simultaneous decline in the rate of deindustrialization and a decrease in the relative rate of employment absorption in sector 9 from 1984 to 1989; from 1989 onwards, deindustrialization speeded up again (at least in Sweden) together with an increasing labour absorption in the welfare services sector.

Finally, for the USA and Japan, the relative growth rates stayed constant over the period (but there remain substantial differences in the structure of employment between the two countries, as mentioned above).

Table 7.2   Results of spline regressions, sector 3

Country	AUS	BEL	DNK	FIN	FRA	GBR	IRL	NLD	NOR	SWE
d1	-0.009	-0.024	-0.024	-0.011	-0.022	-0.023	-0.016	-0.031	-0.032	-0.034
	-8.912	-19.841	-8.243	-2.657	-50.334	-24.853	-8.722	-13.038	-27.339	-11.777
d2	-0.034	0.011	0.018	-0.019	-0.011		0.011	0.023	0.023	0.028
	-6.569	3.643	4.060	-3.600	-8.046		3.626	5.497	3.905	6.249
d3								-0.015		-0.028
								-2.909		-6.297
F-test 1	43.545	13.270	16.484	12.956	64.733		13.145	30.220	15.247	39.044
F-test 2								8.463		39.652
GR1	-0.90	-2.37	-2.35	-1.08	-2.20	-2.30	-1.61	-3.12	-3.18	-3.36
GR2	-4.32	-1.25	-0.59	-2.95	-3.30		-0.48	-0.87	-0.84	-0.56
GR3								-2.39		-3.38
Break 1	1991	1988	1984	1982	1989		1985	1983	1991	1982
Break 2								1990		1989
R^2	0.954	0.980	0.906	0.970	0.998	0.972	0.913	0.979	0.985	0.981
R^2 adj.	0.949	0.978	0.895	0.967	0.998	0.970	0.903	0.975	0.983	0.978
F value	177.306	423.408	81.814	275.579	3929.016	617.672	89.637	243.475	549.966	279.739

226

Country	SP	GRC	ISL	ITA	PORT	USA	AU	JPN
d1	-0.013	-0.007	-0.030	-0.021	-0.005	-0.019	-0.027	-0.008
	-6.818	-4.777	3.052	-8.941	-2.843	-15.575	-29.21	-11.793
d2	-0.011	-0.064	-0.058	0.011	-0.014			
	-3.609	-8.516	-5.300	2.782	-3.790			
d3								
F-test 1	13.023	72.518	28.088	7.737	14.36			
F-test 2								
GR1	-1.35	-0.69	-2.96	-2.11	-0.49	-1.91	-2.69	-0.75
GR2	-2.43	-0.75	-8.76	-1.00	-1.91			
GR3								
Break 1	1984	1990	1980	1985	1987			
Break 2								
R^2	0.979	0.939	0.922	0.934	0.895	0.931	0.979	0.885
R^2 adj.	0.976	0.932	0.913	0.926	0.883	0.927	0.978	0.879
F value	387.887	131.734	101.102	119.737	72.474	242.582	853.232	139.07

Table 7.3   Results of spline regressions, sector 9

Country	AUS	BEL	DNK	FIN	FRA	GBR	IRL	NLD	NOR	SWE
d1	0.013	0.034	0.019	0.021	0.022	0.005	0.018	0.010	0.026	0.022
	12.008	27.597	6.955	29.652	17.349	2.834	12.219	10.991	16.902	16.869
d2	-0.010	-0.028	-0.025		-0.018	0.014	-0.016	-0.009	-0.025	-0.022
	-3.494	-16.569	-6.237		-6.462	2.784	-3.537	-2.551	-7.252	-8.217
d3									0.023	0.021
									5.831	5.925
F-test 1	12.205	274.531	38.904		41.751	7.750	12.507	6.507	52.597	67.519
F-test 2									36.002	35.102
GR1	1.28	3.40	1.86	2.12	2.17	0.47	1.77	1.01	2.59	2.23
GR2	0.32	0.57	-0.67		0.40	1.89	0.18	0.10	0.12	0.05
GR3									2.46	2.19
Break 1	1988	1983	1984		1987	1989	1989	1990	1984	1984
Break 2									1989	1989
R^2	0.938	0.990	0.742	0.980	0.970	0.800	0.932	0.914	0.986	0.982
R^2 adj.	0.931	0.989	0.711	0.979	0.966	0.777	0.924	0.903	0.983	0.979
F value	129.109	844.354	24.378	879.238	271.638	34.037	116.421	89.884	365.766	293.642

Country	SP	GRC	ISL	ITA	PORT	USA	JPN
d1	0.038	0.026	0.018	0.027	0.038	0.003	0.005
	14.375	15.409	19.825	15.384	16.479	4.398	4.778
d2	-0.029	-0.012		-0.015	-0.028		
	-5.019	-2.764		-5.228	-8.106		
d3							
F-test 1	25.187	7.640		27.330	65.714		
F-test 2							
GR1	3.85	2.58	1.79	2.67	3.78	0.29	0.47
GR2	0.91	1.39		1.14	0.95		
GR3							
Break 1	1987	1988		1985	1983		
Break 2							
R^2	0.958	0.968	0.956	0.975	0.975	0.518	0.559
R^2 adj.	0.953	0.964	0.958	0.972	0.972	0.491	0.535
F value	194.858	257.200	393.015	332.250	327.528	19.338	22.831

Table 7.4   Summary of spline regressions

		Sector 3		
		↓	No break	↑
Sector 9	↓	AUS, FRA, IRL, NLD, SP, PORT, GRC		DNK, BEL, ITA
	No break	FIN, ISL	USA, JAP	
	↑	SWE	GBR	NOR

4.   SECTORAL RESTRUCTURING AND UNEMPLOYMENT PERFORMANCE

So far we have not used any unemployment data, as we have referred mainly to the demand side of structural change processes. In this section we use unemployment rates (again from the LFS statistics) and compare these with the rates of structural change.[19] We present some simple regression results to show how the restructuring processes (deindustrialization and tertiarization) are connected to the (un)employment performance of different samples of countries. Figure 7.6 shows the unemployment rates in 1975, 1985 and 1994 for 20 countries. The impact of deindustrialization on unemployment over the period 1973–85 is well documented in Rowthorn and Glyn (1990) where they conclude that changes in the growth rates of industrial employment (that is, the extent and speed of 'deindustrialization') are an important explanatory factor for differences in the unemployment experiences across OECD countries (and is even a better predictor than the growth rate of total employment). This analysis therefore showed that the rise in unemployment since 1973 has a strongly structural character.

Here we want to emphasize this aspect once again, but to add further the absorption capacities of the services sectors, especially of the social and community sector (ISIC 9), the distribution sector (ISIC 6) and the finance and business services sector (ISIC 8). Thus we include employment developments in sectors 3 (manufacturing), 6, 8 and 9. Sector 3 has – as already mentioned – a large influence on unemployment as it is the main labour-shedding sector; sector 6 has played a vital role in the USA and some other countries in the phase of deindustrialization as a labour-absorbing sector, sector 8 experiences the highest growth rates in employment; and sector 9 is the sector which, as has been argued above, may have had different characteristics in the USA and continental European countries as a labour-absorbing sector.

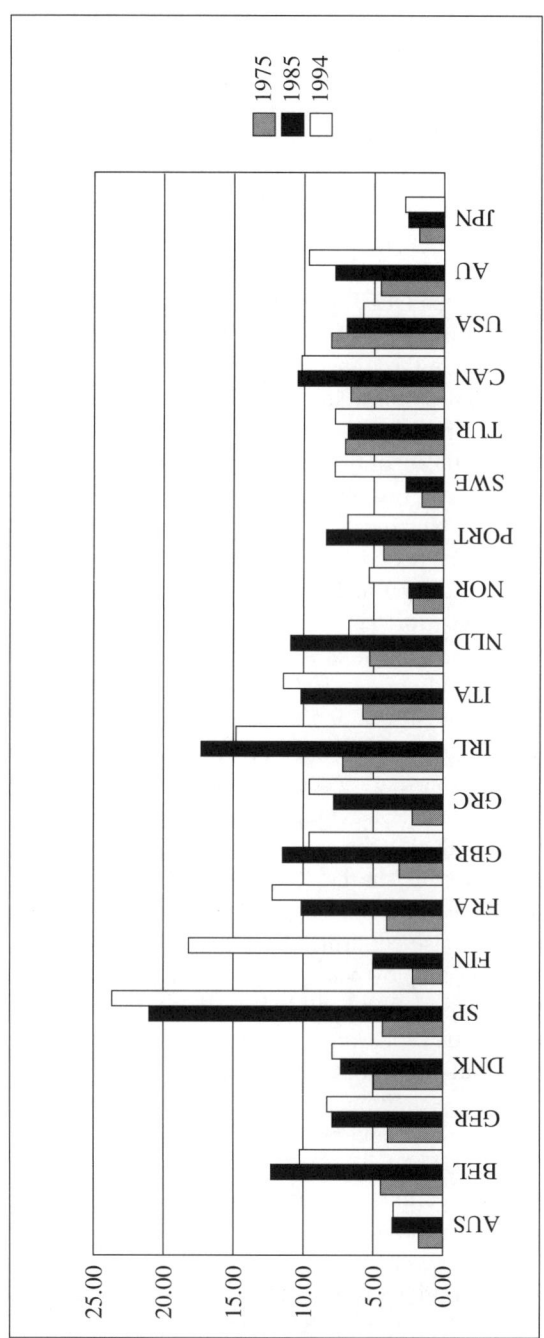

Figure 7.6   Unemployment rates

In the following we report some regression results which attempt to explain changes in the rates of unemployment across OECD economies over the periods 1975–85 and 1985–94 through the employment growth rates of the different labour-shedding and labour-absorbing sectors. We thereby want to examine to what extent the different sectors are responsible for changes in the overall rates of unemployment in the two different periods.

Table 7.5 presents single variable regressions between employment growth in the various sectors and the average annual growth rate of unemployment for the two periods 1975–85 and 1985–94.[20] For the entire sample, including European and non-European OECD economies, we can see that employment growth in sectors 3 and 6 relates negatively and significantly to changes in unemployment rates, while the impact of sectors 8 and 9 is not significant over the first and second periods. In the second period, the impact of the (negative) employment growth in sector 3 is reduced (a lower regression coefficient and only significant at the 10 per cent level).

For the European countries (excluding Great Britain) employment growth in sectors 3, 8 and 9 had a significant negative impact on the growth rate of unemployment over the first period.[21] We can see here that the relation between employment growth in sector 9 and the change in unemployment is indeed highly significant and has, for the period 1975–85, the highest explanatory power of all the four sectors (with an R^2 of about 60 per cent). This changes dramatically for the second period (1985–94) where employment decline in sector 3 and employment growth in sector 6 contribute further and more significantly to the explanation of the cross-country unemployment experience, while employment growth in sector 9 does not contribute to it. A comparison of the results for the total sample as compared to the sample of the European economies without the UK also shows that the deindustrialization factor contributes much more powerfully in the second period to an explanation of the unemployment experience in Europe as compared to the OECD as a whole.

These conclusions are also confirmed by the set of regressions reported in Table 7.6, in which the employment growth rates of all three sectors are considered jointly. Here we left out sector 8, as this sector had no significant impact in the single regressions reported above.[22] Employment growth in sector 9 is the only significant variable (although only at a 10 per cent level) for the group of European economies for the first period and not significant at all for the sample as a whole, and becomes insignificant also for the European economies in the second period. Deindustrialization also contributes powerfully in the second period both for the set of European economies and for the overall sample. Employment growth in sector 6 (a sector which we classified as a low-wage sector) is not very important for the European economies in the first period but is more important for the sample as a whole for both periods and for the European economies in the second period.

Table 7.5   Sectoral growth and unemployment (single regressions)

	Total sample		European countries	
	1975–85	1985–94	1975–85	1985–94

Dependent variable: growth rate of unemployment rate s_u^c

Sector 3	–0.930**	–0.671*	–1.482**	–1.365**
	(–2.565)	(–1.798)	(–2.379)	(–2.653)
R^2	0.291	0.168	0.340	0.390
$\bar{R}^2$	0.247	0.118	0.280	0.335
F value	15.831	1.975	14.284	4.076
Sector 6	–1.645***	–1.215***	–2.569***	–2.240***
	(–2.600)	(–2.960)	(–3.103)	(–5.185)
R^2	0.297	0.354	0.467	0.710
$\bar{R}^2$	0.253	0.314	0.418	0.683
F value	16.022	4.842	18.999	14.614
Sector 8	–1.557	–0.652	–4.299*	–0.800
	(–1.489)	(–0.953)	(–2.198)	(–0.793)
R^2	0.122	0.054	0.305	0.054
$\bar{R}^2$	0.067	–0.005	0.242	–0.032
F value	11.228	0.077	13.298	0.674
Sector 9	0.041	–0.056	–4.822***	–0.538
	(–0.067)	(–0.090)	(–3.687)	(–0.672)
R^2	0.000	0.049	0.609	0.039
$\bar{R}^2$	–0.062	–0.011	0.531	–0.048
F value	8.893	0.718	16.906	0.580

Average annual growth rates

Sector 3	–0.59	–0.41	–0.68	–0.37
Sector 6	–0.02	0.12	–0.03	0.10
Sector 8	0.12	0.23	0.12	0.21
Sector 9	0.29	0.12	0.38	0.14
Unemployment	0.60	0.29	0.71	0.14

Table 7.6   Sectoral growth and unemployment (joint regressions)

	Total sample		European countries	
	1975–85	1985–94	1975–85	1985–94
	Dependent variable: growth rate of unemployment rate s_u^c			
Sector 3	−1.026***	−0.889***	−0.418	−1.178**
	(−3.624)	(−3.679)	(−0.676)	(−3.161)
Sector 6	−1.794***	−1.530***	−1.269	−1.845***
	(−3.638)	(−5.109)	(−1.121)	(−4.740)
Sector 9	−0.577	−0.564	−3.775*	−0.714
	(−1.395)	(−1.531)	(−2.181)	(−1.357)
R^2	0.642	0.717	0.721	0.910
$\bar{R}^2$	0.565	0.657	0.554	0.856
F value	17.116	9.345	9.277	16.148

5.   CONCLUSIONS AND POLICY IMPLICATIONS

This chapter has examined structural employment patterns across the OECD economies and has attempted to contribute towards a 'structural' explanation of the European jobs crisis, especially as it evolved from the mid-1980s onwards. The following are the main results of our analysis.

First, the period since the mid-1970s has seen major shifts in employment patterns across the OECD economies in terms of a decline in the share of employment in industry ('deindustrialization') and an increase in the share of employees in tertiary activities ('tertiarization'); in the southern European economies there was also a significant fall in the shares of employment in agriculture ('deagrarianization').

Second, as compared to the USA, the European economies also experienced a much more dramatic process of deindustrialization after the mid-1980s; we called this the 'lagged deindustrialization thesis'. For a sizeable group of continental European countries, a significant speeding up of the deindustrialization process could be observed some time in the mid-1980s or early 1990s; this was tested by means of spline regressions. This speeding up of deindustrialization could not be observed in the case of the non-European OECD economies or the UK.

Third, another important difference between the USA and Japan, on the one hand, and the majority of the European economies is a sharp slowdown in the rate of employment absorption in the community and social services sector (ISIC sector 9) which employs the largest share of employees of all tertiary sectors. The community and social services sector ('welfare services' for short) showed, on the other hand, much higher rates of employment absorption in continental Europe over the earlier period (1974 to mid-1980s) than it did in the USA, Japan or the UK. Again the significance of a slowdown of employment absorption in the welfare services sector was tested by means of spline regressions.

Fourth, the contribution of this chapter in terms of a 'structural explanation' of the European jobs crisis, particularly from the mid-1980s onwards, is thus based on the above two factors: a continued and sometimes intensified process of deindustrialization in a large group of continental European economies and a dramatic slowdown in the rates of employment absorption in the welfare services sector. Neither of these factors was observed in the case of the UK (which experienced the main phase of deindustrialization earlier), the USA or Japan. For some European countries a somewhat different pattern can be observed. In Denmark, Belgium and Italy both the labour absorption of sector 9 and the labour shedding of sector 3 decreased; Finland and Iceland experienced higher (negative) rates of growth of the employment shares in sector 3 (higher speed of deindustrialization) but no significant decrease in the labour absorption in sector 9. Finally, Sweden and Norway show a similar pattern in the dynamics of sector 9. In both countries there is a sharp decline of labour absorption at the beginning of the 1980s and an increase in 1989. In Norway labour shedding fell dramatically at the beginning of the 1990s. Sweden experienced high labour shedding up to 1982 in sector 3, very low labour shedding out of sector 3 in the 1980s, and again an increase in labour shedding at the beginning of the 1990s.

Finally, other factors played a role in accounting for differences in structural employment patterns. First, the relative growth rates of employment in other tertiary activities (wholesale and retail trade, restaurants and hotels (sector 6) and market services, sector 7 (transport, storage and communication) and sector 8 (financing, insurance, real estate and business services)) differed widely across countries. Second, we were able to show in a regression analysis that cross-country differences in changes in rates of unemployment could be explained by the different countries' experiences in terms of 'deindustrialization' and the patterns of employment growth in the different types of tertiary activities. The distinct patterns of continental European economies as against the UK, USA and Japan (and some of the Scandinavian countries) emerged again.

What are the policy conclusions? Major sectoral shifts in employment patterns can cause severe 'matching problems' on the labour market. The skill structure of the labour force, the demographic and gender characteristics as well as the geographic location of the existing labour force might not match the new requirements of shifts in the structure of labour demand. In addition, legal and institutional features which might have been functional to support growth and employment in an economy with particular structural characteristics might no longer be adequate in a situation with changed structural characteristics. This issue is well known as far as the problem of relocating labour from traditional core manufacturing industries is concerned. In this chapter we have pointed out that the major shifts within the tertiary sector (from the community and social services sector to distribution, recreational and market services) might similarly generate problems of adjustment. In this respect the USA and Japan never had the same unbalanced growth pattern across these different types of tertiary activities which characterized the European economies in the past. It does look as if the 'European model' of fast (relative) expansion of employment in the 'community and social services' sector came to an end at some point in the mid-to-late-1980s and Europe has also moved towards a more balanced pattern of employment growth across the different tertiary serv-ices activities. However, skill requirements, work conditions and wage contracts differ widely in these different tertiary sectors and, hence, it is no wonder that a dramatic change in the structural employment pattern in this area also gener-ates its own adjustment problems.

In a theoretical part of this chapter, we examined why the European model of employment growth in the welfare services could no longer be (politically) sustained and we pointed out some of the differences between the European and the US models of social services provision which have implications for sustainability of growth in this sector: the US model allows much more differentiation in the qualities of services provided and expansion is thus built upon an incentive of producers to exploit consumer rents, while the European model traditionally imposed relative homogeneity in the provision of such services. It was demonstrated that expansion of this sector had to rely on a continuous process of redistribution; unbalanced productivity growth im-plied, furthermore, an increasing burden of such redistribution as a percentage of GDP. The (political) collapse of the 'European model' of welfare services expansion does not imply, by necessity, a full convergence on the US model and a full-scale political and scientific discussion is currently under way on the extent to which a more selective approach towards (mixed public and private) welfare services provision can ensure a sustainable new European model, distinct from both the US and the traditional European models. This chapter did not delve into this discussion, but simply stated that this regime switch did, indeed, start to occur some time in the mid-to-late 1980s and has made its contribution to the 'structural' European jobs crisis.

There are other areas of policy discussion which are relevant to explaining the structural employment patterns which we pointed out in this chapter.

First, the different types of tertiary activities require to different extents low-skill, medium-skill and high-skill employees, second, these activities differ in the degree to which performance can be monitored by employers and, third, the performance relies on job commitments which differs between them. Hence the types of wage and employment contracts will differ and wage, employment and social security regulations but also public training and educational facilities will affect the relative expansion of jobs in the service sectors.

APPENDIX A   EMPLOYMENT EFFECTS OF
DISTRIBUTIVE MEASURES IN THE
EUROPEAN MODEL: A SIMPLE
ANALYTICAL REPRESENTATION

Introduction

In the following we go over the arguments concerning the way redistribution of income affects the growth of output and employment in the European model (see discussion in section 2, above). We characterized the European model of welfare services provision as one in which clear restrictions are imposed on the degree of quality differentiation that can occur in the provision of welfare services. We assume that only a uniform quality of such a service is supplied to all income groups. The additional stylized facts are that in welfare services labour input coefficients are relatively high (welfare services are labour-intensive) and the rate of (labour) productivity growth is rather low. Further, welfare services are relatively skill-intensive.

The further assumption is that income elasticities for welfare services are equal or larger than one for low-income groups. Thus, with two different income groups with $w^l < w^*$ and $w^h > w^*$, a redistribution of income from the high- to the low-income groups would increase demand for welfare services and hence labour-intensive goods and thus overall employment might rise.[23] The further assumption that welfare services employment is more skill-intensive would lead to a shift towards more skill-intensive jobs. Further, with a given (uniform) quality of a particular welfare service, there may also be a satiation level for the high-income groups. This depends on the availability of outside options in service goods and on the supply of high-quality industrial goods as substitutes. Here again, a redistribution away from income groups with high income levels would boost employment levels. The impact of redistributive measures will be shown in more detail in a simple model below.

A second (but, as we shall see, related) potential reason for a demand constraint in the service sector is the increase in the relative price of service goods (Baumol's 'cost disease'). In the above scenario we showed that a redistributive measure (income subsidies to low-income groups or direct price subsidies of welfare services) reduces the demand constraint for a good of homogeneous quality (the constraint on homogeneity in quality justifies the shape of the upper part of the Engels curve). A second step in the argument is to analyse the dynamic implications. Here we rely on the contributions made by Baumol and associates (Baumol *et al.*, 1989) who work out the implications of unbalanced productivity growth in relation to different expenditure structures. As long as real expenditure structures remain somewhat constant, uneven productivity growth (across sectors) combined with rigidities in the relative wage structure across sectors leads to changes in the nominal expenditure shares. A growing share of nominal spending will go into spending on products/services produced by low productivity growth sectors (in our case, welfare services).

The redistributive mechanism characterizing the European model (see section 2, above) implies that subsidies for income groups must rise in proportion to the relative price increases of the welfare services sector so that these income groups are able to pay for the more expensive service goods. Thus, if the subsidy element remains a constant fraction of the price of the welfare services, aggregate subsidies will have to rise as a fraction of total nominal expenditure, as the relative price of services is increasing.

Dynamically, the redistributive measure leads to an increasing subsidy burden in GDP (that is, either a higher public expenditure/GDP ratio or higher lump-sum transfers from high- to low-income groups). This rising burden will lead to political constraint in the expansion of the welfare sector: the unwillingness of (parts of) the population to finance the expansion of this sector. A cap on the rise in transfers will then cause slower employment growth in the service sectors. The welfare services sector will cease to function as an employment absorber, the 'European model' of welfare services growth will collapse and a structural jobs crisis will emerge, until alternative sectors act as employment absorbers (see section 2, above).

The Structure of the Model

The arguments above are now reproduced in a simple analytical framework. First, we present the structure of a multisectoral model, the equilibrium solutions and some comparative-static analysis. On the cost side there is a coefficients matrix $\mathbf{A}$ with the interindustrial input requirements and a (row) vector of labour input coefficients for skill groups and industries $\mathbf{a}'_L = (a^s_{L1}, a^u_{L1},..., a^s_{Li}, a^u_{Li},..., a^s_{LN}, a^u_{LN})$. This vector includes the labour input coefficients a^z_{Li} for each industry $i = 1...N$ and each skill-type of worker $z = s, u$ for skilled and unskilled, respectively.[24] In the following the skill groups also represent the different income groups. Labour demand is then given by $l^z_i = a^z_{Li}q_i$ where q_i denotes output in industry i. Further, the wages of the workers w^z_i are assumed to be fixed exogenously. Nominal prices are assumed to be equal to costs, thus

$$\mathbf{p}' = \omega'(\mathbf{I} - \mathbf{A})^{-1}$$

where $\omega' = (\bar{z}_l\omega^l_1,..., \bar{z}_l\omega^l_N)$. $\omega^z_i = a^z_{Li}w^z_i$ denotes the labour unit costs of each skill group z in each industry i.

In this simple economy only wage income exists, as the profit rate is assumed to be zero. The demand for goods thus consists of demand for interindustrial inputs and the structure of the wage demand. The latter can be described for each income or skill group by a matrix:

$$
\mathbf{D}_{\tilde{L}}^{z} =
\begin{pmatrix}
\alpha_{1}^{z}\dfrac{w_{1}^{z}l_{1}^{z}}{p_{1}} \cdots \alpha_{1}^{z}\dfrac{w_{N}^{z}l_{N}^{z}}{p_{1}} \\
\vdots \quad \ddots \quad \vdots \\
\alpha_{n}^{z}\dfrac{w_{1}^{z}l_{1}^{z}}{p_{N}} \cdots \alpha_{n}^{z}\dfrac{w_{N}^{z}l_{N}^{z}}{p_{N}}
\end{pmatrix}
=
\begin{pmatrix}
\alpha_{1}^{z}\dfrac{\omega_{1}^{z}}{p_{1}} \cdots \alpha_{1}^{z}\dfrac{\omega_{N}^{z}}{p_{1}} \\
\vdots \quad \ddots \quad \vdots \\
\alpha_{N}^{z}\dfrac{\omega_{1}^{z}}{p_{N}} \cdots \alpha_{N}^{z}\dfrac{\omega_{N}^{z}}{p_{N}}
\end{pmatrix}
\begin{pmatrix}
q_{1} \\
\vdots \\
q_{N}
\end{pmatrix}
$$

The nominal income shares, denoted by α_{i}^{z}, are different between income groups and are given exogenously. We assume that the higher-income group has a higher nominal share of expenditure on services than the lower-income group. In a more advanced setting, these shares could be derived from utility functions implying non-linear Engel curves. The quantity system can then be written as

$$
\mathbf{q} = \left(\mathbf{A} + \sum_{z} \mathbf{D}_{\tilde{L}}^{z} \right) \mathbf{q}
$$

where $\mathbf{q}' = (q_{1}, \ldots, q_{N})$. This is a homogeneous system as

$$
\left(\mathbf{A} + \sum_{z} \mathbf{D}_{\tilde{L}}^{z} - \mathbf{I} \right) \mathbf{q} \equiv (\Theta - \mathbf{I})\mathbf{q} = \mathbf{0}.
$$

For non-trivial solutions the condition det $(\Theta - \mathbf{I}) = 0$ has to be fulfilled. This condition is guaranteed, as it can be shown that the rows are linearly dependent. In such a system it is not possible to determine the activity level of the economy, but only the structure of the output.

In this simple setting only stationary solutions are possible as there are no rents to be reinvested or productivity gains which can be used to extend production (the latter will, however, be considered below).[25] We will present the equilibrium solutions with respect to the parameter values used in the simulations.

The simulation model

In this section we present the parameter and starting values for the simulation studies discussed below. There are two sectors, manufacturing (sector 1) and welfare services (sector 2). The first sector uses less labour than the service sector and the service sector is more skill-intensive.[26] There is no saving and all income is spent immediately. The demand structure differs between the two skill groups. For simplicity we assume exogenous nominal shares for the two income groups and also the changes in the shares if income changes because of redistribution (these shifts reflect the non-linearity of the Engel curves discussed above).

The concrete parameter values and resulting equilibrium values (which we use as starting points) are as follows. The input–output coefficient matrix[27] is

$$\mathbf{A} = \begin{pmatrix} 0.25 & 0.00 \\ 0.15 & 0.00 \end{pmatrix}.$$

The vector of labour input coefficients is given by

$$\mathbf{a}'_L = (1, 2, 3, 4).$$

Wages are assumed constant and given exogenously and are assumed equal across industries but different between skill groups, using $w^s = 5$ and $w^u = 2$ as concrete values. The different skill groups represent the two income groups, thus having different nominal shares in consumption, $\alpha_1^s = 0.25$ and $\alpha_1^u = 0.75$. The nominal shares in the services sector are $1 - \alpha_1^z$. The resulting equilibrium prices are

$$\mathbf{p} = (16.6, 23.0).$$

As the level of output is not determined, we set output of sector 1 equal to 1 to determine the output structure

$$\mathbf{q} = (1, 0.841).$$

The resulting labour demand is then $l_1^s = 1$, $l_1^u = 2$, $l_2^s = 2.523$, and $l_2^u = 3.364$.

The structure of output and employment patterns is given in Table 7A.1 (in percentage of total output or total employment, respectively).

Table 7A.1 Equilibrium structure of output and employment

q_1	54.32
q_2	45.68
l^s	39.64
l^u	60.36
l_1	33.76
l_2	66.24

Redistributional effects

What are the effects of redistributing income from the high- to the low-income group? Redistribution in this model set-up does not alter the relative prices, as the wage rates firms have to pay remain constant and productivity levels for now remain constant too. We assume that subsidies to the low-income group are financed by taxing the high-income group, in the way that the budget is balanced; thus $t^s \Sigma_i l_i^s w_i^s = -t^u \Sigma_i l_i^u w_i^u$. For a given subsidy t^u, the high-income group has to be taxed at the rate

$$t^s = -\frac{\sum_i l_i^u w_i^u}{\sum_i l_i^s w_i^s} t^u.$$

We now introduce a subsidy to the low-income groups of 25 per cent and assume that, owing to higher wages, the nominal expenditure share for the manufacturing good is declining to $\alpha_1^u = 0.50$. This shift in nominal shares implies an income elasticity for services larger than 1. The nominal shares of the high-income group remain constant (income elasticity of 1).

The effects of introducing a subsidy of $t^u = 0.25$ on output and employment can be seen from Figure 7A.1. Output of the manufacturing sector is declining, whereas output of services is rising owing to rising demand of services from the low-income groups. The effect depends on the change in the nominal expenditure shares. Redistribution has two effects: first, there is a higher real income in the economy as a whole as the low-income group spends relatively more on the relatively cheaper (manufacturing) goods. This 'real income effect' raises demand and total output, and, potentially, employment. On the other hand, a higher proportion is spent on goods which are produced by less labour (manufacturing), thus depressing employment ('employment effect'). The outcome between these two effects depends on relative prices and the structure of the labour input coefficients. Thus a rise in the nominal expenditure share on services has the effect of reducing the 'real income effect' (that is, depressing output) but also lowers the (negative) 'employment effect' (more is spent on labour-intensive goods). Thus the (exogenous) shift in the nominal shares α_i^z has to be large enough to produce positive effects on employment; this is guaranteed for the parameters used in our simulations. Further, this kind of redistribution only has level effects on output and employment, but no long-term growth effects. The new long-term levels are $q_1 = 0.866$, $q_2 = 0.938$ and, for employment, $l_1^s = 0.867$, $l_1^u = 1.732$, $l_2^s = 2.810$ and $l_2^u = 3.759$. Table 7A.2, which can be compared with Table 7A.1, gives the structure of output and employment in the new steady-state equilibrium. The share of output in sector 1 is declining, whereas the share of services in total output is rising.

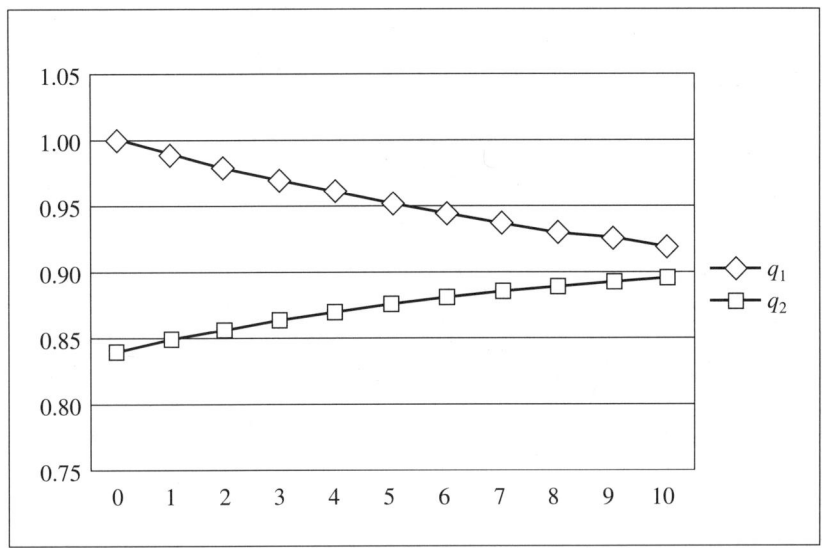

Panel A: Output

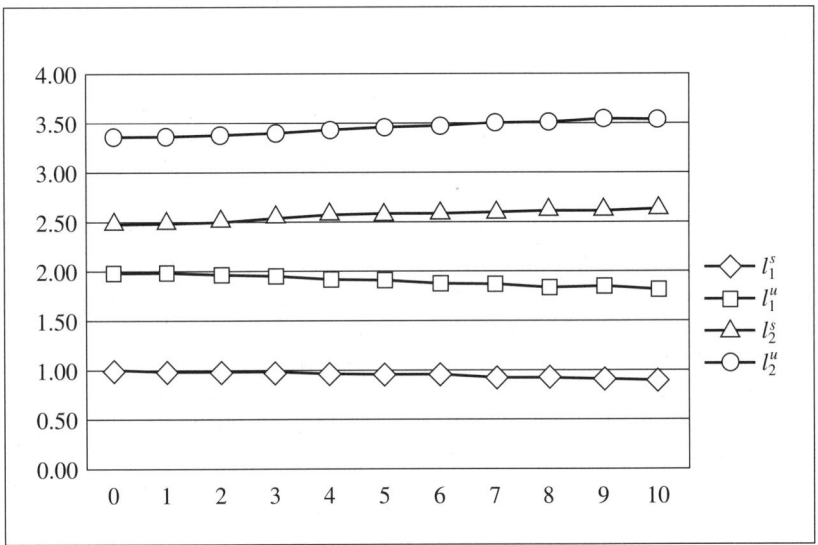

Panel B: Employment

Figure 7A.1   Effects of wage subsidy for low-income groups

*Table 7A.2   Equilibrium structure of output and employment with
              redistribution*

q_1	47.99
q_2	52.01
l^s	40.13
l^u	59.87
l_1	28.35
l_2	71.65

Employment shifts occur mainly between sectors and only to a very small extent between skill groups.

Figure 7A.2 shows the real tax burden b for the high-income group, defined by

$$b = \frac{l_1^{ds} w_1^s + l_2^{ds} w_2^s}{\alpha_1^s p_1 + \alpha_2^s p_2} t^s.$$

The real tax burden b is constant in the long run at a level of -0.127 once the structure of employment reaches an equilibrium as wages and productivity levels (and hence prices) are constant.

Introducing the effects of unbalanced and skill-biased productivity growth

We assume that there is labour-saving technological progress only for un-skilled workers in the manufacturing sector at an exogenous rate of productivity growth. Further, we assume that the gains from productivity growth find expression immediately in higher real incomes and higher real spending which, in turn, is distributed between the sectors in line with changing expenditure structures.[28] As regards prices we assume that they adjust immediately to the lower costs. Output is formulated dynamically as a simple 'supply adjusts to demand' differential equation. Given our information of the demand side of the model (in terms of exogenously fixed nominal expenditure shares), relative price changes imply substitution effects of a Cobb–Douglas type (that is, changes in relative prices generate proportionate changes in relative quantities). Figure 7A.3 presents the simulation results for output and employment without redistribution. Output in sector 1 is rising faster as productivity gains lead to manufacturing goods becoming relatively cheaper, thus expenditures are switching to the manufacturing sector. As a result of this output growth, employment is growing too. Employment for unskilled workers is growing

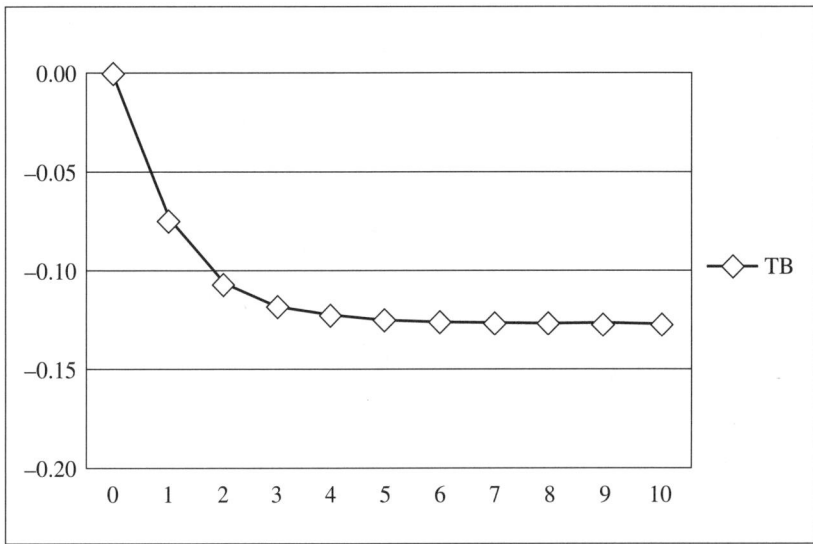

Figure 7A.2   Tax burden

only very little as technological progress is biased against it. Employment in the service sector is growing equally for both skill groups. (We assume no changes in input coefficients in this sector.)

Introducing redistribution as described above into the productivity-enhanced version of the model changes the results slightly. Output of services is rising much faster than without redistribution, whereas the output growth of manufacturing is lower (Figure 7A.4 compared to Figure 7A.3, Panel A). The differences between the scenarios without and with redistribution are best discussed in terms of the growth rates. Table 7A.3 presents the growth rates (in percentages) of output and employment in the two scenarios and the structure of both variables at the end of the simulations.

The first column gives the growth rates of output and employment without redistribution, the second column with redistribution from high- to low-income groups. As one can easily see, in the latter case output growth rises in the services sectors and declines in the manufacturing sector, owing to redistribution. Compared to the initial structure (see Table 7A.1) the output share of services declines in the scenario without redistribution, but rises with redistribution.

This is also reflected in the dynamics of the employment shares. In the first scenario (without redistribution) the share of employment in services is lower at the end than in the second scenario, although it is growing in both scenarios; thus the employment shift due to substitution effects (as relative

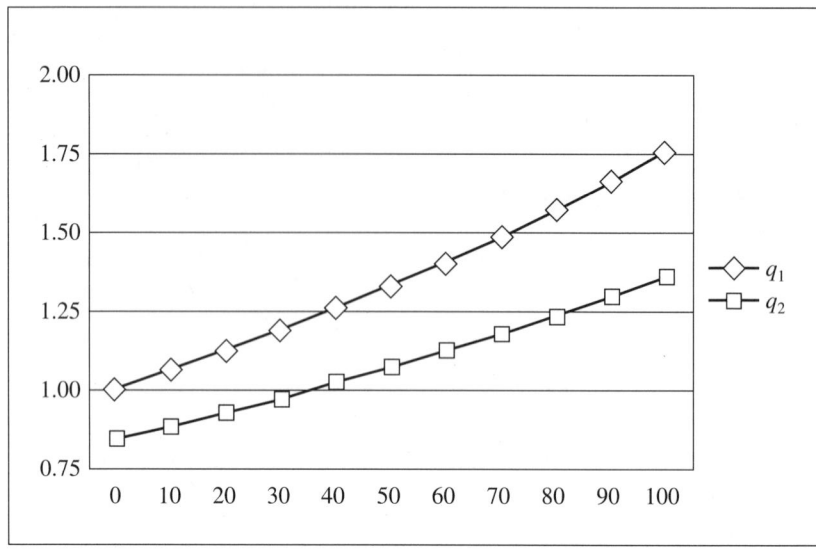

Panel A: Output

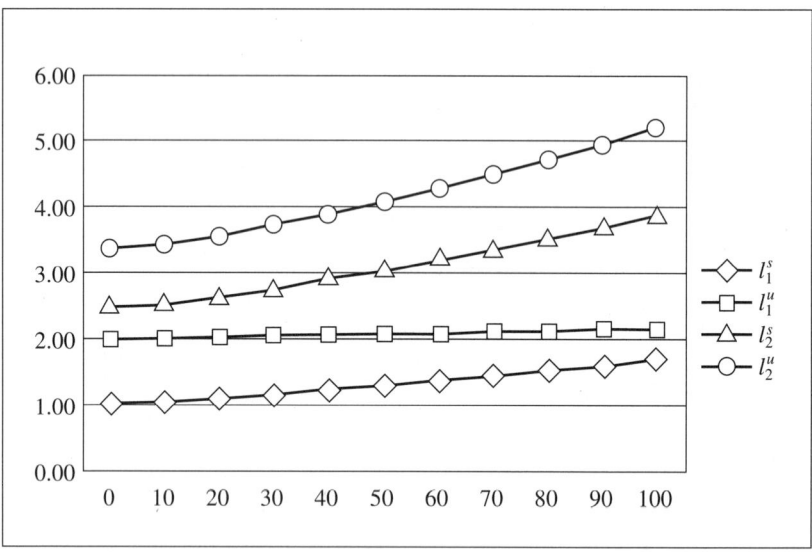

Panel B: Employment

Figure 7A.3   Output and employment with biased technological progress

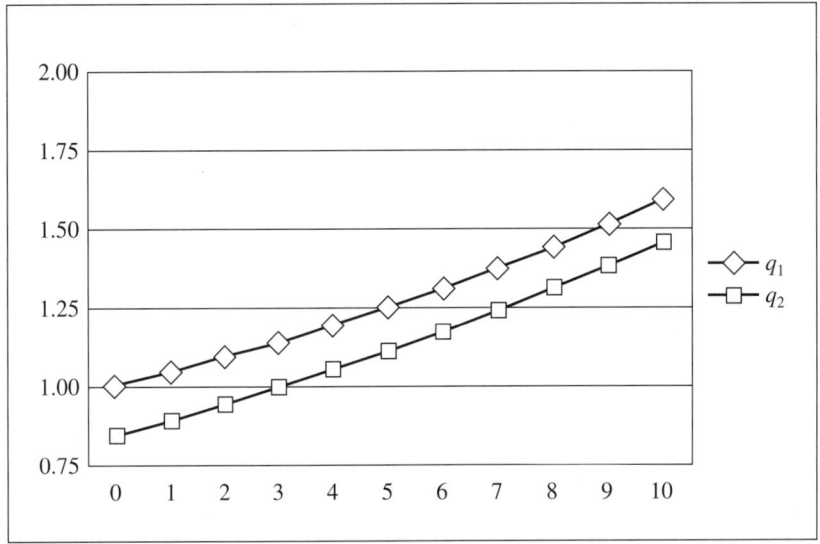

Figure 7A.4   Output with biased technological progress and redistribution

prices change) is lower than the impact of the biased technological progress. But, again, there is no large difference between the two scenarios with respect to the shares of skilled versus unskilled workers.

But owing to changes in relative prices and restructuring of employment, the real tax burden *b* for the high-income groups is rising, as can be seen in Figure 7A.5. A cut in redistribution, that is, lowering the subsidy rate t^u, would then have the opposite effects on output and employment growth presented in Table 7A.3.

Table 7A.3   Effects of technical change and redistribution

	Growth rates		Structure	
	$t^u = 0.00$	$t^u = 0.25$	$t^u = 0.00$	$t^u \, 0.25$
q_1	0.56	0.46	56.54	52.10
q_2	0.48	0.56	43.55	47.90
l^s	0.43	0.46	43.15	43.12
l^u	0.29	0.32	56.85	56.88
l_1	0.24	0.14	29.51	26.26
l_2	0.45	0.53	70.49	73.74

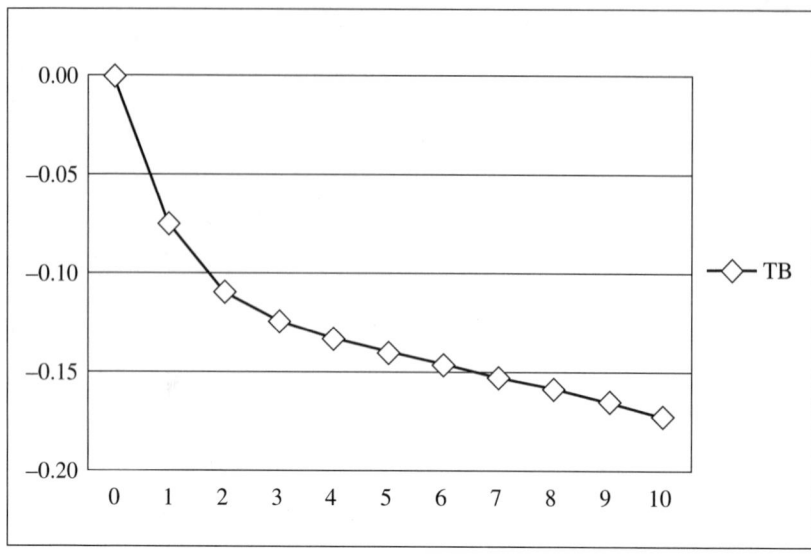

Figure 7A.5   Tax burden with biased technological progress

APPENDIX B   TABLES AND FIGURES

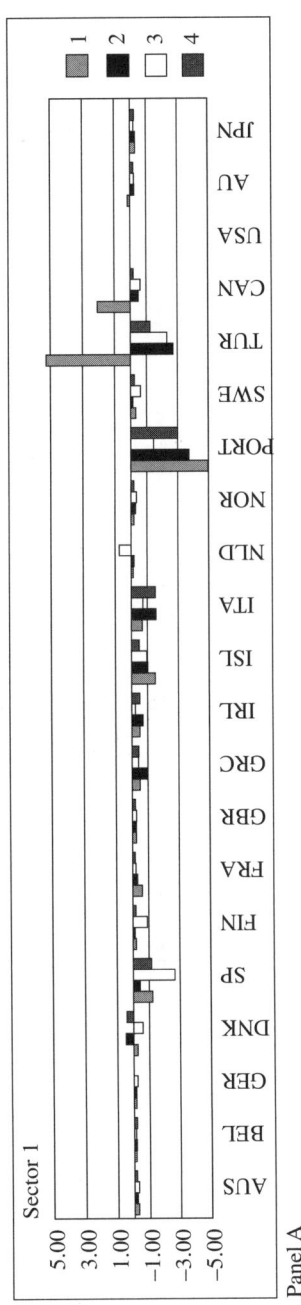

Panel A

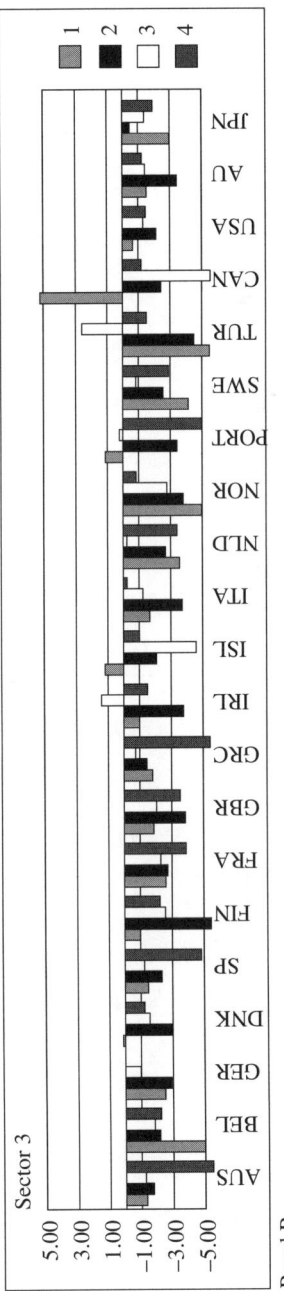

Panel B

Figure 7B.1   Country and sector-specific dynamics of change, sectors 1 and 3

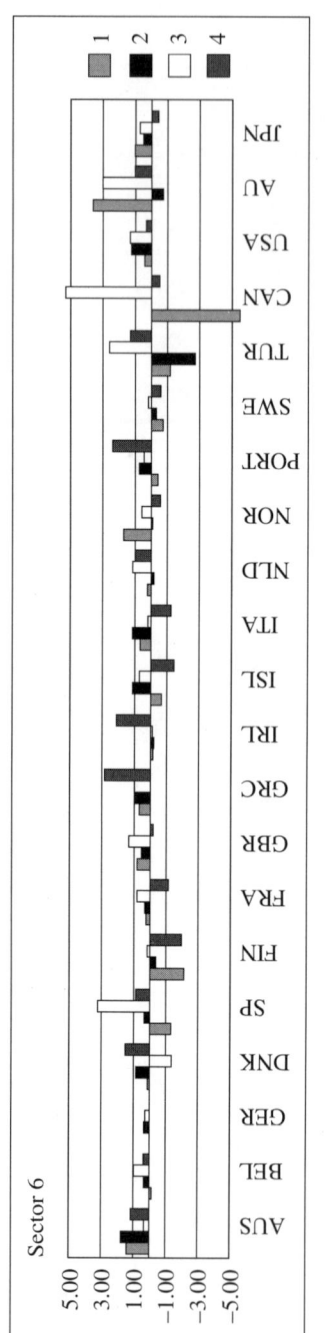

Sector 6

Panel A

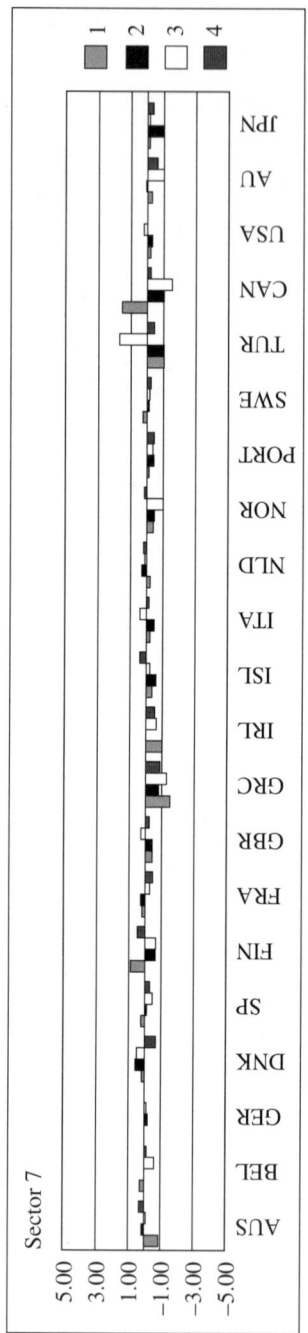

Sector 7

Panel B

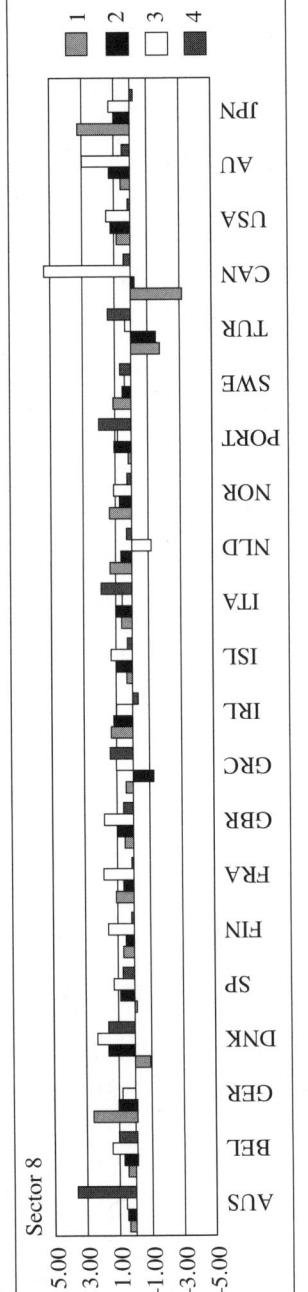

Panel C

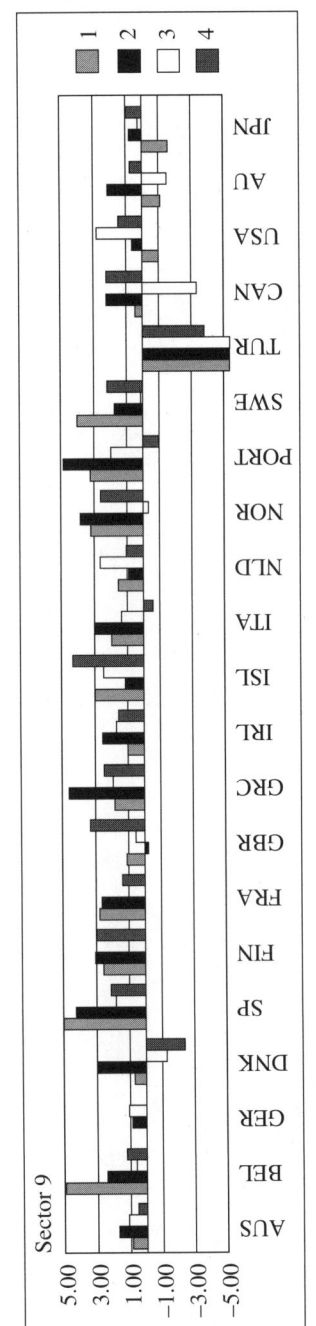

Panel D

Figure 7B.2   Country and sector-specific dynamics of change, sectors 6–9

Sectoral changes and demand

Table 7B.1   Country and sector-specific dynamics of change

Sector 1

	AUS	BEL	GER	DNK	SP	FIN	FRA	GBR	GRC	IRL
1	−0.16	−0.04	−0.04	−0.22	−1.10	−0.09	−0.57	−0.17	−0.44	−0.51
2	−0.07	0.02	−0.01	0.43	−0.32	0.03	−0.23	−0.22	−0.92	−0.66
3	−0.15	0.01	−0.16	−0.44	−2.55	−0.76	−0.12	−0.19	−0.33	−0.11
4	0.02	−0.02	n.a.	0.38	−1.06	−0.06	−0.09	−0.10	−0.41	−0.45

Sector 3

	AUS	BEL	GER	DNK	SP	FIN	FRA	GBR	GRC	IRL
1	−1.20	−4.97	−2.36	0.15	−1.37	−0.81	−2.47	−1.80	−1.76	−0.92
2	−1.61	−2.04	−2.80	−2.91	−2.21	−5.83	−2.64	−3.68	−1.39	−3.57
3	−1.18	−1.73	−0.83	−1.50	−1.13	−2.45	−2.21	−1.93	−0.67	1.40
4	−6.12	−2.05	n.a.	−1.11	−4.77	−2.13	−3.79	−3.35	−5.76	−1.37

Sector 6

	AUS	BEL	GER	DNK	SP	FIN	FRA	GBR	GRC	IRL
1	1.42	−0.07	n.a	0.08	−1.24	−2.07	0.19	0.80	0.66	−0.08
2	1.75	0.33	0.30	0.80	0.29	−0.30	0.26	0.47	0.81	−0.09
3	0.35	0.87	0.27	−1.29	3.18	0.07	0.78	1.30	0.88	−0.02
4	1.13	0.34	n.a.	1.49	0.79	−1.89	−1.09	−0.12	2.76	2.04

Sector 7

	AUS	BEL	GER	DNK	SP	FIN	FRA	GBR	GRC	IRL
1	−0.88	0.14	n.a.	−0.04	0.09	0.80	0.17	−0.24	−1.33	−0.95
2	0.08	−0.01	−0.18	0.47	−0.13	−0.65	0.25	−0.23	−0.65	0.00
3	−0.05	−0.54	−0.05	0.38	−0.44	−0.66	−0.14	0.31	−1.19	−0.46
4	0.25	−0.10	n.a.	−0.64	−0.26	0.34	−0.36	−0.04	−0.73	−0.35

Sector 8

	AUS	BEL	GER	DNK	SP	FIN	FRA	GBR	GRC	IRL
1	0.28	0.38	2.40	−0.84	−0.06	0.49	1.07	0.56	0.39	1.36
2	0.34	0.53	0.79	1.42	0.69	0.38	0.63	0.89	−1.14	1.16
3	0.46	1.28	0.65	2.13	1.17	1.46	1.79	1.76	0.94	0.93
4	3.48	0.79	n.a.	1.43	0.50	−0.19	0.00	0.56	1.32	−0.25

Sector 9

	AUS	BEL	GER	DNK	SP	FIN	FRA	GBR	GRC	IRL
1	0.81	4.80	n.a.	0.70	4.77	2.59	2.76	1.09	1.77	1.04
2	1.75	2.40	0.88	2.88	4.29	3.09	2.62	−0.13	4.58	2.51
3	1.13	0.61	1.04	−1.15	1.79	1.10	0.91	0.58	1.96	1.67
4	0.50	1.27	n.a.	−2.33	2.14	2.90	1.35	3.34	2.46	1.57

Employment growth rates (%)

	AUS	BEL	GER	DNK	SP	FIN	FRA	GBR	GRC	IRL
1	1.63	0.09	1.11	1.86	−1.75	0.24	0.90	0.49	1.64	2.33
2	1.61	−1.26	−0.10	0.90	−2.53	1.74	−0.09	−1.81	1.61	−0.39
3	0.80	0.94	1.19	1.45	3.90	0.67	0.72	1.48	1.59	−0.15
4	2.51	0.42	n.a.	−0.33	−0.51	−4.17	1.23	−0.75	1.34	2.81

Notes: Sub-period 1: 1975–9, Sub-period 2: 1979–84, Sub-period 3: 1984–9, Sub-period 4: 1989–94.

ISL	ITA	NLD	NOR	PORT	SWE	TUR	CAN	USA	AU	JPN
-1.42	-0.59	0.05	-0.07	-4.94	-0.22	6.81	2.04	n.a.	0.13	-0.13
-1.04	-1.45	-0.04	-0.15	-3.59	-0.01	0.00	-0.35	n.a.	-0.08	-0.14
-0.82	-0.65	0.67	-0.21	-1.35	-0.50	-2.19	-0.51	n.a.	-0.08	-0.03
-0.44	-1.43	0.00	-0.08	-2.81	-0.13	-1.10	-0.07	n.a.	-0.02	-0.08

ISL	ITA	NLD	NOR	PORT	SWE	TUR	CAN	USA	AU	JPN
1.19	-1.60	-3.41	-4.91	1.10	-3.98	-12.27	8.54	-0.46	-1.43	-2.79
-1.98	-3.56	-2.51	-3.63	-3.22	-2.38	-4.37	-2.26	-1.93	-3.22	-0.34
-4.44	-1.13	-0.10	-2.66	0.14	-0.73	2.61	-11.51	-1.17	-1.35	-1.25
-0.98	-0.10	-3.24	-0.71	-4.85	-2.76	-1.34	-1.10	-1.33	-1.08	-1.79

ISL	ITA	NLD	NOR	PORT	SWE	TUR	CAN	USA	AU	JPN
-0.63	0.64	0.20	1.67	-0.30	-0.65	-1.09	-5.35	0.37	3.64	1.07
1.07	1.06	-0.09	-0.07	0.70	-0.23	-2.62	0.01	1.21	-0.60	0.53
0.73	0.16	1.09	0.57	0.45	0.16	2.58	6.84	1.32	2.90	0.80
-1.39	-1.22	0.93	-0.50	2.35	-0.51	1.27	-0.44	0.28	1.13	-0.40

ISL	ITA	NLD	NOR	PORT	SWE	TUR	CAN	USA	AU	JPN
-0.40	-0.22	-0.07	-0.25	-0.06	0.18	-1.07	1.41	-0.15	-0.20	-0.08
-0.64	-0.49	0.31	-0.29	-0.43	-0.03	-1.02	-0.99	-0.26	0.03	-0.89
-0.22	0.28	0.01	-0.94	-0.37	-0.10	1.56	-1.56	0.09	-1.02	-0.04
0.29	-0.06	0.19	0.13	-0.42	-0.24	-0.36	-0.19	0.00	-0.59	-0.31

ISL	ITA	NLD	NOR	PORT	SWE	TUR	CAN	USA	AU	JPN
0.33	0.69	1.35	1.36	0.16	1.10	-1.67	-3.07	0.70	0.49	3.21
0.98	1.01	0.63	0.73	1.08	0.62	-1.40	-0.09	1.11	1.21	0.86
1.35	0.62	-1.08	1.13	0.87	0.41	0.37	5.88	1.40	2.80	1.19
0.30	1.91	0.32	0.25	2.01	0.77	1.48	0.31	-0.11	0.40	-0.20

ISL	ITA	NLD	NOR	PORT	SWE	TUR	CAN	USA	AU	JPN
2.98	2.02	1.58	3.19	3.23	4.06	-16.41	0.34	-0.96	-1.09	-1.59
1.13	2.92	0.83	3.89	4.75	1.81	-7.35	2.09	0.49	2.01	0.71
2.49	1.38	2.72	-0.17	2.04	0.12	-7.18	-3.35	2.71	-1.52	0.16
4.41	-0.49	1.09	2.73	-0.90	2.26	-3.66	2.13	1.40	0.68	0.91

ISL	ITA	NLD	NOR	PORT	SWE	TUR	CAN	USA	AU	JPN
2.58	0.34	1.19	2.74	0.42	0.83	3.22	3.58	3.87	0.48	1.53
2.73	-0.12	0.73	0.88	1.98	0.46	2.53	1.12	1.01	1.25	1.90
1.26	0.71	3.95	1.11	2.11	0.63	0.42	2.98	1.67	3.67	1.83
0.71	-0.67	1.23	0.29	0.64	-3.01	2.42	-0.04	1.84	0.20	2.24

NOTES

1. Financial support from the TSER project on 'Technology, Economic Integration and Social Cohesion' is gratefully acknowledged.
2. However, see for example Aghion and Howitt (1998, ch. 4), where the issue of unemployment and growth in a Schumpeterian framework of endogenous growth is discussed.
3. Of course, this is an oversimplified view which does not do full justice to either of these two strands of the literature, but we think that in essence this point can be made.
4. The theoretical (and empirical) discussion which comes closest to the theme of unemployment due to structural change is the discussion on 'technological unemployment'. This topic was first addressed by David Ricardo in his chapter 'On Machinery' (Ricardo, 1881, ch. XXXI) where he investigates the impact of sudden (capital-using) technological process. In doing so he sets up the first model of (what was later called) 'traverse analysis' and highlighted the view of technological unemployment as (transitory) 'capital shortage unemployment'. This kind of unemployment was seen as a medium-term phenomenon, until the capital stock had time to adjust. Ricardo's analysis gave rise to a lively set of contributions in the 1920s and 1930s (particularly by members of the 'Kiel school', such as Burckhardt, Neisser, Lederer and Lowe; see Hagemann, 1990, for an overview of these contributions). Later on two important theoretical contributions, one by Hicks (1973) and the other by Lowe (1976), again addressed this topic, using rather different modelling techniques.

   The spirit of these models is well reflected in a statement by Neisser: 'the capitalist process [is] a race between displacement of labour through technical progress and reabsorption of labour through accumulation ... displacement and reabsorption are two largely independent factors, and it is impossible to predict the outcome of the race between the two on purely theoretical grounds' (Neisser, 1942).

   In the neoclassical literature the problem of unemployment as (partly) derived from major sectoral adjustment problems in the economy as a whole is hardly addressed at all: unemployment is seen as primarily due to rigidities in the price/wage adjustment processes and to coordination problems in labour markets; these are modelled at the microeconomic level without any attention being paid to major developmental patterns at the sectoral level (see, for example, the comprehensive treatment in Pissarides, 1990).
5. Please note that in Panels A–C of Figure 7.1 the dimensions of the vertical axes are not equal.
6. For the EU-North, EU-South and Scandinavia, the average of shares is presented. EU-North consists of Austria, Belgium, Germany (if available), Denmark, France, Italy and the Netherlands; EU-South includes Greece, Ireland, Portugal and Spain, and the Scandinavian countries are Finland, Norway and Sweden.
7. This shift is partly due to a reclassification of certain activities formerly undertaken within manufacturing, and partly due to increased outsourcing of service type of activities (marketing, advertising, accounting, and so on).
8. Of course this can also be seen as a question of the appropriate length of the time horizon of the analysis.
9. For an analytical example of different trajectories, see, for example, Rowthorn and Wells (1987).
10. We have to note here that the indicator we use in this section depends on the choice of beginning and ending of the sub-periods.
11. Owing to data problems the sub-periods for Turkey were 1975–80, 1980–85, 1985–89 and 1989–94, and for the USA the last two sub-periods were 1985–90 and 1990–94.
12. No data available for the USA.
13. We do not report here the results for the absolute levels of the employment series, which are qualitatively more or less similar.
14. For a discussion of the various tests, see Hackl (1989) and Johnston and Dinardo (1997).
15. One could also take some political events (for example; change of government, EU-accession) or international treaties (such as the Maastricht treaty) as indicators for such breaks.

16. Unfortunately, we had to exclude some of the countries because of data problems. The most important ones are Germany and Canada, also Turkey and New Zealand. For some countries we had to include some missing values for one missing year. Although this could of course influence the results, we do not believe that the general pattern would change.

17. The spline regressions can be implemented as a dummy-variable approach; on this, see Johnston (1987).

18. Owing to data problems we had to exclude Australia from the data sample.

19. For this analysis we express the proportion of people employed in sectors 1–9 and of unemployed persons relative to the total labour force (that is, employed plus unemployed persons; not included are self-employed and unpaid family workers): $s_i^c = e_i^c / (\Sigma_i e_i^c + e_u^c)$ with $i = 1...9$ where e_i^c denotes the level of employment in sectors i in country c, and e_u^c the number of unemployed persons; s_i^c thus denotes the employment rate in sector i and s_u^c the unemployment rate.

20. t-values are in brackets. ***, ** and * mean significant at the 1 per cent, 5 per cent and 10 per cent level, respectively.

21. Four European countries (Austria, Denmark, Ireland and the Netherlands) show the negative relationship between unemployment rate and the growth rate in sector 9 at a much lower growth rate of sector 9. We therefore included an intercept dummy for these countries over the first period, which is highly significant.

22. Austria, Denmark, Ireland and the Netherlands were removed from the sample (see note 21).

23. As we shall see below, things are more complicated as shifts in demand towards sectors with lower productivity growth might also imply effects on real incomes which we shall track in the model below.

24. The model could be readily extended to more than two skill groups.

25. See, however, Landesmann and Stehrer (2000) where a more elaborate version of this type of model (including rents) is explored.

26. Ideally, we should expand the model to include a second services sector which could be much less skill-intensive (such as distribution and recreational services). However, the focus is on the welfare services sector which is relatively skill-intensive (see Landesmann and Pichelmann, 1998).

27. The welfare services sector here is assumed to only have direct labour inputs; however, this assumption could easily be dispatched without affecting the qualitative results.

28. This procedure avoids the emergence of an 'effective demand problem', which could arise as a result of the instability of the output system in our current formulation, as the largest eigenvalue of that system is zero in equilibrium.

REFERENCES

Aghion, P. and P. Howitt (1998), *Endogenous Growth Theory*, Cambridge, MA: MIT Press.

Baumol, W. (1987), 'Macroeconomics of unbalanced growth', *American Economic Review*, 53, (June), 941–73.

Baumol, W., S. Blackman and E. Wolff (1985), 'Unbalanced growth revisited: Asymptotic stagnancy and new evidence', *American Economic Review*, 75, (September), 806–17.

Baumol, W., S. Blackman and E. Wolff (1989), *Productivity and American Leadership: The Long View*, Cambridge, MA: MIT Press.

Chenery, H. (1960), 'Patterns of industrial growth', *American Economic Review*, 50, 624–54.

Chenery, H. and M. Syrquin (1975), *Patterns of Development*, London: Oxford University Press.

Clark, C. (1957), *The Conditions of Economic Progress*, New York: Macmillan.

Davis, S., J. Haltiwanger and S. Schuh (1997), *Job Creation and Destruction*, Cambridge, MA: MIT Press.

Fourastié, J. (1949), *Le grand espoir du XXe siècle*, Paris: P.U.F.

Glyn, A. (1995), 'The assessment: Unemployment and inequality', *Oxford Review of Economic Policy*, 11, (Spring), 1–25.

Hackl, P. (1989), *Statistical Analysis and Forecasting of Economic Structural Change*, Berlin: Springer-Verlag.

Hagemann, H. (1990), 'The structural theory of growth', in M. Baranzini and R. Scazzieri (eds), *The Economic Theory of Structure and Change*, Cambridge: Cambridge University Press.

Hansen, B. (1992), 'Testing for parameter instability in linear models', *Journal of Policy Modeling*, 14 (4), 517–33.

Hicks, J. (1973), *Capital and Time*, Oxford: Oxford University Press.

Inman, R. (1985), *Managing the Service Economy*, Cambridge: Cambridge University Press.

Johnston, J. (1987), *Econometric Models*, 3rd edn, New York: McGraw-Hill, ch. 6.

Johnston, J. and J. Dinardo (1997), *Econometric Methods*, 4th edn, New York: McGraw-Hill, ch. 4.

Landesmann, M. and K. Pichelmann (1998), 'Structural employment patterns and the welfare services sector', in M. Landesmann and K. Pichelmann (eds), *Unemployment in Europe*, London: Macmillan.

Landesmann, M. and R. Stehrer (2000), 'Industrial specialisation, catching up and labour market dynamics', *Metroeconomica*, 51 (1).

Lowe, A. (1976), *The Path of Economic Growth*, Cambridge: Cambridge University Press.

Neisser, H. (1942), '"Permanent" technological unemployment', *American Economic Review*, 32, 50–71.

Pasinetti, L. (1981), *Structural Change and Economic Dynamics*, Cambridge: Cambridge University Press.

Pasinetti, L. (1993), *Structural Economic Dynamics*, Cambridge: Cambridge University Press.

Petit, P. (1986), *Slow Growth and the Service Economy*, London: Frances Pinter.

Pissarides, C. (1990), *Equilibrium Unemployment Theory*, Oxford: Oxford University Press.

Poirier, D. (1991), 'The econometrics of structural change: A retrospective view', *Structural Change and Economic Dynamics*, 2 (2).

Ricardo, D. (1881), *On the Principles of Political Economy and Taxation*, 3rd edn.

Rowthorn, R.E. and A. Glyn (1990), 'The diversity of unemployment experience since 1973', *Structural Change and Economic Dynamics*, 1 (1), 57–89.

Rowthorn, R.E. and J. Wells (1987), *Deindustrialisation and Foreign Trade*, Cambridge: Cambridge University Press.

Shaked, A. and J. Sutton (1982), 'Relaxing price competition through product differentiation', *Review of Economic Studies*, 49, 3–13.

PART III

Changes in Organization and Distribution

8. New technologies, organizational change and the skill bias: what do we know?

Eve Caroli[1]

INTRODUCTION

Until recently, economists have known very little about the relations between technology, organization and skills. The current state of knowledge could be as shown in Figure 8.1 and could be summarized by: 'Better live on your feet than die on your knees'. Stated in a more academic way, a firm's performance (in terms of productivity and competitiveness) was to be improved by a high skill level of its workforce, a frontier technology and a flexible organization – with the concept of flexibility to be defined. The question of the relations between technology, organization and skills was not tackled and therefore set up a black triangle carefully avoided by scholars.

One reason for our poor understanding of this topic is that organizational issues have long been disregarded by economic theory. As underlined by Mowery (1990), standard neoclassical theory is a theory of markets. Its main focus has traditionally been on price formation rather than on the organization of production units. The firm is seen as a black box in which inputs are transformed into outputs according to a production function. Moreover, competition is assumed to ensure that all surviving enterprises use the same technology and face similar cost curves. Hence the internal structure is the same for all firms and the black box approach is thereby justified.

The study of organizations has thus been neglected by economists and left to other disciplines. Major advances have been made in sociology (Burns and Stalker, 1961; Woodward, 1965) as well as in management studies (Lawrence and Lorsch, 1967; Pugh, 1973). Scholars have characterized various forms of organization on the basis of detailed empirical analyses in the form of case studies. As we will see in the course of the chapter, they have provided enlightening insights concerning the evolution of organizational forms as well as their determinants. In doing so, they have explicitly tackled the relations between technology and organization.

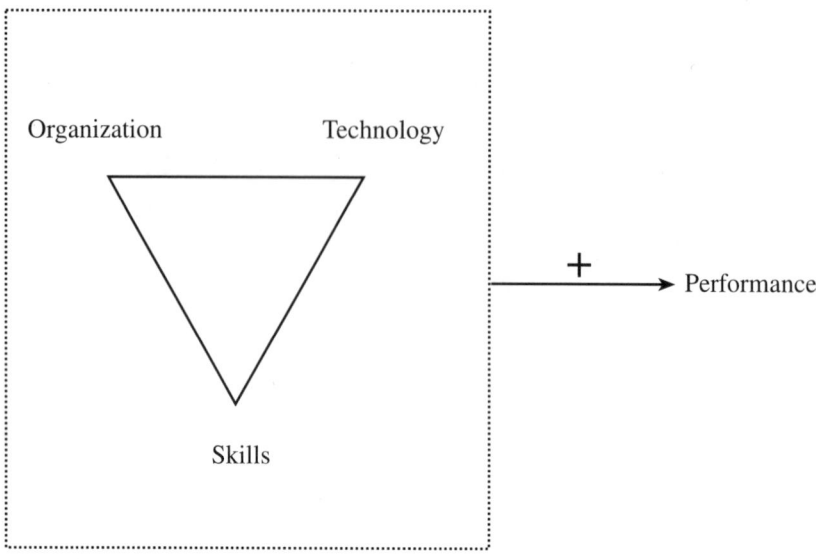

Figure 8.1   Technology, organization and skills: the black triangle

Despite the inadequacy of the standard neoclassical paradigm, an economic literature has slowly developed about firms' organization. The most influential contribution has been made by Alfred Chandler, whose seminal work, *Strategy and Structure*, was published in 1962. Chandler studied the rise of large-scale business enterprises between 1850 and 1920. He suggested that, during this period, a new economic institution was created: the multidivisional firm (M-firm) which replaced the former unified structure (U-firm). Functionally organized divisions were replaced by product-based ones, each of them containing functionally differentiated departments. This shift was accompanied by the rise of a new class of managers. They were both products and developers of the M-firm, which led Chandler to write that their 'visible hand' replaced Adam Smith's 'invisible hand' of market forces.

As far-seeing as it is, Chandler's account of the development of the M-firm is mainly historical (Pugh and Hickson, 1989). One of the few theories of the internal structure of the firm was proposed later on by Oliver Williamson (1975) who emphasized the critical role of transaction costs. Transactions may be conducted either through markets or hierarchies, the trade-off depending on related costs. In this view, the M-firm arose as a rational response to increasing transaction costs in the unitary structure. A second line of theorizing proposed a natural selection setting to analyse the dynamics of organizations. Hannan and Freeman (1977) emphasize that organizations are differentially selected for survival on the grounds of the quality of the fit between their characteristics and the environment. Organizations are hence treated as populations rather than individuals, and the mechanism underlying organizational change is based on variety creation, selection and adaptation. Recently, a more applied literature has developed on information and organization (Sah and Stiglitz, 1985; Milgrom and Roberts, 1990; Radner, 1993). However, until the early 1990s there was virtually no economic study of the relations between technology and organization and no work whatsoever has been done on the linkages between technical change, organizational change and the skill level of the workforce.

Lately, economists have shown a growing interest in this relationship, as a result of the spreading of new information technologies (ITs). Scholars seek to understand what could be the consequences of IT adoption in terms of economic performance and how best to implement these new technologies. Some empirical research has developed in the area. Part of it relies on case studies, many of which have been carried out by the NIESR[2] and compare German and British establishments (see various issues of the *National Institute Economic Review*). More systematic work has also been carried out on American, French and British data,[3] trying to identify the determinants of organizational change as well as its consequences, in particular with respect to skill requirements. A theoretical literature is also developing fast[4] which focuses on the dynamics of organizational change and its relations with human capital accumulation and technical change.

However, this line of research is still in a very early stage of development and much remains to be done before we reach a thorough understanding of the linkages inside the technology–organization–skill triangle. One reason for research not going faster in that direction is the lack of data. Statistical information about organizational forms and their evolution is still scarce, though rapidly growing both in quantity and quality. In France, for example, a number of surveys have been launched which are designed to provide detailed firm-level information on work organization, technologies, skill levels and paying schemes. Some data are already available (see TOTTO, 1987, 1993; REPONSE, 1992, 1998; Enquête COP, 1993) but panel information is still very scarce. One exception is REPONSE (1998), the exploitation of which has just started (see Coutrot, 2000).

The second and more fundamental problem raised by the relations between technology, organization and skills is that of causality. Does technology determine organization or is organization shaping technical choice? How do these two variables depend on the skill level of the workforce? This chapter aims to tackle these issues in a dynamic perspective. We will do so with reference to a precise case: the adoption of new information technologies and the organizational changes that have been going along with it. Getting a more precise view of the relations inside the so-far black triangle is indeed necessary to assess what kind of skills – if any – should be provided in order to ease current technical and organizational changes. As we will see, both are skill-biased but they do not require the same skills, which should be taken into account in the devising of education and training policies.

In our view, some light can be shed on the way causality runs between technological and organizational change by bringing together various types of knowledge built up by different disciplines. As we will show, economists can learn a lot from sociology and management. Taking their analyses into consideration will allow us to put forward a tentative explanatory framework for the way technical and organizational choices have been made by firms over the past two decades. This might look exceedingly simplistic in some respects. However, it permits us to open up the debate about the sequence of firms' decisions and provides a basis for discussion as well as empirical testing.

The chapter is organized as follows. Section 1 characterizes current organizational change and relates it to firms' characteristics, including technical choice. Section 2 builds up a tentative theoretical framework, underlying an asymmetry in the articulation of technical and organizational change in the past 20 years. Section 3 focuses on the last apex of the triangle: skills. We show that they are the key to a successful implementation of new technologies, but that this is as much a consequence of skill-biased organizational change as of skill-biased technical change. We finally analyse the kind of skills required both by technical and organizational changes and emphasize the need for general education and training.

1.   WHAT IS ORGANIZATIONAL CHANGE?

The body of new technologies that have developed over the past 20 years is now well identified by scholars. Whether they are called 'information technologies' (by economists) or 'advanced manufacturing technologies' (by sociologists), their content raises general agreement. They include: computer-assisted design and engineering systems, material resource planning systems, automated material-handling systems, robotics, numerically controlled machines and computer integrated systems. The spreading of these technologies has been paralleled by changes in firms' organization. Delayering, just-in-time and re-engineering are often mentioned, but to date no clear definition of what current organizational restructuring consists of has been provided. Before focusing on its relations with technical change, one needs to define the content and characteristics of current organizational change.

Organizational Change: from a Theoretical to an Empirical Definition

Three different approaches to organizational change can be found in the literature. One is based on ideal types mainly set up by sociologists. It defines current organizational change as a move towards organic structures. A second approach is historical and sees current changes as the end of the Taylorist or Fordist paradigms. The third approach is empirical and defines organizational changes with reference to their content.

The theoretical approach to organizational change

One of the first typologies of organizational forms was proposed by the Aston Group (see Pugh and Hickson, 1976). This group was composed of members of the Industrial and Administration Research Unit founded at Aston in 1961 by Derek Pugh. It included researchers in psychology, sociology, economics and politics, and its research agenda was on organizations structure. They conducted a large number of empirical investigations (case studies), and proposed that an organization should be analysed in terms of its degrees of (a) specialization of functions and roles, (b) standardization of procedures, (c) formalization of documentation, and (d) centralization of authority.

The first three elements determine the structuring of activities, whereas centralization determines the concentration of authority. The cross-classification of both yields the core of the typology proposed by the Aston Group. A high level of structuring of activities and concentration of authority defines *full bureaucracies*. A low level of both characterizes *non-bureaucracies*, with intermediate forms being labelled 'personnel bureaucracies' (low structuring–high concentration) and 'workflow bureaucracies' (high structuring–low concentration). In this framework, current changes in firms' organization could be defined as a move from full to non-bureaucracies. Indeed, as will appear below, they include task integration, more horizontal and informal communication, and some decentralization of decision making.[5] These tend to reduce both the structuring of activities and the concentration of authority.

A more straightforward vision of current changes in organization builds upon the seminal work by Burns and Stalker (1961) and would define organizational change as a shift from a mechanistic to an organic firm structure. Burns (a sociologist) and Stalker (a psychologist) initially studied the conditions for a successful introduction of electronics by traditional Scottish industrial firms. They defined two ideal types of management organization. The *mechanistic* type is adapted to relatively stable environmental conditions. Management tasks are broken into specialisms within which each individual carries out her well-defined task. There is a clear hierarchy of control, with responsibility for coordination and knowledge lying at the top of it. Vertical communication is emphasized and there is a strong insistence on obedience to supervisors' orders. On the contrary, the *organic* structure is adapted to unstable conditions, when new and unfamiliar problems continually arise so that it is impossible to distribute them among the existing specialists' roles. There is a continuous readjustment of individual tasks, and communication occurs at any level. Overall a much higher degree of commitment to the aims of the organization is generated than in the mechanistic structure.

According to Burns and Stalker, these two ideal types are extreme points of a continuum on which most organizations can be placed. Many recent sociological works analyse current organizational changes as a move along this continuum, away from mechanistic structures, towards more organic ones. Zammuto and O'Connor (1992) notice that organic forms are currently spreading and account for their taking over by their greater ability to adjust and react to a rapidly changing environment.

Although grounded on empirical observation, these approaches are in essence theoretical, in that they define organizational change as a move from one ideal type to another.

The historical approach
A more historical perspective has been taken by some economists and can be seen as a middle ground position between the theoretical and empirical views.

Cyert and Mowery (1987) draw a portrait of modern US manufacturing as it arose at the end of the 19th century. The work process was organized along the lines defined by Ford and Taylor. Capital equipment was highly specialized and workers performed a series of relatively unskilled and repetitive activities. Control over the pace of work and the structure of jobs belonged to management. Internal employment regulation was based on seniority rules and job classification, and industrial relations were quite adversarial. This organizational structure is now changing. Firms increasingly rely on external resources for administrative and support services so that 'downsizing' tends to spread. At the same time, collaboration increases among firms, both for product development and manufacturing. More responsibility is allocated to workers, who perform a wider range of tasks. Finally, much of the work control function formerly performed by supervisors and middle managers is vanishing as less hierarchical structures develop.

A similar historical analysis is carried out by Boyer (1991) for several OECD countries. Fordist principles were at the core of the 'old management style' at work since the end of World War II. They encompassed a deep (and deepening) division of labour, centralization of decision making, a high degree of mechanization taking the form of highly specialized equipment, and mass production of standardized goods. This old way has been strongly challenged since the 1970s and a new model of management and work organization started to emerge in the 1990s. This is characterized by a weaker division of labour going along with functional integration, some decentralization of decision making, the development of networking and long-run subcontracting between firms, and the production of small batches of customized goods.

All historical accounts are highly convergent and draw a picture of 'reorganized firms' which is consistent with the one provided by theoretical approaches. Current organizational changes would consist of (a) changes in work organization in the direction of more decentralization of decision making – in particular towards lower layers of the hierarchy – more task integration and less division of labour inside firms, and (b) changes in firms' structure, including less functional differentiation, more outsourcing and more collaboration among firms (networking). Most of these features of organizational change also come out of systematic empirical analysis.

The empirical approach to organizational change

This approach relies on statistical methods and is hence heavily constrained by the (reduced) availability of data on organizational forms. In the following we will refer to a disproportionate number of studies carried out on French data. This is because they have been widely used to characterize the content of current organizational changes, in contrast to US data, which have been used more to study the technology–organization relationship (see below).

As emerging from empirical work, new forms of organization are characterized by a reduced role for hierarchy, more collective work, task/work enlargement, including more horizontal communication, *but* a somewhat lower degree of workers' control over their own activity (see OECD, 1999).

The lessening of the role of hierarchy arises from a series of new work practices. Decentralization of decision making is taking place both between and within production units.[6] Units coordinate their own activity: there is no longer much central planning inside firms. Within units, operatives are given more responsibility[7] and a greater autonomy,[8] in particular for the management of breakdowns and quality controls. This is often accompanied by delayering.[9] A second characteristic of new organizational forms is their strong emphasis on collective work. Whatever form it takes (work teams, quality circles and so on) it is spreading in the USA, the UK and in France.[10] The enlargement of jobs content is the third feature of new organizational forms. Workers rotate among jobs,[11] but, more importantly, in a given occupation, they perform a greater variety of tasks.[12] Functional integration[13] as well as total quality management[14] also contribute to the enlargement of the actual content of jobs.

A greater intensity and variety of communication channels also character-ize 'reorganized firms'.[15] Although workers' autonomy tends to increase as they get more responsibility, a strict prescription of how tasks are to be carried out still persists in many cases.[16] This is referred to as 'controlled autonomy' by Coutrot (1996). Finally, 'reorganized firms' display tighter links with their suppliers and clients.[17]

Moreover, these new work practices tend to be adopted in clusters. In their study of US steel production lines, Ichniowski *et al.* (1997) show that the adoption of human resource management (HRM) practices such as teamwork, flexible job assignment, and intensive labour–management communication are highly correlated. They consider eight different HRM variables and show that two-thirds of the steel lines in their sample are distributed over only four different combinations of these variables. New organizational practices are thus to a large extent complementary, as already pointed out by Osterman (1994).

The empirical approach to organizational change hence draws a precise picture of new forms of organization. The main features are consistent with most of the characteristics emphasized by the theoretical and historical litera-ture. This convergence allows us to borrow from Burns and Stalker (1961) and define current changes in organization as a move towards more *organic* structures characterized by the following:

- a flatter hierarchy – that is, more horizontal organizations – with more responsibility being transferred downstream,
- the development of team working,
- a greater variety of tasks and/or jobs being performed by workers,
- more intensive communication,
- tighter links with other firms and clients.

This definition provides a basis for the analysis of the relations between technical and organizational change. Before going ahead in that direction, one needs to assess how organizational changes correlate with other firms' characteristics and choices, including technological innovation. This will allow us to demonstrate that technical change and organizational change go together, and will then lead us to propose an explanatory framework for their relations.

Reorganized Firms: What do They Look Like?

Towards an ID of reorganized firms
A growing number of studies have investigated the characteristics of reorgan-ized firms. Many focus on the determinants of new work practices, whatever

the date on which they have been introduced in the firm, while a few tackle the process of organizational change itself. The former focus on firms' organization once changed, while the latter aim at uncovering the factors enhancing the introduction of such changes. Both allow us to characterize the main features of firms whose organization is comparatively more organic – according to the definition given above. All studies focus on work practices and organization, rather than on firms' structure and relationships with partners. The reason for this is to be found in the nature of available data.

A number of US data sets provide a wealth of information regarding new work practices as well as other firms' characteristics. This makes it possible to characterize 'reorganized' enterprises. This method was first used by the British Aston Group, which showed that size was a crucial determinant of the organizational structure, with large firms displaying comparatively higher degrees of specialization, formalization and standardization. Lincoln *et al.* (1986) also emphasize the impact of size on work organization. They check whether differences in organization across US and Japanese firms arise from economic or cultural factors. Their conclusion is that cultural factors are important, but do not account for all differences. Economic factors play an important role. They include firm size (large firms tend to have more quality circles), unionization, which enhances formalization, centralization of decision making and technology. Workflow rigidity increases functional specialization, and small batch production reduces the span of control of hierarchy.

Osterman (1994) uses US establishment-level data to study the characteristics of firms that have implemented some – or all – of the following work practices: self-directed work teams, job rotation, employee problem-solving groups and total quality management. Here, again, size appears as a key factor: large establishments are more likely to implement the new practices, with a similar correlation holding at the firm level. Demand conditions also arise as an important determinant. Establishments facing an international market tend to be more innovative than those serving national customers. So are establishments which have adopted a 'high road strategy' in terms of product quality and human resource management. Indeed, profit-sharing schemes and management's efforts to raise workers' commitment have a positive influence on the introduction of any of the four practices.

Ichniowski *et al.* (1997) focus on 36 US steel production lines and study the determinants of the adoption of 'high-performance' work practices. Here, again, firms facing severe competition are found to be more innovative than average with regard to organization. They are also less unionized. An interesting feature is that they display shorter tenure for both managers and workers, thus indicating that the introduction of new work practices tends to be resisted by employees who have long been in the firm.

Studies focusing on organizational change itself put forward similar factors. Using the British WIRS (1984), Machin and Wadhwani (1991) emphasize the role of unionization. Union recognition had a positive influence on organizational change over the period 1981–84. This paradoxical result is accounted for, by the authors, on the grounds of Margaret Thatcher's reforms, which seriously undermined the position of unions, thus allowing unionized firms to catch up with their competitors in terms of organizational change. This result is confirmed – on the same data set – by Caroli and Van Reenen (2001) for 1981–84 and 1987–90, implying that the impact of full power unions on organizational change used to be negative. Using both waves of WIRS, Caroli and Van Reenen also uncover the positive influence of size and demand conditions on organizational change. The same variables also have a positive and significant impact on the flattening of hierarchy (delayering) in French establishments.[18]

Overall, a consistent picture of reorganized firms comes out of the literature. Factors playing a prominent role in the shaping of workplace organization are size, demand conditions, unionization, human resource and quality strategies. Given endogeneity problems in many studies, they cannot yield more than an ID of firms that have introduced new work practices. They are usually larger than average. They face dynamic (or very depressed) demand conditions, thus implying that organizational change is a radical decision and does not come down to fine-tuning adjustment. They have weak unions and have adopted a 'high road' strategy emphasizing the quality of output and human resource management.

New work practices and technological intensity

There is widespread evidence in the literature that information technologies and new work practices are close complements. Using the French dataset REPONSE, Coutrot (1996) displays a strong correlation between technical change and organizational change. Establishments that have introduced new work practices use numerically controlled systems more frequently than average. Similarly, establishments with quality circles use more robotics than average. Similar results are obtained on other French data. Technological innovation – whether radical or incremental, product- or process-oriented – is highly correlated with organizational change, as captured through the intensification of communication and the slackening of hierarchical constraints.[19] Similarly, reorganized firms prove to be using more technically advanced equipment than average – in particular, more robotics and numerically controlled systems.[20]

These results prove robust on US data. Brynjolfsson and Hitt (1998) show that the use of ITs strongly correlates with (a) structural decentralization, that is, the use of self-managing teams, employees' involvement groups and broad job classification, and (b) individual decentralization, that is, more discretion being given to line workers for the methods used in their work. The stock of IT capital and the number of PCs per firm are also correlated with a global measure of decentralization even when controlling for size, industry and workforce composition. This result holds when estimating a full demand equation for ITs. Brynjolfsson and Hitt therefore conclude that information technologies and new work practices do cluster together.

At this point of the analysis we know that current organizational changes lead to more organic firms, with these being characterized by flatter hierarchies, more collective work, enlarged job content and tighter links with other firms and customers. Moreover, organic firms display common features which include a high technological intensity.

Organizational change and the adoption of new technologies thus appear to go hand in hand. In the literature, this assessment is often considered the end of the story. We will take it as a starting point and try and investigate the linkages between the two variables. The next section builds upon work carried out in sociology and management, as well as economics, to propose an explanatory framework for the organizational–technical change relationship.

2. TECHNICAL CHANGE AND ORGANIZATIONAL CHANGE: A TENTATIVE THEORETICAL FRAMEWORK

Current organizational and technical changes[21] go together. Hardly anybody would deny this statement. However, theoretical analyses of the way the two variables relate to each other are difficult to find in the literature. Scholars are usually extremely cautious about a definitely complex matter. We take here a more daring route and put forward a tentative theoretical framework depicting the sequence of technical and organizational choices at the firm level. Our ambition is limited, however, in that we will focus on current technical and organizational choices, that is, those choices leading to the spreading of IT and organic forms of workplace organization. Our theoretical attempt is thus limited in scope, space and time. Despite these limitations, it is in our view necessary, in order to improve our understanding of current microeconomic changes. It also bears important consequences for the devising of adequate training policies, at the macroeconomic level.

In this section, we will first provide a general view of our framework, summarized in Figure 8.2, before coming back to its building blocks and tackling, in particular, the issue of causality.

Organizational and Technical Change: the Case for Coevolution

The main line of our argument is extremely simple. Technological and organizational changes coevolve. Their dynamics do influence each other but are not entirely determined one by the other. This partial independence turns firms' expectations into the main stabilizing factor in the model which may yield persistent mismatch between technologies and organization.

Figure 8.2 provides a general view of our line of argument. At one end of the process, the adoption of new technologies is constrained by the existing set of technological opportunities. It defines the technological frontier which is likely to be modified at any point in time. When adopted, new technologies generate a potential for productivity gains. Here again is a frontier, which determines the maximum increase in productivity which can be expected from implementation. In turn, this potential contributes to shaping actual productivity gains which are finally obtained by firms. The above technological dynamics proves largely independent of that of organization. As we will show below, the feasibility of organizational change – in particular when involving a move towards organic structures – is to a large extent conditional on labour–management relations, as well as cultural and institutional factors. In turn, organizational change contributes to determining actual gains in productivity that can be obtained from the adoption of new technologies.

This sets out the case for coevolution of organizational and technical change. Their dynamics are different but linked through their common impact on actual productivity gains. Here lies the main source of potential instability in the model. The adoption of a technology (or a range of them) creates a need for specific forms of organization and hence, very likely, for organizational change. Indeed, as we will argue below, the actual increase in productivity arising from the choice of a given technology varies considerably according to work organization inside firms. In other words, the gain from the introduction of a specific technology is maximized when it is supported by adequate organizational forms. However, nothing ensures that organizational change will exactly fit the requirements of technical change. Some feedback connections contribute to reducing instability through the direct effects of IT upon organizational change. However, these feedbacks are unlikely to be strong enough to ensure that any technological requirement in terms of organization will be fulfilled at any time. There is no mechanism of automatic equilibration.

In this context, firms' expectations play a major role. They are the only way through which organizational and technical change can be brought in line. Firms will choose whether or not and to what extent they will adopt new technologies on the basis of expected returns. Actual returns directly depend upon organization. So the adoption decision will be taken by firms on the

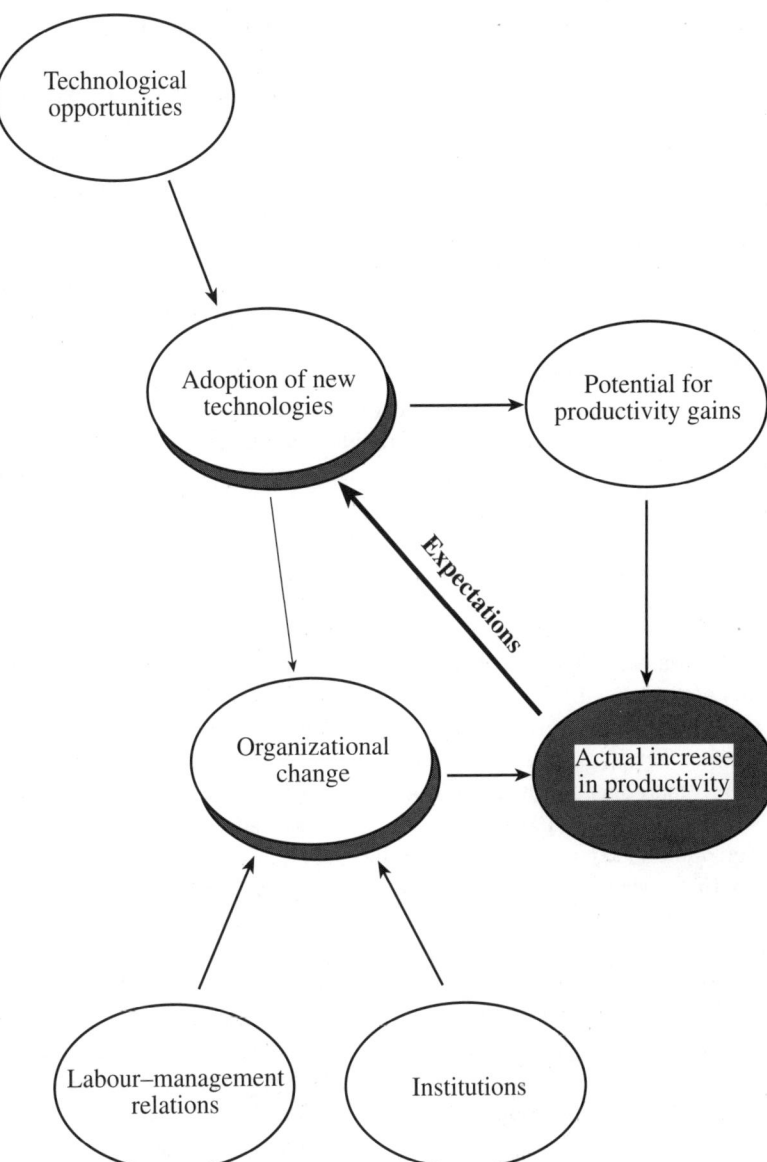

Figure 8.2   Organizational change, technical change: a theoretical framework

grounds of what they expect to be able to do with regard to organizational adjustment. If expectations are perfect, technical and organizational change will automatically fit each other. Firms with adversarial labour relations and a non-supportive institutional environment will expect organizational adjustments to be difficult. They will therefore be reluctant to adopt new technologies. In such conditions, the expected return is low indeed. On the contrary, firms which expect organizational change to be somewhat easier will be more likely to take advantage of technological opportunities. If ever firms' expectations are imperfect (owing to a lack of information or bounded rationality), some mismatch may occur. This cannot be ruled out, given the multiplicity of determinants of organizational change. As stressed above, its feasibility is influenced by factors which are not entirely under firms' control (industrial relations, institutions). As a consequence, their evolution is particularly hard to predict, in particular in a period of rapid economic change.

Now that we have sketched our theoretical framework, let us come back to its building blocks and linkages.

Organizational and Technical Change: the Building Blocks

The core of our model lies in the idea that technical change has a decisive impact upon organizational change. Stated in a static way, this relationship is highly controversial. The idea that organizations might be shaped by technology has been fiercely criticized in the literature. Noble (1984) sharply points to the fact that 'the appearance of automaticity and necessity (in this relation) is false, a product ultimately of our own naïveté and ignorance'. Strauss (1978) and Goffman (1983) stress that the main features of organizations are the outcome of continuously undergoing negotiations and do not derive directly from technological considerations. Similarly, Davis and Taylor (1976) underline the point that organizations embody important social choices and cannot be reduced to a pure technological output. Even Williamson (1975) denies that technology might influence firms' structure: this could only be the case if one single technology should be decisively superior to all others, and if that technology should require one single organization. This being highly unrealistic, Williamson dismisses the idea that organizations might be shaped by technology. His position is quite extreme, but the literature we have just mentioned does question the influence of technology upon organization.

We argue here that this debate arises from the static view adopted by a number of scholars. In a more dynamic perspective, one can show that, though organizational change is far from being entirely determined by technological considerations, the *adoption* of a technology raises some needs in terms of organization, and that, in the end, organizational *change* largely takes place *in response* to technical change. In the present case, the adoption of new forms of workplace organization mainly arose in response to the spreading of IT. However, this response is non-optimal, leading to a dynamic process of the kind, 'I love you, me neither.'

Organizational change as a response to technical change

The recent development of IT both calls for and permits new forms of workplace organization. To that extent, it acts as an impulse to organizational change. According to Zammuto and O'Connor (1992), ITs yield two potential benefits. Productivity gains derive from the incorporation of routine tasks into hardware and software which lowers direct labour costs and work-in-process inventories, and increases machine utilization. In parallel, flexibility benefits arise from the electronic integration of the production process. This shortens lead times and makes it feasible to produce small batches of highly customized goods. Following the authors, flexible (organic) structures make it possible to capture the whole of these benefits, whereas control-oriented (mechanistic) ones do not do such a good job. The latter enable the securing of productivity gains, but not flexibility gains, which cannot be secured when decision making is centralized in the hands of middle management because of the subsequent rigidity in the production process. Moreover, the handling of breakdowns – more frequent with fragile high-tech equipment – is slowed down by the hierarchical structure, thus setting at risk productivity gains themselves. In contrast, an organic structure makes it possible to get both productivity and flexibility gains, thus reducing the risk of failure in the implementation process. Workers and clerks can diagnose problems and suggest solutions so that problem solving is much quicker than when systematic referring to the hierarchy is compulsory. Moreover, firms react more rapidly to unforeseen events and can thus benefit to a larger extent from the integration of the production process. In this view, the full benefit from the introduction of new technologies is more likely to be captured if moving to an organic structure, that is, if introducing adequate organizational change (Thesmar and Thoenig, 2000).

Hage (1997) reaches the same conclusion: organic structures are more effective than mechanistic ones at implementing new technologies because they provide a higher degree of communication and because the management of technology does not belong to a small class of technicians. A greater number of employees are aware of the best way to handle new equipment. As stressed by Greenan and Guellec (1994b), decentralization of decision making and intensification of communication are necessary in order to use new technologies efficiently. Bresnahan *et al.* (2001) provide empirical evidence on that point. They estimate a Cobb–Douglas production function with physical capital, labour and a measure of IT capital. All three factors are significant and remain so when a measure of decentralization of workplace organization is introduced. More interestingly, when IT and organization interact and are introduced into the regression, they have a strong and significant effect on productivity. Decentralized firms tend to get a higher output from their IT investment. So the development of information technologies calls for new work practices in order for the return on technological adoption to be maximized.

To some extent, at the same time as the development of technological opportunities calls for new forms of workplace organization, it makes them possible. Old technologies used to generate rigidities, thus supporting the centralization of decision making. In contrast, ITs are flexible. They do not constrain so much the production process, thus permitting some decentralization of decision making. Similarly, they embody performance control in hardware and software, thus making it possible to delayer and reduce the role of hierarchy (Greenan and Guellec, 1994b). Moreover, ITs reduce the cost of lateral communication and/or increase the capabilities of line workers to perform information-processing tasks. This reduces the benefit from hierarchical decision making, thus making it more profitable to decentralize authority (Radner, 1993; Van Zandt, 1997). From the point of view of both technical and economic efficiency, the newly available information technologies constituted a strong impulse to the spreading of more organic forms of workplace organization.

Sociology as well as management theory teach us that, to some extent, organizations do respond to technological impulsion. This view was first expressed by the contingency theory (Lawrence and Lorsch, 1967) which states that (a) there is no one best way to organize, (b) any way of organizing is not equally effective, and (c) the best way to organize depends on the nature of the environment to which the organization must relate (Galbraith, 1973). One of the key elements of this environment is technology, which is defined by its complexity/diversity, that is, the number of different items that must be dealt with by the organization, by its degree of uncertainty/unpredictability and by the interdependence of the items involved in the work process. As technology evolves – with the development of IT – diversity, uncertainty and interdependence tend to increase, thus posing increasing information-processing demands on the organizational structure. The challenge is then to select the structural arrangement appropriate for the information-processing requirements of the task to be performed. The underlying assumption of the contingency paradigm is that organizations with internal features best matching the demands of their environment are more effective, hence more likely to survive.

This view is close to the one taken by the natural selection approach. It underlines that environments differentially select certain types of organizations for survival on the basis of the fit between organizational and environmental characteristics (Hannan and Freeman, 1977). Both the contingency and selection paradigms emphasize that changes in organization take place in response to changing requirements arising from the technological environment.

This view is supported by the history of past and present organizational changes. Cyert and Mowery (1987) recall that the organizational structure of modern US manufacturing arose at the end of the 19th century in response to innovations in the production process. This is also Chandler's view. According to him, the transition from the U-firm to the M-firm could occur only in response to external pressure. It took the form of expanding urban markets and new mass production technologies for manufactured goods. In this environment, the M-firm took over the U-firm. In line with this argument, and despite Williamson's strong opposition, the transaction cost theory also accounts for the development of multidivisional firms on the grounds of their greater ability to manage international transfers of technology (Scott, 1990). Finally, the current spreading of 'empowered', 'decentralized' work practices demonstrates that organizational change does react to the impulse of information technologies. Depending on the country, the period and the exact work practice that is considered, the proportion of firms having implemented some organizational changes ranges from 30 per cent to 60 per cent. As stated by Aoki (1990): 'The tendency towards the delegation of decision making to the lower levels of organisational hierarchies, where economically useful on-the-spot information is available, as well as the non-hierarchical communication among operating units, is becoming a more discernible phenomenon on a world wide scale, wherever conditions permit.'

Overall, the successful implementation of IT requires specific organizational changes, namely the implementation of new work practices corresponding to more organic forms of workplace organization. So far, organizational change appears to have responded, at least to some extent, to that technical impulse. The next question is then: is this response optimal? Does organization fully adapt to technological requirements? To these questions, we will answer *no*, because technical and organizational evolution are partly independent.

Organizational change as a non-optimal response to technical change
The actual implementation of new work practices is far from being entirely determined by technological considerations. Firms' decisions in this matter also depend on the quality of labour–management relations as well as on cultural and institutional factors.

Labour–management relations are crucial to the implementation of organizational changes, in particular when these are to go in the direction of more organic structures. The very implementation of changes has to be negotiated between unions and management. This negotiation is, of course, considerably tougher when labour–management relations are highly suspicious. This is well illustrated by Carr (1994) who studied the introduction of team working at Vauxhall Motors at Luton, UK. The negotiation process lasted for more than two years because unions saw team working as the last (and smartest) attempt made by management to undermine their position. Beyond the introductory phase, employees' involvement is crucial when moving to organic structures. Workers are to get more responsibility and perform a wider range of tasks. They also have to be part of communication networks within the firm, as well as workgroups of various types. In such an organization, their commitment is a key component of success. Osterman (1994) stresses that human resource management practices are crucial to the building of trust. He underlines the role of insider preference as opposed to external hiring, pay systems rewarding workers for success (profit and gain sharing) and job security as ways to enhance employees' involvement. Boyer (1991) goes one step further and argues that a higher commitment cannot be secured without an explicit advantage to wage earners.

So the extent to which work organization can actually and successfully be changed heavily depends on the quality of industrial relations inside firms. It can be improved by appropriate human resource management, but remains to a large extent dependent on relations with unions which are – obviously – not under perfect control of the firm. As Burns and Stalker (1961) put it, an organization is not a pure formal authority system. It is also a cooperative and political structure.

A second determinant of the feasibility of organizational change is firms' cultural and institutional environment. As already mentioned, most authors who have dismissed the influence of technology upon organization have done so on the grounds that organizations are also shaped by a large series of other factors. Many insist on the social dimension of organizations. As an illustration, Perrow (1967) puts forward the radical change which took place in the organization of mental hospitals as the way in which society used to see mentally ill people evolved. Organizations are thus partly shaped by their social and cultural environments. In a more economic perspective, Scott (1990) underlines the point that institutions play a key role. He defines them as a set of rules and regulations, belief systems and legal frameworks and argues that they are determining factors of organizational change. Changes which would be desirable from firms' point of view may then be slowed down or even stopped by unsupportive institutional factors. Lincoln *et al.*'s (1986) results support this view. Part of the organizational differences across US and Japanese firms are due to country-specific factors. In particular, it emerges that technology plays a much smaller role in the shaping of work organization in Japan than in the USA. The authors account for this result on the grounds of a unique set of institutions and a strong cultural tradition in

Japan. This is also the view taken by Gerwin and Tarondeau (1984), who conclude that political and institutional factors play a major role in the devising of work practices, on the basis of a comparative study of US and French motor plants. Finally, Boyer (1998) strongly emphasizes the need for a supportive institutional set in order for the new mode of organization to survive.

Labour–management relations, and cultural and institutional determinants thus contribute to generating a dynamics of its own for organizational change. Given that it is independent from technical change, there is no mechanism ensuring that the global dynamics will be stable.

Back to Coevolution

Our statement of technological and organizational coevolution can be summarized as follows. First, individual firms decide on the adoption of new information technologies, given their expectations on future returns. This is conditional on what they expect to be able to do in terms of organizational change, knowing that it is strongly influenced by factors they do not control. Second, when ITs are adopted, firms seek to adjust their work practices towards more organic forms of workplace organization. In doing so, they are constrained by external factors (relations with unions, cultural and institutional environments) so that organizational change ends up exactly in line with technical change if, and only if, expectations have been perfect in the first step.

At this stage, our coevolution statement is but a theoretical hypothesis, with the idea that technological impulsion has been crucial for organizational change, lying at the core of it. In the present case, this seems particularly likely, given the different time horizons of both types of changes. The adoption of ITs has been a very rapid phenomenon. As underlined by Brynjolfsson and Hitt (1998), the share of computers in US real capital stock has displayed a hundredfold increase since 1970. In 1987, there was one PC for 30 employees in large US firms, as compared with one for six in 1994. The diffusion of new information technologies has thus taken place in a short time. In fact, one could say that, in order for new technologies to be adopted, managers just need to order the corresponding equipment. In contrast, the time scale of organizational changes is long. This is mainly due to very strong organizational inertia, in particular when organic forms of workplace organization are at stake.

A number of theoretical works have emphasized that, when the new work practices to be implemented are complementary to one another, organizational inertia is bound to be important. Milgrom and Roberts (1990) show that, in such circumstances, organizational choices are discrete, which prevents firms from moving slowly from one organization to the other. Athey and Schmutzler (1995) go one step further and show that, when organizational policies are complementary, the profit function exhibits multiple local equilibria. Local searches involving changes in one or two policies are therefore likely to be detrimental to profit. As a consequence, incumbent HRM practices are likely to persist, especially if the firms' information about the performance of new policies is incomplete. At the empirical level, Ichniowski and Shaw (1995) provide some evidence of organizational inertia. When estimating an equation with a combination of new work practices on the left-hand side, the year when the firm was set up comes as a key determinant. The newer the plant, the more decentralized the work practices in use. Consistently, when regressing the current number of new work practices on its past value, the authors find a coefficient close to one. Although the number of new work practices has increased over time, the predominant mechanism for their adoption has been the birth of new plants. Within plants, practices do not change very much.

Firms' organization thus appears to be quite inert. This is no surprise in a context in which a whole cluster of new HRM practices need to be implemented at once in order for them to be beneficial. Moreover, the new work practices require workers' involvement, which requires in turn that their implementation be negotiated. Overall, the combination of a strong organizational inertia and a significant correlation between IT and new 'decentralized' work practices supports the idea of a technological impulsion behind organizational *changes*. This does not prevent a subsequent retroaction of decentralized work practices upon technological adoption.[22] However, in such a relation, organization is assumed to be given. In a dynamic perspective, organizational changes appear to be fostered by technological change.

This sequence of decisions should definitely be tested empirically. Given that our statement deals with *changes* in technology and organization, rather than with levels of the corresponding variables, empirical testing requires dynamic data. Some panel data exist for technology indicators, but this was not the case with organization until recently. Consequently, the debate about the sequence of choices between technical and organizational choices is still open.

Though limited in its present shape, this debate is worth having, in that it puts organizational change to the fore. So far, this has been hidden behind technological considerations, in particular in the literature related to the skill bias. However, it could very well be the case that the two types of change do not require the same amount, or the same kind, of skills. If this were to be the case, disentangling the technology–organization–skill triangle would be a prerequisite to a proper devising of education and training policies.

3. SKILLS: LAST BUT NOT LEAST

In the description of IT adoption decisions given above, skills are impressively ... missing! However, they come out as the bottom line of success. Those skills we consider here are all capacities acquired or enhanced in the education and training system. Employees may obviously embody innate abilities, or abilities acquired in their family environment. In this section, we will focus on those capacities that are likely to be enhanced by appropriate education and training policies. We call them 'skills' in what follows.

Skills as the Bottom Line of Success

The large number of comparative case studies carried out by the NIESR widely display that the education and training (that is, the skill) level of the workforce is *the* key determinant to a successful implementation of new technologies.

Most of the studies compare German and British plants in similar industries and try to account for the wide differences in their technological strategies. Steedman and Wagner (1987) study the wood furniture industry (production of fitted kitchens) and observe a huge difference in the type and technological intensity of equipment used in the German and British samples (respectively, nine and eight plants). The production of fitted kitchens is a highly sophisticated industry in Germany. All plants use linked machinery and numerically controlled equipment. This allows them to produce a wide variety of products and to shorten delivery periods. Production is then more flexible in the sense in that it adapts more rapidly to consumers' needs. It also relies on lower manning levels, thus enhancing productivity. In contrast, British plants still employ large numbers of unskilled operatives to feed and unload machines. They refuse to implement linked machinery, arguing that this would hamper flexibility, that is, in their view, the ability to interrupt the production process at any time. This is indeed necessary to rush through special small batches that have not been planned, and to cope with numerous machine malfunctions and breakdowns. The two strategies yield very different results in terms of productivity with British plants lagging behind German ones by some 60 per cent.

Steedman and Wagner (1989) and Mason *et al.* (1994) also display large cross-country discrepancies in technological strategies in the clothing manufacturing and food processing (biscuit) industries. In both cases, German firms produce high value-added goods. They cluster at the top end of the quality ladder, producing small batches of highly differentiated products. In contrast, British plants still produce long runs of standardized goods. In the clothing industry, this ends up with British products being sold at half the price of German ones. This strategy could prove unstable in the middle run, given the harsher competition of low-wage countries at the bottom end of the market.

Another example is provided by Mason and Wagner (1994) who display higher innovativeness in German than in UK engineering, leading to faster productivity growth in German plants. According to the authors, this is due to a much stronger link between the shopfloor and technical support departments in Germany, which permits a steady flow of incremental innovations.

Such examples could be multiplied to infinity. In all cases, the low-tech route taken by UK firms is attributed to the low skill level of the British labour force. In the wood furniture industry, German plants display 90 per cent of craft employees with a three-year vocational qualification, as compared to less than 10 per cent in British plants. Similarly, all German foremen have a basic craft qualification, whereas British ones usually do not, and prove technically not so much at ease as their German counterparts. According to Steedman and Wagner (1987), the comparatively low skill level of British operatives and foremen accounts for the low-tech strategy of firms. Linked machinery requires a flexible organization in order to be effective and this cannot be achieved when skills are scarce on the shopfloor. Moreover, the introduction of high-tech equipment requires some technical understanding by foremen which, again, requires a minimum amount of skills. When this is lacking, a low-tech strategy comes out as a rational choice. In the clothing and food processing industries, German plants can produce high value-added customized goods thanks to a well-trained labour force. In biscuit manufactures, apprentice-trained bakers can produce top-quality grades and Meister-trained supervisors can collaborate systematically with engineering departments on the incremental innovation process. In contrast, the low supply of skilled operatives and maintenance personnel creates a serious obstacle to UK manufacturers moving towards more differentiated products. Hence they rationally specialize in the production of long series of standardized goods. The same goes for engineering. Highly-skilled German supervisors can communicate easily with technicians, thus enhancing continuous process innovation. UK foremen cannot do the same owing to their low level of technical understanding.

Overall, NIESR studies show that differences in the adoption of high-tech equipment and 'high road' quality strategies across UK and German firms are due, to a large extent, to differences in the skill level of the labour force that is available. Therefore skills appear as the bottom line of success in implementing new technologies. Whether firms dispose of a highly-educated, well-trained workforce will largely determine their choosing of a high-track or low-track technological road.

This is usually interpreted in a purely technological way: new equipment would require higher-skilled workers. In other words, technical change would be skill-biased. A large econometric literature has developed on this issue and accounts for the rising skilled–unskilled wage gap and the upgrading of the in-firm occupational structure on the grounds of technical change. If the relative price of skilled workers is steadily going up (as has been the case in the USA and the UK over the past 20 years) despite a regular increase in supply, it must be the case that the demand curve is shifting out. Macroeconometric studies show that the major part of this shift is due to intrasectoral rather than intersectoral reallocations and that the rise in the share of skilled workers – both in employment and in the wage bill – is correlated with indicators of technological change (see Berman *et al.*, 1994; Machin and Van Reenen, 1998). At a more micro level, the skilled–unskilled wage gap appears to be correlated with investment in R&D and with the use of technically advanced equipment (Krueger, 1993; Chennells and Van Reenen, 1997). Here, again, that IT requires a higher skill level in the workforce often comes out as a natural conclusion.[23]

However, to date, none of these studies controls for organizational changes. It could therefore be the case that the source of rising skill requirements is organizational as much as technical change. If this were to be true, a large part of what is currently seen as skill-biased technical change would actually be skill-biased organizational change. This would be important for empirical studies, implying that the skill bias on technical change is systematically overestimated in the microeconomic literature. Moreover, it would be practically very important if organizational and technical change did not require identical skills.

Organizational Change, Technical Change and the Skill Bias

In this section, we will take a completely agnostic view and investigate the skill requirements arising both from IT adoption and from the spreading of a more organic firm structure.[24]

The 'upskilling/deskilling' effect of new technologies has been much debated in the sociological literature over the past 15 years, with economists lately tackling the issue. All suggest that the impact of technical change upon jobs content (and related skill requirements) is actually ambiguous. In the short run, the introduction of new technologies calls for higher-skilled employees because these have a comparative advantage in dealing with unforeseen events. This argument was first put forward and tested by Bartel and Lichtenberg (1987). According to them, when a technology is new, it is to a large extent unknown, which generates a high level of uncertainty. Given that skilled workers have a comparative advantage both in learning and in coping with uncertainty, skill requirements increase in the early phase of implementation. As experience on the technology grows, their comparative advantage decreases. In this view, the skill-biased nature of new technologies would be a transitory effect, likely to vanish as experience on them increases. This is actually what comes out of a series of case studies focusing on office automation (Cyert and Mowery, 1987): the impacts of new technologies on skill requirements do change as the former spread. In clerical positions, the initial effect has been some enlargement of job content because employees have become less specialized and have absorbed new computer-related tasks. But, as computer technologies developed, the number of high-skill opportunities went down. This is just one case study, but it suggests that long-run skill requirements associated with new ITs might well differ from short-run ones.

And, indeed, in the long run, the evolution in jobs content is quite ambiguous. An upskilling 'tradition' emphasizes that technical change tends to eliminate 'noxious physical labour' (Cappelli and Daniel, 1997). Howell and Wolff (1992) also insist that new technologies require fewer motor skills and more cognitive ones, thus raising educational requirements on the part of workers. This is consistent with Bresnahan *et al.* (2001) displaying a positive and significant impact of the stock of ITs per worker upon firms' investment in human capital – defined as training and screening on the basis of education. Another argument in favour of upskilling is not so often mentioned and has to do with the fragility of linked machinery and numerically controlled systems. This requires expertise in the use of machines, hence calling for some (re)training of workers. However, a strong deskilling tradition puts forward opposite arguments: technological change would reduce the breadth of skills required from workers because more and more complex tasks are performed by technologically sophisticated equipment (Flynn, 1985) and because workers tend to lose control over their own activity (Shaiken, 1986).

Overall, evidence regarding the skill requirements of ITs is fragmentary. In contrast, changes in work organization towards more organic structures definitely upgrade the skill content of jobs. The reasons for this are threefold. As demonstrated in section 1, current organizational changes increase workers' responsibility. This has a positive impact on skill requirements, in particular for operatives (Greenan and Guellec, 1994b), but also modifies the roles, hence skills, required from supervisors and technicians (Fréchou and Greenan, 1995). Employees at the bottom end of the hierarchical structure perform a wider range of tasks. This requires multi-skilling and mental abilities enhancing the capacity to adapt.

Interpersonal abilities also become more important owing to the rising need for communication and coordination (Zammuto and O'Connor, 1992). Overall, new organizational forms unambiguously require flexible, self-directed participants (Scott, 1981), which in turn raises skill requirements. Indeed, in the French COP survey, where the question is directly asked, more than 70 per cent of reorganized firms acknowledge that changes in work organization have strongly increased the skill level required from operatives, technicians and supervisors, and 80 per cent of them provide some training in response to this evolution.

Systematic empirical studies, though still scarce, provide similar results, with some of them even suggesting that organizational change might be more skill-biased than technical change. On US data, Bresnahan *et al.* (2001) show that, when an index of decentralization in work organization is introduced in the regression, the impact of the stock of ITs upon human capital investments drops sharply. Cappelli and Daniel (1996) displays a positive impact of organizational changes on rising skill requirements for production jobs. Similarly, Greenan (1996b) shows, on French data, that changes in skill requirements – as reported by firms – are more strongly correlated to organizational than to technical change. Moreover, when focusing on changes in the occupational structure, Greenan shows that technical change has no significant impact, whereas moving towards a flexible organization increases the share of managers at the expense of lower categories. Similarly, Caroli and Van Reenen (2001) show on UK and French data that both organizational and technical change reduce the share of unskilled manual workers inside firms, with the strongest impact coming from organizational change. Though still exploratory, these studies support the idea that organizational change might be even more strongly skill-biased than technical change.

Our theoretical framework can thus be completed as follows (see Figure 8.3). The supply of skills is crucial to the adoption of new technologies (a) through a direct effect on implementation – shortening of the initial learning period and possibly better use of sensitive equipment – and (b) through its determining impact on organizational change when this takes the form of a move towards more organic firms.

The fact that not only technical change, but also organizational change might be skill-biased raises the question of the nature of skills that both require. Investigations in this direction should prove of great interest for the devising of educational and training policies.

General versus Technical Intermediate Skills: What do We Need?

Technical and organizational changes do not require identical skills. Empirical (case study) evidence regarding new technologies suggests that, if any, their long-run use requires specific technical skills. Indeed, the main source of potential need for higher skills lies in the technical complexity of new systems. Numerically controlled devices, robots and computers are fragile and workers should be trained to use (and repair) them properly. This holds for operatives, as for supervisors and technicians, and can be coped with by providing adequate technical training. Such deepening of technical training is

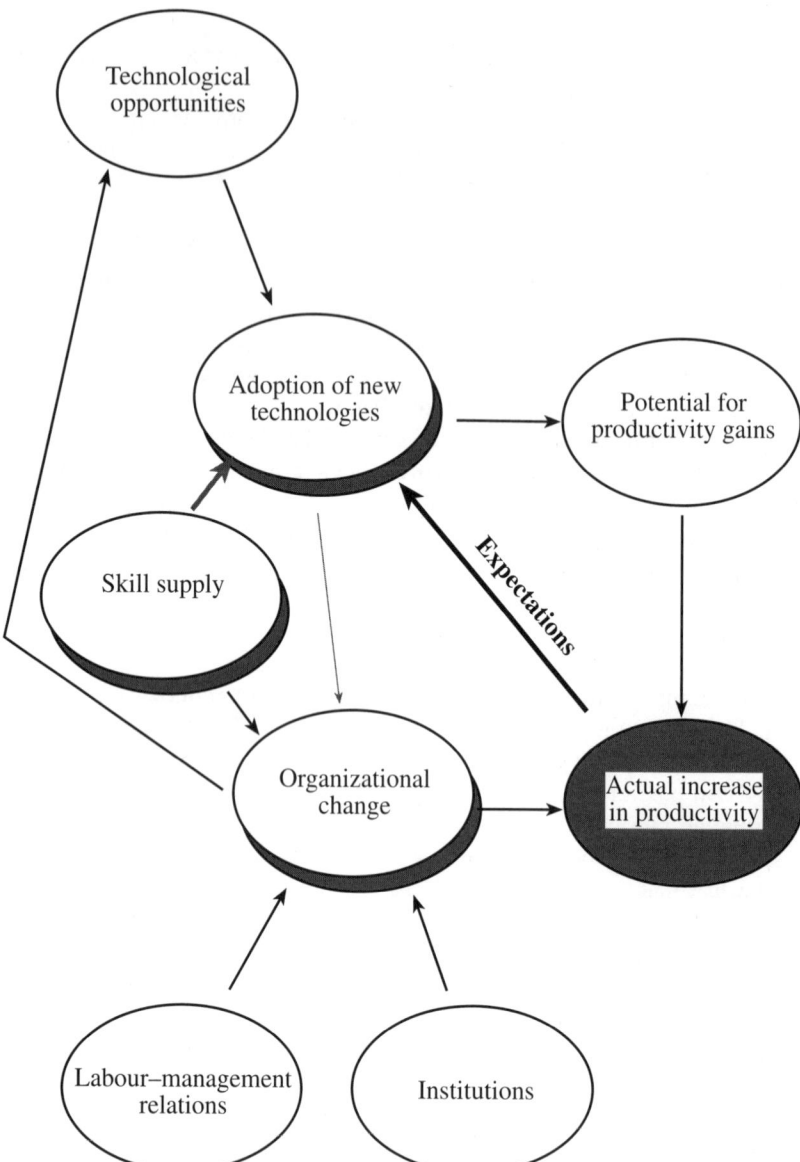

Figure 8.3   Organizational change, technical change and skills

strongly advocated by Hage (1997), who takes it to be the priority for education and training policies in the years to come.

Our analysis of the skill-biased nature of organizational change leads us to disagree with Hage and underline the need for a 'big push' in general – rather than technical – education and training. As stressed above, the spreading of new forms of organization deeply modifies job content at the office and shopfloor levels. Workers need to be multi-skilled, to adapt rapidly to changing tasks. They must be able to diagnose problems and solve them quickly and to display a high level of interpersonal skills. More importantly, they must be able to learn. In all industrialized countries, such skills are acquired through general education and training. Beyond basic knowledge, this provides synthesis abilities together with a basis for continuous – and fast – learning.

The improvement of general knowledge is particularly important at the bottom end of the hierarchical structure, that is for operatives and low-level clerks, as well as supervisors and technicians. According to Steedman *et al.* (1991), 'intermediate' skills (at the craft, foremen and technical levels) are the ones most strongly needed. The reason for this asymmetry lies in the nature of organizational change itself. Owing to the flattening of hierarchies and the spreading of horizontal communication, changes in work organization particularly affect the content of intermediate jobs, hence putting special pressure on intermediate skills. Steedman and Wagner (1987) underline the point that craft qualifications and foremen training are key elements of German superiority in the wood furniture industry. In the clothing industry, UK firms' specialization at the bottom end of the market is due to the low skill level of operatives and maintenance personnel and, in the food processing industry, bakers' and supervisors' qualifications are the problem. The same goes for engineering, so that Steedman *et al.* (1991) stress that intermediate skills are crucial for the smoothing of the production process.

The need for more general knowledge at the bottom end of the occupational structure is clearly displayed on French data by Caroli and Van Reenen (2001). Plants which delayer have a significantly higher propensity to provide general training. This correlation is particularly strong for unskilled manual workers. Similar estimations for specific training display a negative correlation for unskilled workers. It is positive for other occupational layers, but coefficients are in all cases lower than for general training. These results support the idea that organizational change is skill-biased and that requirements are particularly strong in terms of general training and at the bottom end of the occupational structure.

As a conclusion, we would like to stress that the need for more general skills at the intermediate level calls for special education and training policies. Although French data show that firms do invest in general training, this should be strongly supported by adequate initial education and training.

Several reasons for this can be put forward. It is now established that people learn faster when younger. With regard to general knowledge, it is therefore more efficient to concentrate efforts on initial training rather than waiting until people have grown up into adults. This underlines the need for education and training policies aimed at improving the general knowledge content of school curricula. Moreover, precisely because it is not specific, general training raises problems of non-cooperative behaviour on the part of firms. For fear of labour poaching, they may be reluctant to provide the adequate level of general knowledge. This shifts the burden onto the education system and emphasizes the need to improve the general content of initial training.

As we have seen, there is a special need for higher intermediate skills. At the moment, in many countries, this need is not met. This leads to overeducation phenomena that are particularly costly. Steedman *et al.* (1991) stress that, in France and in the UK, people with technician qualifications are employed at the shopfloor level as supervisors, or even operatives. According to the authors, this 'drawing-down' process is due to the lack of craft and supervisory skills. Indeed, it does not take place in Germany where the dual system of apprenticeship provides firms with an adequate supply of skilled manual workers. Mason (1996) displays similar underutilization of employees' competencies in the UK financial services. The proportion of graduate entrants to the labour market increased sharply in the 1980s (from 13 per cent to 30 per cent between 1983 and 1993). They amounted to 13 per cent of the British workforce in 1993, as compared to 9 per cent in 1983. In the financial services, this trend has been accompanied by a rising share of these graduates being employed in clerical jobs where a graduate diploma used not to be necessary. According to Mason, this partly compensates for the lack of intermediate skills. Mason and Finegold (1997) display exactly the same phenomenon in the US food processing industry: a significant proportion of engineering graduates are actually employed at the shopfloor level, where they compensate for the lack of craft and supervisory skills.

Drawing down of higher-skilled workers thus appears as a widespread solution to the lack of intermediate skills. However, this is a very costly one. If resources devoted to education and training are limited – which they are everywhere – they would be more efficiently used if shared between higher and further education. Indeed, a higher level of knowledge at the intermediate level does not necessarily mean higher attendance or staying-on rates at university. This might be more efficiently achieved through improvements of the general content of secondary school curricula.

CONCLUSION

The aim of this chapter was to shed some light on the relationships inside the technology–organization–skill triangle, using a more varied toolbox than is usually the case.

Current organizational changes take the form of a move towards more organic forms of workplace organization. This includes more responsibility being awarded to workers at the lower layers of the hierarchy, a more collective organization of work, more horizontal communication and a greater variety of tasks being performed by each employee. Moreover, new forms of workplace organization tend to be associated with a high technological intensity: technical and organizational change go hand in hand.

With the help of studies coming from the economics literature, but also from sociology and management, we have highlighted one possible sequence of choices between technical and organizational change. Recent evidence regarding the spreading of information technologies and new work practices suggests that organizational changes have largely taken place in response to technological evolution. This is due to the combination of ITs calling for, as well as permitting, new forms of workplace organization *and* of a strong organizational inertia. In a dynamic perspective, this ends up in technical change proving a crucial impulse to organizational evolution. However, organizational change is far from being an optimal response to technical change. The implementation of new HRM practices heavily depends on labour–management relations and on cultural as well as institutional factors. This leads to the coevolution of technical and organizational changes.

Disentangling the roles of technical and organizational changes in current microeconomic evolution is particularly important in the current debate on the so-called 'skill bias'. Recent econometric studies indicate that organizational change is actually as much skill-biased as technical change. Moreover, the tentative analysis carried out in this chapter suggests that organizational changes call for more general education, whereas technological changes could presumably be coped with by relying on specific training. In short, organizational change would be *knowledge-biased*, whereas technical change is essentially skill-biased.

A lot of work remains to be done so as to confirm this view. However, if the contrast between the needs of technical and organizational changes were to be supported by further empirical evidence, this would bear heavy consequences for the devising of education and training policies in OECD countries.

NOTES

1. I am grateful to Paul David, Pascal Petit and Luc Soete for helpful comments. Useful remarks have also been offered by participants in seminars at CEPREMAP and TSER workshops on 'Technology and the Future Employment of Europe', in Paris, and 'The Economics of Science and Technology: Microfoundations and Policy', in Urbino, Italy. The usual disclaimer applies.
   Correspondence: INRA-LEA, 48 Boulevard Jourdan, 75014, Paris, Tel: 33 1 43 13 63 66, Fax: 33 1 43 13 63 62, Email: Eve.Caroli@ens.fr
2. British National Institute for Economic and Social Research.
3. Osterman (1994), Brynjolfsson and Hitt (1995, 1996, 1998), Bresnahan *et al.* (1999), Askenazy (1998), Greenan (1996a, 1996b), Caroli and Van Reenen (2000).
4. Acemoglu (1999), Lindbeck and Snower (1997), Caroli *et al.* (2001), Thesmar and Thoenig (2000).
5. Greenan and Guellec (1994a).
6. See Greenan (1996a) who uses the French COP data set focusing on new work practices in manufacturing.
7. See Caroli and Van Reenen (2001) who use the British WIRS (1984, 1990) and the French REPONSE (1992) surveys of industrial relations and new work practices both in manufacturing and services.
8. See Fréchou and Greenan (1995) on COP data and Coutrot (1996) on REPONSE.
9. See Cappelli and Daniel (1997) who use the US EQW survey (1994) on investment, work organization and the educational level of the workforce; also Greenan (1996a) and Coutrot (1996).
10. See Osterman (1994) using the US survey on innovative work practices. Also Cappelli and Daniel (1997) and Coutrot (1996).
11. Osterman (1994), Greenan and Mairesse (1999).
12. Caroli and Van Reenen (2001).
13. Greenan (1996a).
14. Osterman (1994), Cappelli and Daniel (1997) and Coutrot (1996).
15. See Bué (1989) on the French TOTTO survey on technical and organizational changes. Also Coutrot (1996).
16. Caroli and Van Reenen (2001).
17. Fréchou and Greenan (1995).
18. See Caroli and Van Reenen (2001) using data from REPONSE.
19. Greenan and Guellec (1994b).
20. See Fréchou and Greenan (1995) and Greenan (1996b) on COP data.
21. What we call technical change in the following is actually process innovation. We do not consider product innovation since we are interested in the process of *adoption* of new technologies.
22. See Bresnahan *et al.* (2001) who show that firms with a comparatively more decentralized workplace organization tend to invest more than average in information technologies.
23. See Autor *et al.* (1998) for a discussion of the extent to which technical change proves biased against less skilled workers.
24. In order to do so, we rely on existing empirical evidence. Unfortunately, to date, this remains quite scarce. Let us hope that the growing interest displayed by economists in organizational issues will foster empirical research in this area

REFERENCES

Acemoglu, D. (1999), 'Changes in Unemployment and Wage Inequality: An Alternative Theory and Some Evidence', *American Economic Review*, 89, 1259–78.

Aoki, M. (1990), 'The Participatory Generation of Information Rents and the Theory of the Firm', in M. Aoki, B. Gustafsson and O. Williamson (eds), *The Firm as a Nexus of Treaties*, London: Sage Publications, 26–52.

Athey, S. and A. Schmutzler (1995), 'Product and Process Flexibility in an Innovative Environment', *RAND Journal of Economics*, 26(4), 557–74.

Autor, D., L. Katz and A. Krueger (1998) 'Computing Inequality: Have computers changed the labor market?', *Quarterly Journal of Economics*, 113, 1169–213.

Bartel, A. and P. Lichtenberg (1987), 'The Comparative Advantage of Educated Workers in Implementing New Technologies', *Review of Economics and Statistics*, 64(1), 1–11.

Berman, E., J. Bound and Z. Griliches (1994), 'Changes in the Demand for Skilled Labor Within U.S. Manufacturing: Evidence from the Annual Survey of Manufactures', *Quarterly Journal of Economics*, 109(2), 367–97.

Boyer, R. (1991), 'New Directions in Management Practices and Work Organisation. General Principles and National Trajectories', *Couverture Orange*, CEPREMAP no. 9130.

Boyer, R. (1998), 'Evolution des Modèles Productifs et Hybridation: Géographie, Histoire et Théorie', *Couverture Orange*, CEPREMAP, no. 9804.

Bresnahan, T., E. Brynjolfsson and L. Hitt (2001), 'Information Technology, Workplace Organization and the Demand for Skilled Labor: Firm-level Evidence', NBER working paper 7136; *Quarterly Journal of Economics*, forthcoming.

Brynjolfsson, E. and L. Hitt (1995), 'Information Technology as a Factor of Production: The Role of Differences Among Firms', *Economics of Innovation and New Technology*, 3(4), 183–200.

Brynjolfsson, E. and L. Hitt (1996), 'Paradox Lost? Firm-level Evidence on the Returns to Information Systems Spending', *Management Science*, April.

Brynjolfsson, E. and L. Hitt (1998), 'Information Technology and Organizational Design: Evidence from Micro Data', mimeo, MIT.

Bué, J. (1989), 'Les Différentes Formes de Flexibilité', *Travail et Emploi*, 41, 21–7.

Burns, T. and G. Stalker (1961), *The Management of Innovation*, London: Tavistock.

Cappelli, P. (1996), 'Technology and Skill Requirements: Implications for Establishment Wage Structures', *New England Economic Review*, 139–54.

Caroli, E. and J. Van Reenen (2001), 'Skill-biased Organizational Change: Evidence from a panel of British and French establishments', working paper INRA-LEA 00-06; *Quarterly Journal of Economics*, forthcoming.

Caroli, E., N. Greenan and D. Guellec (2001), 'Organizational Change and Skill Accumulation', *Industrial and Corporate Change*, 10(2), 479–504.

Carr, F. (1994), 'Introducing Team Working – a Motor Industry Case Study', *Industrial Relations Journal*, 25(3), 199–209.

Chandler, A. (1962), *Strategy and Structure*, Boston: MIT Press.

Chennells, L. and J. Van Reenen (1997), 'Technical Change and Earnings in British Establishments', *Economica*, 64, 587–604.

Coutrot, T. (1996), 'Les Nouveaux Modes d'Organisation de la Production: Quels Effets sur l'Emploi, la Formation, l'Organisation du Travail?', *Données Sociales*, INSEE, 209–16.

Coutrot, T. (2000), 'Innovations et Gestion de l'Emploi', *Premières Synthèses DARES*, 2000.03.

Cyert, R. and D. Mowery (1987), *Technology and Employment. Innovation and Growth in the U.S. Economy*, Washington: National Academy Press.

Davis, L. and J. Taylor (1976), 'Technology, Organization and Job Structure', in R.

Dubin (ed.), *Handbook of Work, Organization and Society*, Skokie, IL: Rand McNally.

Flynn, P. (1985), 'The Impact of Technological Change on Jobs and Workers', paper prepared for the US Department of Labor, Employment Training Administration.

Fréchou, H. and N. Greenan (1995), 'L'Organisation de la Production dans l'Industrie: des Changements Profonds', *4-pages Sessi*.

Galbraith, J. (1973), *Designing Complex Organizations*, Reading, MA: Addison-Wesley.

Gerwin, D. and J.C. Tarondeau (1984), 'La Flexibilité dans les Processus de Production: le Cas de l'Automobile', *Revue Française de Gestion*, 46, 37–46.

Goffman, E. (1983), 'The Interaction Order', *American Sociological Review*, 48, 1–17.

Greenan, N. (1996a), 'Innovation Technologique, Changements Organisationnels et Evolution des Compétences', *Economie et Statistique*, 298, 15–34.

Greenan, N. (1996b), 'Progrès Technique et Changements Organisationnels: leur Impact sur l'Emploi et les Qualifications', *Economie et Statistique*, 298, 35–44.

Greenan, N. and D. Guellec (1994a), 'Coordination Within the Firm and Endogenous Growth', *Industrial and Corporate Change*, 3(1), 173–98.

Greenan, N. and D. Guellec (1994b), 'Organisation du Travail, Technologie et Performances: Une Etude Empirique', *Economie et Prévision*, 113–14, 39–57.

Greenan, N. and J. Mairesse (1999), 'Organizational Change in French Manufacturing: What do we learn from firm representatives and from their employees?', NBER Working paper no. 7285.

Hage, J. (1997), 'Organization Innovation: an Overview of Research in the Area', mimeo, Center for Innovation, University of Maryland.

Hannan, M. and J. Freeman (1977), 'The Population Ecology of Organizations', *American Journal of Sociology*, 82, 929–64.

Howell, D. and E. Wolff (1992), 'Technical Change and the Demand for Skills by US Industries', *Cambridge Journal of Economics*, 16(2), 127–46.

Ichniowski, C. and K. Shaw (1995), 'Old Dogs and New Tricks: Determinants of the Adoption of Productivity-enhancing Work Practices', *Brookings Papers on Economic Activity, Microeconomics*.

Ichniowski, C., K. Shaw and G. Prennushi (1997), 'The Effects of Human Resource Management Practices on Productivity: A Study of Steel Finishing Lines', *American Economic Review*, 87, 291–313.

Krueger, A. (1993), 'How Computers have Changed the Wage Structure: Evidence from Micro-Data, 1984–1989', *Quarterly Journal of Economics*, 108, 33–60.

Lawrence, P. and J. Lorsch (1967), *Organization and Environment*, Boston: Harvard University Press.

Lincoln, J., M. Hanada and K. McBride (1986), 'Organizational Structures in Japanese and US Manufacturing', *Administrative Science Quarterly*, 31, 338–64.

Lindbeck, A. and D. Snower (1997), 'Reorganization of Firms and Labour Market Inequality', *American Economic Review*, 86, 315–21.

Machin, S. and S. Wadhwani (1991), 'The Effects of Unions on Organisational Change and Employment', *Economic Journal*, 101, 835–54.

Machin, S. and J. Van Reenen (1998), 'Technology and Changes in Skill Structure: Evidence from Seven OECD Countries', *Quarterly Journal of Economics*, 113(4), 1215–44.

Mason, G. (1996), 'Graduate Utilisation in British Industry: The Initial Impact of Mass Higher Education', *National Institute Economic Review*, 93–103.

Mason, G. and D. Finegold (1997), 'Productivity, Machinery and Skills in the United States and Western Europe', *National Institute Economic Review*, 85–98.

Mason, G. and K. Wagner (1994), 'Innovation and the Skill Mix: Chemicals and Engineering in Britain and Germany', *National Institute Economic Review*, 148, 61–72.

Mason, G., B. Van Ark and K. Wagner (1994), 'Productivity, Product Quality and Workforce Skills: Food Processing in Four European Countries', *National Institute Economic Review*, 147, 62–81.

Milgrom, P. and J. Roberts (1990), 'The Economics of Modern Manufacturing: Technology, Strategy and Organization', *American Economic Review*, 80(3), 511–28.

Mowery, D. (1990), 'Technology and Organizations: An Economic/Institutional Analysis', in P. Goodman and L. Sproull (eds), *Technology and Organization*, San Francisco: Jossey-Bass.

Noble, D. (1984), *Forces of Production: A Social History of Industrial Automation*, New York: Knopf.

OECD (1999), *Employment Outlook*, Paris: OECD.

Osterman, P. (1994), 'How Common is Workplace Transformation and Who Adopts It?', *Industrial and Labour Relations Review*, 47(2), 173–88.

Perrow, C. (1967), 'A Framework for the Comparative Analysis of Organizations', *American Sociological Review*, 32, 194–208.

Pugh, D. (1973), 'The Measurement of Organization Structures: Does Context Determine Form?', *Organizational Dynamics*, 19–34.

Pugh, D. and D. Hickson (1976), *Organizational Structure in its Context: The Aston Programme I*, Aldershot: Gower.

Pugh, D. and D. Hickson (1989), *Writers on Organization*, 4th edn, London: Penguin Books.

Radner, R. (1993), 'The Organization of Decentralized Information Processing', *Econometrica*, 61(5), 1109–46.

Sah, R. and J. Stiglitz (1985), 'Human Fallibility and Economic Organization', *AEA Papers and Proceedings*, 75, 292–7.

Scott, W. (1981), *Organizations: Rational, Natural and Open Systems*, Englewood Cliffs, NJ: Prentice-Hall.

Scott, W. (1990), 'Technology and Structure: an Organizational Level Perspective', in P. Goodman and L. Sproull (eds), *Technology and Organization*, San Francisco: Jossey-Bass.

Shaiken, H. (1986), *Work Transformed: Automation and Labor in the Computer Age*, New York: Holt, Rinehart & Winston.

Steedman, H. and K. Wagner (1987), 'A Second Look at Productivity, Machinery and Skills in Britain and Germany', *National Institute Economic Review*, 122, 84–95.

Steedman, H. and K. Wagner (1989), 'Productivity, Machinery and Skills: Clothing Manufacture in Britain and Germany', *National Institute Economic Review*, 128, 40–57.

Steedman, H., G. Mason and K. Wagner (1991), ' Intermediate Skills in the Workplace: Deployment, Standards and Supply in Britain, France and Germany', *National Institute Economic Review*, 136, 60–76.

Strauss, A. (1978), *Negotiations: Varieties, Contexts, Processes and Social Order*, San Francisco: Jossey-Bass.

Thesmar, D. and M. Thoenig (2000), 'Creative Destruction and Firm Organization Choice', *Quarterly Journal of Economics*, 115(4), 1201–37.

Van Zandt, T. (1997), 'Decentralised Information Processing in the Theory of Organizations', in Sertel Murat (ed.), *Contemporary Economic Development Reviewed, Volume 4: The Enterprise and its Environment*, London: Macmillan.

Williamson, O. (1975), *Markets and Hierarchies: Analysis and Antitrust Implications*, New York: Free Press.

Woodward, J. (1965), *Industrial Organization: Theory and Practice*, Oxford: Oxford University Press.

Zammuto, R. and E. O'Connor (1992), 'Gaining Advanced Manufacturing Technologies' Benefits: the Roles of Organization Design and Culture', *Academy of Management Review*, 17(4), 701–28.

9. Unemployment and labour market flexibility: a misplaced question?

Donatella Gatti[1]

1. INTRODUCTION

A major puzzle in current economic debate stems from the differences in observed rates of unemployment across developed countries. Economic analysis has generally focused on divergence in labour market organization (and flexibility) as the most plausible institutional explanation of this phenomenon (see, for instance, Layard *et al.*, 1991). Labour market flexibility, in turn, is related to structural rigidities characterizing the wage bargaining system or the social security system. Therefore, from this perspective, the rise of (direct and indirect) workers' social protection and the improvement in working conditions are seen as essential causes of increasing unemployment rates throughout western Europe (Layard *et al.*, 1991; OECD, 1986). Hence, high unemployment must be understood as a 'fair price' that Europe has to pay in order to be able to keep, as a counterpart, a high workers' protection legislation (CEPR, 1996).

In contrast to this, a growing body of literature has been devoted to defining a more comprehensive approach to the comparative analysis of institutional arrangements and performances (see Aoki, 1986; Boyer, 1987; Soskice, 1990; Streeck, 1996). The main idea underlying contributions in this field is that a variety of (forms of) capitalism exists which defines several institutional models respectively characterizing existing industrialized countries. The main elements of these institutional settings involve both the internal organizational features of the firm and the whole environmental structure (that is, relationships with other firms and the 'regime of competition', nature and size of product markets, and structure of both labour and capital markets).

More particularly, in recent works special attention has been paid to the study of national differences in workers' competence formation (training systems) and in firms' organization (see, for instance, Aoki, 1994). These two elements, in fact, seem to be tightly interlinked and strongly complementary; at the same time, striking differences in national experiences are observed as far as the nature of skills as well as firms' organization are concerned. This seems to indicate that they can both be taken as fundamental elements in the definition of opposite ideal-types of institutional settings.

The claim of this chapter is that both the nature of workers' skills and firms' organization play an important role in shaping the structure of workers' incentives. In particular, the two factors are likely to affect effort determination, in an efficiency wage framework. Therefore, from this perspective, we expect that both factors will eventually affect the macroeconomic outcome in terms of equilibrium rate of unemployment. This crucial idea is developed in the chapter through two stages.

In a first part of the work, a theoretical framework is proposed to address the theoretical analysis of cross-country differences in the equilibrium rate of unemployment. This framework is based on the definition of an institutional setting encompassing (a) the structure of the firm, (b) the system of education or vocational training, and (c) the nature of the labour market. In particular, the notion of the 'mode of organization' of the firm will be developed as a starting point of the analysis.

Building on a stylization of different institutional configurations in the second part of the chapter, a theoretical model is proposed that formalizes the choice of firms about employment and wages, given the institutional constraints. Following an approach proposed by the radical American economists (see Bowles, 1985), the unemployment–wage equilibrium is derived through the definition of an optimal incentive scheme for the firm. This implies addressing two related trade-offs shaping firms' choice – threat versus positive incentives, and internal versus external discipline – whose resolution eventually will lead to the definition of the equilibrium configuration of wage and the 'opportunity rate' (see below). The final equilibrium configuration, including the unemployment and vacancy rates, is then obtained by imposing the equality of inflows and outflows from the unemployment set. This is done through an endogenous determination of the separation and job-finding rates.

The equilibrium value of the unemployment and vacancy rates turns out to be affected by the nature of the microinstitutional setting (as defined above). This gives some important insights concerning differences in the equilibrium rate of unemployment among developed countries: a major role could actually be played by the mode of organization of firms and the nature of workers' competence, rather than labour market flexibility. In fact, by distinguishing on the basis of the nature of both firms' organization and the vocational training system, it is possible to identify several microinstitutional stylizations respectively characterizing the French, German, Japanese and US models. This allows a tentative ranking of different microinstitutional settings (and corresponding 'real models') as regards their sustainable employment performance.[2]

Compared to previous studies (Gatti, 1997), this approach allows a widening of the scope of our cross-country comparison, progressing from the Japan–US models' opposition to the analysis of a larger set of countries. Moreover, the model allows the determination of both the unemployment rate and the vacancy rate, thus partially contributing to the current debate on cross-country differences in Beveridge curves.

The chapter is organized as follows. Section 2 presents a theoretical framework based on the definition of several microinstitutional settings. In section 3, a model of the choice of the firm concerning its system of control upon effort is presented, with a central role being played by the nature and shape of the prevailing institutional setting. We then focus more directly on the determination of the wage–opportunity equilibrium configuration and, finally, we derive the flow equilibrium condition for unemployment and vacancy and show that the equilibrium rate of unemployment is strongly affected by the nature of the microinstitutional setting.

2.   THE MICROINSTITUTIONAL SETTING

A microinstitutional setting will be characterized according to three different dimensions: the organization of the firm; the system of education or vocational training; and the nature of the labour market. Here we take up the first two aspects; the third will be dealt with later, when issues of effort and workers' mobility are discussed.

Generally speaking, both firms' organization and workers' competence are crucial in the analysis of unemployment since they can be expected to have a strong influence on workers' incentives. The mode of organization defines and shapes the nature of the mechanisms implemented by the firm in order to improve workers' effort. If we think about it in an efficiency wage framework, for instance, it is clear that each firm has to make a fundamental choice between different incentive devices, namely, direct monitoring and workers' cost of job loss (see below). The actual form of the effort function will reasonably be modified according to the nature of the firm, which means that the equilibrium level of the efficiency wage is likely to be affected, too. From a macroeconomic point of view, the main consequence is that the equilibrium rate of unemployment will eventually be modified. The nature of skills is also a crucial factor in this respect. In fact, one main characteristic of competence is the degree of transferability of skills. This is also a crucial feature affecting workers' potential external mobility– and their cost of job loss. If skills are non-transferable, workers will hardly be able to take them outside the originating firm, and this makes their cost of job loss significantly higher than in the case of highly transferable skills. Therefore the structure of the educational system as a whole has a crucial influence on workers' incentives and, once again, on the equilibrium level of wage and unemployment.

In the following, we will analyse in more detail these two aspects of firms' organization and workers' competence in a comparative perspective, starting with training and competence. In this respect, we initially take up the traditional distinction (first proposed by Becker, 1962) between general and specific training. Unlike most theoretical speculations about economic consequences of different typologies of training, what we are interested in is the relationship between nature of training and workers' external mobility. This is a crucial factor – also shaping the operation of labour markets – that has been given little attention in recent theoretical and empirical work.[3]

In what follows, competence will be assumed to range from firm-specific to generic.[4] Firm-specific competence should be associated with a reduced or absent external transferability, in the sense that firm-specific workers will not regard themselves as being able to find comparable employment conditions elsewhere. The contrary holds for workers with generic competence.

Let the parameter $1-l$ grasp the (average) degree of competence's transferability characterizing the labour force in a given economic system (l is then the proportion of non-transferable competence to total workers' competence). In other terms, all workers will be assumed to be characterized by the same average parameter; this means that workers in a given economic system have homogeneous competence. Quite a rich literature now exists studying the main features of vocational training systems and competence formation in developed countries. We can therefore propose a (tentative) cross-country ranking of the value taken by l in each institutional context, building on results from recent comparative studies on vocational training systems.

Following Soskice and Hancké (1997) we first have to single out institutional contexts where vocational training systems deliver generally accepted diplomas (as in Germany). In this case, in fact, workers acquire competences that are more likely to be transferable, namely, through these diplomas. In institutional contexts where this is not possible (the USA, Japan and, to a lesser extent, France), transfer of competence is more difficult and mostly depends on the actual contents of training.

Concerning Europe, Marden's cross-country comparative analyses on the structure of labour markets confirm this first insight on competence transferability: occupational labour markets characterize the German experience while internal labour markets are more predominant in France (see Marsden, 1990). However, Maurice and Soskice–Hancké both underline the central place of general education and diplomas in the French training system (see Maurice, 1993). This allows workers to (partially) transfer their general competence from one firm to another.

Several recent works allow us to extend this comparison to Japan and the USA. These works underline the highly firm-specific nature of vocational training systems in these two countries, as compared to European countries, and Germany in particular (see Aoki, 1994; Berg, 1994; Maurice, 1990; Streeck, 1996). Streeck, for instance, states: 'Japanese skills are not occupational, in that there are no socially defined bundles of intrinsically related work capacities ... Rather than belonging to an occupation, workers in the core Japanese industry belong to a company; they are not electricians but "Toyota men".' On the same lines, concerning the USA Berg observes: 'the most striking contrast between the two countries [Germany and the USA] is the qualitative differences in the training offered to workers'. In particular, 'the German enterprise-based occupational training system provides broad, general skills training that is more sophisticated than the narrow firm-specific skills provided in the US'.

We can sum up the arguments presented above by means of a corresponding ranking of the parameter $1-l$, assumed to characterize the degree of competence's transferability in each institutional context. According to the above arguments, we can write:[5]

$$l(\text{Jap}) > l(\text{US}) > l(\text{Fr}) > l(\text{Ger}).$$

We have thus illustrated the aspect of our microinstitutional analysis concerning workers' competence. We can now return to the comparative analysis of firms' organizational structures. We tackle this issue by proposing the notion of 'mode of organization'.[6] A mode of organization (MoO) is a complex system of rules that should provide the firm with an answer to the following questions of (a) how to control work intensity and effort, and (b) how to manage information and communication flows. This can be done through the definition of a specific system of work organization and knowledge distribution inside the firm. The importance of jointly analysing these two aspects is also stressed by Coriat and Dosi: 'we have tried to show that the explanation of a particular set of routines [inside the firm] can be traced back to the coevolution between corporate patterns of knowledge distribution and mechanism of coordination and governance' (Coriat and Dosi, 1995). Let me quickly define these different configurations taken, respectively, by work organization and knowledge distribution.

Concerning work organization, two polar solutions can be conceived, namely hierarchical and horizontal organization of work. In the former case, the 'communication network' inside the firm is vertically oriented: central authority is in charge of all kinds of decisions and guarantees internal coordination. The reverse holds when work organization is horizontal: in this case, decision making is a matter of individual workers' choices and decentralized coordination. A similar reasoning holds regarding knowledge distribution. An opposition can be assumed between concentrated and shared knowledge, which concerns the way a firm's 'know-how' is distributed among workers. With a concentrated distribution, the stock of knowledge owned by the firm is localized at the central authority (or experts) level. Workers do not share firms' stock of knowledge and their competence is determined outside the firm (see Aoki, 1986). Conversely, when knowledge is shared, the know-how is spread all around the firm through a process that is generally called 'workers' involvement'.

Indeed, a hierarchical (horizontal) organization of work is likely to be associated with a concentrated (shared) distribution of knowledge. Therefore, combining these two factors actually leads to the previous definition of two opposite MoOs: a centralized MoO with a hierarchical work organization and concentrated knowledge; and a decentralized MoO with a horizontal work organization and a shared knowledge base (see also section 3 on this point). An extensive literature is now available concerning the prevailing nature of firms' MoO across different countries (see, for example, Kristensen (1996), Lazonic and West (1995), Milgrom and Roberts (1990), Lorenz (1992)): it is a commonly shared view between scholars that Japanese and German firms mostly present the main characteristics of a decentralized mode of organization, while US and French firms prevalently show features of a centralized MoO.

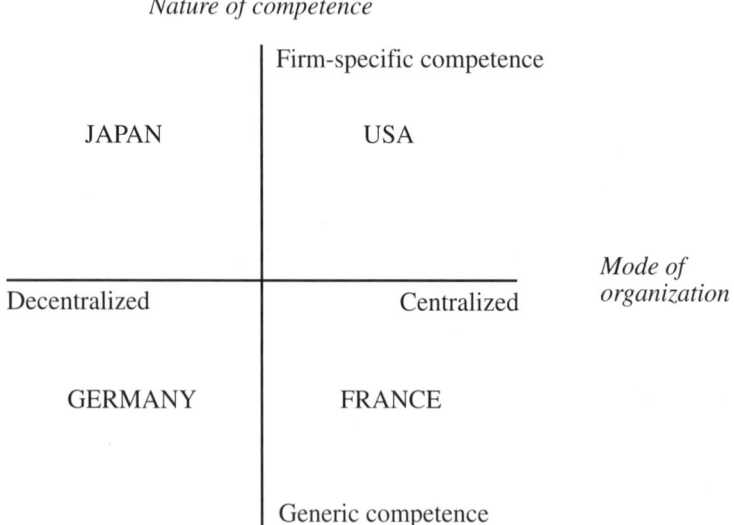

Figure 9.1   Vocational training systems

If we consider simultaneously the typologies proposed for vocational training systems and the MoO, we can describe different microinstitutional settings according to their position along two axes: the 'nature of competence axis' (from firm-specific to generic competence) and the 'mode of organization axis' (from centralized to decentralized MoO). Figure 9.1 proposes an application of this exercise, trying to position the microinstitutional settings prevailing in the USA, Germany, France and Japan.

The following sections will explore how the institutional features presented above influence the choice made by firms about their incentive scheme and eventually affect the equilibrium unemployment rate. We first present the general set-up of the model and then solve it alternatively under the different parametric assumptions characterizing each of the previous microinstitutional settings.

3. THE MODEL

Let us consider the decision problem faced by firms when determining the optimal level of production and wage. This generally consists of maximizing profits, that is:[7]

$$\pi = \text{profit} = Q - (w + s) \cdot Lp, \tag{9.1}$$

where Q is quantity produced, w is real wage, Lp is hours of work hired, s is real cost of supervising resources per hour of work.

The introduction of s allows us to consider the twofold role played by the authority respectively in the domain of coordination between production units and control upon work intensity: $s = \text{coord} + \text{monit}$, where coord is coordination cost per hour of work, monit is monitoring cost per hour of work.

The production function is:

$$Q = h \cdot (e \cdot Lp), \quad h \leq 1, \tag{9.2}$$

where h is coordination efficiency, e is effort per hour of work. For the sake of simplicity, we assume that the production function shows constant return to scale on labour: this does not constitute a major assumption and makes the following analysis much clearer.

In particular, the variable h is assumed to grasp the possibility of a (partial) output waste due to lack of coordination between production units. Contrary to other models, in this framework workers' effort and ability to coordinate are both determined by the nature and efficiency of implemented incentive schemes, and the nature of this incentive scheme is conditioned by firms' organizational structure. We clarify this point by modelling explicitly effort and coordination functions.

Labour Market and the Effort Function

As far as work intensity is concerned, we assume that firms want to make sure that workers provide a satisfactory level of effort. It is then necessary to implement some devices in order to guarantee this level of effort. In this respect, we take up an efficiency wage approach developed by the American radical economists (see Bowles, 1985).

In particular, we can assume that the intensity of workers' effort corresponds to the part of the working day when workers are actually working: $e \in [0,1]$, where $e = 1$ grasps the case where workers work all day long. Effort can then be increased in two different ways: by increasing 'monitoring intensity' (m) over workers, or by increasing workers' 'cost of job loss' (w_c). Bowles shows that the maximization of workers' anticipated income, given the probability of being fired if discovered not working, leads to the definition of an 'effort function' (e) which is the optimal response of workers to the firm's incentives strategy:[8]

$$e = \text{effort per hour of labour hired} = e(w_c, m; r), \qquad (9.3)$$

with r = external factors affecting effort; $\partial e / \partial w_c > 0$, $\partial e / \partial m > 0$.

This function translates what Bowles shows in his general approach, namely that the optimal level of effort selected by workers depends positively on the probability of being monitored (linked to the intensity of monitoring) and the cost of job loss. The main external variable (with respect to the firm's decision) influencing effort determination is in fact firing costs; that is, labour markets' rigidity: this affects the *ex post* probability of being fired if a worker is discovered not working. Therefore, r = index of barriers to firing (higher r means lower barriers to firing).

Concerning monitoring, we assume that the available 'monitoring technology' is such that a maximum monitoring intensity is reached when the amount of monitoring resources per hour of work is equal to (M). As we will see, the firm will not necessarily choose this maximum amount of monitoring, since monitoring is an expensive activity; m = intensity of monitoring = monitoring resources per hour of work/M. Therefore, as we have defined it, the actual monitoring intensity inside the firm is such that $0 < m < 1$.

As far as the 'cost of job loss' is concerned, following Bowles we can define it as the income loss that workers incur when they lose their job.

Conditional to the fact that they are fired, we can model workers' income loss as the difference between wage (w) and alternative expected income (w_d):

$$\text{cost of job loss } = w_c = w - w_d, \tag{9.4}$$

where w is wage; w_d is alternative income in case of firing.

The expression for w_d is given by the average between the alternative wage that workers can earn if rehired and the unemployment benefits they get if they do not find new jobs. The average is calculated on the basis of the probability that workers have to find new jobs (when they are fired). This depends on the transferability of workers' competence ($1-l$), and the global availability of job opportunities.

Let us define a variable p called the opportunity rate. This variable is supposed to grasp global available opportunities for workers. We assume:

$$p = p(u,v), \quad \text{such that} \quad 0 < p < 1, \tag{9.5}$$

where u is unemployment rate, v is vacancy rate. The unemployment/vacancy rates are supposed to grasp the global availability of job opportunities inside the economic system. The higher (lower) the unemployment (vacancy) rate, the fewer the alternative job opportunities globally available. However, it should be considered that the degree $1-l$ of competence transferability also plays a role in job matches: when the proportion l of non-transferable to total workers' competence increases, the probability of getting re-employed if fired reduces. Therefore the higher the (average) degree of competence transferability, the higher the probability of workers finding (if fired) new jobs comparable to their original ones. In formal terms, we can say that the probability of finding a new job (if fired) is given by the product $p \cdot (1-l)$.

We can sum up the operation of matching process (for fired workers) on the labour market as in Figure 9.2. In the case of a successful match, workers find a new job which fits their competence; then they are paid a wage (w_v) that will be equal (*ex post*) to their previous one. In the alternative case, workers only earn unemployment benefits.

From Figure 9.2 we can deduce the expression of the alternative income (w_d) that workers should expect if they are fired:[9]

$$w_d = p \cdot (1-l)w_v + (1 - p \cdot (1-l))\underline{w}, \tag{9.6}$$

where w_v is alternative market wage, $\underline{w}$ is unemployment benefits.

Substituting (9.6) into (9.4), we can easily obtain the definition of the cost of job loss, and then we have completely defined the second control device that firms can implement to obtain a positive level of effort.

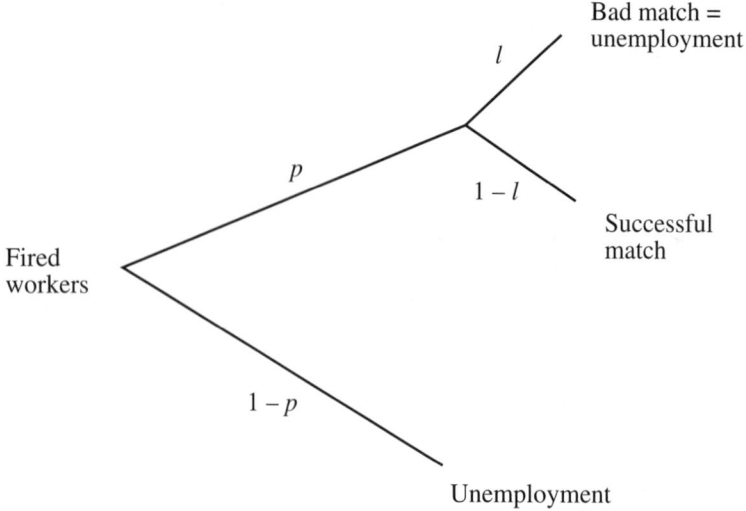

Figure 9.2　The matching process

Finally, we should consider that one crucial difference between centralized and decentralized MoO concerns the role played by the central authority inside the firm. In fact, in a centralized firm, supervising resources have to be used for two different functions – direct monitoring (m) and coordination between production units (t) – while in a decentralized firm workers are capable of coordinating autonomously. We can sum up our claim concerning the difference between centralized and decentralized firms, as far as supervision is concerned, as follows:

$$\text{centralized MoO} \Rightarrow t > 0 \Rightarrow s = m \cdot M + t,$$

$$\text{decentralized MoO} \Rightarrow t = 0 \Rightarrow s = m \cdot M.$$

This difference concerns the cost and nature of coordination inside the firm, while the effort function (and monitoring) are similar in the two types of firms. We now turn to the definition of the coordination function.

Coordination and Knowledge Sharing

This section analyses how the coordination function (h) can be specified according to the nature of firms' MoO. We start with the common assumption that a given amount of communication per hour of work (t) is needed inside the firm to ensure coordination and avoid wasting (potential) output.[10] We can write:

$$h = h(t), \partial h / \partial t > 0. \tag{9.7}$$

Two options are then possible (corresponding to different MoO):

1. *Centralized mode of organization*: the MoO is characterized by a vertical organization of work and concentrated knowledge; coordination is therefore assured by hierarchy. This makes it possible to fully exploit the specialization of tasks: no loss of output should occur. Therefore we can put $h(t) = 1$, adding a cost $(\bar{t} \cdot Lp)$ to production costs.
2. *Decentralized mode of organization*: the MoO relies on decentralized coordination among workers implemented through a process of knowledge sharing. Intensity of this knowledge sharing can be (partially) determined by choosing the amount of 'socialization activities' (teamwork, quality circles and so on) undertaken. Let us define: t = amount of socialization activities per hour of work; $h(t)$ = decentralized coordination ability.

Thus firms can choose the optimal degree of decentralization, taking into consideration that, in order to further socialization activities, supervision is always needed. This means that part of a firm's monitoring resources must be diverted from direct control of production effort. In fact, we will assume that decentralized coordination needs time and resources to be implemented and that these resources should be taken out of a firm's stock of internal resources (namely, monitoring resources). Following the approach proposed by Greenan and Guellec (1994), we will assume that an amount $t \cdot (m \cdot M)^\delta$ of a firm's monitoring resources $(m \cdot M)$ are necessary in order to obtain a degree of coordination $h(t)$. Therefore resources left for direct monitoring are such that *ex post* monitoring intensity is equal to $m - t \cdot (m \cdot M)^\delta / M$.

The firm has to choose an optimal amount of t, deciding the appropriate allocation of monitoring resources between control and socialization. To do this, each firm solves the following maximization problem:

$$\text{Max}_t \ \pi = Q - (w + s) \cdot Lp,$$

where $Q = h(t) \cdot e[w_c; r; (m - t \cdot (m \cdot M)^\delta / M)] \cdot Lp.$

Profit maximization turns out to be a separable problem, since choice about t is independent of the costs structure. Therefore we can first solve maximization for t, and subsequently treat the general profit-maximization problem. Choice about t consists, then, in:

$$\text{Max}_t \ h(t) \cdot e[w_c; r; (m - t \cdot (m \cdot M)^\delta / M)] \cdot Lp.$$

In order to obtain explicit results, we propose the following specifications for coordination and effort functions:[11]

$$h(t) = t^\varepsilon ; \qquad (9.8)$$

$$e(w_c; m) = r \cdot (w_c)^a \cdot (m - t \cdot (m \cdot M)^\delta / M)^b . \qquad (9.9)$$

The solution to the maximization problem is given by $\partial Q / \partial t = 0$, which means:

$$\varepsilon \cdot t^{(\varepsilon-1)} \cdot (m - t \cdot m^\delta \cdot M^{\delta-1})^b - t^\varepsilon \cdot b \cdot (m - t \cdot m^\delta \cdot M^{\delta-1})^{(b-1)} \cdot (m^\delta \cdot M^{\delta-1}) = 0.$$

Then we obtain:

$$t = [\varepsilon / (\varepsilon + b)] \cdot [1 / (m \cdot M)^{(\delta-1)}], \qquad (9.9a)$$

and

$$h(t) \cdot e(w_c; m) = h \cdot r \cdot w_c^a \cdot m^{b - \varepsilon(\delta-1)}, \qquad (9.10)$$

where

$$h = M^{-\varepsilon(\delta-1)} \left(\frac{\varepsilon}{\varepsilon + b} \right)^\varepsilon \left(\frac{b}{\varepsilon + b} \right)^b < 1.$$

Therefore, if $\delta > 1$,[12] the process of knowledge sharing determines a negative feed-back of monitoring on coordination. The coefficient of monitoring in the production function now turns out to be $b' = b - \varepsilon \cdot (\delta - 1) < b$. This result shows that a decentralized firm is characterized by a lower reliance on monitoring (lower elasticity of the effort function with respect to monitoring) than a centralized firm. This is coherent with reported stylized facts about Japanese firms (as typical examples of decentralized MoO), where monitoring has a significantly weaker role than in most centralized firms, owing to the crucial role played by workers' autonomy and responsibility (Aoki, 1994; Okuno, 1984; Marsden, 1996). In fact, the activities by which workers get to share their knowledge are generally of an informal nature, and are not performed under explicit control by the authority. Quoting Okuno, 'a senior worker frequently helps junior workers in the same work line learn special skills needed for the job .... Yet on-the-job-training is not assigned as a part of the senior worker's job, nor does he receive any extra reward for his efforts' (Okuno, 1984).

The assumptions that we make concerning MoO parametric specification are summed up in Table 9.1.

Table 9.1 A parametric characterization of the MoOs

	Mode of organization	
	Centralized	Decentralized
Knowledge distribution	$h = 1$	$h = h(m)$
Work organization	$t > 0$	$t = 0$
	$(s = m + t)$	$(s = m)$

Choice of an Optimal Incentive Scheme

We now turn to the solution of firms' maximization problem. First, the firm has to determine the optimal combination of direct monitoring and cost of job loss in order to obtain a given level of effort. Once this combination is determined, the firm has to find the optimal level of wage. As we have seen (eqs (9.1) and (9.2), a firm's profits and production function are:

$$\pi = Q - (w + s) \cdot Lp, \text{ with } Q = h \cdot F(e \cdot Lp) = h \cdot e \cdot Lp.$$

Moreover, we will assume that the effort function has the same specification as in (9.9):

$$e = r \cdot w_c^a \cdot m^b.$$

The coordination function h is defined as in the previous section: in a centralized MoO (C), coordination is perfectly assured by hierarchy which means that $h = 1$; in a decentralized MoO (D) coordination is instead decentralized and the process of knowledge sharing leads to a function $h(m)$ influencing the role of monitoring inside the effort function (the specification of effort and h is therefore redefined as in eq. (9.10)).

Finally, we make standard assumptions about coefficients a and b: $0 < a < 1$, $0 < b < 1$, in order to guarantee concavity of $e(w_c, m)$ in its arguments. Maximization stands as follows:

$$\underset{m, w, Lp}{\text{Max}} \ \pi = h \cdot e \cdot Lp - (w + s) \cdot Lp. \tag{9.11}$$

Substituting respectively expressions (9.5) or (9.6) for the cost of job loss into the profit function and considering the corresponding definition of s

given above (Table 9.1), we can finally resolve (9.11). After simplifying Lp on both sides, we obtain the following first-order conditions:

$$\partial\pi/\partial m = 0 \quad \Rightarrow h\cdot(\partial e/\partial m) = M, \tag{9.12a}$$

$$\partial\pi/\partial Lp = 0 \quad \Rightarrow h\cdot e = w + s, \tag{9.12b}$$

$$\partial\pi/\partial w = 0 \quad \Rightarrow h\cdot(\partial e/\partial w) = 1. \tag{9.12c}$$

Combining conditions (9.12a) and (9.12c) gives the definition of the optimal relationship between the two control devices (monitoring and cost of job loss):

$$w_c/m = M\cdot a/b = 1, \tag{9.13}$$

where we fix, in order to simplify notations, $M = b/a$.[13] Substituting into the effort function (9.9) we obtain:

$$e^* = r\cdot(w_c)^{a+b}. \tag{9.14}$$

This expression allows the determination of the level of effort as a function of the cost of job loss (and therefore wage). In order to rule out any possibility of perverse results due to increasing returns to scale in the effort function, we assume: $a + b \le 1$.

In order to determine the equilibrium configurations of wages and opportunity rates, we have to consider now the two remaining first-order conditions. To do that, we first combine conditions (9.12b) and (9.12c) to obtain the wage curves associated with different MoO; then we impose the constancy of the cost of job loss at the equilibrium (condition (9.12c)).

Equilibrium Wage/Opportunities Configurations

Combining conditions (9.12b) and (9.12c), resolving and substituting (under the *ex post* equilibrium assumption that $w = w_y$) for the definitions of the cost of job loss respectively corresponding to a decentralized (D) and a centralized (C) firm, eventually we obtain the following optimal wage curves:

$$w(C) = \frac{a\cdot t + (1-b)\cdot[1 - p\cdot(1-l)]\cdot w}{[1 - p\cdot(1-l)]\cdot(1-b) - a}, \tag{9.15}$$

$$w(D) = \frac{(1-b')\cdot[1 - p\cdot(1-l)]\cdot w}{[1 - p\cdot(1-l)]\cdot(1-b') - a}, \tag{9.16}$$

where $b' = b - \varepsilon \cdot (\delta - 1) < b$.

We can see from the above that both wage curves establish a positive relationship between wage and the opportunity rate $p(u,v)$. Moreover, we can easily show that a maximum opportunity rate exists that ensures positive wages $(< \infty)$:

$$p_{max} = \frac{1 - b - a}{(1 - l) \cdot (1 - b)}. \tag{9.17}$$

Two points are worth making concerning expression (9.17). First, since $b' < b$, the maximum opportunity rate associated with decentralized firms is higher than the one associated with centralized firms. The reason for this is that a lower value of the monitoring coefficient b (as in decentralized firms) reduces the scope for the effort-enhancing mechanism relying on external discipline (that is, linked to the threat of being fired), thus reducing, *ceteris paribus*, the intensity of effort. This must be counter-balanced by an external shift of the wage function (reduction of wage for any given level of the opportunity rate) which also means that a higher opportunity rate is compatible with a positive wage.

Second, we can see from (9.17) that the maximum opportunity rate is a decreasing function of the average transferability of workers' competence (1–l). This results from the role played by firm-specific competence as an alternative disciplinary device replacing the external mechanism relying on unemployment and vacancy rates. When competence becomes less transferable, workers' opportunity to find a new job is reduced for any given level of the opportunity rate.

Building on these observations and on assumptions previously made concerning the cross-country ranking of parameters b and l (see section 2), we can easily show that $p$max(Jap) > $p$max(Ger), and $p$max(US) > $p$max(Fr), since $b' < b$.

More generally, considering also the role of l, we can easily see that $p$max $(D) > p$max(C) if:

$$l(C) - l(D) < a \cdot (b - b') / [(1 - b) \cdot (1 - b')]. \tag{9.18}$$

According to the proposed ranking for l (section 2), that is, l(Jap) > l(US) > l(Fr) > l(Ger), condition (9.18) on $p$max(D) is always satisfied for Japan (since in this case $l(D) > l(C)$), while the result is uncertain for Germany. Since i(US) > l(Fr), condition (9.18) for Germany is more easily verified against France than against the USA, so that we can expect the following tentative ranking to hold:

$$p\,\text{max(Jap)} > p\,\text{max(US)} > p\,\text{max(Ger)} > p\,\text{max(Fr)}. \tag{9.19}$$

Considering now condition (9.12c) leads us to define a complementary relationship between wage and opportunity rate, which allows us to determine the equilibrium configuration for these two variables. Condition (9.12c) states that the cost of job loss has to be constant at equilibrium. This condition can actually be rewritten as $h \cdot (\partial e / \partial w) = \Rightarrow w_c^* = B(I)$, where $I = D,C$ respectively indicates decentralized and centralized MoO. Considering that B values are $B(D) = [a \cdot h \cdot r]^{1/(1-a-b')}$, $B(C) = [a \cdot r]^{1/(1-a-b)}$, we can show that the inequality $B(D) < B(C)$ generally holds, since $h < 1$.

Substituting the definition of the cost of job loss for $wc(I)$ into the above condition, we obtain the following complementary positive relationships between the wage and the opportunity rate:

$$p(I) = \frac{w - \underline{w} - B(I)}{(1-l) \cdot (w - \underline{w})}. \tag{9.20}$$

In order for the predicted value of p to be positive, the following condition must hold: $w > w\text{min} = \underline{w} + B(I)$, stating that a minimum wage level exists (even assuming away unemployment benefits). Actually, a maximum wage level also exists corresponding to $p = 1$. In this case, in fact, $w(I) = w\text{max}(I) = \underline{w} + B(I)/l(I)$.

The maximum wage level always increases when competence specificity (l) decreases (so it is higher for France compared to the USA, and for Germany compared to Japan). We can also show that $w\text{max}(D) > w\text{max}(C)$, if $B(D) \cdot l(C) > B(C) \cdot l(D)$.

For the above condition to hold, we need either $l(D)$ to be very low or $l(C)$ to be very high. Therefore we can reasonably assume the condition to hold for Germany (low $l(D)$). Concerning Japan, the above condition is more easily verified against the USA than against France (because $l(\text{US}) > l(\text{Fr})$). Depending on this, the final equilibrium configurations for centralized and decentralized MoO are respectively defined considering the wage curve and the corresponding complementary relationship (9.20). The result is illustrated in Figure 9.3, which shows the wage curves as well as the opportunity curves corresponding to the different microinstitutional settings that have been singled out. On the horizontal axis we consider the complement to one of the opportunity rate p; therefore countries showing higher equilibrium opportunity rates are to be found closer to the origin.

The figure shows different equilibrium values of the opportunity rate depending on the assumptions made concerning the parameters that characterize the microinstitutional settings. As we can see from the analysis above, the parameter r grasping the role of firing barriers does not affect the position of the wage curves above while it does affect the opportunity curves (9.20). This means that deregulation (that is, an increase of r) leads to an upward shift of the opportunity curves, thus pushing the equilibrium opportunity rate toward

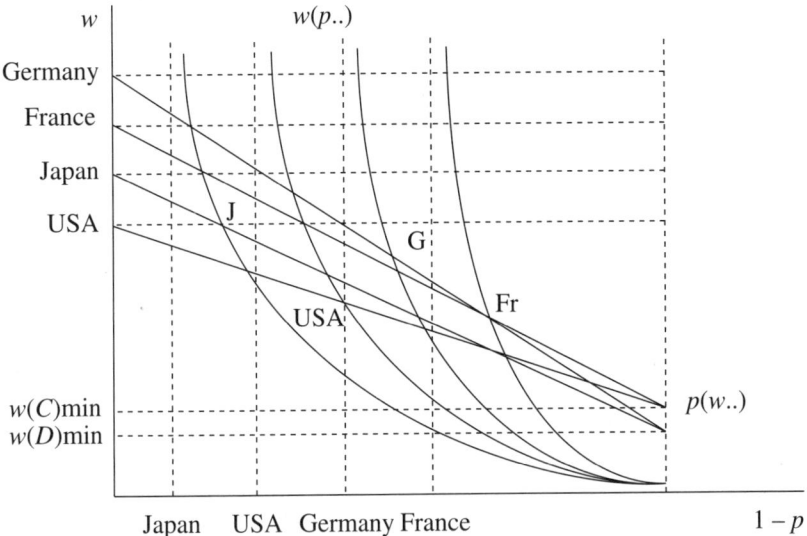

Figure 9.3   Equilibrium opportunity rate

p_{max}. However, it also appears that labour market deregulation alone does not yield a modification of the maximum level of the opportunity rate (p_{max}) that is determined by the wage curves.

4.   EQUILIBRIUM UNEMPLOYMENT

The Equilibrium Level of the Opportunity Rate

Until now we have retained a general specification of the opportunity rate. However, we know that the opportunity rate actually depends on unemployment and vacancy rates. Therefore we should consider that the above result concerning the equilibrium opportunity rate actually defines a first equilibrium relationship between unemployment and vacancy:

$$p(u,v) = p^*, \tag{9.21}$$

with $\partial u / \partial v > 0$. This relationship will be called the 'opportunity locus'. In order to simplify calculations, we assume the opportunity rate to be characterized by the following functional form:

$$p = v \cdot (1 - u). \tag{9.22}$$

This is a minor (and reasonable) assumption that has the merit of allowing us to derive explicit results concerning the role of microinstitutional settings on equilibrium unemployment rates.

Given the above specification, we can easily see that the relative position (in the (u, v) space) of the opportunity locus corresponding to different microinstitutional settings actually depends on the optimal p value derived in previous sections, that is: $p^*(\text{Jap}) > p^*(\text{US}) > p^*(\text{Ger}) > p^*(\text{Fr})$.

Given this first relationship between unemployment and vacancy rates, we still have to define a complementary one in order to characterize the global equilibrium configuration for all variables. This complementary condition is the flow equilibrium condition necessary to ensure equality of flows inside and outside the unemployment set. In fact, the efficiency wage equilibrium that was described in the previous sections does not prevent firms from firing workers who are discovered not working. In fact, we can easily show that the equilibrium level of effort is $e^* = r^b \cdot [h \cdot a \cdot r^b]^{(a+b)/1-a-b} < 1$, which means that the optimal choice of effort intensity leads workers to work only a part ($e^* < 1$) of the working day. Monitoring then allows the firm to discover (and possibly fire) workers when they are not working.

In the following section, we will derive the aggregate unemployment–vacancy relationship implied by the equilibrium flows condition on the labour market, under the hypothesis of an endogenous determination of both the separation and job finding rates.

Flows Equilibrium

In order to determine the equilibrium unemployment rate, we have to consider a complementary relationship ensuring the equilibrium on the labour market. Therefore a flows equilibrium condition is required to impose equality of flows in and out of the unemployment set. We define d as separation rate, and f as job finding rate. The flows equilibrium condition is then $L \cdot d = (N - L) \cdot f$, where L is employed people, and N is the labour force. From this we obtain:

$$u = (N - L) / N = d / (d + f). \qquad (9.23)$$

As we can see, an increase of the job finding rate makes unemployment go down, while an increase of the separation rate pushes unemployment up.

We now define the separation and job finding rates. Since all workers are identical and have the same probability of finding a job when entering the labour market (either for the first time or after being fired), the job finding rate can actually be understood as the probability of finding a new job (when fired). As we have seen in section 3, this probability is determined by the opportunity rate combined with the degree of specificity of workers' human capital:

$$f = p \cdot (1-l), \tag{9.24}$$

Substituting (9.22) and (9.24) for p into (9.23) we obtain the following equilibrium conditions:[14]

$$u \cdot v = d/(1-l). \tag{9.25}$$

We now turn to the definition of the separation rate. In our model, separations from the firm occur only when a worker is discovered not working. Therefore the probability of being discovered not working (and fired) is a function of the quantity of resources allotted to monitoring activities as well as of firing barriers.[15] In particular, γ = probability of being fired = $\gamma (r \cdot m)$, γ' > 0. The separation rate is then:

$$d = \text{separation rate} = \gamma(m), \tag{9.26}$$

which is endogenously determined inside the model by the quantity of resources affected to monitoring activities.

From (9.26), and considering that the optimal level of (m) is such that (eq. (9.13)) $m^* = w_c^*$, with (from condition (9.12c)):

$$w_c^*(I) = B(I), \quad I = D \text{ or } C, \tag{9.27}$$

where $B(D) = [a \cdot h \cdot r^b]^{1/(1-a-b')}$, $B(C) = [a \cdot r^b]^{1/(1-a-b)}$, we can deduce that

$$\gamma = \gamma(B(I)). \tag{9.28}$$

Hence the equilibrium condition (9.25) implies a negative relationship between the vacancy rate and the unemployment rate: $u = u(v; b, l)$, $\partial u / \partial v < 0$.

Building on this framework, it is possible to interpret differences in the equilibrium rate of unemployment across developed countries as a consequence of the varying nature of microinstitutional settings.

Since $d = \gamma$, we can substitute (9.28) for γ into condition (9.25). Considering that at the equilibrium the level of monitoring is the same as the level of the cost of job loss (eq. (9.13)) and that the latter is given by eq. (9.27), we can easily obtain an explicit formulation of condition (9.25). We finally arrive at the following unemployment/vacancy curve:

$$u(I) = \frac{\gamma(r \cdot B(I))}{v \cdot (1 - l(I))}. \qquad (9.29)$$

As we can see, the position of the $u(v)$ curves crucially depends on parameters identifying the different microinstitutional settings. In particular, a higher equilibrium value of monitoring (that is, higher $B(I)$) determines a higher separation rate and therefore pushes the Beveridge curve outwards; at the same time, a better transferability of workers' competence (that is, lower l) increases the job finding rate, thus shifting down the Beveridge curve.

Building on these results and considering our assumptions on parameters b and l (see section 2) and previous results regarding $B(I)$, we can show that the minimum unemployment rate (corresponding to a vacancy rate's value $v = 1$) is such that $u$min(Ger) < $u$min(Fr), $u$min(Jap) < $u$min(US).

This allows us to determine the relative position of the Beveridge curves corresponding to different microinstitutional settings (see Figure 9.4). Combining eq. (9.26) with the corresponding (Beveridge) curve allows us to define the final equilibrium configurations for the unemployment and vacancy rates. The figure presents a graphical representation of the equilibrium configurations across different microinstitutional settings.

We can use Figure 9.4 to study the impact of labour market deregulation on the equilibrium configurations. Labour market deregulation is grasped by an increase of the parameter r. As we have seen in section 3, this leads to an

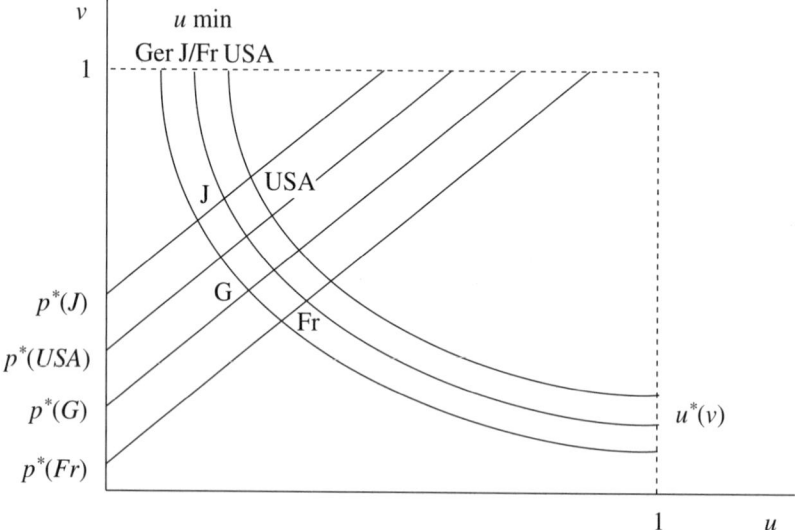

Figure 9.4   Equilibrium unemployment rate

increase in the equilibrium value of the opportunity rate (p^*). At the same time, it can easily be seen from eq. (9.29) that it also pushes the Beveridge curves upwards. Therefore, according to this model, the final effect of deregulation on the unemployment rate is likely to be uncertain, while this kind of measure certainly has a role in increasing the vacancy rate.

5. CONCLUSIONS

This chapter has built on an efficiency wage framework to model firms' choice on wage and effort as an 'institutionally biased' optimization. In order to do this, a stylization of different microinstitutional settings has been proposed and the notion of 'mode of organization' introduced. This allows us to take into account the influence of microinstitutional parameters on firms' behaviour.

The proposed stylization deals with varying microinstitutional settings across developed countries (France, Germany, Japan and the USA) and interprets them as a consequence of two factors: the nature of workers' competence (parameter *l*) and the MoO of firms (parameter *b*). On the basis of this approach, a model was presented combining an efficiency wage determination of wage/opportunity equilibrium configurations and a flow equilibrium condition for unemployment and vacancy rates. Results concerning the equilibrium unemployment/vacancy configurations were derived in the chapter, which point to a major role played by the mode of organization of firms and the nature of workers' competence, rather than labour market flexibility, as crucial determinants of cross-country differences in the equilibrium rate of unemployment. These results throw some light on the structural causes underlying European high unemployment experience and on possible determinants of Japan's exceptional employment performance.

In particular, this theoretical exercise allows us to deduce some important policy implications.

1. Labour market deregulation is not the only way out of a 'high unemployment equilibrium', and it is an insufficient measure as long as differences in firms' organization and workers' competence exist and persist.
2. A wider 'institutional reorganization' needs to be carried out in order to modify the equilibrium level of structural unemployment in European countries.
3. This reorganization should be focused on both the mode of organization of firms and the nature of workers' competence (for example, development of more firm-specific competence).

In this respect, the empirical evaluation of the actual strength (in different countries) of the results provided by the model could be an important step further in the direction of a wider and deeper understanding of the problem of structural unemployment.

NOTES

1. I am very grateful to Eve Caroli, Daniele Checchi, Giovanni Dosi and Youri Kaniovski for their suggestions and critical comments on a preliminary version of this chapter. I am also indebted to Masahiko Aoki, Robert Boyer, Jérôme Bourdieu, Bob Hancké and participants in the TSER project on 'Technology, Economic Integration and Social Cohesion' and for their suggestions. A special thanks to David Soskice for his precious analytical comments on the model presented in the chapter. The usual caveats apply.
2. This is actually a crucial point in our approach. Our concept of equilibrium rate of unemployment is meant to grasp the sustainable level of unemployment inside the economy, compatible with labour market equilibrium. It can easily be shown (see Carlin and Soskice, 1990) that in an open economy framework the equilibrium rate of unemployment actually defines a minimum sustainable rate of unemployment, while the actual rate is also determined by other factors (demand factors).
3. For an interesting empirical study on British data, concerning the intensity of labour mobility associated with different forms of apprenticeship, see Booth and Satchell (1996).
4. The definition of 'generic competence' encompasses both low, generic skills and high, general skills. In other words, the distinction between the quality of skills will not be taken into account; we will focus only on the distinction in the nature of competence.
5. The parameter l, as I have defined it, is the proportion of non-transferable competence to total workers' competence. The same kind of ranking for l, based on a comparative analysis of individual countries' training systems, can be found in Soskice and Hancké (1997). Inequalities given in the text can actually be weakened (for example, considering only two groups of countries: high versus low firm-specific competence countries) with no substantial modification in my subsequent analysis. Strong inequalities are given just to simplify exposition.
6. A more comprehensive analysis of the nature and characteristics of the mode of organization of the firm can be found in Gatti (1997).
7. Given the assumption of perfectly competitive markets, the relative product price P has been directly assumed equal to 1.
8. The specification of the effort function has a sociological justification relying on the conflicting nature of the labour–capital relationship. Under the same assumption, Bowles (1985) obtains a similar specification considering explicitly workers' utility maximization.
9. The retained alternative income definition is derived under two crucial assumptions: there is no distinction in the nature of the jobs offered by firms, and (b) workers acquire their competence after being hired the first time: it is therefore impossible for firms to make any distinction between specific and generic workers upon the first hiring. These assumptions are rather common ones in matching models literature. In the present context, they allow me to define the alternative income simply as an aggregate average.
10. A similar approach is generally taken up in recent models of firms' organization (see Greenan and Guellec, 1994).
11. The specification taken up is a generalization of the functional form proposed by Bowles and Boyer (1990).
12. This is actually a standard hypothesis, being m^δ, a cost and therefore generally a convex function.
13. In order to make sure that $M > 1$ it is then sufficient to assume that $b > a$.

14. There is actually a second root which is always equal to 1, so we will not consider it as a possible solution for the equilibrium unemployment rate.
15. As Bowles (1985) points out, this probability is also influenced by the actual value of effort: when monitoring increases, effort increases too, and the probability of being fired is reduced. However, it can be shown that if $e^* < 1 / (1 + b)$ then $\partial\gamma /\partial m > 0$ in spite of the positive effect on effort. In the text we take up the assumption that this condition is satisfied and just focus on a simpler version of the model, where γ is a linear function of m. However, results are completely unchanged if we consider the more complex version where m also affects effort and γ is a non-linear function of monitoring, provided that the condition $e^* < 1 / (1 + b)$ still holds.

REFERENCES

Aoki, M. (1986), 'Horizontal vs. Vertical Information Structure of the Firm', *American Economic Review*, 76, 972–83.

Aoki, M. (1994), 'Sur certains aspects des conventions dans l'entreprise', in A. Orléan (ed.), *L'analyse économique des conventions*, Paris: Economica, pp. 281–305.

Becker, G. (1962), 'Investment in Human Capital: a Theoretical Analysis, *Journal of Political Economy*, 70(5), 9–49.

Berg, P.B. (1994), 'Strategic Adjustment in Training: A Comparative Analysis of the U.S. and German Automobile Industries', in L.M. Lynch (ed.), *Training and the Private Sector; International Comparisons*, London: University of Chicago Press, pp. 77–107.

Booth, A.L. and S. Satchell (1996), 'On apprenticeship qualifications and labour mobility', in A.L. Booth and D.J. Snower (eds), *Acquiring Skills. Market failures, their symptoms and policy responses*, Cambridge: Cambridge University Press, pp. 285–302.

Bowles, S. (1985), 'The Production Process in a Competitive Economy: Walrasian, Neo-Hobbesian and Marxian Models', *American Economic Review*, 75, 16–36.

Bowles, S. and R. Boyer (1990), 'Labour Market Flexibility and Decentralisation as Barriers to High Employment? Notes on Collusion, Centralised Wage Bargaining and Aggregate Employment', in R. Brunetta and C. Dell'Aringa (eds), *Labour Relations and Economic Performance*, London: Macmillan, pp. 325–53.

Carlin, W. and D. Soskice (1990), *Macroeconomics and the Wage Bargaining*, London: Oxford University Press.

CEPR (1996), *Unemployment: Choices for Europe*, Paris.

Coriat, B. and G. Dosi (1995), 'Learning How to Govern and Learning How to Solve Problems: On the Co-Evolution of Competences, Conflicts and Organisational Routines', IIASA Working Paper, WP-95-06, Laxenburg.

Gatti, D. (1997), 'Flexible technology, unemployment and effort. The role of the organization of the firm', IIASA Report, IR-97-004.

Greenan, N. and D. Guellec (1994), 'Coordination within the Firm and Endogenous Growth', *Industrial and Corporate Change*, 3(1), 173–97.

Hollingsworth, J.R. and R. Boyer (eds) (1997), *Contemporary Capitalism, the Embeddedness of Institutions*, Cambridge: Cambridge University Press.

Kristensen, P.H. (1996), 'Variations in the nature of the firm in Europe', in R. Whitley and P.H. Kristensen (eds), *The Changing European Firm*, London: Routledge, pp. 3–20.

Layard, R., S. Nickell and R. Jackman (1991), *Unemployment, Macroeconomic Performance and the Labour Market*, Oxford: Oxford University Press.

Lazonic, W. and J. West (1995), 'Organizational Integration and Competitive Advantage: Explaining Strategy and Performance in American Industry', *Industrial and Corporate Change*, 4, 229–70.

Lorenz, E.H. (1992), 'Trust and the Flexible Firm: International Comparisons', *Industrial Relations*, 31, 455–72.

Marsden, D. (1990), 'Institutions and Labour Mobility: Occupational and Internal Labour Markets in Britain, France, Italy and West Germany', in R. Brunetta and C. Dell'Aringa (eds), *Labour Relations and Economic Performance*, London: Macmillan, pp. 414–38.

Marsden, D. (1996), 'Employment Policy Implications of New Management Systems', *Labour*, 10, 17–61.

Maurice, M. (1993), 'La formation professionelle en France, en Allemagne et au Japon: trois types de relation entre l'école et l'entreprise', *Entreprise et Histoire*, 3, 317–26.

Milgrom, P. and J. Roberts (1990), 'The Economics of Modern Manufacturing: Technology, Strategy and Organization', *American Economic Review*, 80, 511–28.

OECD (1997), *Employment Outlook*, Paris: OECD.

Okuno, M. (1984), 'Corporate Loyalty and Bonus Payment: an Analysis of Work Incentives in Japan', in M. Aoki (ed.), *The Economic Analysis of the Japanese Firm*, Amsterdam: North-Holland, Elsevier Science Publishers.

Soskice, D. (1990), 'Reinterpreting Corporatism and Explaining Unemployment: Co-ordinated and Non-co-ordinated Market Economies', in R. Brunetta and C. Dell'Aringa (eds), *Labour Relations and Economic Performance*, London: Macmillan, pp. 170–211.

Soskice, D. and B. Hancké (1997), 'De la construction des normes industrielles à l'organisation de la formation professionelle. Une approche comparative', in M. Möbus and E. Verdier (eds), *Les diplômes professionnels en Allemagne et en France*, Paris: l'Harmattan, pp. 245–62.

Streeck, W. (1996), 'Lean Production in the German Automobile Industry: A test case for Convergence Theory', in S. Berger and R. Doore (eds), *National Diversity and Global Capitalism*, London: Cornell University Press, pp. 138–70.

10. Sweeping the chimney before kindling the fire as a workable option for employment policy

Adriaan van Zon, Huub Meijers and Joan Muysken

1. INTRODUCTION

One of the most pressing problems of today is the uneven distribution of the burden of unemployment across different skill groups. Low-skilled workers especially are more often unemployed than others (OECD, 1994). They are also unemployed for longer periods of time (Muysken and Ter Weel, 1997). The unevenness of this distribution seems to indicate that having different skills determines to some extent one's employment/unemployment opportunities. However, policy prescriptions meant to alleviate the problems associated with the uneven distribution of unemployment are primarily based on (conceptual) models which effectively ignore skill differences between workers. Policy recommendations involving wage flexibility or reductions in costs of hiring and firing, for example, seem to be based on the notion that, essentially, skills are not the distinctive elements in defining the employment opportunities of an individual. Instead, individual skill services, being overpriced or being too costly to hire and/or fire, are often regarded to be the real cause of disparities in unemployment rates by skill. In this chapter we will argue that this neglects the fact that people with high-level skills have intrinsically more employment opportunities than people with low-level skills. This is because low-skilled people may find it (too) hard to fill high-level jobs.[1] High-skilled people, on the other hand, may be expected to be able to perform, in principle at least, in both low-level jobs and high-level jobs. In addition, it may well be possible that high-skilled workers are more efficient than low-skilled workers on low-level jobs. Given these asymmetries between low-skilled and high-skilled workers' employment opportunities, it follows directly that the number of low-level jobs is an upper limit to low-skilled employment, while the number of high-skilled jobs is a lower limit to high-skilled employment.[2] Hence, in the absence of supply bottlenecks, high-skilled employment cannot fall below the number of high-level jobs, whereas low-skilled employment may even fall to zero, when high-skilled workers take over low-level jobs.[3] While, therefore, the creation of a high-level job necessarily favours high-skilled employment, employment prospects of the low-skilled are not necessarily improved by the creation of low-level jobs. The creation of high-level jobs draws high-skilled workers into those jobs, possibly leaving low-level job vacancies in the process. This gives low-skilled workers the opportunity to fill these vacancies. Thus, by 'sweeping the employment chimney' (that is, the creation of high-level jobs), unemployed low-skilled workers may simultaneously be drawn into the 'employment fire' at the low-skilled end of the chimney. In the Netherlands, this is known as the 'chimney effect'. By contrast, an increase in the amount of fuel present at the low end of the chimney through the creation of low-level jobs (this is the trickling down effect mentioned in note 3) may actually result in more smoke rather than a bigger fire. The latter depends to a large extent on the general condition of the employment chimney itself, as we intend to illustrate in this chapter.

An additional asymmetry between different skill types of labour lies in the possibility that low-skilled workers may be low-skilled because it is difficult for them to learn enough to become high-skilled. With firms having become 'leaner and meaner' over the last decade, this 'learning inability' may provide incentives for entrepreneurs to hire workers who are flexible, in the sense that they could easily learn different types of tasks, instead of hiring people who can only perform a limited number of tasks. These learning asymmetries between low-skilled and high-skilled workers may also be part of the explanation of the uneven distribution of unemployment across skills.[4]

The above suggests that there are a number of ways to increase the employment prospects of low-skilled workers. The first is by increasing the number of low-level jobs, the second by decreasing the relative labour costs of low-skilled workers, thus increasing the willingness of entrepreneurs to allocate low-skilled workers instead of high-skilled workers to low-level jobs, and the third by increasing the number of jobs which only high-skilled workers can take. A fourth option would be to raise the skill level of low-skilled workers to some extent by means of (re-)training programmes. The effectiveness of the first possibility depends in part on the availability of high-skilled workers for low-level jobs. The effectiveness of the second measure is limited to the extent that such a change in relative wage costs at first tends to increase the unemployment of high-skilled workers while decreasing the unemployment of low-skilled workers. However, when wage costs per unit of output fall, a net expansion of jobs may be expected to occur and therefore a positive net effect on overall employment. Otherwise, the bias in unemployment will just shift against high-skilled workers.[5] The third measure structurally decreases the availability of high-skilled workers for low-level jobs, and hence increases employment prospects for the low-skilled, almost as a beneficial side-effect. The fourth measure, insofar as it would be feasible, would be highly desirable, for it would structurally diminish the constraints put on employment opportunities by the existence of substitution asymmetries between skills.

In this chapter we will describe a model which combines most of the notions regarding asymmetries in employment perspectives between low-skilled workers and high-skilled workers mentioned above. We will look at the influence of asymmetries in substitution possibilities between skills, but also of asymmetries in learning capabilities between skills. To this end we specify a model of labour demand defined in terms of jobs, which, given the supply of labour which is heterogeneous in the skill dimension, requires matches to be made between the skills supplied and the jobs under consideration. The model is highly neoclassical in nature, except for the asymmetries mentioned above. The skill allocation model to be described in this chapter is part of a more general production structure which has an explicit vintage production structure (see MASTER, 1997 for more details). In this chapter, however, we focus exclusively on the skill allocation model and the way in which it allows the chimney effect to do its work.

The organization of this chapter is as follows. In section 2 we describe the principal outlines of the skill allocation model (SAM). It summarizes the more detailed model description present in MASTER (1997). Section 3 describes the results we have obtained with a first empirical implementation of the skill allocation model for Germany and the Netherlands. Section 4 describes the outcomes of two *ex post* simulation experiments conducted for those countries. These experiments were performed in order to highlight the differences between the working of the chimney effect and standard policy methods to favour employment prospects of low-skilled workers. Section 5 contains a summary and some concluding remarks.

2.   A MODEL OF SUBSTITUTION ASYMMETRIES BETWEEN JOBS AND SKILLS

Introduction

High-skilled people may in general be expected to perform better in low-level jobs than low-skilled people in high-level jobs. Hence there is an intrinsic bias against allocating low-skilled people to high-level jobs. Nonetheless, some kind of substitution between higher skills and lower skills should be possible (in principle at least), since there is empirical evidence to this effect.[6] Our approach differs from the foregoing approaches in that we distinguish between substitution possibilities between jobs as well as between skills, because these are intrinsically different. In our set-up, a job is a set of tasks which need to be performed. This requires the people engaged in that job to have a certain *minimum* skill level in order to be able to perform the tasks in question. These specific minimum skill requirements define the level of the job. Therefore, generally speaking, high-skilled people can be hired for more job levels than low-skilled people. This provides an asymmetry in employment opportunities for high-skilled and low-skilled people, which, given a certain lack of compensating asymmetries in wage formation, leads to a bias in employment opportunities in favour of high-skilled people.

Ex ante Job Demand and *ex post* Skill Demand

We assume that the design of a production process entails the definition of certain packages of tasks for low- and high-skilled people. We assume, furthermore, that technical constraints on combining these tasks are such that a decrease of the number of high-level jobs must be compensated by an increase in the number of low-level jobs, and vice versa.[7] But, once a job combination has been chosen and implemented in the design of machinery and equipment, it cannot be altered: the job composition of employment is assumed to be 'putty' *ex ante* and 'clay' *ex post*. We also assume that the capital stock depreciates with rate δ, and so does the number of jobs embodied in the capital stock. We furthermore assume that the efficiency of a high-skilled worker on a low-level job relative to the efficiency of a low-skilled worker on that job is given by the constant number e'. This implies that e' low-skilled workers on a low-level job could be replaced by one high-skilled worker: we assume therefore that the actual allocation of skills to jobs is essentially 'putty' *ex post*.[8] The corresponding substitution possibilities between high- and low-skilled workers on a low-level job are given by equation (10.1), which combines the substitution and relative efficiency assumptions mentioned above:

$$L_t = L^* \cdot e^{-\delta t} - e' \cdot (H_t - H^* \cdot e^{-\delta t}). \qquad (10.1)$$

The framework sketched above can be represented as in Figure 10.1. In this figure, H^* and L^* denote high-level and low-level jobs, respectively, while H and L denote high- and low-skilled workers, respectively. *Ex ante* job combinations (H^*, L^*) are described using a standard 'unit job isoquant', while the feasible *ex post* combinations of high- and low-skilled workers associated

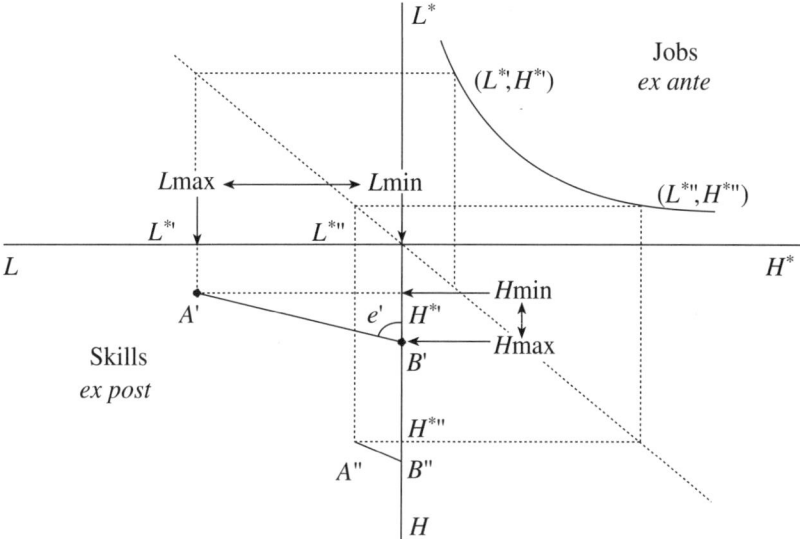

Figure 10.1 Ex ante *job substitution and* ex post *skill substitution*

with a certain job combination are given by linear relations in the *L,H* quadrant, as described by (10.1).

In Figure 10.1, job combinations (L^*, H^*) are given in the north-east quadrant. The 45-degree lines in the north-west and south-east quadrants are used to obtain the skill-substitution possibilities between *L* and *H* for any given job combination (L^*, H^*) as depicted in the south-west quadrant. Note that in the latter quadrant, the lines through the points A', B' and the points A'', B'' are parallel lines. The angle of these lines with respect to the horizontal is equal to $-1/e'$ (see equation (10.1)). The *ex ante* choice on the job combination determines the minimum and maximum employment opportunities for both low- and high-skilled workers. For instance, in the case of the job combination $(L^{*'}, H^{*'})$, low-skilled employment is bounded between *L*min and *L*max, whereas high-skilled employment is bounded between *H*min and *H*max. Note, furthermore, from this figure, that the line $A''B''$ represents smaller employment opportunities for low-skilled workers than the line $A'B'$ due to the fact that the number of low-level jobs in job combination $(L^{*''}, H^{*''})$ is much smaller than in job combination $(L^{*'}, H^{*'})$. Note also that the intercepts B' and B'' represent the maximum number of high-skilled workers associated with a certain job combination; that is, B' and B'' are the number of high-skilled workers fully occupying both low-level and high-level jobs. From now on we define this maximum number of high-skilled workers to be H'. Finally, it should be noted that the choice of a certain *point* in the L^*,H^* quadrant defines a *feasible choice set* in the L,H quadrant as given by (10.1).

Assuming that entrepreneurs can determine the job layout of their production process (that is, (H^*,L^*)) only when new capacity is installed, it follows that the expected present value of the operating cost of one labour efficiency unit installed at time zero declines over time with the discount rate r and with the rate of technical depreciation δ.[9] Hence the expected present value of total operating cost of one labour efficiency unit which is installed at time zero is given by:

$$T = \int_0^\infty \left(w_{H,t}^e \cdot H_t + w_{L,t}^e \cdot L_t \right) \cdot e^{-rt} dt, \tag{10.2}$$

where $w_{H,t}^e$ and $w_{L,t}^e$ are the expected values of the wage rate for high-skilled and for low-skilled workers at time t.

The Allocation of Skills in the Absence of Skill Supply Constraints and Adjustment Costs

In the absence of supply constraints and adjustment costs, and given a job combination (L^*,H^*), the choices of L_t and H_t can be modelled by solving:

$$\text{Min } F_t = \int_0^\infty \left(w_{H,t}^e \cdot H_t + w_{L,t}^e \cdot L_t \right) \cdot e^{-rt} \tag{10.3}$$

$$s.t. \ L_t = L^* e^{-\delta t} - e' \cdot (H_t - H^* e^{-\delta t}).$$

Given the linear nature of the problem, $(H = H', L = 0)$ is the solution to (10.3) when $w_H^e / w_L^e < e'$, while $(L = L^* \cdot e^{-\delta \cdot t}, H = He^{-\delta \cdot t})$ is the solution when the opposite is the case.[10] Using the shorthand notation $w = w_H^e / w_L^e$ and $\hat{w} = \hat{w}_H - \hat{w}_L$, for given and constant expected exponential growth rates $\hat{w}_H$ and $\hat{w}_L$, there are now four distinct possibilities to consider, depending on whether $w < e'$ or $w > e'$ and $\hat{w} < 0$ or $\hat{w} > 0$. For instance, when $w < e'$ and $\hat{w} < 0$ (case 1) the optimal solution will always be $(H = H',L = 0)$. But when $w < e'$ and $\hat{w} > 0$ (case 2) there will be a moment when the other solution $(H = H^*,L = L^*)$ will be chosen. A switch between states will also occur in case 3, which starts out with $w > e'$ and $\hat{w} < 0$. Case 4 will always have $(H = H^*,L = L^*)$ since $w > e'$ and $\hat{w} > 0$. The four cases are depicted in Figure 10.2.

In cases 2 and 3 a switch from the one corner solution of (10.3) to the other will take place at time t^*.[11] t^* itself is easily determined from the condition that, at the moment of switching,

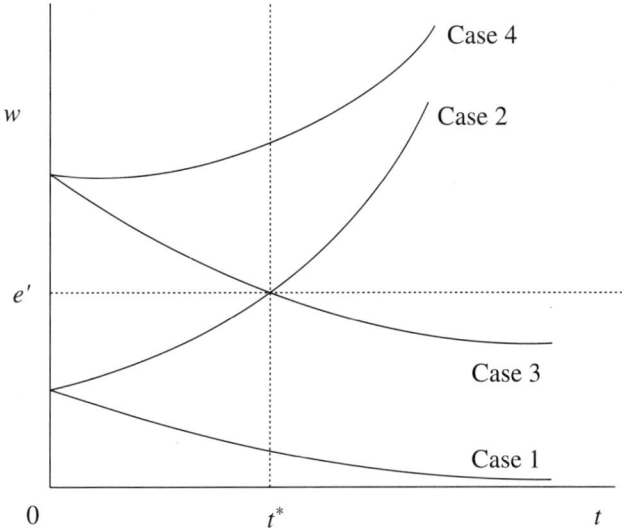

Figure 10.2   Expected wage growth and skill switching

$$w_{t^*} = \frac{w_{H,t^*}}{w_{L,t^*}} = \frac{w_{H,0}}{w_{L,0}} \cdot e^{(\hat{w}_H - \hat{w}_L)t^*} = e' \Rightarrow t^* = \frac{\ln\left(e' / \left\{\frac{w_{H,0}}{w_{L,0}}\right\}\right)}{\hat{w}_H - \hat{w}_L}. \quad (10.4)$$

Using equation (10.4), the solution of the original problem over an infinite horizon can be defined as the solution to the problem of minimizing unit operating costs over two consecutive horizons, namely $0 - t^*$ and $t^* - \infty$. Thus the minimum expected present value of operating costs of a low-level job can be obtained directly for all four cases.

Choosing jobs

The solution to the allocation problem defined by the minimization of (10.3) subject to (10.1) is easily obtained by substituting (10.1) and (10.4) into (10.3). The solution of the objective function for case 1 can now be written as:

$$T_1 = \int_0^\infty w_{H,0} \cdot H' \cdot e^{(\hat{w}_H - \rho)t} dt = -\frac{w_{H,0}}{\hat{w}_H - \rho} \cdot (L^* / e' + H^*) = \alpha_1 \cdot H^* + \beta, \quad (10.5)$$

where α_1 and β_1 are implicitly defined by (10.5), and T_1 denotes the value of the objective function given by (10.3) for case 1, as depicted in Figure 10.2.

The term ρ is equal to the rate of technical decay plus the discount rate. Note that α_1, $\beta_1 > 0$. Note also that the objective function T_1 is linear in L^* and H^*. Hence maximization of T_1 by choosing (L^*, H^*) constrained by the unit job isoquant implicitly defined by $\phi(L^*, H^*) = 1$ provides values for L^*, and H^* in terms of e', relative wages, relative growth rates in wages and the parameters of the job-substitution function $\phi(L^*, H^*)$. As was the case with T_1, $T_2 - T_4$ can also be written as $T_i = \alpha_i \cdot H^* + \beta_i \cdot L^*$.[12]

The αs and βs can readily be interpreted as the expected present value of the minimum cost of operating high- and low-level jobs conditional on expectations with regard to the growth of relative wages. The solution to the job-composition problem is therefore implicitly given by the standard first-order condition for a cost-minimization problem:

$$\left(\frac{\partial \phi}{\partial H_k^*}\right) \Big/ \left(\frac{\partial \phi}{\partial L_k^*}\right) = \frac{\alpha_k}{\beta_k}, \tag{10.6}$$

where L_k^* and H_k^* reflect the optimum values of L^* and H^* when case k (for $k = 1$ to 4 as depicted in Figure 10.2) is expected to hold. Note that (10.6) is totally comparable with the first-order conditions of a static cost-minimization problem, and hence we can conclude that a fall in the ratio α_k / β_k corresponds to a rise in the relative expected present value of the operating cost of a low-level job. This in turn leads to a decrease in the slope of the iso-cost lines and hence to a movement down the unit job isoquant (see Figure 10.1) and to a decrease in the L^* / H^* ratio.

The optimum choice of a job combination as described above, can easily be understood by 'folding' Figure 10.1 along the 45-degree line in the north-west and south-east quadrants, so that the job dimension and the skill dimension in Figure 10.1 would come to overlap completely. The result of this 'folding' is depicted in Figure 10.3.

In this figure the points A', A'', B' and B'' correspond to the same points in Figure 10.1. The lines through $A'B'$ and through $A''B''$ now both have slope $-e'$, however. The vertical and horizontal axes measure two dimensions at the same time. Points on the straight lines with slope $-e'$ are points in the L, H dimension, while job combinations on the unit job isoquant through points A', A, A'' are actually points in the L^*, H^* dimension. Point A on the unit job isoquant has slope $-e'$ by construction.

It is fairly easy to show from the parameterization of the α / β ratios (see MASTER, 1997) that case 1 will have the lowest value of the L^* / H^* ratio, while case 3 will have a larger L^* / H^* ratio than case 4. Case 2 in turn will have a larger L^* / H^* ratio than case 1; that is: $(L^* / H^*)_3 > (L^* / H^*)_4 > (L^* / H^*)_1 = e'$ and $(L^* / H^*)_2 > (L^* / H^*)_1 = e'$. Case 1 is especially interesting, since the corresponding L^* / H^* ratio depends on e' only. This is caused by the fact

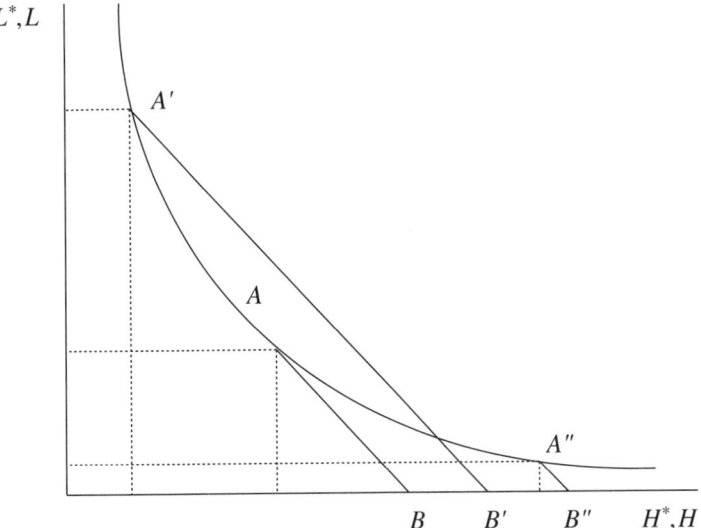

Figure 10.3   Ex ante *job choices*

that e' provides a limit to the marginal rate of substitution of high-level jobs for low-level jobs which is determined by economic considerations. Indeed, it does not pay to let the marginal rate of job substitution along the unit job isoquant fall below e', because the corresponding solution to the cost-minimization problem ($L = 0$, $H = H'$) would in that case result in an increase in the operating cost of a low-level job. This follows immediately from Figure 10.3, where a job combination to the right of point A on the unit job isoquant obviously generates a corresponding point on the horizontal axis to the right of point B. Because in this case the solution to the cost-minimization problem would be ($L = 0$, $H = H'$), it follows that any point on the horizontal to the right of B represents higher total expected wage costs than point B itself. Hence, from a cost-minimization point of view, any point to the right of B is inferior to B itself. The same conclusion obviously applies to points on the unit job isoquant to the right of point A. For a more detailed discussion of these results, see MASTER (1997).

The Allocation of Skills and Learning Asymmetries

We now assume that the allocation of a specific skill to a job gives rise to adjustment costs, which we link to learning cost. When a person is allocated to a job, we assume that he needs to spend some time learning all the tricks of the trade. The time spent is a given fraction of 'normal working hours' for a person of a similar skill level already trained for the job. We assume, furthermore, that a trainee does not produce anything while learning. Nonetheless, the trainee gets paid for his time. This implies that the learning cost per person is proportional to the wage rate which applies to the skill level of the person in question. The corresponding factor of proportion depends both on the skill level of the person in question and on the type of job this person is trained for. In addition to this, we assume that, each time a job is filled, training costs will have to be incurred. Obviously, people who have been trained already for a certain job do not have to be trained again, as long as they remain allocated to the job they have been trained for. Consequently, the experience they have in a certain job brings about an additional asymmetry in the costs of operating a job by either experienced workers or inexperienced workers of the same skill type.

Let the training cost of an allocation of skill i to job type j, as a fraction of the relevant wage rate w_i, be represented by $\xi_{i,j}$. Furthermore, let us consider the case where $\hat{w} > 0$. Then, eventually, the allocation $(L \to L^*)$ will be the cost-minimizing allocation, because the relative wage rate of high-skilled workers will become infinitely high, while the relative efficiency e' of high-skilled workers *vis-à-vis* low-skilled workers in low-level jobs will remain the same. The question arises whether low-skilled workers (L) should be allocated to a low-level job (L^*) from the outset, or high-skilled workers (H) should be allocated to that low-level job first. As in the no-adjustment case, we assume that this depends on the present value of operating the job over an infinite period of time for given expectations of the growth rate of relative wages and for given values of the learning cost parameters.

For experienced workers initial matching costs can be disregarded. However, for a switch in the future, we assume that entrepreneurs will always expect to use inexperienced workers, in order to be on the 'safe side'. Hence, the present value of the cost of 'manning' and operating a low-level job by starting with the allocation $(H \to L^*)$ first and then switching to the final allocation $(L \to L^*)$ at time t^* can be written as a function of t^* and the various expected present values of the wage costs associated with the different skills to be allocated to the low-level job. As we show in MASTER (1997), this relation can be used to obtain the optimum switching time t^* for an *allocation sequence* which starts out with the allocation $(H \to L^*)$ and which ends with the allocation $(L \to L^*)$ by minimization of the expected costs with respect to t^*, giving:

$$t^* = \frac{\ln\left(\left\{ \frac{e'}{w_0} \right\} \cdot \{ 1 + (\rho - \hat{w}_L) \cdot \xi_{L,L^*} \} \right)}{\hat{w}}. \tag{10.7}$$

t^* depends positively on the (effective) rate of discount ρ (which includes the rate of technical decay).[13] A rise in either of the parameters would decrease the relative importance of future costs in the present value of total cost per low-level job. Hence the incentive to switch would diminish, thus increasing t^*, *ceteris paribus*. Likewise, a rise in the adjustment cost of the final allocation $(L \rightarrow L^*)$ would tend to postpone the moment of switching, that is increase t^*, while a rise in the rate of growth of the wage rate of the skill used in the final allocation would tend to increase the unit cost associated with that allocation, and hence the moment of switching may again be expected to be postponed. This is in line with the fact that $\partial t^* / \partial \hat{w}_L > 0$, for $t^* > 0$. Note that the introduction of switching costs postpones the moment of switching (c.f. equation (10.4)). From equation (10.7) it is clear that t^* is positive only when $w_0 < e' \cdot (1 + (\rho - \hat{w}_L) \cdot \xi_{L,L^*})$, since $\hat{w} > 0$ by assumption. If t^* is positive, then an entrepreneur who had started out with the allocation $(H \rightarrow L^*)$ would be happy to stick with that allocation up to time t^*. If $w_0 < e' \cdot (1 + (\rho - \hat{w}_L) \cdot \xi_{L,L^*})$, t^* would be negative, and the optimum moment of switching from $(H \rightarrow L^*)$ to $(L \rightarrow L^*)$ would lie in the past, implying that under the circumstances it would have been wise to choose the final allocation $(L \rightarrow L^*)$ from the outset. But even if $t^* > 0$, it may be wise to start with the allocation $(L \rightarrow L^*)$ rather than the allocation $(H \rightarrow L^*)$, for minimization of the expected present value of operating costs with respect to t^* really assumes that learning costs associated with the allocation $(H \rightarrow L^*)$ have already been sunk. This would obviously be true in the case of experienced workers, but false in the case of inexperienced workers. In order to resolve this issue, a direct comparison must be made between the expected operating costs associated with both feasible allocation sequences $(L \rightarrow L^*, L \rightarrow L^*)$ and $(H \rightarrow L^*, L \rightarrow L^*)$.

The above analysis can be repeated for the case where $\hat{w} < 0$, which implies that ultimately the allocation $(H \rightarrow L^*)$ instead of $(L \rightarrow L^*)$ will be the less costly one. The analysis is completely analogous to the one above, and it is therefore not repeated here.

Linear Programming and the 'Chimney Effect'

In order for the chimney effect to work, it is necessary to introduce supply constraints which take account of asymmetries in substitution possibilities between high- and low-skilled workers. In those circumstances, an increase in the number of high-level jobs would require an increase in the employment of high-skilled workers on those jobs. The inflow of high-skilled workers into those high-level jobs would in part have to be met by an outflow of high-skilled workers from low-level jobs. The employment holes this outflow would leave in low-level jobs create the employment opportunities on which the low-skilled workers ultimately depend. In order to add supply constraints to the existing labour demand framework, we can use the technique of linear programming, as we have done before in van Zon and Muysken (1992). In this way, we are able to account explicitly for both job-demand constraints and skill-supply constraints.

Definition of the Linear Programming Problem

The demand side described so far ignores supply constraints. Consequently, the allocation of skills to jobs referred to first-best allocations only. But explicit supply constraints which become binding may force entrepreneurs to use second-best allocations instead. In order to be able to cover both first-best and second-best allocations, we assume that entrepreneurs solve a linear programming problem in which they try to meet job requirements as well as skill supply constraints, while minimizing the mismatch[14] between skill levels and job levels on low-level jobs.

The definition of the linear programming problem is now quite straightforward. Let $E^i_{j,k}$ denote the employment of skill j with experience type i on job k. The index i has two values, experienced and inexperienced with respect to the match in question. The distinction between experienced and inexperienced workers is important because of the fact that the allocation of an inexperienced worker with a certain skill level implies having to bear the training/matching costs associated with that particular combination of skills and jobs, while this is not the case with experienced workers. Consequently, it is important to distinguish between experienced and inexperienced workers on the supply side too. In this respect we simply assume that the experienced workers available for allocation in the present period are those which were allocated to the job under consideration in the previous period, adjusted for a constant exit rate. Supply from other sources (like the educational system or the unemployment pool) which is available for employment is inexperienced by assumption.

Given these supply constraints, we can now force the solution of the linear programming problem to be as close as possible to the first-best allocation sequence in the following way. First, we introduce the notion of the quasi-job-utilization rate q with $0 \leq q \leq 1$. When $q = 1$, the skills allocated to low-level jobs and high-level jobs are exactly sufficient to generate the labour efficiency units in those jobs as implied by the composition and level of labour demand by job. At the same time, consistency between *ex ante* and *ex post* considerations would require a particular allocation of a skill to a certain job to be the one which follows from the *ex ante* cost considerations, unless supply constraints for the skill in question prove to be binding. Hence the assignment of skills to jobs at any moment in time can be seen as the implementation of the first parts of the various allocation sequences associated with the different jobs under consideration, within the constraints set by the supply of the various skills involved.[15]

Let J_k^* denote the optimum job levels H^*, L^*, where k denotes a job level; that is the number of people with a corresponding skill level needed to fill the job. Let $P_{j,k}^i$ denote the wage cost (including the corresponding training cost for inexperienced workers) associated with a non-zero value of first- and/or second-best allocation sequences starting with the employment of skill j and experience type i on type k jobs. V denotes the total labour costs associated with the allocation sequence which results in *maximum costs*.[16] Then the following linear programming (LP) problem can be defined:

$$\text{MAX } F = q \cdot V - \sum_j \sum_k \sum_i P_{j,k}^i \cdot E_{j,k}^i$$

$$s.t. \quad \sum_i \sum_j e'_{j,k} E_{j,k}^i \geq q \cdot J_k^* \quad \forall k$$

$$E_{j,k}^x \leq S_{j,k}^x \quad \forall j,k \tag{10.8}$$

$$\sum_i \sum_k E_{j,k}^i \leq S_j \quad \forall j$$

$$q \leq 1$$

where S_j denotes the total supply of skill j at time t, $S_{j,k}^i$ is the total supply of skill j with experience type i with respect to job k; $e'_{i,j}$ is the relative efficiency of skill level i on job level j. Note that $e'_{j,k} = 0$ for skill levels j less than the job level k, and $e'_{i,j} = 1$ for all j, by definition.

The first constraints are the job demand constraints. The employment of available skills should generate the required efficiency units of labour in each job. The second set of constraints is the set of skill supply constraints. These constraints state that the total use of experienced workers cannot exceed the available supply. The third set of constraints states that the total use of skills may not exceed total supply including both experienced and inexperienced workers. The last constraint, in combination with the first, requires employment to be not larger than needed. From (10.8) it follows that maximization of F requires both q to be equal to 1 and a cost-minimizing allocation sequence to be chosen, as it should.

Given the solution to this LP problem, total employment by skill can now readily be obtained by aggregating over all experience categories and all jobs. The only things which need to be defined are the expected present value of the costs of the relevant allocation sequences per unit of the skill which is the first one in each sequence, that is the $P_{j,k}^i$s.

The Valuation of Alternative Allocation Sequences

In valuing a certain allocation sequence, we assume that entrepreneurs do not expect to face supply constraints in the future, unless they are faced by supply constraints in the present. But, even if the latter is the case, they are assumed to act as if these supply constraints would remain binding for Θ years only. Non-binding supply constraints are assumed to remain non-binding forever, until proven wrong. Furthermore, future matches/switches, as opposed to present matches, are assumed to involve inexperienced workers in all cases.

Given expectations regarding the growth rate of relative wages by skill, the optimum switching moment t^* for each possible allocation sequence starting with a certain skill can be calculated. If t^* proves to be positive, then the allocation sequence in question may prove to be first-best. If not, then the allocation sequence must be second-best, because switching to the second part of the allocation sequence at time zero, that is, starting out with the second part of the allocation sequence, as indicated by $t^* \leq 0$, would lead to a lower expected present value of the costs of the allocation sequence under consideration. Note, however, that this second-best allocation sequence would only be chosen when supply constraints would force entrepreneurs to actually do so. Hence, in this particular case, entrepreneurs expect to stay at least Θ years with this particular skill–job combination, after which they could decide to return to the unconstrained optimum allocation sequence, if that would prove to be worthwhile, or they could decide to stick with the 'forced' allocation instead. The latter could actually prove to be the most profitable thing to do, since 'returning' to the unconstrained allocation sequence entails the 'sinking' of matching costs in the future. We now assume that the cost to be assigned to the first part of the allocation sequence is equal to the minimum of these two options, 'sticking' and 'returning'.

When we wish to evaluate the expected present value of the costs of a certain allocation sequence starting out with an experienced worker, the corresponding costs of the same allocation sequence starting out with an inexperienced worker can simply be obtained by adding matching costs to the cost of an experienced worker. Hence all possible matches can be defined, knowing the expected present value of allocating an experienced worker to a certain job, as part of a certain allocation sequence. These present value costs of the various allocation sequences are described in more detail in MASTER (1997).

Asymmetries in Matching Behaviour

One of the problems associated with the linear programming approach taken here is that small changes in relative wages could bring about large changes in the optimum composition of employment. The introduction of a distinction between experienced workers and inexperienced ones 'causes' entrepreneurs to have a tendency to stick to a certain allocation once it has been made in the past, for whatever reason (either because it was an optimum allocation in the first place, or a second-best allocation due to supply shortages), because learning costs will be sunk in the process. However, it is still possible that large changes in the skill composition of employment occur when wage differentials more than offset learning costs.

In order, therefore, to smooth this transition between different skill compositions of employment, we assume, just as van Zon and Muysken (1996) have done, that entrepreneurs differ with respect to the value of their switching thresholds (perceived switching/matching/learning costs in this case). In fact, we assume that the population of entrepreneurs can be subdivided into 'fast switchers' (characterized by relatively low perceived adjustment costs) and 'slow switchers'. The relative class sizes depend on relative wages, expected growth rates and so on (for more details, see MASTER, 1997). Because entrepreneurs are by assumption distributed over those classes, aggregate employment must be obtained by aggregation over the corner solutions belonging to each of the three classes of entrepreneurs, using the number of entrepreneurs in each class as weights. While this approach avoids 'bang-bang' behaviour, the valuation of the present value of the costs associated with the relevant allocation sequences for each class becomes much more intricate. For a detailed description, see MASTER (1997).

3. THE ESTIMATION OF THE MODEL

Introduction

In this section we present the estimation results for the model described in the previous section. Because of the relatively limited amount of data we had at our disposal, it proved to be necessary to simplify the overall substitution structure of the model to some extent: we specified the job isoquant to be given by a Leontief structure, while leaving the rest of the model intact. This implies that, measured in efficiency units, the job composition of employment has been assumed to be more or less fixed, with the exception of the influence of exogenous trend terms.

It should be stressed that the results used here are for illustrative purposes only, because more work is needed to extend the database in order to create the degrees of freedom necessary for sound empirical estimation of the parameters of the model. Despite the obvious shortcomings in this respect of both the data and our empirical work so far, the results are interesting enough by themselves for us to present them.

Data and Classification Issues

Data

The data we have had at our disposal are from Eurostat. However, they cover only very few years. In fact, for Germany we have available employment and unemployment data for the years 1988–91, while data for the Netherlands are available for a longer period (1979–89), owing to additional data from the labour force surveys of the Netherlands Central Bureau of Statistics. An extensive overview of the data and the preliminary calculations we have performed on them is presented in MASTER (1997).

The employment data are available at a sectoral level for seven sectors with indices A, B, E, F, M, N and T, respectively: A being the advanced sector, B the building and construction sector, E the energy sector, F the financial and other services sector, M the material services sector, N the non-market sector and T the traditional industries sector. For each worker we know the type of job she or he is doing, with indices A, P and T, respectively: A stands for administrative/commercial job, P stands for personal services and T represents technical jobs. Finally, we also know the level of skill of each worker, where we distinguish three skill levels: low, medium and high.

As regards the wage data, these are only available by sector and by skill level. This is unfortunate, because it excludes the possibility of specifying an alternative to the present model in which wages are paid in accordance with the job level of the match under consideration rather than with the skill level.[17] Although one might regard this as a shortcoming of the model, one should consider that a change in perspective towards wages being paid by job level rather than by skill level necessarily refocuses the explanation of the skill composition of employment on the supply of skills rather than the demand for skills.[18] However, this is not the approach taken in most macro-economic models, nor is it the approach taken in the present chapter.

Classification issues

The analysis presented above can easily be extended to more than two job levels/job types and skill levels. Indeed, in the empirical implementation of the model for Germany and the Netherlands, we have actually distinguished between three job levels (high, medium and low) as well as three different skill levels (also high, medium and low), with indices H, M and L respectively. The naming of the job and skill levels suggests a one-to-one correspondence which is not supported by the data: one would expect jobs of a certain level to be filled predominantly by skills of the same level, while employment of skills with levels lower than the job level would have to be ruled out on the basis of our a priori notions about substitution asymmetries. Again, this a priori notion is not supported by the data.

There are several reasons why such a one-to-one correspondence might be lacking. First of all, the educational classification covers only the level of education attained during normal schooling years: any additional learning/schooling is ruled out. This would provide a direct explanation of the possibility of observing matches between skills and jobs, where skill levels apparently fall short of job levels. Secondly, if both the job classification and the skill classification are obtained by aggregation over more detailed classifications which do have a one-to-one correspondence, the correspondence between skill levels and job levels may be broken when the classes over which job levels and skill levels are aggregated are not the same. Since we have only fairly aggregated data at our disposal in the first, we should not really expect to find a one-to-one correspondence. However, the unfortunate consequence is that the data we have at our disposal are not consistent with the model we have presented above.

In order to resolve this issue, we assume that when a person with a low skill level is observed to be working in a job with a higher level, a wrong classification has been made. Hence that person should be reclassified. The problem is that we do not know whether a misclassification is due to an error on the job side of the classification or on the skill side. For the reclassification of the misclassified person we assume that both errors are present in proportion to the relative frequencies of the adjoining classes in the job skill classification.[19] This reclassification introduces a discrepancy between the 'official' data and the data which are consistent with the model, which is an unfortunate situation. However, by assuming that the number of people who are wrongly classified from the perspective of the model are a more or less constant fraction of the 'new' elements in the modified classification, we can 'deconstruct' the outcomes of our model by applying the reclassification procedure to the model outcomes in reverse. In the absence of more detailed information regarding effective skill levels, this is obviously a second-best solution to the problem of how to handle data inconsistencies, but a comparison of our results with the observed data remains possible in principle.

In addition to these manipulations and reclassifications, we have smoothed the employment figures and unemployment figures obtained in this way, because the changes over time in the distribution of the employment figures are fairly irregular. In fact, we have obtained the shares of employment and unemployment by skill level and skill type by means of a two-year moving average for the years from and including 1989. In order to obtain employment figures by skill level, skill type and job level and job type, we have retrapolated the distributions of employment and unemployment from their 1988 values in the case of Germany. In the case of the Netherlands, we have obtained figures for the years which are not covered by the CBS 'Enquête beroepsbevolking' data by means of straightforward linear interpolations of the distributions of employment and unemployment in both skill and job dimensions in the years adjoining the gaps in the data.

Estimation Results

Because of the linear programming nature of the matching process, it is not possible to use standard methods to estimate the various relative efficiency parameters and switching cost parameters from the data described above. Instead, we use a genetic estimation procedure which is based on Goldberg (1989). This genetic search algorithm encompasses the linear programming problem which was described in section 2, and combines both data with respect to job demand measured in (quasi-) efficiency units and sectoral supply constraints. The objective of this search algorithm is to find a parameter vector that minimizes an objective function. This objective function is defined as the sum of the squared residuals between generated optimum employment and observed employment by skill/job type match. We also add a penalty to this sum when the calculated quasi-rate of labour utilization falls below one. The procedure is described in detail in MASTER (1997).

Estimation results for Germany and the Netherlands

For both Germany and the Netherlands the estimated relative efficiencies are relatively close to 1.25, as is shown in Figure 10.4. A notable exception is the relative efficiency of medium-skilled workers in low-level jobs for the German E sector, which appears to be about 1.6 for all three job types. This reflects the exceptionally high relative wages for medium-skilled workers *vis-à-vis* low-skilled workers in the E sector (also roughly 1.6). For the Netherlands, the relative efficiency of high-skilled workers in medium-level jobs is relative high for all three job types in the M sector, whereas the relative efficiency of the high-skilled workers in medium-level jobs is very low for the personal jobs in the T sector. Moreover, the relative efficiencies are more equal for Germany than for the Netherlands. In general, however, the relative efficiencies are roughly equal to relative wage rates – both for high skills relative to medium skills and medium skills relative to low skills. Because we have transformed job demand measured in man-years into job demand measured in efficiency units by using relative wages by skill as a relative efficiency indicator, this is more or less what one could expect. However, the class sizes of different switching classes depend on relative wages and efficiencies too. Hence there are other influences of relative wages on the estimation results beside the ones through the job demand constraint and the valuation of the present value of the costs of the various allocation sequences.

Another remarkable result is that the learning costs as a fraction of the wage costs by skill are very different for different sectors and different job

Germany

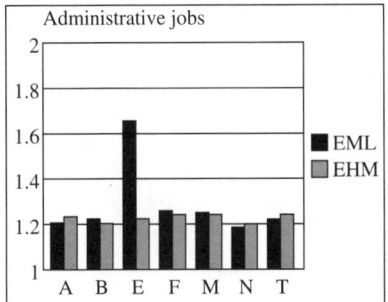

Netherlands

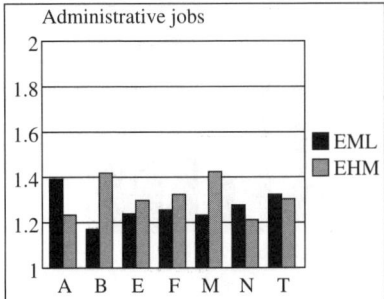

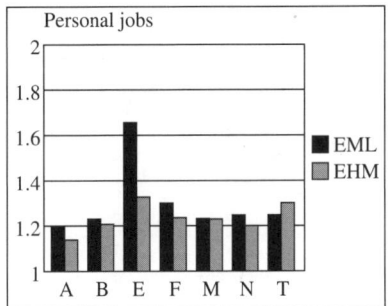

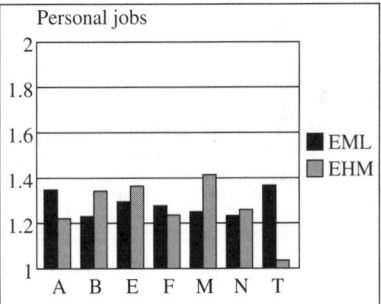

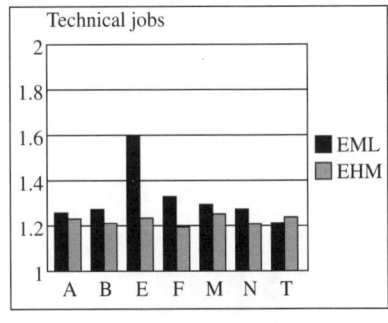

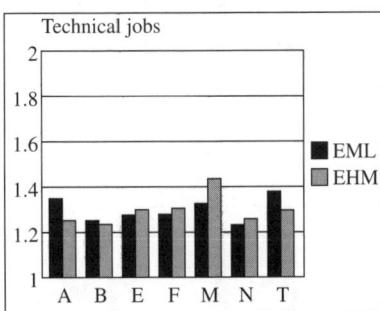

Note:
EML: relative efficiency of medium-skilled people on low-level jobs.
EHM: relative efficiency of high-skilled people on medium-level jobs.

Figure 10.4   Parameter estimates of the relative efficiencies

Germany

The Netherlands

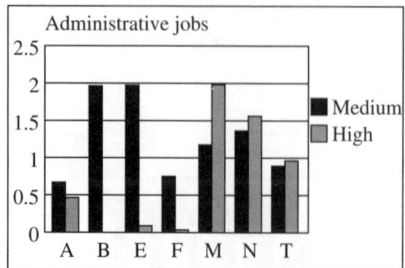

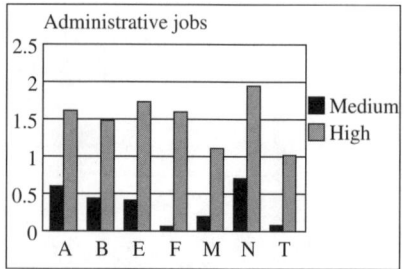

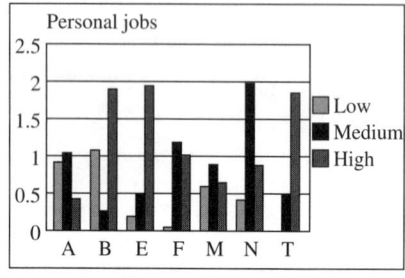

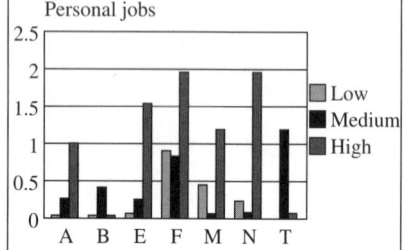

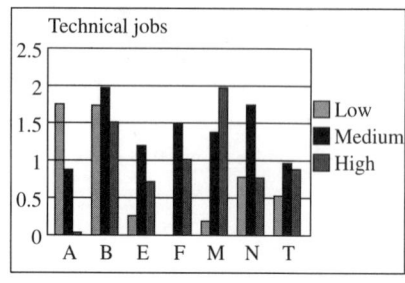

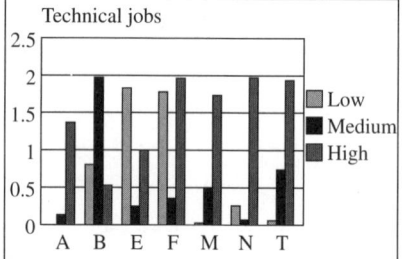

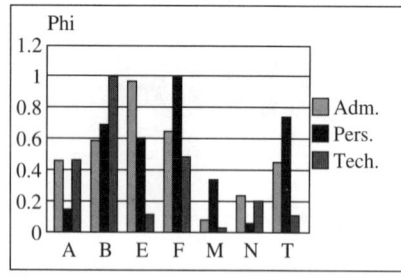

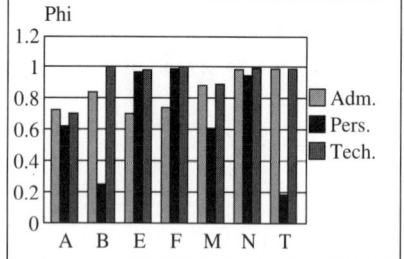

Figure 10.5   Parameter estimates of the learning costs ($\xi_{L,L}$, $\xi_{M,M}$, $\xi_{H,H}$, and ϕ)

types.[20] The estimated upper bounds[21] on these learning cost parameters are presented in Figure 10.5 for both countries. One can roughly conclude that there is a tendency for learning costs to be relatively low for low-skilled workers for all sectors in personal jobs, and for most technical jobs.[22] This indicates that, apparently, given changes in relative wage rates, supply conditions and so on, the skills needed to perform satisfactorily in low-level technical jobs in the German advanced sector A and the building and construction sector B and in the Dutch energy sector E and financial services sector F are more difficult to acquire than the skills necessary for low-level technical jobs in other sectors. For the other sectors one can roughly conclude that as far as technical jobs are concerned, the costs of acquiring the skills necessary for medium-level jobs are often higher than the costs of acquiring the skills necessary for low-level jobs and for high-level jobs. Although this may seem to be hard to understand at first, it may indicate that the medium-level jobs especially determine the sector specificity of the jobs in question, where low-level jobs only need 'eye–hand coordination', in the words of Romer (1990), while high-level jobs require the execution of functions which are less sector-specific, like management, computer engineering and so on. This conclusion also holds for administrative jobs, with the exception of the M and N sectors, for medium-level and high-level administrative jobs.[23]

As regards the asymmetries in learning costs for higher-skilled persons on job levels below their skill levels, there is again a fairly wide dispersion in the results. This can be seen from the results for the parameter phi, which represents the fraction by which the learning costs of medium-skilled workers on medium-level jobs should be multiplied to obtain the learning costs of the medium-skilled workers on low-level jobs. The same fraction is used for high-skilled workers on medium-level jobs. The results show that, for the Netherlands, the learning costs for these workers are not that different from the learning costs of the medium-skilled workers on medium-level jobs and high-skilled workers on high-level jobs. For Germany, the results are different where, for instance, the learning costs of high- and medium-skilled workers on medium- and low-level jobs are almost zero in the M and N sectors.

In general, we may conclude that the estimates are roughly as expected. First, the relative efficiencies seem to be much more reliable than the learning cost parameters. This is only natural, since we have much more direct information about relative efficiencies (in the form of relative wages) than for the other parameters. In addition to this, the estimation period is very short, while at the same time only a limited number of estimation runs have been performed, owing to the time-consuming nature of the estimation process. Nonetheless, the results we have obtained so far, seem to indicate that learning costs are an important determinant of the matches between skills and jobs, since a zero value of all learning parameters is included in the permissible range of variation of the parameters under consideration.

4. SOME SIMULATION EXPERIMENTS

The model described in section 2 has been used to perform two *ex post* simulation experiments for Germany and the Netherlands. These experiments are meant to illustrate the influence of the chimney effect on employment and to emphasize some interactions which might be relevant from a policy point of view. It should be stressed here that these are 'stand-alone' experiments: the simulation results obtained here do not reflect induced substitution effects between labour and capital (those between jobs have been ruled out already because of the Leontief structure we have had to impose). Such effects would obscure the illustration of the working of the chimney effect, which is the subject of this chapter. Nonetheless, it should be stressed here that these induced substitution effects may alter the results, depending on the nature of the experiment. Where relevant, we will indicate how the results would be influenced by these induced substitution effects. So, in the simulation experiments, the total demand of labour in terms of efficiencies is given.

The first experiments we have performed are concerned with the creation of low-, medium- and high-level jobs for all job types we have distinguished. As will be recalled from section 2, we have specified a Leontief job demand structure, measured in efficiency units, where the shares of the job types in total employment are constant (apart from an exogenous linear trend term) and where the shares of job levels in the various job types are constant too (again apart from a linear trend term). This means that the job demand $J_{t,l}$ can be written as: $J_{t,l} = a_{t,l} \cdot b_t \cdot E$, where the subscript t denotes the job type, and l denotes the job level; $a_{t,l}$ denotes the share of job-type t and level l in total jobs of type t, while b_t denotes the share of all jobs of type t in total job demand E, all measured in efficiency units. Note that we have dropped the sectoral dimension here for ease of exposition. The experiment we have conducted pertains to an increase in $a_{t,l}$ by 0.01, for each of the different job levels.[24] The first experiment is concerned with an increase in all low-level jobs, the second with all high-level jobs and the third with all medium-level jobs. Note that the replacement of $a_{t,l}$ by $a_{t,l} + 0.01$ increases $J_{t,l}$ by the same absolute amount in all cases, that is by $0.01 \cdot b_t \cdot E$. Hence, if the additional jobs were manned by people with the same skill level as the job level under consideration, employment of a particular type would increase by the same number of people. Obviously, the latter does not have to be the case, since higher-skilled people may take the jobs of lower-skilled people, depending on the situation on the various (skill) labour markets. With regard to induced substitution effects, it should be noted that this experiment increases, *ceteris paribus*, effective labour costs per unit of output. Consequently, a decrease in the labour/capital ratio would be expected, thus decreasing the original expansion of jobs somewhat. This would still leave an absolute increase in the number of jobs which had expanded in the first place, but it would also lead to an absolute decrease in the demand for other jobs.

A second experiment is concerned with a reduction in the wage costs of low-skilled workers, by means of a 15 per cent subsidy on wage costs. While this may seem to be a large amount, this will only lead to modest changes in the total wage sum by sector of industry and hence to modest induced substitution effects between labour and capital. This is due to the fact that the share of the low-skilled in total employment is fairly low, as opposed to their share in unemployment.

Simulation Results for Germany

The results of the experiment in which we have increased the number of low-skilled jobs, as described above, are summarized in Figures 10.6 and 10.8. This can be compared to the results when we increase the number of all high-level jobs instead (cf. Figures 10.7 and 10.9). In all Figures 10.6 to 10.11 a reference run is presented too. This is the broken line: it shows the outcome of the model when the estimated parameter values are used. As is discussed in MASTER (1997), the reference run generally shows a very close fit to the observed data.[25] From Figure 10.6 we see that, owing to the increase in low-level jobs, the rate of unemployment falls by about 0.7 percentage points. However, in the case of an increase in the number of high-level jobs, the drop in the overall rate of unemployment is of the order of 1 per cent, as can be seen from Figure 10.7.

The reason for this asymmetry in the results can be understood when the underlying changes in employment are studied in more detail. The increase in the number of low-level jobs does not affect employment of high-skilled workers (see Figure 10.8a), while employment of both low-skilled workers and medium-skilled workers are positively affected. In fact, the increase in employment for medium-skilled workers is much larger than for low-skilled workers. This points to two things. First of all, there are many more medium-skilled workers occupying a low-level job than there are low-skilled workers doing so. Secondly, although low-level jobs are created in this experiment, this is primarily beneficial for medium-skilled workers.[26]

When we increase the number of high-level jobs we see that, contrary to the previous experiment, employment of high-skilled workers is now positively affected, although still fairly small (Figure 10.9a). Moreover, the increase in employment of the low-skilled is far higher than the corresponding in-

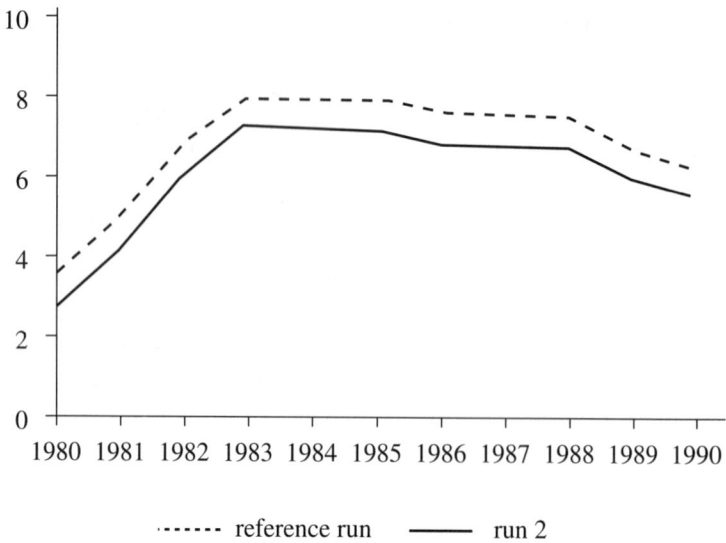

Figure 10.6   Unemployment change due to an increase in low-level jobs

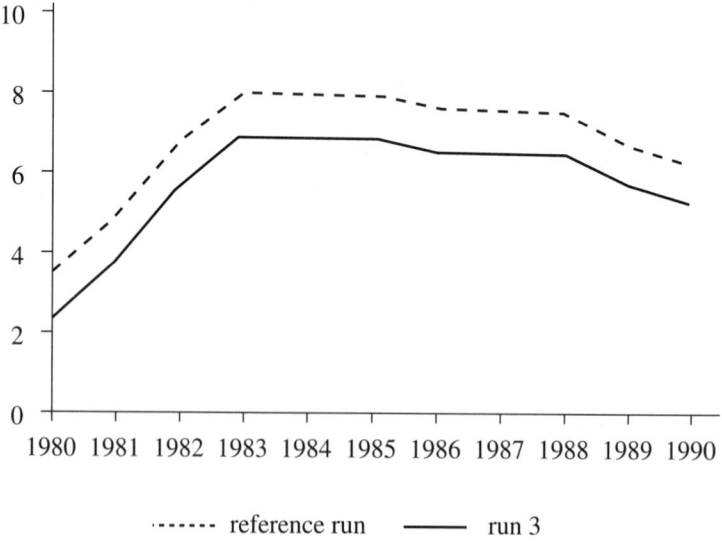

Figure 10.7   Unemployment change due to an increase in high-level jobs

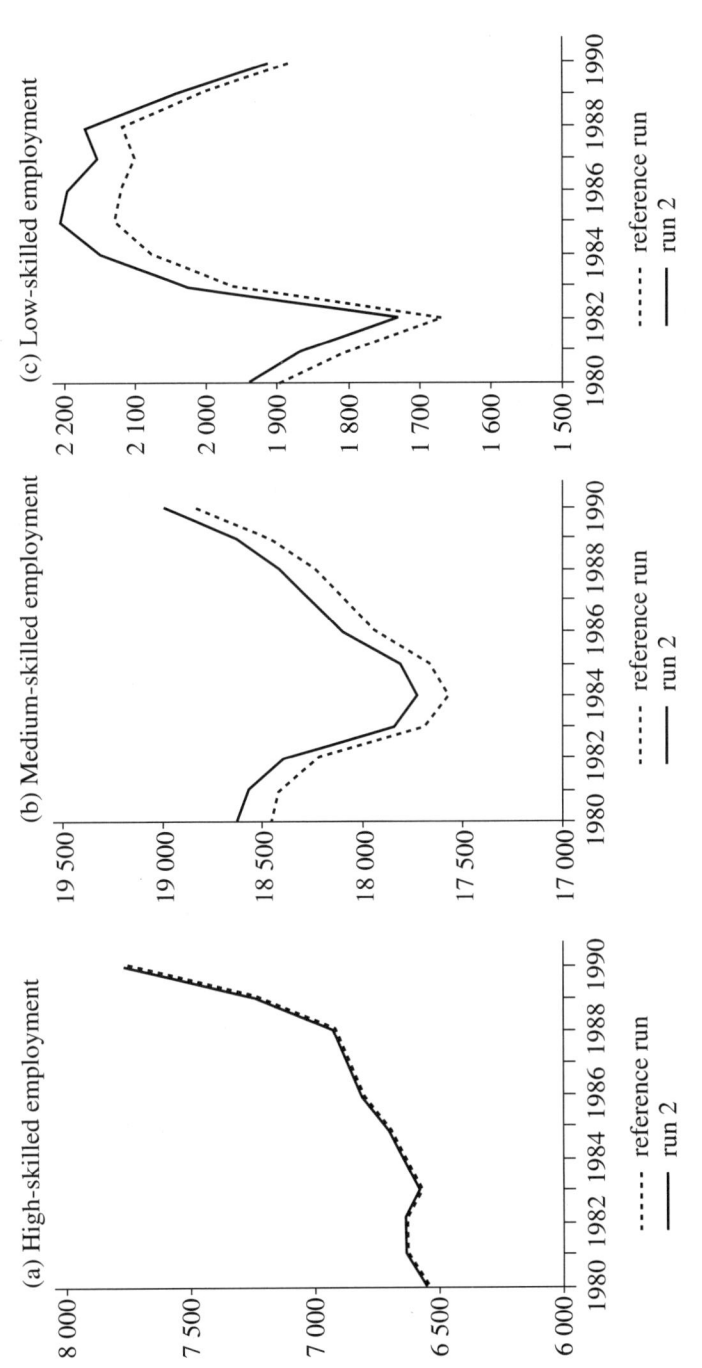

(a) High-skilled employment

(b) Medium-skilled employment

(c) Low-skilled employment

····· reference run
—— run 2

Figure 10.8   An increase in low-level jobs

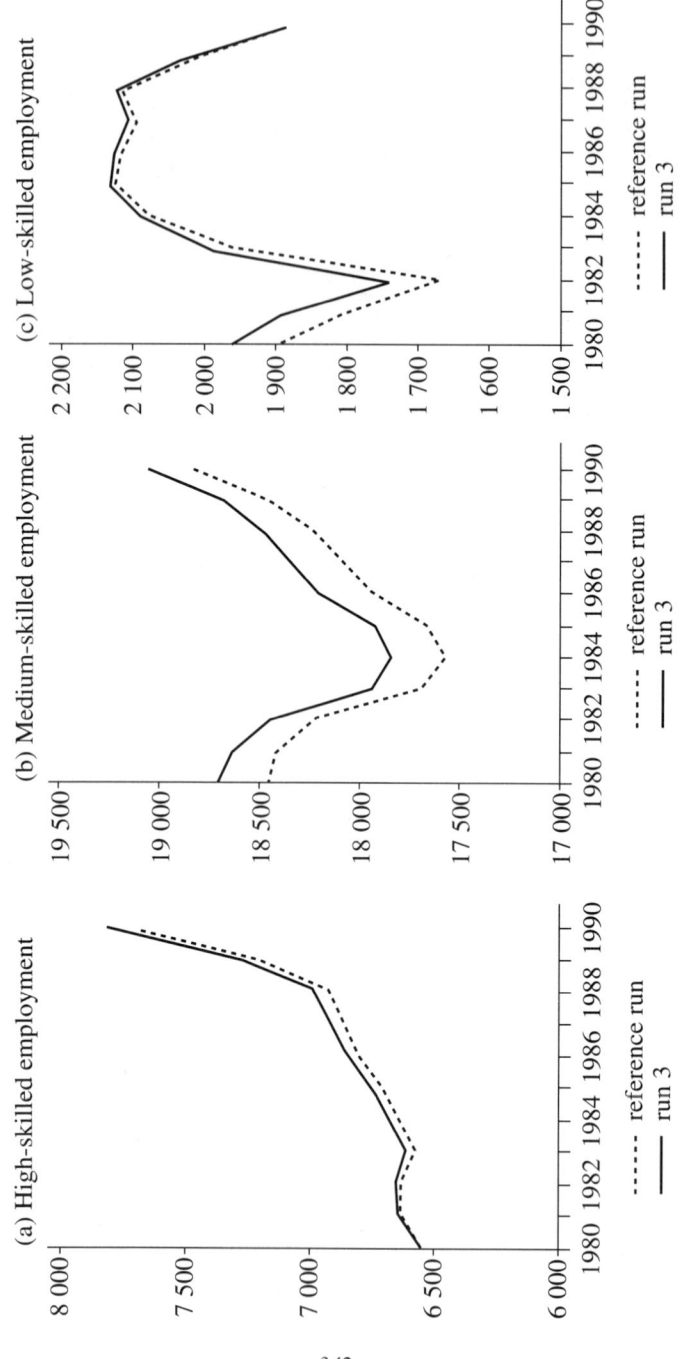

Figure 10.9 An increase in high-level jobs

crease in employment of the high-skilled (Figure 10.9c), especially in the beginning of the experimental period. A similar change occurs to employment of medium-skilled workers (see Figure 10.9b). The conclusion is that we see here an example of the chimney effect at work. Total employment of high-skilled workers cannot change much, because unemployment of high-skilled workers is relatively low already. Hence the additional high-level jobs need to be filled using high-skilled workers previously employed on medium-level jobs. The vacant medium-level jobs in turn are filled using medium-skilled workers, who need to be drawn in part from the pool of unemployed, but also from low-level jobs. Here, therefore, we see an example of the creation of high-level jobs having a beneficial effect on low-skilled employment. However, medium-skilled workers profit most. This is only natural, since these are the ones who are first in line to fill the employment gaps left by high-skilled workers moving into high-level jobs, while the low-skilled workers are only second in line if there are supply shortages of medium-skilled workers.

We do not present the outcomes of the experiment in which we increased the number of medium-level jobs. In MASTER (1997) we explain that this raises both the employment of low-skilled and that of the medium-skilled workers, because the pool of unemployed medium-skilled workers is not enough by itself to supply the extra amount of labour needed to fill the additional medium-level jobs. And, although high-skilled workers could be used to fill medium-level jobs, this does not happen, because high-skilled workers are already in relatively short supply.

Figures 10.10 and 10.11 contain the results of a 15 per cent subsidy on wage costs for all low-skilled workers. Since total employment measured in efficiency units is fixed by assumption (because of the 'stand-alone' character of the experiments), this means that we should expect low-skilled workers to be hired for low-level jobs more frequently than before. However, they would replace medium-skilled workers who were employed in low-level jobs, hence the net effect on total employment should be very limited. But even though employment in efficiency units is fixed a priori, this is not necessarily the case with employment measured in physical units, because one medium-skilled worker can replace roughly 1.25 low-skilled workers, while one high-skilled worker replaces about 1.25 medium-skilled workers, as shown by the estimates of the relative efficiencies in section 3.[27] This implies that a reshuffling of employment for a given job type can still have net positive effects, even though employment in efficiency units is constant. This is actually what is shown in Figure 10.10, which shows a drop in the aggregate rate of unemployment due to the subsidy. The subsidy favours the employment of low-skilled workers on low-level jobs, which makes the previously employed medium-skilled workers more or less redundant. They enter the labour market, and some of them succeed in driving out the high-skilled workers who

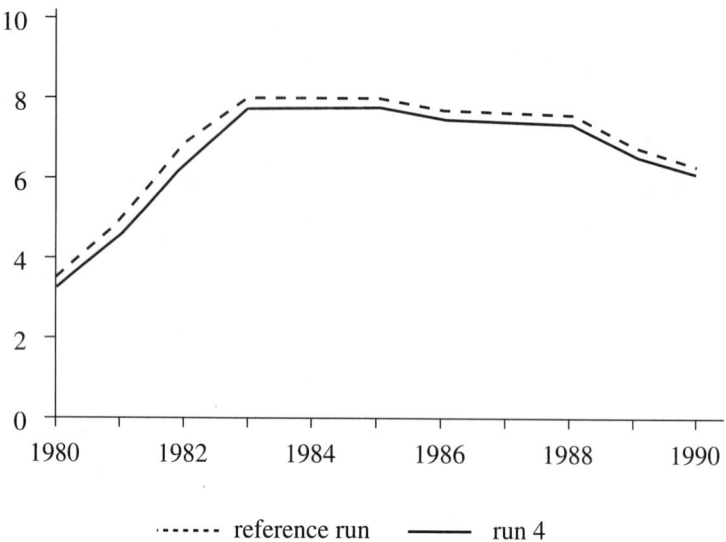

Figure 10.10   Unemployment change due to a decrease in low-skilled wage costs

took up medium-level jobs, owing to the existence of previously binding constraints with respect to the supply of medium-skilled workers. We see then that a subsidy on the employment of low-skilled workers does indeed lead to positive employment effects at the aggregate level, although medium-skilled employment and even high-skilled employment fall.

Results for the Netherlands

For the Netherlands we have performed the same experiments as for Germany. We will not present them in detail, but refer to MASTER (1997) instead. The results are very similar as far as the low-skilled jobs are concerned. But the results for the increase in high-skilled jobs are very different, because the rate of unemployment rises at first, which may seem to be somewhat odd. The reason is that an increase in the high-level job intensity of labour demand for a given supply of high-skilled workers may actually lead to a quantitative shortage of the supply of high-skilled workers, which, by assumption, are the only ones which can take those high-level jobs. Since high-skilled workers were already in short supply, the supply constraints even become so binding that total demand for high-level jobs cannot be filled for the given sectoral distribution of high-skilled workers. Consequently, owing to the Leontief character of the job structure of employment, the demand for

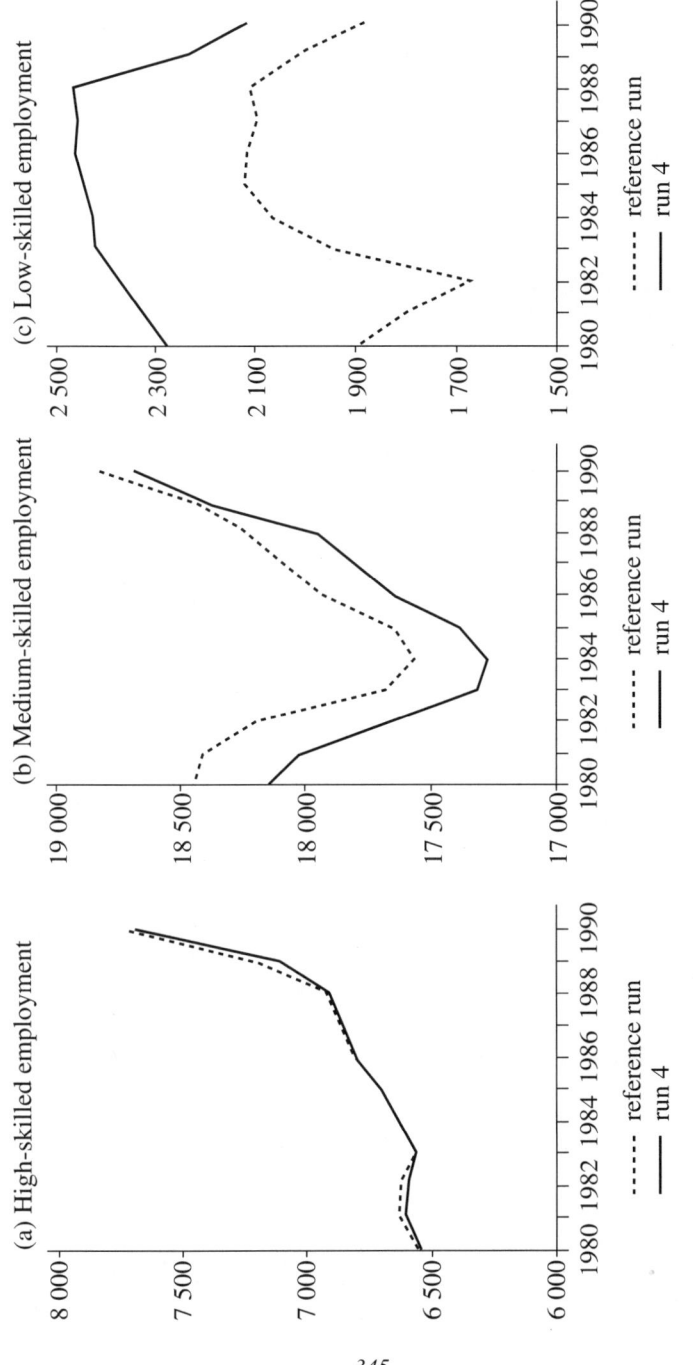

Figure 10.11   Employment effects due to a decrease in low-skilled wage costs

low- and medium-level jobs is negatively affected too (there is no use in hiring low- and medium-skilled workers, when the functions which high-skilled workers should perform cannot be executed because of a quantitative shortage of the supply of high-skilled workers). Only when job-to-job movements help to alleviate these quantitative supply shortages does the aggregate rate of unemployment drop below the base-run value.

In the experiment with a 15 per cent subsidy on low-skilled wage costs, we see a very minor net positive effect on unemployment. This indicates that supply constraints on medium-skilled workers are more binding in the Netherlands than in Germany; that is, even though the relative efficiency estimates are of the same order of magnitude in both Germany and the Netherlands, the substitution effects *ex post* are more limited owing to the existence of supply constraints.

5.   SUMMARY AND CONCLUSION

In this chapter we have presented a model of the allocation of labour of three different skill levels (low, medium and high) to jobs which require certain minimum skill levels (also low, medium and high). This means that low-skilled people cannot perform medium- and high-level jobs, while medium-skilled people can take both medium-level and low-level jobs. High-skilled people could perform well in medium- and high-level jobs. Furthermore, we assumed that people with a certain skill level may perform more efficiently on certain jobs than people with a skill level which exactly matches the job level. This means that from a demand point of view a match between skills and jobs depends on relative wage costs by skill. Consequently, mismatches refer to non-cost-minimizing matches between job and skill levels rather than to direct differences between job levels and the skill levels of the people associated with those jobs. Additional asymmetries between skills are introduced with respect to learning abilities, which favour the employment of groups with relatively low learning costs. These asymmetries between skills are important, because they stress the fact that a match on the labour market will only take place when the person in question has the ability to (learn to) perform satisfactorily in a certain job. Consequently, reductions in wage costs only increase the relative demand for certain skills for certain jobs if the skills in question are suited to those jobs: reductions in wage costs alone are not sufficient to further employment of certain skill groups.

Apart from the demand for skills which follows from the job composition of labour demand, we have integrated the supply of skills in a linear programming framework. Thus we are able to show how the creation of high-level jobs can improve the employment prospects of low-skilled people by reducing the displacement of low-skilled people by medium-skilled people on low-level jobs. The mechanism present in the model sees to it that the increase in the demand for high-skilled people caused by an increase in high-level job demand (owing to our assumptions regarding allocation asymmetries, high-skilled workers are the only ones who can take on high-level jobs) shifts high-skilled workers out of medium-level jobs, medium-skilled workers into medium-level jobs and out of low-level jobs, while low-skilled workers can move into the job positions freed by medium-skilled workers. We have seen this mechanism at work in a number of experiments we have conducted for Germany and the Netherlands.

In the experiments for Germany in particular, we have seen the chimney effect at work. We saw that the creation of high-level jobs may lead to the employment of medium-skilled workers and low-skilled workers in job openings which arise as a result of job-to-job movements of high-skilled workers from medium-level jobs to high-level jobs. We also saw that a subsidy on low-skilled wages, even when labour demand in efficiency units was constant, may actually raise the level of total employment, because medium-skilled workers are driven out of low-level jobs, where those workers took the place of 1.25 low-skilled workers on average. This implies that the full employment multiplier associated with the creation of a high-level job may be as high as 1.55; that is, 1.55 people may find work owing to the creation of one high-level job. By contrast, the creation of one additional low-level job will lead to additional employment of, at most, one worker.

In the experiments for the Netherlands we saw that the creation of high-level jobs may actually cause the occurrence of unemployment for medium- and low-skilled workers, instead of generating positive employment effects as it did in the case of Germany. The reason is that the logic of the model allows supply constraints of high-skilled workers to become so binding that total job demand cannot be fulfilled within the limits set by those constraints, and, given the Leontief structure of job demand which we have postulated, the demand for medium- and low-level jobs is negatively affected too. This underlines the point that the creation of high-level jobs is not a panacea for low-skilled unemployment problems. Rather, its effectiveness in this respect depends on high-skilled workers being in relatively high demand, but not too high at the same time, so as to avoid the negative impact on overall job demand of an 'absolute' excess demand for high-skilled workers.[28]

In general, then, we conclude that the experiments we have conducted with the model underline the relevance of both the chimney effect and the 'trickle down' effect, as well as the importance of displacement in determining overall employment opportunities for the various skills under consideration. The experiments also show how asymmetries in substitution possibilities may actually lead to job creation multiplication effects which, however, depend very much on the existence of qualitative binding supply constraints in the higher-skill echelons and the simultaneous absence of binding quantitative supply constraints in these same echelons.

NOTES

1.   This argument is used by Kettunen (1994) in his job competition model.
2.   This is corroborated by the fact that, particularly in the lower segment of the labour market, overschooling occurs to a large extent. For the Netherlands this is documented in Groot and Maasen van den Brink (1996) and Oosterbeek and Webbink (1996).
3.   The implied crowding-out effect – or as Drèze and Malinvaud (1993) put it 'trickling down effect' – is not found for the Netherlands by Van Ours and Ridder (1995). However, their results are criticized by both van der Laan (1996) and de Beer (1996), who do find crowding-out effects.
4.   Gelderblom *et al.* (1997) observe that education may be related to the opportunity to find a job, which in turn may lead to crowding out of low-skilled persons.
5.   Note, however, that differences in relative efficiencies of different skills in a certain job may actually lead to net changes in unemployment measured in man-years, while employment measured in efficiency units could remain the same. We will come back to this later.
6.   Many writers have investigated the possibility of direct substitution between skill categories. A survey for the USA is provided by Hamermesh and Grant (1979), while Hebbink (1991) and Broer and Jansen (1989) provide some results for the Netherlands. Kugler *et al.* (1990) do the same for Germany. Mincer (1989) provides additional results for the USA. The general conclusions which emerge from these studies are, first, that capital and high-skilled labour are complements, and second, that low-skilled labour and the capital/high-skilled labour complex are substitutes.
7.   These are the 'standard' substitution assumptions in neoclassical production theory. We also assume that the implied 'job aggregator function' or 'job substitution function' obeys the usual restrictions on the signs of the first and second partial derivatives.
8.   We are referring here to substitution possibilities between low-skilled and high-skilled workers on low-level jobs only, since, by assumption, high-level jobs can only be taken by high-skilled workers.
9.   The influence of δ is already accounted for in the definition of L and H in terms of L^* and H^* (see equation (10.1)).
10.   From now on we drop time subscripts, unless this could lead to confusion.
11.   Although this has not been indicated in Figure 10.2, t^* can of course be different for cases 2 and 3.
12   For a detailed description, see MASTER (1997).
13.   In van Zon and Muysken (1996), switching costs are introduced by means of the notion of a switching threshold Δ. That is, entrepreneurs switch from $(H \rightarrow L^*)$ to $(L \rightarrow L^*)$ only when $w > e'(1 + \Delta)$ and $\hat{w} > 0$. Note that Δ can directly be interpreted in terms of matching costs, since the switching threshold approach and the matching costs approach taken here are equivalent when $\Delta = (\rho - \hat{w}_L) \cdot \xi_{L'L}$.
14.   Note that the term 'mismatch', used here, usually refers to the a priori notion that the best match is the one where the skill level is exactly equal to the job level. From a (learning) cost-minimization point of view, this is not necessarily the case, because of the possibility of 'skill switching'. In this context, therefore, a mismatch refers to a non-cost-minimizing allocation of skills to jobs, which is due to the existence of supply constraints.
15.   Note that, because the linear programming problem is essentially a static optimization problem, the intertemporal effects of particular allocations are not taken into account (like, for instance, the fact that the inexperienced workers of today are the experienced workers of tomorrow). In the case of intertemporal optimization under uncertainty regard-

ing the level of demand for output, other interesting aspects of asymmetries in learning costs and substitution possibilities by skill become more relevant. In particular, one can envisage a situation where a bias would be expected in the demand for skills towards the ones which could be used in as many jobs as possible at intertemporal costs which are as low as possible (including expected matching costs). This provides an intertemporal cost minimization interpretation to the notion of a 'flexible' worker.

16. V therefore represents the operating cost of the worst-case allocation sequence.

17. Although the approach we have taken here is consistent with the neoclassical idea that wages reflect the marginal productivity of workers and that the marginal productivity of a worker is higher, *ceteris paribus*, the higher his skill level, one could also consider situations where high-skilled workers on low-level jobs are paid lower wages than the same workers on high-level jobs.

18. A certain match between skills and jobs, and especially the employment of high skills on low jobs, depends explicitly on the willingness of the persons under consideration to be employed on those jobs. See also Borghans and Heijke (1998) and Robinson (1997).

19. In order to clarify the reclassification procedure, let $f_{j,s}$ represent the two-dimensional frequency distribution of skills over jobs, where j represents the job level and s the skill level. If $f_{j,s} > 0$ for some $s < j$ (where increases in s and j correspond to increases in skill and job levels), then a new frequency distribution $f_{j,s}^i$ is obtained by distributing the value $f_{j,s}^i$ over the adjoining elements $f_{j-1,s}^i$ and $f_{j,s+1}^i$, in proportion to the values of these elements; i denotes a stage in the distribution process. The process continues until all 'wrong' elements have disappeared. With respect to the employment of high-skilled workers on low-level jobs, we assume that the observation is due to a mismeasurement of the job level. Consequently, we reclassify this entry as high-skilled workers on a medium-level job.

20. A word of warning is necessary here, since the administrative job type does not have any positive employment entries in the low-level administrative jobs, both for Germany and the Netherlands. This means that the job demand constraint, in combination with a positive present value price of an allocation to such a low-level administrative job, forces the estimated employment entries to be equal to zero for any value of relative efficiencies and relative wages. Consequently, one should disregard the results obtained for low-skilled administrative jobs.

21. As one may recall, we assume entrepreneurs to be uniformly distributed over these learning costs parameters between a zero lower bound and the estimated upper bounds presented here.

22. The exception is technical jobs in the A and B sectors for Germany and in the E and F sectors for the Netherlands.

23. Note that, in the case of administrative jobs, we may disregard the estimates for low-level jobs. See also note 20.

24. Note that this increase by one percentage point in $a_{t,l}$ is not matched by an average decrease of 1 percentage point elsewhere, so that the interpretation of a_t as an employment share is actually violated. The 0.01 increase in $a_{t,l}$ therefore represents a net increase in the corresponding skill intensity of labour demand.

25. There are a few exceptions to this rule, however. For example, the drop in low-skilled employment in 1982 as shown in Figure 10.8c is due to a relatively bad performance of the model in those years. This drop is not in the data themselves, but because the employment share of low-skilled workers is so low, such a drop can be compensated within the model by only a relatively minor increase in employment of medium-skilled workers, which is indeed what happens.

26. It should be stressed here that our conclusion may be exaggerated because of the transformations on the data we have had to perform in order to make them compatible with our model.

27. Note that this implies that we essentially have a low-skilled employment creation multiplier which is associated with the creation of one high-level job with a maximum value of about 1.55 which would be realized when this high-level job was filled by a high-skilled

worker previously occupying a medium-level job, which was then taken by a medium-skilled worker who previously occupied a low-level job.
28. Note that the ensuing increase in high-skilled wages would lead to substitution of capital for labour and to substitution of medium-skilled workers for high-skilled workers on medium-level jobs. This would mitigate the pressures of excess demand for high-skilled workers to some extent, thus reducing the negative impact of excess demand for high-skilled workers on the development of overall employment.

REFERENCES

Beer, P. de (1996), 'Laag opgeleiden: minder kans op een baan, meer kans op ontslag', *Economisch Statistische Berichten*, 908–12.

Borghans, L. and H. Heijke (1996), 'Flexibility and the structure of the Dutch labour market', ROA10, Maastricht.

Borghans, L. and H. Heijke (1998), *Flexibility and Structure of the Dutch Labour Market, Towards a Transparent Labour Market for Educational Decisions*, Aldershot: Ashgate, 119–50.

Broer, D.P. and W.J. Jansen (1989), 'Employment, Schooling and Productivity Growth', *De Economist*, 137, 425–53.

Drèze, J.H. and E. Malinvaud (1993), 'Growth and Employment, the Scope of European Initiative', mimeo, CORE/Collège de France, Louvain-la-Neuve/Paris.

Gelderblom, A., J. de Koning and J. Odink (1997), 'Loont Studeren?', *Economisch Statistische Berichten*, 500–504.

Goldberg, D.E. (1989), *Genetic Algorithms in Search, Optimization and Machine Learning*, Reading, MA: Addison-Wesley.

Groot, W. and H. Maassen van den Brink (1996), 'Overscholing en verdrininging op de arbeidsmarkt', *Economisch Statistische Berichten*, 74–7.

Hamermesh, D.S. and J.H. Grant (1979), 'Econometric Studies of Labor–Labor Substitution and Their Implications for Policy', *Journal of Human Resources*, 14, 518–42.

Hebbink, G.E. (1991), 'Employment by Level of Education and Production Factor Substitutability', *De Economist*, 139, 379–401.

Kettunen, J. (1994), 'The Effects of Education on the Duration of Unemployment', *Labour*, 8(2), 331–52.

Kugler, P., U. Muller and G. Sheldon (1990), 'The Labor Market Effects of New Technologies – an Econometric Study for the Federal Republic of Germany', in R. Schettkat and M. Wagner (eds), *Technological Change and Employment Innovation in the German Economy*, Berlin: Walter de Gruyter.

Laan, L. van der (1996), 'Substitution in the regional labour markets in the Netherlands', KNAG/Faculty of Geographical Sciences Utrecht/Economic Geographical Institute Rotterdam.

MASTER (1997), 'Final Report to the Commission of the EC', MERIT, Maastricht.

Muysken, J. and B. Ter Weel (1997), 'Does overeducation reduce unemployment', Maastricht University.

OECD (1994), *The OECD Jobs Study: Evidence and Explanations*, Paris.

Oosterbeek, H. and D. Webbink (1996), 'Over scholing, overscholing en inkomen', *Economisch Statistische Berichten*, 240–41.

Robinson, P. a. M.M. (1997), 'Qualifications and the labour market in Britain: skill-biased change in the demand for labour or credentialism?', CEPR discussion paper 330, CEPR, London.

Romer, P. (1990), 'Endogenous Technological Change', *Journal of Political Economy*, 71–102.

Van Ours, J.C. and G. Ridder (1995). 'Job Matching and Job Competition: Are Lower Educated Workers at the Back of Job Queues?', *European Economic Review*, 39(9), 1717–31.

Zon, A.H. van and J. Muysken (1992), 'Matching Skill Supply and Demand: A Linear Programming Approach', MERIT Research Memorandum, 92–005.

Zon, A.H. van and J. Muysken (1996), 'MASS: A Model of Asymmetric Skill Substitution', MERIT Research Memorandum, 2/96–016.

11. Determinants of sectoral average wage and employment growth rates in a specific factors model with production externalities and international capital movements

Ivo De Loo and Thomas Ziesemer[1]

1. INTRODUCTION

Sectoral wages are the average of the wages for skilled and unskilled labour. Explaining their development has recently led to some controversies (see Freeman, 1995). The major problems discussed are why wages for skilled and unskilled labour diverge in the USA and why unemployment has been heavily concentrated on low-skilled workers in Europe. These shifts can also be observed in newly industrialized countries (NICs) (see Richardson, 1995). The wage determination question, however, is of broader interest.

Many economists using closed or open economy growth models would explain wage growth mainly as a consequence of technical progress. Labour market economists would tend to emphasize (sectoral) supply and demand with little weight on international aspects (see Richardson, 1995). Trade economists would tend to ignore the supply of labour when using the Stolper–Samuelson theorem. However, in a multisectoral world of international trade and capital movements it is tempting to take a broader perspective. Consequently, one may ask the question, what is the relative importance of the major determinants of (average) wage growth and employment – international trade or factor movements, technological change or labour market developments – once one integrates all of them into one framework? In this chapter we try to answer this question with regard to the USA and six European countries (where wage inequality seemingly has changed much less than in the USA). The inequality issue will not be addressed in this chapter. We analyse average wages.

Lawrence and Slaughter (1993) and Krugman (1994) have argued that international trade would have an impact on wages, if any, via changes in the

352

terms of trade. However, they indicate that the terms of trade of the USA are almost unchanged and therefore changes in wages must be due to technical change. This argument leaves us with several open issues.

1. Results may be different for countries other than just the USA.
2. Results may change if we do not argue in terms of a two-sector model but at a more disaggregated level, because some of us will remember that in continental Europe the shipbuilding sector did shrink in the 1970s, the motor industry was faced by increased competition from Japan in the early 1980s, and the European consumer electronics sector lost ground in the 1980s and 1990s. Ultimately, protectionists lobby at the sectoral or even the firm level, and not at the macro level.
3. Once international capital movements are taken into account, not only the terms of trade but also interest rates become an exogenous variable for a (model of a) country and their changes should have an impact on wage growth according to economic theory.

How did the literature treat these three issues? The only contribution on average wages so far is Lawrence and Slaughter (1993). Some other insights are gained from the wage inequality debate by Lücke (1997), who has looked at data for Germany and the UK, and Oscarsson (1997) for Sweden. Seemingly, for many other countries this has not been done (within an international trade framework). Oliveira Martins (1994), using an industrial economics rather than an international trade approach, also looks at several countries.

Leamer (1996) sees the point of relevance for single sectors too, mentioning apparel and textiles in the USA. Krugman and Lawrence (1993) acknowledge that Japan did threaten US textiles in the 1960s and semiconductors in the 1990s. Leamer (1993) takes international capital movements into account when making theoretical scenarios but not when running estimations. Wood (1995), as well as Sachs and Shatz (1994), also look at several sectors and international capital movements. However, they do not have an integrating framework but rather look at all aspects, separately running regressions that give some intuition on their idea that international trade, technology and international capital movements are all important. Thus it seems to be worthwhile to investigate all of these points more closely.

Most of the wage inequality debate in international economics has been conducted in terms of Heckscher–Ohlin (HO) models (see Sachs and Shatz, 1994; Baldwin and Cain, 1997; Lücke, 1997; Oscarsson, 1997). Krugman and Obstfeld (1997) give a justification for this choice: although labour may not be mobile between sectors because its skills are specific to one sector only, reschooling could achieve the desired mobility after some time, which would justify the mobility assumption of the Heckscher–Ohlin model. Against this we wish to propose that, before reschooling, labour is specific to one (or several) sector(s) and after reschooling it is specific to different sectors or just one. We prefer to capture this with a specific factors model that has an exogenously changing labour supply for each sector and allows for sectoral differences in wages, whereas the HO model does not (see Leamer, 1994). Also most of the literature uses the Stolper–Samuelson theorem for the analysis (see Leamer, 1994; Richardson, 1995; Baldwin and Cain, 1997; Lücke, 1997, Oscarsson 1997), which makes the latter heavily dependent on the empirical validity of the zero profit conditions in every sector or period.[2] Using the cost-minimization part of a specific factors model with production externalities for perfect and imperfect competition and international capital movements can avoid this drawback. It provides a simple way to include the supply of labour, technical change, international trade and factor movements in one framework. Yet it does so at the cost of slightly exaggerating the immobility aspect of labour (which is now restricted to merely one sector). Other alternatives to the Stolper–Samuelson approach are presented in Francois (1996).

To allow for the treatment of more sectors motivated under point (2) we will construct a multisectoral, specific factors model in section 2. The inclusion of international capital movements brings in interest rate changes in accordance with the motivation of point (3). In section 3, some remarks on the data and analysis techniques are made. Section 4 contains our main findings, after which section 5 will discuss the policy conclusions which may be drawn from them. Finally, section 6 addresses the limitations of our approach and gives some guidelines for further research.

2.   MODEL DESCRIPTION

The details of our model are as follows. For each product i we assume the following production function for n identical firms to be responsible for the generation of variable costs, where Y indicates output, K capital, L labour, A technology and $Q = nY^i$ is sectoral output:

$$Y^i = (K^i)^{\alpha^i} (A^i)^{\theta^i} (L^i)^{\beta^i} Q^{\eta^i}$$

α, β and η are elasticities of the production of capital, labour and technology. η indicates production externalities which can have any sign. If the sum of α and β is smaller, larger than or equal to one, we have decreasing, increasing or constant returns to scale at the firm level and therefore upward- or downward-sloping or constant cost functions (for given technology A). We do not exclude any of these cases a priori.

From *cost minimization* we get (with w as the wage rate and r as the interest rate):

$$w^i = \lambda^i F^i{}_{L^i}$$

$$r = \lambda^i F^i{}_{K^i}.$$

λ is the Lagrange multiplier of the technology constraint, whose economic interpretation is marginal costs. Subscript indices K or L indicate a partial derivative with respect to K or L. The three equations given above allow us, together with the definition for sectoral output, Q, to find a solution for the value of the Lagrange multiplier. Dropping the index i, we get:

$$\lambda = \left(\frac{r}{\alpha}\right)^a Y^b A^c \left(\frac{w}{\beta}\right)^d Q^{-\eta b}$$

with

$$a = \frac{\alpha}{\alpha+\beta}, \quad b = \frac{1-\alpha-\beta}{\alpha+\beta}, \quad c = \frac{-\theta}{\alpha+\beta}, \quad d = \frac{\beta}{\alpha+\beta}.$$

In the case of perfect competition, marginal costs equal prices given by the world market (under the small country assumption) and marginal productivity conditions can therefore be rewritten as:

$$w^i = p^i F^i{}_{L^i}$$

$$r = p^i F^i{}_{K^i}.$$

Rewriting the marginal productivity conditions in growth rates, using the Cobb–Douglas form of production functions, and eliminating the term for capital yields an equation for several sectors in different countries (we do not write down a country index):

$$\hat{w}^i = \gamma_1 \hat{p}^i + \gamma_2 \hat{r} + \gamma_3 \hat{A}^i + \gamma_4 \hat{L}^i$$

with

$$\gamma_1 = \frac{1}{1 - \dfrac{\alpha^i}{1-\eta}} > 0, \quad \gamma_2 = -\frac{\dfrac{\alpha^i}{1-\eta}}{1 - \dfrac{\alpha^i}{1-\eta}} \leq 0,$$

$$\gamma_3 = \frac{\dfrac{\theta^i}{1-\eta}}{1-\dfrac{\alpha^i}{1-\eta}} \geq 0, \ \gamma_4 = \frac{\dfrac{\beta^i+\alpha^i}{1-\eta}-1}{1-\dfrac{\alpha^i}{1-\eta}}$$

In this model, the terms of trade are exogenous in the case of perfect competition and the small country assumption. These assumptions are made in most of the related literature. With perfect capital movements, the real interest rate, r, is given by the world market at each moment in time. Technology is exogenous by assumption and so is labour input because of the assumption that it is specific to each sector. Alternatively, we could have had employment as an endogenous variable and wages as an exogenous one. Then the equation would seek to explain employment of a sector in a country.[3]

The right-hand side of the above equation captures all variables that play a role in the debate on real wages. International trade is captured by changes in the terms of trade, technology is contained and international capital movements are represented by changes in the interest rate. Finally, factor supply is included, which could not be done in a Stolper–Samuelson approach using the zero profit assumption.

An estimate of this equation at the firm level would give us a result for $\alpha/(1-\eta)$, the elasticity of production of capital of a sector in a country corrected for the production externality, from either γ_1 or γ_2. Therefore we have to impose or test the constraint, $\gamma_1 + \gamma_2 = 1$, when doing the estimation. Having found a value for $\alpha/(1-\eta)$ we can deduct the value of $\beta/(1-\eta)$ from γ_4 and that of $\theta/(1-\eta)$ from γ_3. The question whether or not we have increasing returns to scale can be answered by looking at γ_4. If it is less than, more than or equal to zero, we have decreasing, increasing or constant returns to scale in labour and capital at the sectoral level, including the production externalities. However, only if the previous coefficient restriction is accepted may we draw such a conclusion, for then we may suspect that the definitions of the other coefficients hold too. Moreover, the assumption of perfect competition is only justified if we have non-increasing returns to scale at the firm level, which means that $(\alpha + \beta - 1) \leq 0$. It will turn out below that the estimations yield a positive sign for the labour variable, although it can have any sign. With non-increasing returns at the firm level this requires a positive externality that is large enough to make the numerator positive but also leaves the denominator positive. With the positive denominator the signs given for all terms in the previous formula follow.

In the case of increasing returns to scale at the firm level, we have to resort to imperfect competition and endogenous prices. Therefore we must give up the small country assumption, because price determination by domestic firms and prices given by the world market are mutually exclusive concepts (see Helpman and Krugman, 1989). If a sector is faced by a constant-elasticity demand function, $p^i = B^i Y^{i\phi} M_{eu}^{i\delta} M_{neu}^{i\varepsilon}$, with ϕ as an inverse of the price elasticity, M_{eu} as import quantities of competing products from the EU, M_{neu} as their non-European equivalent, B as a shift parameter which captures all other demand effects (such as effects of other imports coming into the country) and each product being produced by only one firm (as it would under monopolistic competition), profit maximization will yield $p^i = \lambda^i/(\phi^i + 1)$. Prices are now an endogenous variable because marginal costs (λ) are endogenous, for they depend on output and wages. A division between European and non-European trade is made because competition from the Asian NICs has been of special interest in the recent debate. If trade has an impact we would expect $\delta, \varepsilon < 0$.

Equating prices from the first-order conditions with those of the demand function yields:

$$B^i Y^{i\phi} M_{eu}^{i\delta} M_{neu}^{i\varepsilon} = \lambda^i /(\phi^i + 1).$$

Taking growth rates of this equation, the marginal productivity conditions and the expression for λ gives us four linear equations for four endogenous variables: the growth rates of wages (w), capital (K), marginal costs (λ) and sectoral output ($Q = nY^i$). The exogenous variables are the growth rates of A, B, L, r, M_{eu} and M_{neu}. Parameters are $\alpha, \beta, \theta, \eta, a, b, c, d, \delta, \varepsilon$ and ϕ. Solving the system for the growth rate of wages yields:

$$\hat{w}^i = e_0 + e_2\hat{r} + e_{1a}\hat{M}_{eu}^i + e_{1b}\hat{M}_{neu}^i + e_3\hat{A}^i + e_4\hat{L}^i$$

with

$$e_0 = \frac{-\hat{B}[\dfrac{\eta(2\alpha+2\beta)-1}{1-\alpha-\beta}+1]}{\eta[1-\alpha-\dfrac{\beta}{1-\alpha-\beta}]+\alpha(\phi+1)-1},$$

$$e_{1a} = \frac{-\delta[\dfrac{\eta(2\alpha+2\beta)-1}{1-\alpha-\beta}+1]}{\eta[1-\alpha-\dfrac{\beta}{1-\alpha-\beta}]+\alpha(\phi+1)-1},$$

$$e_{1b} = \frac{-\varepsilon[\dfrac{\eta(2\alpha+2\beta)-1}{1-\alpha-\beta}+1]}{\eta[1-\alpha-\dfrac{\beta}{1-\alpha-\beta}]+\alpha(\phi+1)-1},$$

$$e_2 = \frac{\alpha[\eta\frac{\alpha+\beta}{1-\alpha+\beta}+(\phi+1)]}{\eta[1-\alpha-\frac{\beta}{1-\alpha-\beta}]+\alpha(\phi+1)-1},$$

$$e_3 = \frac{-\theta[\eta\frac{\alpha+\beta}{1-\alpha+\beta}+(\phi+1)]}{\eta[1-\alpha-\frac{\beta}{1-\alpha-\beta}]+\alpha(\phi+1)-1},$$

$$e_4 = \frac{-\eta(1-\alpha-\beta)+1-(\alpha+\beta)(\phi+1)}{\eta[1-\alpha-\frac{\beta}{1-\alpha-\beta}]+\alpha(\phi+1)-1}.$$

In this equation, compared to that of the perfect competition case, imports are the exogenous variable that replace prices. The exogenous shift variable B can go either way. If it is decreasing, competition is increased. Then the demand function is shifted towards lower prices. As these coefficients are anything but easy to overlook, it is useful to consider first the case of no externalities, $\eta = 0$. In this case the numerator of e_4 becomes $1 - (\alpha + \beta)(\phi + 1) > 0$, which must be positive because its negative value is exactly equal to the second-order condition of the monopolist. Therefore the denominator with no externalities would have to be negative, leaving us with a negative coefficient. However, the empirics below give us mostly positive coefficients. Therefore it is most convenient to make the following assumptions under which the second-order conditions are not violated and the empirics can be understood. Let us assume the sufficient conditions that $\eta > 0$, $2(\alpha + \beta) - 1 > 0$ and $(1 - \alpha - \beta)$ is positive and sufficiently small to make the denominators of all coefficients negative. Then the second-order condition of the monopolist is not violated, e_4 is negative for sufficiently small externalities and positive for large externalities and therefore it can have any sign. Moreover, all the terms in square brackets in the numerators of e_{0-4} are positive if $(\phi + 1) > 0$ (as required by the first-order condition of the monopolist). The expected coefficients for both import terms (e_{1a} and e_{1b}) and the interest rate (e_2) are non-positive and that of technical change (e_3) is non-negative; e_0 and e_4 can have any sign.

3. DATA AND ECONOMETRIC METHODS

The estimated equations have been derived from the firms' rules for cost minimization and profit maximization and then have been aggregated into sectoral equations under the assumption of a given number of identical firms. We have data at the sectoral level.

Having constructed a model that differs from those of standard international trade models in textbooks by the production externality and the international capital movements, we have to relate a non-monetary model to data that stem from a monetary world. This requires dividing the data for wages and sectoral prices by the gross domestic product (GDP) price level of the country in question. Moreover, nominal interest rates have to be deflated by subtraction of the growth rate of the GDP deflator. We start from national nominal interest rates because, in spite of our assumptions, it is not clear that national capital markets are perfect. Although we have not modelled capital market imperfections explicitly, national rates seem to be the more adequate data.

We will test for structural breaks. The question whether employment drives wages or wages drive employment will be 'answered' using Granger causality tests. The regression equations will be estimated by ordinary least squares (OLS),[4] without the aforementioned coefficient restrictions (at least initially). This technique is applied so that a heteroscedasticity-consistent covariance matrix arises.[5] A description of the data can be found in the appendix to this chapter. At this point, only the choice of R&D expenditures as a proxy of technical change will be elaborated upon.

Basically, there are two sets of indicators that can serve as a proxy of technical change: R&D data and patent statistics. However, both have their drawbacks. R&D data are an input measure of the innovation process. Not all R&D inputs lead to innovations, and also the efficiency with which inputs are used influences the amount of successful R&D efforts. Thus more R&D expenditures do not necessarily imply more innovative activities. On the other hand, patent statistics are an output measure of the innovation process. Not all innovations are patented, and not all patents are put to effective and/or commercial use.[6] Moreover, the propensity to patent differs among countries.[7] In addition, neither R&D expenditures nor patent data refer exclusively to process innovation as our model does. At least product innovations for consumers should (but cannot) be excluded. Another problem associated with using R&D statistics as a technology indicator is that series containing labour or capital data will mostly include, to some extent, labour and capital used as an input to R&D. Thus adding R&D as a separate factor in the analysis could create a sort of 'double-counting'. However, there is mixed evidence on the question whether and how far the consequences of this reach. For example, while Schankerman (1981) and Hall and Mairesse (1992) state that corrections for double-counting should be made,[8] Verspagen (1995) finds only very limited effects. We will not touch upon this issue either, assuming the bias that arises because of double-counting to be negligibly small (which seems reasonable, given that the capital variable drops out of the regression equations).[9]

Nevertheless, the decision to use R&D expenditures as a proxy of technical change was mainly motivated by data availability, which was greater for R&D data.

4.   RESULTS AND INTERPRETATIONS

In the first sub-section below we will discuss the estimation results for the perfect competition model. The results for the imperfect competition model are examined in the second sub-section.

The Case of Perfect Competition

We begin with the basic regression output.[10] At first, a constant term is included in the regressions to capture the mean effect of (possibly) missing variables (such as additional productive factors). We expect γ_1 and γ_3 to have a positive sign, γ_2 to have a negative one, whereas γ_4 and γ_0 (which will be used to denote the constant term) can have either sign.[11] The constant term is (statistically) significant at the 5 per cent level for all of Germany, almost all of Italy (except for textiles, footwear and leather products and the basic metals sector), whereas it is only significant for total manufacturing and wood, cork and furniture in France, the French, British and Spanish paper and printing industry and the Spanish chemical industry. For the Netherlands and the USA, a rather mixed picture emerges (with chemicals, total manufacturing, stone, clay and glass and paper and printing being the significant sectors for the Netherlands and food, drink and tobacco, basic metals, total manufacturing, wood, cork and furniture and other manufacturing industries for the USA). Reasons for this outcome may be that labour market aspects (such as changes in union power, falling real values of the minimum wage, an upgrading of skills and compensation policies of firms), incomplete capacity utilization, developments in the non-traded sector, or additional production factors (such as land and natural resources) are at work (which are all not present in our model).

Many of the variables do have their expected signs to some degree, but are often not significant, as is typical of the whole literature discussed above. An exception is the labour variable, which is generally both positive and significant (only the British food, drink and tobacco and other manufacturing industries have a negative coefficient). This might point to increasing returns at the sectoral level.[12]

It is likely that there are structural breaks underlying the results. Such breaks may stem especially from the movement from negative to positive real interest rates at the beginning of the 1980s.[13] For Great Britain, Germany, Italy and the Netherlands, such a change in sign occurs in 1981. In France it occurs in 1980, whereas in Spain and the USA a change in sign of the real interest rate takes place in 1976 and 1986, respectively. Moreover, the high dollar value of 1985 may have induced another structural break. To test for these notions, a Chow break test[14] is applied to both the aforementioned year of the sign change of the real interest rate and the dollar value.[15]

Only a limited number of breaks is found. They arise for total manufacturing and wood, cork and furniture in France, the chemical industry in France and the UK, fabricated metals products in France and Italy, leather products in Italy and the UK, and the German other manufacturing industries. Of these sectors, three seem to have been affected by the dollar value of the mid-1980s: total manufacturing in France and the two British sectors.[16] It was decided to let the estimation period for all the aforementioned sectors start in either 1980, 1981 or 1986 instead of (mostly) 1974 and to redo the estimation.

Redoing the estimations leads us to the conclusion that structural breaks do not seem to be at the heart of the unexpected signs and large sizes of some of the variables in our model. Factors that remain are the significance of the constant term (in some equations) and the fact that we have not yet imposed the coefficient restriction derived in the theoretical part. We now leave out the constant term for those sectors for which it is statistically insignificant at the 5 per cent level and then test whether the proposed restriction is in place.[17]

The omission of the constant term alters our results somewhat (leading, among others, to several smaller (yet more significant) values of the labour variable),[18] but the overall results are quite similar to the ones already reached. Besides, we see that at the 5 per cent significance level, the coefficient restriction can be accepted only twice. We find significant results for the British chemical industry and Dutch fabricated metal products. Only for these sectors, if we get plausible estimates for $\alpha/(1 - \eta)$ and $\beta/(1 - \eta)$, can we say something about the presence of increasing returns. We can do so by checking whether $(\alpha + \beta)/(1 - \eta) - 1 \leq$ or ≥ 0 holds. It is unlikely that plausible estimates arise for both these sectors, because not all coefficients have the expected sign: for example, for fabricated metals products in the Netherlands, the interest variable turns up with a positive coefficient. In fact, inferring values for $\alpha/(1 - \eta)$, $\beta/(1 - \eta)$ and $\theta/(1 - \eta)$ does lead to estimates for these two sectors $\alpha/(1 - \eta)$ equals either –0.010 or 0.019, $\beta/(1 - \eta)$ –0.083 or 1.71 and $\theta/(1 - \eta)$ –0.011 or 0.430).[19] Negative values indicate $\eta > 1$), which may seem to be somewhat implausibly high. Together with the theoretical part, this may indicate the presence of an aggregation problem (or omitted variables).

Of all variables, labour is for a large part significantly different from zero, whereas, especially for the price and interest variables, there are many unexpected entries as far as sign and significance are concerned. However, statements about increasing or decreasing returns to scale cannot be made any more, since the restriction that would give rise to such an outcome is not accepted. It can only be said that a significant and mostly positive relationship exists between sectoral wage growth and employment growth in almost all sectors and countries under consideration.[20] One might suggest that specific factors matter, although the less plausible results for the other variables possibly overstate the importance of such a conclusion.

The question remains in what direction the relationship between employment and wages holds. Do wages determine employment or does employment determine wages? Tentatively, this question will be 'answered' by means of Granger causality tests. These tests[21] examine whether the occurrence of a certain event (variable) X precedes the occurrence of another event (variable) Y over a certain period of time. Stated differently, it is tested whether variable Y is temporally dependent upon variable X. Thus it is not causality in a strict sense that is analysed here: it is the order in which events happen that matters.[22] Besides, Granger causality is like a two-way street: only when X Granger causes Y, and Y does not (at the same time) Granger cause X, may we say that there is temporal dependence of Y upon X. More specifically, the following model is estimated:

$$Y_t = \delta_0 + \sum_p \alpha_p Y_{t-p} + \sum_q \beta_q X_{t-q} + \varepsilon_t,$$

where p,q are predetermined lag orders, and ε_t is a random disturbance term.

The null hypothesis that X does not Granger cause Y is that $\beta_q = 0 \forall q$ (while, simultaneously, Y should not Granger cause X: $\alpha_p = 0 \forall p$). The size of p and q is mostly agreed upon a priori on theoretical grounds. Here we will assume, letting Y_t denote sectoral wage growth and X_t the corresponding growth of labour, that p and q range from one to three. Tests were carried out with both one and two lags, but this did not alter our basic results very much.[23]

Employment Granger causes wage growth in a limited number of cases: only for the British fabricated metal products and food, drink and tobacco are significant results found (at a 5 per cent level of significance). However, wages determine employment growth more often: for three British sectors (chemicals, textiles, footwear and leather products and basic metals) this turns out to be the case. Two other significant results emerge, namely for Spanish leather products and for the Italian paper and printing industry. For wood, cork and furniture in Germany and total manufacturing in Great Britain, there are statistically significant relationships in both directions: wages determine employment and, by the same token, employment determines wages. Nevertheless, the conclusion in these cases is the same as for all sectors not mentioned: the Granger causality test is inconclusive.

It is quite interesting that, when significant results are found, they occur most often for Great Britain. There seems to be no apparent reason for this outcome, however.

In the five cases where wages Granger cause employment growth, the estimation is redone, with wage growth now being an explanatory variable and labour growth the dependent one. As far as the value of the coefficients is concerned, this simply means rewriting the equations already estimated.[24] However, the fit does change, as does the significance of the coefficients. Tests for structural breaks have to be redone too. Also, in the first stage, a constant term is included in the regression equation. For the sectors for which it does not differ significantly from zero at the 5 per cent level of significance,[25] it is dropped and the modified model is re-estimated.

Note that the desired sign of the explanatory variables switches when moving from wage growth to labour growth as the dependent variable. Only for the relationship between wages and labour does it remain the same. Even then, there are many wrong signs to be found.[26] The interest variable does not have the correct sign for any sector. The technology variable has the wrong sign for three sectors: chemicals in the UK and textiles, footwear and leather products in both the UK and Spain. The price variable has the wrong sign for the two non-UK sectors. Thus there is no sector for which all variables have the desired sign. Therefore no new insights on coefficients are created here either.[27] However, below we will report on all reversed causality cases independent of the outcome of the Granger causality analyses.

For those sectors where all coefficients have the expected sign,[28] it might be illuminating to examine how far the explanatory variables attribute to the explanation (of variation in) the dependent variable.[29] This means conducting a sort of 'growth accounting' exercise.

There are 17 sectors for which we found the expected signs. None of them is located in Germany. All sectors are shown in Table 11.1. Except for chemicals, most industries in the table are the more traditional ones. The basic procedure we follow for the 17 sectors where all variables have the expected signs is to take the regression coefficients of corresponding B3 and pre-multiply them by the means of the corresponding explanatory variables (calculated as an average of the entire estimation period).[30] Then this figure is divided by the mean of the dependent variable over the same period and multiplied by 100 to arrive at percentages. Finally, to obtain country figures, unweighted means of these percentages are taken for all sectors in Table 11.1 within a certain country.

If we leave out total manufacturing,[31] and check the relative importance of all variables in explaining wage growth in a certain country in the way described above, we reach the results presented in Table 11.2. For Italy and the Netherlands, no results at the country level are calculated because of the relevance of just one sector.

Table 11.1   Sectors with correct expected signs

Country	Sector
USA	Chemicals
	Basic metals
	Paper and printing
	Wood, cork and furniture
France	Chemicals
	Stone, clay and glass
	Wood, cork and furniture
Great Britain	Chemicals
	Food, drink and tobacco
	Paper and printing
	Other manufacturing industries
Netherlands	Total manufacturing
Italy	Textiles, footwear and leather
Spain	Basic metals
	Food, drink and tobacco
	Total manufacturing
	Wood, cork and furniture

Perhaps the first impression Table 11.2 gives rise to is that a large part of the explanation of wage growth is attributed to both the constant term (in the UK) and the residual (in France). This implies that for these countries a significant part of wage behaviour is not captured by our model, as discussed above.[32]

Table 11.2   Relative importance of explanatory variables in explaining per country wage growth (per cent)

	Variables					
	Constant	Technology	Capital	Trade	Labour	Residual
USA	19.7	16.9	−3.1	−3.5	57.5	12.3
France	0.0	28.8	30.9	−3.0	79.7	−36.4
UK	40.7	23.8	−18.3	31.8	16.6	5.4
Spain	0.0	22.5	19.8	−6.7	55.8	8.7

Note: In regressions *without* a constant term the residuals do not necessarily have to sum to zero. Therefore, a certain weight is assigned to them in these cases.

However, it does not mean that we cannot draw any (at least, preliminary) conclusion from the table. It is evident that, for most countries here, a large part of wage growth is determined by employment growth: labour supply is a dominating factor in three countries (all but the UK). In the UK, a substantial part is contributed by terms of trade changes.[33] The UK is also the only country where terms of trade are more influential than technology. Looking at the overall results, we may conclude that technology is a more important factor than trade in determining (national) wages in three countries. Labour supply is an even more important factor. Again, specific factors seem to matter. This raises the question of what we can see at the sectoral level.[34]

We can derive from Table 11.3a below for the whole period under consideration (and from Table 11.3b for the 1980s onwards – the results of which will be indicated in the text in brackets) that 12 (12) out of the 17 sectors included in the 'growth accounting' exercise have negative terms of trade growth, indicating that there may be an international problem. In seven (seven) sectors we have falling and in ten (ten) we have increasing wages (according to the last column). In only four (four) sectors do R&D expendi-

Table 11.3a   *Growth rates of explanatory and dependent variables over the estimation period given by sample (SMPL)*

Sector	SMPL	Technology	Capital	Trade	Labour	Wages
USAZ35	74–93	0.0455	0.3643	–0.0019	0.0063	0.0179
USAZMB	74–93	–0.0049	0.3643	0.0015	–0.0286	–0.0222
USAZOP	74–93	0.0522	0.3643	0.0052	0.0108	0.0179
USAZOW	74–93	0.0209	0.3643	–0.0056	0.0000	0.0046
FRAZOG	74–91	0.0215	0.5437	0.0010	–0.0244	–0.0037
FRAZ35	80–91	0.0458	–0.2758	–0.0079	–0.0060	0.0030
FRAZOW	80–91	0.0779	–0.2758	–0.0051	–0.0235	–0.0082
GBRZLF	74–92	–0.0886	0.2590	–0.0047	–0.0212	0.0046
GBRZOO	74–92	–0.0019	0.2590	0.0359	–0.0253	0.0209
GBRZOP	74–92	–0.0022	0.2590	0.0060	–0.0101	0.0107
GBRZ35	86–92	0.0692	0.4908	–0.0210	–0.0070	0.0264
NLDZMT	74–93	0.0187	–0.2920	–0.0095	–0.0123	0.0028
ESPZMB	80–91	0.0173	–9.5446	–0.0339	–0.0399	–0.0323
ESPZMT	80–91	0.1139	–9.5446	–0.0089	–0.0138	–0.0078
ESPZOW	80–91	0.5935	–9.5446	–0.0047	–0.0189	–0.0173
ESPZLF	80–91	0.1110	–9.5446	–0.0188	–0.0085	0.0092
ITAZLX	81–94	0.3001	0.1873	–0.0176	–0.0197	–0.0107

Note: See appendix for a list of abbreviations used.

Table 11.3b   Growth rates of explanatory and dependent variables over the estimation period given by SMPL, 1980s onwards

Sector	SMPL	Technology	Capital	Trade	Labour	Wages
USAZ35	81–93	0.0508	0.1435	–0.0153	0.0037	0.0143
USAZMB	81–93	–0.0311	0.1435	–0.0244	–0.0377	–0.0371
USAZOP	81–93	0.0500	0.1435	0.0083	0.0108	0.0214
USAZOW	81–93	0.0179	0.1435	–0.0023	0.0052	0.0091
FRAZOG	80–91	0.0234	–0.2758	–0.0023	–0.0279	–0.0129
FRAZ35	80–91	0.0458	–0.2758	–0.0079	–0.0060	0.0030
FRAZOW	80–91	0.0779	–0.2758	–0.0051	–0.0235	–0.0082
GBRZLF	81–92	–0.0581	0.1612	0.0040	–0.0276	0.0052
GBRZOO	81–92	–0.0646	0.1612	0.0546	–0.0285	0.0417
GBRZOP	81–92	–0.0018	0.1612	0.0034	–0.0138	0.0153
GBRZ35	86–92	0.0692	0.4908	–0.0210	–0.0070	0.0264
NLDZMT	81–93	0.0209	0.0287	0.0030	–0.0072	0.0034
ESPZMB	81–91	0.0116	–10.2546	–0.0367	–0.0449	–0.0410
ESPZMT	81–91	0.1085	–10.2546	–0.0162	–0.0138	–0.0134
ESPZOW	81–91	0.6157	–10.2546	–0.0124	–0.0182	–0.0265
ESPZLF	81–91	0.0940	–10.2546	–0.0200	–0.0063	0.0016
ITAZLX	81–94	0.3001	0.1873	–0.0176	–0.0197	–0.0107

tures have a negative growth rate. R&D therefore has a positive effect on wages in both periods. Interest rates have risen in ten (ten) sectors and therefore have decreased wage growth.[35] With three (three) exceptions, labour supply has fallen and therefore – given the positive sign of the correlation – there is decreased wage growth.[36]

From Table 11.4a (and Table 11.4b for the more recent period of the 1980s, the results of which are again given in brackets), which shows us a similar table as Table 11.2 but then at the sectoral level, it follows that in four (seven) out of the 17 combination of countries and sectors the terms of trade have a larger impact than technology. This means that technology matters more often over the whole period but terms of trade changes are more influential in the recent period. Out of these four (seven) sectors, two (four) have falling terms of trade. Thus at the sectoral level international trade is quite important. These two sectors are located in Spain (basic metals) and the USA (wood, cork and furniture). In the 1980s more Spanish sectors have terms of trade losses but there is also one additional sector in the USA (basic metals). However, of the two (four) sectors one (four) have falling wages.

Table 11.4a *Relative importance of dependent and explanatory variables in explaining per sector wage growth (per cent)*

Sector	Constant	Technology	Capital	Trade	Labour	Residual
USAZ35	0.0	53.3	−2.5	−0.4	30.0	19.7
USAZMB	−38.8	1.0	2.8	−0.4	135.4	0.0
USAZOP	0.0	4.7	−2.5	3.0	65.1	29.6
USAZOW	117.7	8.8	−10.0	−16.0	−0.5	0.0
FRAZOG	0.0	−72.8	86.4	−13.6	251.7	−151.7
FRAZ35	0.0	198.8	49.5	−6.4	−121.7	−20.1
FRAZOW	0.0	−10.8	−12.3	8.1	188.8	−73.7
GBRZLF	0.0	−17.7	−6.5	−5.9	107.4	22.7
GBRZOO	0.0	−0.9	−6.4	120.4	7.6	−20.8
GBRZOP	162.8	−2.0	−12.2	24.0	−72.5	0.0
GBRZ35	0.0	115.8	−48.2	−11.2	23.8	19.9
NLDZMT	422.7	64.4	8.7	−6.3	−389.5	0.0
ESPZMB	0.0	−1.6	−3.5	7.3	109.9	−12.2
ESPZMT	0.0	−67.4	−3.4	42.8	151.4	−23.4
ESPZOW	0.0	−12.5	−1.1	8.0	90.4	15.2
ESPZLF	0.0	81.5	63.9	−35.5	−32.8	23.0
ITAZLX	0.0	−5.3	3.2	4.8	109.0	−11.8

The more recent period therefore is (much) less favourable (in terms of losses) than the whole period and the terms of trade are catching up with technology in importance.[37]

In 11 (12) of the 17 national sectors, labour has the strongest impact; in only four (four) cases is it technology and in two (one) is it trade. In the more recent period, labour has become even more important than it already was over the entire period. When counting variables that rank second we find six (six) times technology, twice (six times) trade, three times (twice) labour and six (three) times capital movements by interest changes. The overall impression therefore is that labour supply matters most, technology second and trade and interest rates last (in that order), but in the more recent period terms of trade have completely caught up with technology. All evaluations have been made without taking the constant term or the residual into account.

A similar exercise can be carried out by switching the roles of wage and labour growth in the regression equation and redoing the entire analysis up to this point.[38] Then we would find ten sectors where all variables have the expected signs (five of which had not been included before), as shown in

Table 11.4b   Relative importance of dependent and explanatory variables in explaining per sector wage growth (per cent), 1980s onwards

Sector	Constant	Technology	Capital	Trade	Labour	Residual
USAZ35	0.0	74.4	−1.3	−4.1	21.9	9.0
USAZMB	−23.2	3.7	0.7	4.1	106.5	8.2
USAZOP	0.0	3.8	−0.8	4.0	54.4	38.7
USAZOW	59.2	3.8	−2.0	−3.4	56.9	−14.5
FRAZOG	0.0	−22.8	−12.6	8.8	83.0	43.6
FRAZ35	0.0	198.8	49.5	−6.4	−121.7	−20.1
FRAZOW	0.0	−10.8	−12.3	8.1	188.8	−73.7
GBRZLF	0.0	−10.3	−3.6	4.4	124.2	−14.7
GBRZOO	0.0	−14.7	−2.0	91.8	4.3	20.6
GBRZOP	113.9	−1.2	−5.3	9.6	−69.4	52.4
GBRZ35	0.0	115.8	−48.2	−11.2	23.8	19.9
NLDZMT	353.2	60.0	−0.7	1.7	−190.3	−123.9
ESPZMB	0.0	−0.8	−2.9	6.3	97.6	−0.2
ESPZMT	0.0	−37.6	−2.1	45.6	88.7	5.4
ESPZOW	0.0	−8.5	−0.7	13.8	56.8	38.6
ESPZLF	0.0	387.6	385.6	−212.0	−136.2	−325.1
ITAZLX	0.0	−5.3	3.2	4.8	109.0	−11.8

Table 11.5   Sectors with correct expected signs when the roles of labour and wages are interchanged

Country	Sector
USA	Chemicals
	Fabricated metal products
	Food, drink and tobacco
	Textiles, footwear and leather products
	Basic metals
	Total manufacturing
	Wood, cork and furniture
France	Wood, cork and furniture
Great Britain	Paper and printing
Spain	Other manufacturing industries

Table 11.5. Note that the majority of the sectors (seven out of ten) are located in the USA.

It is interesting to see what proportion of labour growth is explained by the other variables, as done before. At the national level a result can only be presented for the USA, for there is too limited a number of sectors available for the other countries. Table 11.6 lists the relevant statistics.

Table 11.6   Relative importance of explanatory variables in explaining per country wage growth (per cent)

	Variables					
	Constant	Technology	Capital	Trade	Wages	Residual
USA	71.1	523.1	−209.9	−167.1	−3 251.2	3 133.9
USA (excl. ZOW)	85.3	−9.1	−2.3	−16.4	20.1	22.5

Looking at Table 11.6, the large percentages we find for all variables besides the constant term indicate that there may be an outlier between the sectors at hand. This is indeed the case for wood, cork and furniture. Dropping this sector yields the result that the most important variable in determining labour growth in the USA is wage growth (leaving aside the constant term and the residual). Trade, technology and capital (in that order) all play a less important role.

Again, at the sectoral level some insights can be gained by analysing both the periods starting from the 1970s and the 1980s. Therefore, in Tables 11.7a and 11.7b the growth rates for the variables under consideration are shown for these periods.

A general conclusion that can be drawn when comparing the two tables is that the period of the 1980s is less favourable in many respects: for example, more sectors suffer from adverse terms of trade (eight instead of seven) and wage growth (four instead of three). Although R&D growth is greater in the 1980s than in the 1970s for some sectors (paper and printing in the UK, chemicals in the USA and other manufacturing industries in Spain), mostly it is less than in the earlier period.

Table 11.8a (and Table 11.8b for the 1980s, the results of which will again be presented in brackets) is a sectoral version of Table 11.6 (but now for all countries). We can derive that, in three (five) of the ten sectors, terms of trade have a larger influence on labour growth than technology. Technology thus matters more over the whole period, but since the 1980s the terms of trade have caught up in importance. This conclusion is in accordance with the one we obtained above.

Table 11.7a   Growth rates of explanatory and dependent variables over the estimation period given by SMPL

Sector	SMPL	Technology	Capital	Trade	Labour	Wages
GBRZOP	74–92	–0.0022	0.2590	0.0060	–0.0101	0.0107
USAZ38	74–93	0.0126	0.3643	–0.0157	–0.0036	0.0062
USAZLX	74–93	0.0320	0.3643	–0.0265	–0.0199	–0.0145
USAZMB	74–93	–0.0049	0.3643	0.0015	–0.0286	–0.0222
USAZMT	74–93	0.0187	0.3643	–0.0085	–0.0045	0.0047
FRAZOW	80–91	0.0779	–0.2758	–0.0051	–0.0235	–0.0082
USAZ35	74–93	0.0455	0.3643	–0.0019	0.0063	0.0179
USAZLF	74–93	0.0294	0.3643	0.0023	–0.0020	0.0048
USAZOW	74–93	0.0209	0.3643	–0.0056	0.0000	0.0046
ESPZOO	80–91	0.5407	–9.5446	–0.0489	–0.0129	0.0017

Table 11.7b   Growth rates of explanatory and dependent variables over the estimation period given by SMPL, 1980s onwards

Sector	SMPL	Technology	Capital	Trade	Labour	Wages
GBRZOP	81–92	–0.0018	0.1612	0.0034	–0.0138	0.0153
USAZ38	81–93	0.0081	0.1435	–0.0216	–0.0123	0.0001
USAZLX	81–93	0.0470	0.1435	–0.0207	–0.0197	–0.0104
USAZMB	81–93	–0.0311	0.1435	–0.0244	–0.0377	–0.0371
USAZMT	81–93	0.0158	0.1435	–0.0137	–0.0083	0.0013
FRAZOW	80–91	0.0779	–0.2758	–0.0051	–0.0235	–0.0082
USAZ35	81–93	0.0508	0.1435	–0.0153	0.0037	0.0143
USAZLF	81–93	0.0241	0.1435	0.0102	–0.0019	0.0033
USAZOW	81–93	0.0179	0.1435	–0.0023	0.0052	0.0091
ESPZOO	81–91	0.5679	–10.2546	–0.0707	–0.0107	–0.0067

In nine (eight) of the sectors wages have the largest impact. Technology comes first once (zero times), whereas capital and terms of trade hold the first position zero (zero) and zero times (twice) respectively. The influence of the terms of trade on labour growth has thus grown over time. When counting the variables that rank second, we get five (five) times technology, four (three) times trade, one (one) capital movements and zero times (once) wages. It will by now not come as a surprise that such results are achieved. Wage growth is the most dominant factor in explaining labour growth, with technology in second place, trade third and capital last. The roles of tech-

Table 11.8a   *Relative importance of dependent and explanatory variables*
in explaining per sector labour growth (per cent)

Sector	Constant	Technology	Capital	Trade	Wages	Residual
GBRZOP	112.9	−0.6	−1.6	20.1	−30.8	0.0
USAZ38	337.3	8.0	−14.7	−62.8	−167.9	0.0
USAZLX	54.9	5.1	−0.7	−22.3	63.1	0.0
USAZMB	34.5	−0.8	−1.8	0.2	67.9	0.0
USAZMT	205.8	11.6	−8.5	−9.1	−99.8	0.0
FRAZOW	0.0	8.7	3.8	−0.1	38.9	48.8
USAZ35	0.0	−120.0	5.8	1.7	256.0	−43.5
USAZLF	0.0	62.1	−0.2	1.0	−118.9	155.9
USAZOW	0.0	3 184.2	−1 247.8	−920.2	−19 607.3	18 691.2
ESPZOO	0.0	45.4	5.3	−44.5	−12.4	106.3

Table 11.8b   *Relative importance of dependent and explanatory variables*
in explaining per sector labour growth (per cent), 1980s
onwards

Sector	Constant	Technology	Capital	Trade	Wages	Residual
GBRZOP	82.6	−0.4	−0.7	8.4	−32.2	42.3
USAZ38	99.7	1.5	−1.7	−25.6	−0.7	26.9
USAZLX	55.6	7.5	−0.3	−17.7	45.6	9.2
USAZMB	26.2	−4.0	−0.5	−2.1	86.4	−6.0
USAZMT	110.7	5.2	−1.8	−7.8	−14.8	8.5
FRAZOW	0.0	8.7	3.8	−0.1	38.9	48.8
USAZ35	0.0	−229.0	3.9	23.8	349.8	−48.5
USAZLF	0.0	53.6	−0.1	4.8	−86.1	127.8
USAZOW	0.0	−11.4	2.1	1.6	163.5	−55.7
ESPZOO	0.0	57.7	6.8	−77.8	59.0	54.3

nology and trade switch when we look at the more recent period of the 1980s.

We have already stated that the perfect competition version of our model leaves something to be desired, for increasing returns at the firm level cannot be excluded from consideration. Yet, despite its deficiencies, it is clear that the model's results are quite robust: specific factors do indeed seem to matter for wage and/or labour growth, whereas the influence of the terms of trade on the results has risen over time (when set against the role of technology).

Nevertheless, it is equally clear that there still is a need to analyse an imperfect competition version of the model.

The Case of Imperfect Competition

The approach that is followed in the case of imperfect competition is very similar to the one followed in the perfect competition case. We first ran OLS regressions[39] on the basic model, which, from a theoretical point of view, already contains a constant term. As explained in section 2, we expect e_{1a} and e_{1b} to have a negative sign, while e_0 and e_4 can have either sign; e_3 is highly likely to be positive.

We see that the constant term (e_0) differs significantly from zero at the 5 per cent significance level only 19 times. These are all positive entries. Negative entries turn up only ten times. A similar conclusion holds with respect to the coefficients of the import variables (e_{1a} and e_{1b}): the coefficient for EU imports differs significantly from zero five times, which are all positive entries but two. It has the desired sign 29 times. Non-European imports turn up significantly nine times (of which three entries are negative), with a total of 27 negative signs. Thus non-European imports, including those of the Asian NICs (may) have a substantial impact on wage growth in some sectors. Given the construction of e_{1a} and e_{1b}, they should have the same sign. This happens only 22 times (with five cases in which they are both negative). [40]

The imperfect competition version of our model thus picks up some factors that were (unjustly) left out at the perfect competition stage. To see whether the fit can be improved even further, we tested whether there are structural breaks underlying the results. Thus Chow structural break tests were applied for (mostly) the years 1981 and 1986 to check whether such breaks were indeed present.[41] Structural breaks were found for nine sectors: four British (chemicals, food, drink and tobacco, leather products and total manufacturing), one American (basic metals), two Italian (chemicals and fabricated metal products) and two Dutch ones (basic metals and total manufacturing). For only three of them did we also find structural breaks in the perfect competition case.[42] The estimation for these sectors was redone, with the estimation period now mostly starting in 1986.[43] The results are split into two groups: the cases where employment growth determines wage growth and the cases where wage growth determines employment growth.[44]

Some changes occur for the aforementioned sectors. We find both correct sign switches (for example, in the case of the interest variable for Italian chemicals, the European import variable for total manufacturing in the Netherlands and its non-European counterpart for basic metals in the USA and the Netherlands), and incorrect ones – sometimes even within the same sectors (for example, in the case of the interest and non-European import variable for

total manufacturing in the Netherlands and the latter variable for total manufacturing in the UK). So, overall, no very new insights on coefficients are
created here either.

Nevertheless, it remains quite difficult to be more specific about the results
without actually knowing the values of the parameters α, β, δ, ϵ, θ, η, ϕ and
$\hat{\beta}$. This would be possible if we could solve the system of equations we get
for e_{0-4} numerically. We would then start with a system of six equations for
eight unknowns, which can only be solved if we could reduce the number of
unknowns to six or lower. We found no way of doing so. Therefore the only
thing we can do is to perform another 'growth accounting' exercise using the
latest estimates. First, however, we have to determine which sectors should
be included in such an exercise.

We may recall from section 2 that there are several conditions which have
to hold in order to fulfil the requirements of the model. Most of them are
about (combinations of) single parameters (such as $\eta > 0$ and $2(\alpha + \beta) - 1 >$
0), which cannot be checked so easily. By looking at the coefficients of the
two import variables (e_{1a} and e_{1b}), however, we can indirectly see whether δ,
$\epsilon < 0$. If we take the 62 sectors for which labour growth explains wage
growth, we find negative coefficients for e_{1a} and e_{1b} on only five occasions.
Furthermore, we can impose certain constraints on the parameter values so
that e_3 is non-negative, e_2 is non-positive and e_0 and e_4 can take on any value
(cf. section 2). Of the five sectors we got when checking the sign of the
import variables, we therefore have three sectors left: total manufacturing in
the USA, fabricated metal products in the Netherlands, and food, drink and
tobacco in Spain. For these three sectors a similar 'growth accounting' exercise as in the previous section can be conducted. No such exercise will be
conducted at the national level, since we have too few observations available
within each country to do so.

Growth rates at the sectoral level are presented in Tables 11.9a (for the
whole period) and 11.9b (for the period starting from the 1980s). Again, they
yield the same conclusion reached in the perfect competition case: the period
of the 1980s is less favourable in many respects. Most variables have lower
growth rates in the later period (except for fabricated metal products in the
Netherlands, where five out of the six variables have higher growth rates).

The final growth accounting results for both periods are given in Tables
11.10a and 11.10b. What we see (the results for the 1980s are in brackets) is
that wage growth is dominated by labour growth in all but one sector (food,
drink and tobacco in Spain, in both periods). There EU trade (non-EU trade)
is the most important. In answering the question whether technology or trade
(EU and non-EU) drives wage growth most, we find technology once (once)
and trade twice (twice). When looking at variables that rank in second place,
we find technology once (once), trade once (twice) and capital movements

Table 11.9a   Growth rates of explanatory and dependent variables over the estimation period given by SMPL

Sector	SMPL	Technology	Capital	EU trade	Non-EU trade	Labour	Wages
USAZMT	74–92	0.0227	0.3942	0.0434	0.0671	–0.0047	0.0041
NLDZ38	74–92	0.0153	–0.2976	0.0354	0.0677	–0.0099	0.0037
ESPZLF	80–91	0.1110	–9.5446	0.1244	0.0275	–0.0085	0.0092

Table 11.9b   Growth rates of explanatory and dependent variables over the estimation period given by SMPL, 1980s onwards

Sector	SMPL	Technology	Capital	EU trade	Non-EU trade	Labour	Wages
USAZMT	81–92	0.0218	0.1724	0.0395	0.0551	–0.0090	0.0001
NLDZ38	81–92	0.0177	0.0464	0.0440	0.0778	–0.0061	0.0033
ESPZLF	81–91	0.0940	–10.2546	0.1363	0.0556	–0.0063	0.0016

Table 11.10a   Relative importance of dependent and explanatory variables in explaining per sector wage growth (per cent)

Sector	SMPL	Constant	Technology	Capital	EU trade	Non-EU trade	Labour	Residual
USAZMT	74–92	200.5	50.2	–14.6	–13.7	–9.8	–112.7	0.0
NLDZ38	74–92	544.7	19.8	21.0	–74.4	–116.5	–294.6	0.0
ESPZLF	80–91	125.5	64.3	99.8	–99.9	–66.8	–22.9	0.0

Table 11.10b   Relative importance of dependent and explanatory variables in explaining per sector wage growth (per cent), 1980s onwards

Sector	SMPL	Constant	Technology	Capital	EU trade	Non-EU trade	Labour	Residual
USAZMT	81–92	9 514.0	2 287.5	–302.2	–589.7	–384.0	–10 226.1	–199.5
NLDZ38	81–92	625.4	26.3	–3.8	–106.0	–153.6	–206.5	–81.8
ESPZLF	81–91	705.2	306.1	602.7	–614.8	–760.5	–95.3	–43.3

once (zero times). Overall, the importance of technology seems to have been rather steady over time, the influence of trade has increased slightly, whereas labour matters most here too. Again the constant term and the residual have not been taken into consideration.

The results for the five sectors for which the role of labour and wages has been interchanged on the basis of Granger causality tests do not change the conclusions reached above. No sector fulfils all the previous requirements. Although the small number of sectors for which all requirements are met may cast some doubt on the validity of our entire model, we should not forget the fact that we are still – and always will be – faced with an aggregation problem from the assumption of identical firms within one sector which clearly may have distorted our results (even if the model was correct in itself). Yet altering the role of wage and labour growth in all equations may be an interesting route to follow, for in that way we can check the robustness of some of the conclusions reached previously.[45] In doing so, we find expected signs of the parameters for eight sectors: fabricated metal products in the USA, the UK and the Netherlands, food, drink and tobacco in Spain and the USA, total manufacturing in the USA, chemicals in Italy, and stone, clay and glass in Spain. We will perform another 'growth accounting' exercise for these sectors. Tables 11.11a and 11.11b list the relevant growth rates.

From the above growth rates we can conclude that, for the eight sectors included, the recent period of the 1980s is less favourable for most variables: they have lower growth rates more often than over the entire period (with the exception perhaps of the two trade variables). In particular the mean wage growth rate has fallen for all sectors, except fabricated metal products in the UK in the last period (where we considered the same sample period). Thus we can reinforce the conclusions already made before. This raises the ques-

Table 11.11a   Growth rates of explanatory and dependent variables over the estimation period given by SMPL

Sector	SMPL	Technology	Capital	EU trade	Non-EU trade	Labour	Wages
USAZ38	74–92	0.0193	0.3942	0.0531	0.0850	–0.0031	0.0063
USAZLF	74–92	0.0339	0.3942	0.0172	0.0053	–0.0026	0.0040
USAZMT	74–92	0.0227	0.3942	0.0434	0.0671	–0.004	0.0041
GBRZ38	86–92	–0.0537	0.4908	0.0241	0.0315	–0.0150	–0.0024
ITAZ35	74–92	0.0520	0.6333	0.0442	0.0773	–0.0018	0.0156
NLDZ38	74–92	0.0153	–0.2976	0.0354	0.0677	–0.0099	0.0037
ESPZLF	80–92	0.1110	–9.5446	0.1244	0.0275	–0.0085	0.0092
ESPZOG	80–92	0.0470	–9.5446	0.0675	0.1094	–0.0221	–0.0159

Table 11.11b   *Growth rates of explanatory and dependent variables over the estimation period given by SMPL, 1980s onwards*

Sector	SMPL	Technology	Capital	EU trade	Non-EU trade	Labour	Wages
USAZ38	81–92	0.0183	0.1724	0.0461	0.0799	−0.0121	−0.0003
USAZLF	81–92	0.0308	0.1724	0.0118	−0.0011	−0.0028	0.0020
USAZMT	81–92	0.0218	0.1724	0.0395	0.0551	−0.0090	0.0001
GBRZ38	86–92	−0.0537	0.4908	0.0241	0.0315	−0.0150	−0.0024
ITAZ35	80–92	0.0487	0.2304	0.0003	−0.0164	−0.0066	0.0108
NLDZ38	80–92	0.0177	0.0464	0.0440	0.0778	−0.0061	0.0033
ESPZLF	81–91	0.0940	−10.2546	0.1363	0.0556	−0.0063	0.0016
ESPZOG	81–91	0.0341	−10.2546	0.0728	0.1121	−0.0219	−0.0200

Table 11.12a   *Relative importance of dependent and explanatory variables in explaining per sector labour growth (per cent)*

Sector	SMPL	Constant	Technology	Capital	EU trade	Non-EU trade	Labour	Residual
USAZ38	74–92	480.2	114.4	−44.4	−54.4	−215.9	−179.9	0.0
USAZLF	74–92	136.1	66.7	−6.5	−28.6	−7.6	−60.1	0.0
USAZMT	74–92	264.4	52.0	−20.3	−28.0	−101.0	−67.1	0.0
GBRZ38	86–92	663.5	−407.5	−102.1	−27.3	−34.4	7.9	0.0
ITAZ35	74–92	499.7	124.4	−6.5	−1.6	−46.3	−469.8	0.0
NLDZ38	74–92	164.5	3.9	6.5	−35.6	−14.9	−24.4	0.0
ESPZLF	80–92	262.9	79.5	62.5	−191.7	−35.4	−77.8	0.0
ESPZOG	80–92	63.6	16.9	15.3	−62.3	−0.3	66.8	0.0

Table 11.12b   *Relative importance of dependent and explanatory variables in explaining per sector labour growth (per cent), 1980s onwards*

Sector	SMPL	Constant	Technology	Capital	EU trade	Non-EU trade	Labour	Residual
USAZ38	81–92	121.9	27.5	−4.9	−12.0	−51.5	2.4	16.6
USAZLF	81–92	125.0	55.6	−2.6	−18.2	1.4	−27.1	−34.2
USAZMT	81–92	138.2	26.1	−4.6	−13.3	−43.4	−0.7	−2.2
GBRZ38	86–92	663.5	−407.5	−102.1	−27.3	−34.4	7.9	0.0
ITAZ35	80–92	133.8	31.2	−0.6	0.0	2.6	−87.4	20.4
NLDZ38	80–92	269.4	7.3	−1.7	−72.3	−28.0	−34.8	−40.0
ESPZLF	81–91	355.2	91.0	90.7	−283.8	−96.7	−18.7	−37.7
ESPZOG	81–91	64.4	12.4	16.6	−68.0	−0.3	85.2	−10.4

tion of what we find when analysing the impact of all variables on wage growth (in percentages). Tables 11.12a and 11.12b give an indication.

Wage growth has the largest impact on labour growth in two (two) out of the eight sectors. Trade explains a larger percentage of labour growth than technology in five (five) sectors. Given that non-EU trade sometimes even exhibits the most explanatory power (in three (three) cases during both periods), we may sustain the hypothesis that competition from the Asian NICs may have had a substantial impact over the years. There are not many differences when comparing the period of the 1970s and that of the 1980s as far as shifts in variables and their explanatory power with regard to labour growth are concerned.

Combining the sectors we worked with above with the 22 non-overlapping sectors we found in the perfect competition case, we have a group of 26 sectors (excluding overlap) for which either a perfect or an imperfect competition approach gives expected signs. The evidence that can be obtained from the imperfect competition model is that the main conclusions of the perfect competition model are endorsed. Specific factors are important and the role of both technology and trade in explaining wage and/or labour growth changes (with trade being influential in both periods) has clearly come forward.

5. POLICY CONCLUSIONS

Protectionism or compensation mechanisms are probably the first policy instruments firms and sectoral institutions point to when trying to counterbalance the (financial) effects from losses from trade. From a model point of view, the effects of such measures are difficult to determine, for under perfect competition and the small country assumption protectionism is damaging. However, even from a theoretical perspective mechanisms like protectionism do seem somewhat short-sighted, for firms are not encouraged (enough) to strengthen their international competitive position over time, which may easily find them falling behind more and more (leading to rising compensation from governments). Case study evidence also indicates that protectionism may have adverse effects. For example, one policy action has been the Trade Adjustment Assistance Program in the USA. Sachs and Shatz (1994) show that the sectoral distribution of compensation from that programme are strongly correlated with the underlying sectoral distribution of employment losses (so that losing sectors are compensated adequately – which might make losing more attractive). Moreover, in our analysis of both the perfect and imperfect competition cases we found a total of nine sectors that have a negative effect from adverse terms of trade or import movements and decreasing wages. These sectors are listed in Table 11.13.

Table 11.13   Sectors that have adverse effects from trade and decreasing wage growth

Country	Sector
USA	Textiles, footwear and leather products
	Basic metals
France	Stone, clay and glass
	Wood, cork and furniture
Italy	Textiles, footwear and leather products
Spain	Basic metals
	Total manufacturing
	Other manufacturing industries
	Wood, cork and furniture

Given the aforementioned policy measures and looking at these nine sectors, we could ask the crucial question whether income policies for the short run and R&D subsidies for the long run would be a better means than protectionism of helping sectors to cope with negative trade effects. As international trade has gained in importance since the 1980s, this question has become more urgent for several sectors (especially for leather products in the USA and other manufacturing industries in Spain). However, we should also study the forces playing within each of these sectors separately before reaching a definite conclusion.

Of course, tax reductions for the less skilled or low-income brackets are one variation on income policies that would invoke a problem at the household level rather than at the firm level. However, it should be clear that behind the given interest rate there is a critical issue of interest rate determination and behind the given sectoral labour supply and wages there are labour market imperfections. Given the dominance of the labour supply variable in both the perfect and the imperfect competition version of our model, it seems reasonable to search for a diagnosis and a solution to trade problems in the labour market sphere (for example, by ensuring a better match between the skills necessary to perform a certain job and the skills workers have, or by easing the hiring and firing conditions of workers). Here specific factors have turned out to be a robust variable that is more important than both technology and trade.

6. LIMITATIONS AND SUGGESTIONS FOR FURTHER RESEARCH

The major drawback of a trade-theoretic approach is that international trade models are not related to models explaining unemployment, and vice versa. This is the reason why economists currently have to choose between a closed economy labour market imperfections approach and a trade approach. The integration of the two must be left for further research (provided that major intertemporal changes in the labour market situation occur). Moreover, owing to the simplifying assumption of constant price elasticities of demand and therefore of mark-ups over marginal costs, we cannot include their change across the business cycle without considerably complicating the model.

An incentive for further research from our analysis follows from three results. First, in the perfect competition case the constant term in our model was absent but the empirics tell us we should have one (thus indicating that there are possibly other explanatory variables that should have been included in the model). Second, the model would predict relations between the coefficients, but the corresponding constraint has been rejected by statistical tests. Third, we could move from the firm level to the sectoral level only under the assumption of identical firms, which in all likelihood is fairly unrealistic. This aggregation problem will (probably) remain even when switching to different types of production functions with constant or variable elasticities of substitution. Yet, even if our model and estimation results are rather too crude to give a 'robust' answer to the question of what factor drives sectoral wage growth most strongly (technology or trade), our results do have their relevance. In particular, the supply of specific factors turned out to matter in both models (with the results being very robust in that respect) and the changing role of international trade (becoming more important than technology in the 1980s according to the perfect competition model) has been clearly illustrated.

NOTES

1. Parts of this chapter have been presented at the ESF conference, 'Economic growth in closed and open economies', Lucca, September 1997, the TSER group seminars on technology and employment, Paris, October 1997 and May 1998, and the conference, 'Unemployment in Europe', Maastricht, October 1997. We especially would like to thank Bruno Amable, Donatella Gatti, Huw Lloyd-Ellis, Erik de Regt, Giovanni Russo, Luc Soete and Winfried Vogt for their comments. The usual disclaimer applies.
2. Note that the estimation of Jones's (1970) dynamic version of the zero profit conditions uses data on factor shares (see Baldwin and Cain, 1997), which consist of a cost term in the numerator and revenue terms in the denominator. If we (empirically) have zero profits on average across time, we might guess from a business cycle perspective that there are

losses in recessions and positive profits in booms. This yields higher than average values of cost shares in recessions and lower values in booms. In time-series estimates this may bias the results, in particular in view of the possibility that capital and labour shares may be affected unequally because of the irreversibility (or costly reversibility) of the investment of capital which makes it difficult to reduce its cost in a recession.

3. In the standard partial equilibrium labour market diagram, an increase in the labour supply would decrease wages. However, the increase in employment has an indirect effect via the marginal productivity of capital, which is increased by higher employment and therefore more capital is attracted from the world market. With the increase in capital, labour demand also increases, which will increase wages. Under increasing (decreasing) returns to scale, the indirect demand effect is stronger (weaker) than the direct supply effect.

4. Applying non-linear least squares (NLS) or maximum likelihood (ML) (while simultaneously imposing the coefficient restriction derived in the theoretical part) would have been an option, were it not that we would then be implying that the coefficient restriction already holds a priori. Thus, given the reservations expressed above, OLS seems to be preferable. Pooling data (across sectors, countries or both) would have been an option too, but it was dropped when relatively few interpretable results emerged. See also notes 19 and 40.

5. White's method (1980) is used to achieve this.

6. Scherer (1983) and Griliches (1990) examine more closely the points in favour and against using either R&D or patent statistics as an indicator of technical change.

7. Cf. Scherer (1983) and Feldman and Florida (1994). See Caniëls (1998) for European evidence of this.

8. With the estimated return to R&D being downward biased.

9. In the appendix, additional remarks on this subject are made.

10. Three sectors were excluded because of missing R&D data: the Dutch and Italian wood, cork and furniture sectors and the Dutch other manufacturing industries.

11. The sign of γ_4 depends on the presence of increasing returns to capital and labour at the sectoral level; see section 2. The fact that it can take on either sign is illustrated by Efendioglu and von Tunzelmann (1998) and Spiezia and Vivarelli (1998).

12. An alternative interpretation that is somewhat independent of our model could be that the economy is moving along an upward-sloping labour supply curve – a view found in the work of Bovenberg (see Bovenberg, 1995, for details).

13. From a model point of view, the period characterized by positive real interest rates is the only one of interest, because only then does the model hold. It is assumed, however, that when no structural breaks are found, the influence of negative real interest rates on the regression results is negligibly small.

14. See Chow (1960) for details.

15. It is reckoned that econometrically more sophisticated methods exist to assess points in time at which structural breaks occur (see, for example, Gallant and Fuller, 1973). However, we concentrate on the years which we assume to be the most influential.

16. The British chemical industry is also affected by the sign switch of the real interest rate at the beginning of the 1980s.

17. All regressions were also carried out with a time variable included. This variable was always insignificantly different from zero at a 5 per cent significance level (which is not that surprising since we are working with series expressed in first differences).

18. The technology variable now has the desired sign more often and (especially) becomes more significant. This might point to the fact that R&D expenditures are rather a flawed indicator of technical change. However, putting the technology variable into the residual would then again seem too drastic an action, for it would, in a statistical sense, lead to omitted variable bias.

19. If all coefficient definitions given in the theoretical part are substituted into the regression equation and the model is re-estimated by means of NLS, these $\alpha/(1 - \eta)$, $\beta/(1 - \eta)$ and $\theta/(1 - \eta)$ estimates follow. Of course, it would have been preferable to solve the system numerically. This did not yield any result, for then it is implicitly assumed that the

imposed coefficient restriction holds *exactly*, whereas our test examines whether it holds *within a certain margin*.

20. Exceptions (with respect to significance) are all British sectors except textiles, footwear and leather products, stone, clay and glass, paper and printing and wood, cork and furniture, the French food, drink and tobacco, stone, clay and glass and other manufacturing industries, food, drink and tobacco in the Netherlands and Spain and the Dutch basic metals sector.

21. First introduced by Granger (1969). Sims (1972) and others provided tests (mostly) along the same lines, but the Granger causality test is the one most commonly used.

22. See Eels (1991) for a more elaborate analysis.

23. Do note that it is short-run causality that we test for here. If we had wanted to detect long-run causality, the existence of a cointegration equation between labour and wage growth should have been proven. Given that we found no evidence thereof when testing (possibly because of the somewhat limited number of observations available), the current approach is chosen.

24. Note that it is not necessary to test the validity of the derived coefficient restriction again, for the same reason (as long as the estimation period remains the same). Here we have to perform this test anew for two sectors: chemicals and leather products in the UK. For the latter sector, the coefficient restriction is accepted, so new estimates for $\alpha/(1 - \eta)$, $\beta/(1 - \eta)$ and $\theta/(1 - \eta)$ can be generated. See also note 27.

25. As turned out to be the case for British and Spanish textiles, footwear and leather products and the Italian paper and printing industry.

26. This is not surprising since coefficients that already had the wrong sign when wages were taken as the dependent variable will have the same now too (as long as the constant term remains either absent or present and the estimation period remains the same).

27. This is why we find no reasonable estimates for $\alpha/(1 - \eta)$, $\beta/(1 - \eta)$, $\theta/(1 - \eta)$ in the case of leather products in the UK (where we did accept the coefficient restriction): $\alpha/(1 - \eta)$ equals $2.93.10^{-3}$, $\beta/(1 - \eta)$ 1.50 and $\theta/(1 - \eta)$ –0.128.

28. At first, we will only look at cases where employment determines wage growth. If we had reversed the position of the wage and labour variables in the regression equation and redone the entire analysis up to this point, we would have ended up with ten sectors to work with (instead of the 17 we have now); see Table 11.5 below. Compared to the 17 sectors we find here (see Table 11.1), we have an overlap for five sectors: three American (chemicals, basic metals and wood, cork and furniture), one French (chemicals) and a British one (paper and printing). So if we include both relationships (where labour growth determines wage growth and vice versa), we would have 22 sectors to continue with. Later we will consider the other five.

29. Ideally, we would have preferred looking at variables that both have the expected sign and are statistically significant. However, this is not the case for any sector (as in all of the international trade literature). Since we do want to give an indication whether either terms of trade or technology drives wage growth most, the present approach is opted for.

30. Alternatively, we could have taken medians or calculated an average based on just the first and last period. However, given the way in which the OLS estimates are obtained, calculating means over the entire estimation period is to be preferred.

31. It is an aggregate across all other sectors and including it would create a bias.

32. Which was to be expected, given our previous results.

33. Leaving aside the constant term.

34. A similar exercise was carried out for the period starting (mostly) from 1980 or 1981. There, we looked at sectors which had (by and large) falling growth rates of wages. With the *same* regression coefficients (which, in a rough sense, is a valid approach, for structural breaks have already been taken into account), we found results that were almost identical to the ones obtained in Table 11.2. However, the results for total manufacturing in the Netherlands and food, drink and tobacco in Spain became worse, with, respectively, –123.9 per cent and –325.1 per cent of wage growth now being attributed to the residual. On the contrary, we found improved results for France, where technology now emerged as the most prominent factor in wage determination. Moreover, in Spain capital became the

most important explanatory factor. Yet, overall, labour still turned out to be the most influential factor in national wage formation. More results at the sectoral level are discussed (and shown) later on. A similar approach will be followed in the case of imperfect competiton.

35. The extremely high value for the mean growth rate of the Spanish interest rate is due to an outlier in 1986. Possibly, this outlier is caused by the alliance of Spain to the EU (and it was therefore explicitly taken along in our exercise).

36. In seven (seven) cases the growth rates of L and w have opposite signs, but have positive regression coefficients. The inclusion of other explanatory variables and interaction effects between them play an important role in this 'switch' in sign.

37. This conclusion is independent of the fact that the residual sometimes explains a large part of wage growth (for example for wood, cork and furniture in France). Even if we had included a constant term in the regressions for these sectors (and checked whether all variables had the correct signs), the economic interpretation of the results would have remained virtually the same.

38. No regression outputs or intermediate results for the reversed relationship are included in the main text. However, they can be obtained from the authors upon request. One may claim that, since wages (which now appear as an explanatory variable) are determined endogenously, the use of instrumental variables is advisable in order to reach more accurate regression results. Since the outcome of instrumental variables techniques is highly dependent on the number and quality of the instruments included in the analysis, we feel that the present approach has certain advantages. A pooling exercise was carried out here, too, finally supporting this feeling. The fact that no definite answer could be given to the question of the existence of a long-run relationship between wage and labour growth makes using this 'reversed causality' equation for *all* sectors an interesting route to follow (to consider all possible relationships). See also note 23.

39. NLS and ML regressions were also tried (with either the results from the OLS regressions or zero as starting values), but this yielded hardly any result (convergence only occurred when running NLS regressions from zero. If we had continued with these figures, the results presented below would remain roughly the same).

40. The 'conflict' in sign between the two import variables may lead us to the conclusion that even the current specification leaves something to be desired. Do note, however, that this sign 'conflict' may be due to multicollinearity: the two import variables have a correlation that is mostly larger than 60 (and often exceeds 80). Although there are solutions to multicollinearity (for example, dropping one of the collinear variables), this is not an option in the present context, for it would imply an explicit change of the theoretical variable.

41. Why these years were chosen was explained at the beginning of section 4.

42. These sectors are the British chemicals and leather products industries, and fabricated metal products in Italy.

43. Exceptions are the British food, drink and tobacco sector and the Italian chemicals and Dutch basic metal industries. For these sectors, the estimation period started in 1981 instead of 1986.

44. The results of Granger causality tests we carried out are still valid. For the sectors where we found structural breaks, this implies that in two British cases (chemicals and leather products) labour is taken as the independent variable and wages as an explanatory one, and the entire estimation procedure has to be redone (including testing for structural breaks). So, in effect, the structural break tests that were carried out change the results for only seven sectors instead of nine.

45. No intermediate results will be shown here either, but they are available from one of the authors upon request. See also notes 33 and 39.

REFERENCES

Baldwin, R.E. and G.G. Cain (1997), 'Shifts in U.S. Relative Wages: The Role of Trade, Technology and Factor Endowments', NBER WP 5934.

Bovenberg, L. (1995), 'Environmental Taxation and Employment', *De Economist*, 143, 2.

Caniëls, M.C.J. (1998), 'The Geographic Distribution of Patents and Value Added across European Regions', MERIT Research Memorandum 98-004.

Chow, G.C. (1960), 'Tests of Equality between Sets of Coefficients in Two Linear Regressions', *Econometrica*, 28, 591–605.

Eels, E. (1991), *Probabilistic Causality*, Cambridge: Cambridge University Press.

Efendioglu, U.D. and G.N. Von Tunzelmann (1998), 'Technology and Employment in Postwar Europe: Short-Run Dynamics and Long-Run Patterns', TSER Working Paper, May.

Feldman, M.P. and R. Florida (1994), 'The Geographic Sources of Innovation: Technological Infrastructure and Product Innovation in the United States', *Annals of the Association of American Geographers*, 84, 210–29.

Francois, J.F. (1996), 'Trade, Labour Force Growth and Wages', *Economic Journal*, 106, 1586–609.

Freeman, R.B. (1995), 'Are your Wages set in Beijing?', *Journal of Economic Perspectives*, 9, 15–32.

Gallant, A.R. and W.A. Fuller (1973), 'Fitting Segmented Polynomial Regression Models whose Joint Points have to be Estimated', *Journal of the American Statistical Association*, 68, 144–7.

Granger, C.W.J. (1969), 'Investigating Causal Relations by Econometric Models and Cross-spectral Methods', *Econometrica*, 44, 424–38.

Griliches, Z. (1990), 'Patent Statistics as Economic Indicators: A Survey', *Journal of Economic Literature*, 28, 1661–707.

Hall, B.H. and J. Mairesse (1992), 'Productivity and R&D at the Firm Level', NBER Working Paper 3956.

Helpman, E. and P.R. Krugman (1989), *Trade Policy and Market Structure*, Cambridge: MIT Press.

International Financial Statistics Yearbook 1990 (1990), Washington, DC: International Monetary Fund Publication Services.

International Financial Statistics Yearbook 1995 (1995), Washington, DC: International Monetary Fund Publication Services.

Jones, R.W. (1970), 'The Role of Technology in the Theory of International Trade', in R. Vernon (ed.), *The Technology Factor in International Trade*, New York: NBER.

Krugman, P.R. (1994), 'Competitiveness: A Dangerous Obsession', *Foreign Affairs*, March/April.

Krugman, P. and R. Lawrence (1993), 'Trade, Jobs and Wages', NBER WP W4478.

Krugman, P.R. and M. Obstfeld (1997), *International Economics*, 4th edn, Reading, MA: Addison-Wesley.

Lawrence, R.Z. and M.J. Slaughter (1993), 'International Trade and American Wages in the 1980s: Giant Sucking Sound or Small Hiccup?', *Brookings Papers on Economic Activity*, 2, 161–226.

Leamer, E.E. (1993), 'Wage Effects of a U.S. Mexican Free Trade Agreement', in P.M. Garber (ed.), *The Mexico–U.S. Free Trade Agreement*, Cambridge, MA: MIT Press.

Leamer, E.E. (1994), 'Trade, Wages and Revolving Door Ideas', NBER Working Papers Series, no. 4716.

Leamer, E.E. (1996), 'Wage Inequality from International Competition and Technological Change: Theory and Country Experience', *American Economic Review*, AEA Papers and Proceedings, May.

Lücke, M. (1997), 'European Trade with Lower-income Countries and the Relative Wages of the Unskilled: An Exploratory Analysis for West Germany and the UK', Kiel University WP 819.

Oliveira Martins, J. (1994), 'Market Structure, Trade and Industry Wages', *OECD Economic Studies* No. 22, Spring, 131–54.

Oscarsson, E. (1997), 'Trade and Relative Wages in Sweden 1968–91', Department of Economics, Stockholm University Research Memorandum, January.

Richardson, J.D. (1995), 'Income Inequality and Trade: How to Think, What to Conclude', *Journal of Economic Perspectives*, 9, 33–55.

Sachs, J.D. and H.J. Shatz (1994), 'Trade and Jobs in U.S. Manufacturing', *Brookings Papers on Economic Activity*, 1, 1–84.

Schankerman, M. (1981), 'The Effects of Double-counting and Expensing on the Measured Returns to R&D', *Review of Economics and Statistics*, 63, 454–8.

Scherer, F.M. (1983), 'The Propensity to Patent', *International Journal of Industrial Organization*, 1, 107–28.

Sims, C.A. (1972), 'Money, Income and Causality', *American Economic Review*, 62, 540–55.

Spiezia, V. and M. Vivarelli (1998), 'Growth, Technical Change and Employment: Some Evidence from G-7 Manufacturing Industries', TSER Working Paper, May.

Verspagen, B. (1995), 'R&D and Productivity: A Broad Cross-Section Cross-Country Look', *Journal of Productivity Analysis*, 6, 117–35.

White, H. (1980), 'A Heteroscedasticity-Consistent Covariance Matrix Estimator and a Direct Test for Heteroscedasticity', *Econometrica*, 48, 817–38.

Wood, A. (1994), *North–South Trade, Employment and Inequality*, Oxford: Clarendon Press.

Wood, A. (1995), 'How Trade Hurt Unskilled Workers', *Journal of Economic Perspectives*, 9, 57–80.

APPENDIX: DATA DESCRIPTION

All data except the Spanish, data on long-term interest rates and data on technical change are taken from the OECD's ISDB database. Employment data contain the number of employees, excluding the self-employed. Wages include all payments made to wage and salary earners (which also exclude the self-employed), including social security payments. Both sectoral and national prices are also calculated from the ISDB database, via value added at market prices (with 1985 as a base year). Technically speaking, it would have been preferable to use value added at factor costs to construct price levels, for this would exclude taxes and subsidies, which may differ between countries. Only for Great Britain was value added available at factor costs in the database (and subsequently used). All variables are expressed in national price levels.

Interest data are taken from the *International Financial Statistics Yearbook* published by the IMF (from 1990 and 1995 publications). The long-term government bond yield was taken as a proxy for the long-term interest rate (as suggested by the IMF itself: see the *International Financial Statistics Yearbook 1995*, pp. xv–xvi).

As a proxy of technical change, R&D expenditures are used. (One may claim that, because R&D personnel are included in the labour variable, our regression results are biased, since we are also using R&D expenditures as a separate variable. However, this only means that there may be some collinearity between the technology and labour variable, which is justified from a theoretical point of view. Regression results do not become biased because of collinearity. See also section 3 of the main text.) These data are taken from the OECD's ANBERD database. For Spain, employment, wage and sectoral price levels are calculated from the OECD's STAN database. Spanish employment figures do include the self-employed. R&D data are again taken from ANBERD, whereas both national price levels (the GDP deflator) and the interest rate data are taken from the *International Financial Statistics Yearbook*. All import data (for all countries) come from the OECD's BITRA database.

The sectors included in the analysis are the two-digit ISIC sectors 31 to 39, which define total manufacturing (ISIC sector 30). In the remainder, we will denote these sectors by means of an abbreviation. These abbreviations are shown in Table 11A.1.

The country codes used are shown in Table 11A.2.

In the regression analyses three sectors were dropped because of missing R&D data: the Dutch and Italian wood, cork and furniture sector, and the Dutch other manufacturing industries.

Table 11A.1   Sector classification and abbreviations

ISIC code	Abbreviation	Sector description
30	ZMT	Total manufacturing
31	ZLF	Food, drink and tobacco
32	ZLX	Textiles, footwear and leather
33	ZOW	Wood, cork and furniture
34	ZOP	Paper and printing
35	Z35	Chemicals
36	ZOG	Stone, clay and glass
37	ZMB	Basic metals
38	Z38	Fabricated metal products, machinery and equipment
39	ZOO	Other manufacturing industries

Table 11A.2   Country codes

Country	Country code
USA	USA
Former West Germany	DEU
France	FRA
Great-Britain	GBR
Netherlands	NLD
Italy	ITA
Spain	ESP

12. Modelling the link between skill biases in technical change and wage divergence through labour market extensions of Krugman's North–South model

Adriaan van Zon, Mark Sanders and Joan Muysken

1. INTRODUCTION

A growing problem in the OECD area over the recent past has been the position of low-skilled workers relative to their high-skilled counterparts. The employment and income perspectives of the former have deteriorated significantly over the past two decades owing to a drop in the relative demand for their services. Economic theory has tried to come to grips with the empirical observation that the effects of this drop in low-skilled labour demand differ widely between Europe and the USA. In fact, there are two distinct empirical observations which led us to the problem we wish to address in this chapter. First of all, the OECD (1994) and several independent academic researchers (Brauer and Hickok, 1995) have observed that in the Anglo-Saxon parts of the world wages for low-skilled workers have deteriorated in relative but in some cases even in absolute terms.

To explain this phenomenon economists have formed two hypotheses. The first attributes the effect to the factor price equalizing effects of increased trade with low-wage countries (Leamer, 1994, 1995; Burtless, 1994; Lawrence and Slaughter, 1993). The second hypothesis emphasizes the possibility of technological change causing a divergence between the productivity of high- and low-skilled workers and thus causing a drop in relative demand for low-skilled labour, together with wage divergence in line with productivity. Some notable references are Krugman (1995), Jackman (1995), Howell (1995) and Agenor and Aizenman (1996). The OECD (1994), however, emphasizes that this relative wage deterioration for the low-skilled is almost absent in continental Europe, although Europe and the USA have been exposed to roughly similar trade and technology circumstances over the past decades.

In Europe, therefore, the drop in demand must have a different cause. Some European writers (Drèze and Malinvaud, 1993; Groot and Maassen van de Brink, 1996; Laan, 1996; Beer, 1996) have suggested the crowding-out hypothesis, which claims that low-skilled workers are 'pushed' out of their jobs by their high-skilled colleagues. This hypothesis rests on the assumption that asymmetries in employment opportunities exist between skills, and it explains the observed deterioration in employment and unemployment rates in Europe (OECD, 1994). The latter are remarkably stable over time in the USA, Canada and Australia, although the assumed asymmetries in employment opportunities should also exist in these countries. We conclude, therefore, that the similarities between the technology and trade circumstances seem to contradict the disparity between the experience of Europe and the USA with regard to wage divergence and changing employment perspectives. The challenge then is to formulate a model which is able to cope with this apparent contradiction.

In this chapter we will present a model based on Krugman's North–South model (Krugman, 1979), that generates two distinct labour market regimes, each of them supporting one of the two broad classes of empirical observations described above. We show that countries with identical production structures and identical employment opportunity asymmetries, but (slightly) different attitudes with respect to innovative activity and labour market flexibility, may find themselves in, or moving towards, either a wage divergence regime, or a stable relative wage-cum-crowding-out regime, depending on the nature of the attitudes and the initial conditions. In constructing the model, we start from the stylized fact that employment opportunities for high- and low-skilled people are asymmetric: high-skilled people generally can execute the tasks which low-skilled people perform, but this is not necessarily the case the other way around. Moreover, if the production of new goods and services is high-skill biased, at least in the initial stages of production, we can imagine a situation where innovative activity makes high-skilled people more scarce, while making low-skilled workers relatively less scarce at the same time (after all, product innovation means finding new products rather than old ones). In short, the process of product innovation aggravates the fundamental asymmetry between low- and high-skilled workers' employment opportunities mentioned above through impulses from the demand side. Given these employment asymmetries, we are able to show that wage divergence and crowding-out are two 'natural' states of the model.

In the debate on the causes of asymmetries in employment in Europe, however, economists seem to have a tendency to put the blame on the existence of price rigidities and distortions in the European labour markets. It is commonplace to argue for more flexible labour market institutions and to call in the help of market forces to equilibrate supply and demand. In our model a lack of labour market flexibility may indeed cause Europe and the USA to converge upon different types of equilibria explaining the divergence between Europe and the USA with respect to wages and employment by skill, but it is not clear that the suggested remedies actually help us very much.

The outline of the rest of the chapter is as follows. Section 2 will be devoted to a description of the Krugman (1979) model. In section 3 we add our labour market extensions, as well as a 'reduced-form endogenization' of technical change. Section 4 discusses the conditional predictions of the model regarding various possible regimes of wage divergence, while section 5 contains some concluding remarks.

2. THE KRUGMAN NORTH–SOUTH MODEL[1]

Krugman assumes two countries, North and South, which differ from each other in the sense that North can produce new goods after inventing them, while South can produce those goods only after North has invented them. North therefore innovates, while South imitates: there is an intrinsic asymmetry in time in North's and South's technological capabilities.

The Demand for Goods

The Krugman model assumes that consumers show 'love of variety'; that is, a good being different from other goods already available will always contribute in a positive way to utility. Krugman simplifies this notion by assuming a symmetric additively separable utility function of the form:

$$U = \int_0^n c_i^\alpha di, \tag{12.1}$$

where U represents total utility, and $0 - n$ is the 'width' of the continuous range of varieties produced and consumed; c_i is the level of consumption of variety i; α is a constant parameter which is positive and smaller than one. Note that the utility function is completely symmetric in the consumption of every single variety. Moreover, it is concave in the consumption of the varieties. If, as is the case in the Krugman model, prices of all varieties would be the same, the concavity in combination with the symmetry of the utility function ensures that utility would be maximized by distributing the total consumption budget evenly over all varieties. This follows directly from the first-order condition with regard to utility maximization:

$$\frac{c_i}{c_j} = \left(\frac{p_i}{p_j}\right)^{\frac{-1}{1-\alpha}}$$

(12.2)

for all $0 \le i, j \le n$. Equation (12.2) describes the relative demand for each variety, which depends negatively on the corresponding relative price.

The Supply of Goods

Each variety is produced in a situation of profit maximization under imperfect competition on the product markets (varieties are heterogeneous by assumption) and perfect competition on the labour market. Labour is the only factor of production distinguished in the model. The production functions for each variety are identical and linear homogeneous, which implies:

$$\frac{x_i}{x_j} = \frac{l_i}{l_j}$$

(12.3)

for all $0 \le i, j \le n$, and where x_i represents the level of supply of variety i, while l_i denotes the amount of labour used in sector i. With perfect mobility between sectors (each sector produces one variety), labour would earn the same wage everywhere, and, with labour productivity set equal to one for all varieties, profit maximization would ensure that the prices of all varieties would be strictly proportional to the common wage rate (this follows from Amoroso–Robinson, which applies in the case of imperfect competition on the goods market):

$$\Pi_i = p_i c_i - w l_i \Rightarrow \frac{\partial \Pi_i}{\partial p_i} = 0 \Rightarrow c_i \left(\frac{\partial c_i}{\partial p_i}\frac{p_i}{c_i} + 1 - \frac{w}{p_i}\frac{\partial l_i}{\partial c_i}\frac{\partial c_i}{\partial p_i}\frac{p_i}{c_i}\right)$$
$$= 0 \Rightarrow p_i = \frac{w}{\alpha}$$

(12.4)

where we have made use of (12.2) and (12.3), and where we have assumed equilibrium between the demand for goods and the supply of goods. Equation (12.4) indicates that the profit-maximizing prices of all varieties will be the same, in which case the demand for all varieties will be the same as well (cf. (12.2)). Therefore labour will be equally distributed over the production of all varieties (cf. (12.3)).

Krugman now assumes that there are varieties which can be produced both by the North and by the South ('old' varieties) and varieties which can only be produced by the North ('new' varieties). If wage costs per worker in the South would be lower than in the North, the latter would have an incentive to concentrate on the production of 'new' varieties, and to trade these for 'old' varieties produced by the South. The question Krugman then raises is who will be producing what, and which consequences this will have for relative wages earned in the North and in the South.

Technology

In the Krugman model, technical change takes the form of product innovation in the North and imitation by the South. Krugman assumes that the rate of innovation is exogenous:

$$\hat{n} = \mu_0, \tag{12.5}$$

where a 'hat' over a variable denotes its proportional rate of change over time; μ_0 is the constant and exogenous proportional rate of innovation. With respect to imitation, Krugman assumes that the rate of change in the number of 'imitated' varieties (n^i) is proportional to the number of 'new' varieties, that is, the number of varieties not yet imitated. This implies for the proportional rate of change in the number of imitations that:

$$\hat{n}^i = \mu_1 \frac{n - n^i}{n^i}, \tag{12.6}$$

where μ_1 is the propensity to imitate. It is easy to show that equations (12.5) and (12.6) define a steady-state value for the ratio of 'new' varieties and 'imitated' varieties:

$$\hat{n} = \hat{n}^i \Rightarrow \mu_0 = \mu_1 \frac{n - n^i}{n^i} \Rightarrow \frac{n}{n^i} = 1 + \frac{\mu_0}{\mu_1}. \tag{12.7}$$

If this ratio exceeds its steady-state value, then equation (12.6) indicates that the rate of imitation will increase, which will lower the ratio of 'new' and 'old' varieties again, until the ratio has reached its steady-state value. If the ratio is lower than its steady-state value, the rate of imitation will decrease, thus raising the value of the ratio of 'new' and 'old' varieties again. Equation (12.6) therefore describes an adjustment process which converges to the steady state given by (12.7).

Production and Trading Regimes

In Krugman's model, the North has two options. The first is to produce both 'old' and 'new' varieties to satisfy internal demand, while the second is to concentrate on the production of 'new' varieties only and trade part of total output for 'old' varieties produced by the South. The latter option would be chosen if the opportunity cost of obtaining 'old' varieties through trade would be lower than the opportunity cost of producing 'old' varieties by the North itself. This would be the case if the relative world price of 'new' varieties *vis-à-vis* 'old' varieties would be larger than one.[2]

Because of the linear and identical production technologies and the symmetry in the utility function, all goods produced by the South would have the same price. This also goes for all goods produced by the North. Hence, if the North ended up producing both 'old' varieties and 'new' varieties, both types of varieties would have the same price as well. However, if the North ended up producing 'new' varieties and the South produced 'old' varieties, then relative prices might change in favour of 'new' varieties, thus increasing the terms of trade in favour of the North.

Let L^N and L^S be the supply of labour in the North and in the South, respectively. Assume now that the North and the South have specialized completely in the production of 'new' varieties and 'old' varieties, respectively. Assume furthermore that North and South have identical consumer preferences. Then the average world supply of goods will be given by:

$$\frac{x_N}{x_S} = \frac{L^N/(n-n^i)}{L^S/n^i}, \tag{12.8}$$

where the subscripts and superscripts N and S refer to North and South, respectively. Relative world demand would still be given by equation (12.2). The steady-state long-run market-clearing relative price for 'new' varieties would therefore be given by equating (12.8) and (12.2) and substituting (12.7):

$$\frac{p_N}{p_S} = \left(\frac{L^N}{L^S}\right)^{-(1-\alpha)}\left(\frac{n}{n^i}-1\right)^{1-\alpha} = \left(\frac{L^N}{L^S}\right)^{-(1-\alpha)}\left(\frac{\mu_0}{\mu_1}\right)^{1-\alpha}. \tag{12.9}$$

If $p_N/p_S>1$, then North will gain from trade. Moreover, North will gain more from trade if its rate of innovation is relatively high, and if the propensity to imitate is low. Because of (12.4), equation (12.9) also describes the development of relative wages in the North. Hence a relatively large supply of labour in the North *vis-à-vis* the South increases the supply of 'new' varieties and therefore lowers the terms of trade for the North, but also relative wages.

3. LABOUR MARKET AND TECHNOLOGY EXTENSIONS

We now transform the Krugman model in three ways. First, we interpret North as the 'high-tech' production sectors in a country which produces 'new' varieties of goods. These new varieties of goods can be produced using high-skilled workers only. South represents the sectors that are able to produce 'low-tech' varieties of goods. In the production of those goods low-skilled workers and high-skilled workers can be used as perfect substitutes.[3] Secondly, we assume that a high-skilled worker can replace ε low-skilled workers in the production of 'low-tech' goods. We therefore assume that high-skilled workers are completely mobile within the economy, whereas low-skilled workers are not. This contrasts sharply with the Krugman model, where workers are immobile between countries, but completely mobile between sectors within a country. We thus introduce employment opportunity asymmetries between high-skilled and low-skilled workers. These asymmetries are associated with the relative importance of 'high-tech' and 'low-tech' production technologies within a country.[4] Third, we assume that the rate of imitation present in the Krugman model now reflects the behaviour of the marginal entrepreneur who switches from a 'high-tech' to a 'low-tech' production technology. We will actually link the speed of adoption of 'low-tech' production technologies, the equivalent of the imitation rate in the Krugman model, to the profitability of such technologies relative to the 'high-tech' production technologies. At this stage, we do not provide the micro foundations for this behaviour, but focus on the consequences for employment and price and wage divergence of this technology adoption behaviour instead.

Labour Market Extensions

We now assume that high-skilled labour can move between the 'high-tech' and the 'low-tech' sectors. In that case, one would observe the simultaneous employment of high-skilled workers in both sectors only if the wages earned by a high-skilled worker in both sectors are the same. Assuming that wages earned by skill in the 'low-tech' sector reflect intrinsic productivity differences between different skills, we can now readily distinguish between two specialization cases. The case of complete specialization with respect to employment is associated with the situation in which wages for high-skilled workers in the 'high-tech' sector exceed wages for high-skilled workers in the 'low-tech' sector. In the case of incomplete employment specialization, wages for high-skilled workers are the same in both the 'low-tech' and the 'high-tech' sectors. In this case, one may observe employment of high-skilled workers in both sectors.

The case of complete specialization generates wage and price results which are completely analogous to the results obtained in the basic Krugman model. We will find high-skilled workers employed solely in the 'high-tech' sectors, and only low-skilled workers employed in the 'low-tech' sectors. This implies that relative prices for high-tech varieties and low-tech varieties are given by:

$$\left(\frac{p_H}{p_L}\right)_{CS} = \left(\frac{w_H}{w_L}\right)_{CS} = \left(\frac{H^*}{L^*}\right)^{-(1-\alpha)}\left(\frac{n^H}{n^L}\right)^{1-\alpha}, \qquad (12.10)$$

where H^* reflects the supply of high-skilled workers in a country, L^* is the supply of low-skilled workers, and n^H and n^L are the number of 'high-tech' varieties and the number of 'low-tech' varieties, respectively. The subscript cs refers to a labour market regime of complete specialization. Because of (12.4), relative wages are equal to relative prices. Equation (12.10) implies, as in Krugman's model, that an increase in the relative number of 'high-tech' varieties would drive up the relative price of those varieties, and hence the relative wages earned by 'high-skilled' workers. In the complete specialization regime, relative profits for the marginal entrepreneur who considers a switch from a 'high-tech' to a 'low-tech' production technology for a certain variety are given by:

$$\left(\frac{\Pi_H}{\Pi_L}\right)_{CS} = \left(\frac{p_H(1-\alpha)(H^*/n^H)}{p_L(1-\alpha)(L^*/n^L)}\right) = \left(\frac{H^*}{L^*}\right)^{\alpha}\left(\frac{n^H}{n^L}\right)^{-\alpha}, \qquad (12.11)$$

where we have used (12.10) and (12.4). Equation (12.11) states that an increase in the number of high-tech varieties would decrease the relative profitability of continuing to produce relatively new varieties using high-tech production technology. The reason is that such an increase would drive up wage costs of high-skilled workers, thus providing an incentive to switch to low-tech technologies. In the case of complete specialization, the relative employment in the 'high-tech' sectors (H^e) and in the 'low-tech' sectors (L^e), both measured in efficiency units, is given by:

$$\left(\frac{H^e}{L^e}\right)_{CS} = \frac{H^* - HL}{L^* + \varepsilon HL} = \frac{H^*}{L^*}, \qquad (12.12)$$

where HL denotes the number of high-skilled workers employed in the 'low-tech' sectors. In the complete specialization case, HL is equal to zero.

The case of incomplete specialization can readily be derived from the condition that in such circumstances the wage rate earned by a high-skilled worker in the 'high-tech' sectors should be equal to the wage rate earned by a low-skilled worker in the 'low-tech' sector, times the relative productivity of a high-skilled worker in that sector:[5]

$$
\left(\frac{w_H}{w_L}\right)_{IS} = \varepsilon = \left(\frac{p_H}{p_L}\right)_{IS} = \left(\frac{H^e}{L^e}\right)^{-(1-\alpha)}\left(\frac{n^H}{n^L}\right)^{1-\alpha} \Rightarrow \left(\frac{H^e}{L^e}\right)_{IS}
$$

$$
= \left(\frac{n^H}{n^L}\right)\varepsilon^{-1/(1-\alpha)}
$$
(12.13)

Equation (12.13) shows that, in the incomplete specialization case, the employment in efficiency units of high-skilled labour in the high-tech sector will increase, with an increase in the number of high-tech varieties, as expected. An increase in the relative efficiency of high-skilled workers will increase the relative profitability of using high-skilled workers in the low-tech industry. This in turn will decrease the relative employment of high-skilled workers in the high-tech industry and increase low-tech employment in efficiency units. Using (12.4) and (12.13), relative profits in the case of incomplete specialization would be given by:

$$
\left(\frac{\Pi_H}{\Pi_L}\right)_{IS} = \left(\frac{p_H(1-\alpha)(H^e/n^H)}{p_L(1-\alpha)(L^e/n^L)}\right) = \varepsilon^{-\alpha/(1-\alpha)}.
$$
(12.14)

Technology Extensions

As stated above, we now assume that the rate of change in the number of low-tech production[6] technologies is driven by relative profits. The idea is that entrepreneurs have two options to produce a certain variety. A variety can be produced using a high-tech production technology which requires high-skilled workers only. After a while, the same variety could also be produced using a low-tech production technology.[7] Entrepreneurs now have to decide whether to stick to their high-tech production technology (that is, the technology they had to use at the moment the new variety arrived) or to switch to a low-tech production technology. We assume that their decisions depend in part on the relative profitability of those two options. We would expect the number of entrepreneurs engaging in such a switch to be positively influenced by the relative profitability of low-tech production technologies. We furthermore assume, as in Krugman, that the rate of change is proportional to the number of high-tech production technologies present. We therefore essentially restate μ_1 in equation (12.6) and obtain:

$$
\hat{n}^L = \xi\left(\frac{\Pi_H}{\Pi_L}\right)^{-\theta}\left(\frac{n^H}{n^L}\right),
$$
(12.15)

where ξ and θ are positive and constant parameters. Note that equation (12.15) is completely comparable to equation (12.7), when the profit elasticity θ would be equal to zero. In that case $\xi = \mu_1$. For ease of exposition, we define $z = n^H / n^L$. In order for z to be constant in the steady state, we should have $\mu_0 = \hat{n}^L$, as before (cf. equation (12.7)). This implies that the steady-state value of z, that is, $\bar{z}$, is given by:

$$\bar{z} = \left(\frac{n^H}{n^L} \right) = \left(\frac{\mu_0}{\xi} \right) \left(\frac{\Pi_H}{\Pi_L} \right)^{\theta}. \tag{12.16}$$

Note that $\bar{z}$ depends positively on relative profits. Because relative profits are different for the two specialization cases, we will also have different steady-state values for z:

$$\bar{z}_{CS} = \left(\left(\frac{\mu_0}{\xi} \right) \left(\frac{H^*}{L^*} \right)^{\alpha\theta} \right)^{1/(\alpha\theta+1)} \tag{12.17a}$$

$$\bar{z}_{IS} = \left(\frac{\mu_0}{\xi} \right) \varepsilon^{-\alpha\theta/(1-\alpha)}, \tag{12.17b}$$

where (12.17a) and (12.17b) are obtained from (12.16), by substituting (12.11) and (12.14), respectively.

Finally, equation (12.13) can be used to find the critical value of z, that is z^*, for which there is a transition between the cases of complete and incomplete specialization. Since at the moment of the transition we should have $HL = 0$ (cf. equation (12.12)), equation (12.13) implies:

$$z^* = \left(\frac{H^*}{L^*} \right) \varepsilon^{1/(1-\alpha)}. \tag{12.17c}$$

Any value of $z \geq z^*$ indicates that complete specialization will occur, while any value of $z < z^*$ indicates that incomplete specialization will prevail.

4.   LABOUR MARKET REGIMES AND WAGE DIVERGENCE: EUROPE AND THE USA COMPARED

Using equations (12.17), it is easy to distinguish a number of constellations of technology and labour market 'parameters' which influence the occurrence of wage divergence in their own way. Basically, there are two classes of parameter constellations: those where $z_{CS} > z_{IS}$ (constellation A) and those

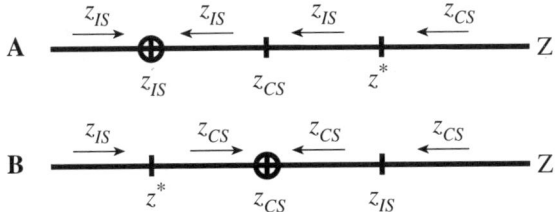

Figure 12.1   Labour market specialization regimes

where $z_{CS} < z_{IS}$ (constellation B). However, it can be shown that the structure of the model implies that, if $z^* \leq z_{CS}$, it must be the case that $z_{CS} \leq z_{IS}$. Likewise, if $z_{CS} \leq z^*$, it automatically follows that $z_{IS} \leq z_{CS}$. In constellation A we have a case of 'globally stable' incomplete specialization. In constellation B we have a case of 'globally stable' complete specialization. Both cases are depicted in Figure 12.1.

In this figure, the relevant attractors are marked with a small circle, while the various threshold values and attractors of z are marked by means of a small vertical line through the horizontal. The direction of an arrow indicates the movement of a point z on the line segment containing the arrow. The text above a line segment denotes which attractor is relevant for points on the line segment under consideration. In order to understand how the adjustment processes works, consider a point on the right-most line segment of case A. Since the point is to the right of z^*, we must be in the complete specialization regime with the z_{CS} attractor. This is indicated by z_{CS} directly above the line segment containing the point z. In case A, the relevant attractor for z lies to the left of z^*, and z will start to move in that direction in accordance with (12.15). But once z has arrived to the left of z^* we are in the incomplete specialization regime, and z_{IS} becomes the relevant attractor. Hence z continues to move until it arrives at z_{IS}. Actually, any point z in case A will move to z_{IS}. Likewise, in case B any point z will move towards the z_{CS} attractor.

Suppose now that a country can be categorized by means of its 'scores' on the labour market 'parameter' H^* / L^* and the technology parameters ξ and θ. A country with a relatively high participation of low-skilled workers on the labour maket will have a relatively low value of H^*/L^*, while a country with relatively high technological capabilities and/or a keen eye for profitable technological adoption opportunities will have relatively high values for ξ and θ. This has immediate consequences for the probability of a certain variant being relevant for a specific country.

For constellation A to be relevant for a country, we require $\bar{z}_{CS} < \bar{z}_{IS}$. Using equation (12.17b) to be able to drop the term μ_0 / ξ from equation (12.17a), we find:

$$\bar{z}_{CS} = \bar{z}_{IS}^{\,1/(\alpha\theta+1)} \varepsilon^{\theta\alpha/((1-\alpha)(\alpha\theta+1))} (H^* / L^*)^{1/(\alpha\theta+1)}. \qquad (12.18)$$

Equation (12.18) defines $\bar{z}_{CS}$ as a concave function of $\bar{z}_{IS}$. Moreover, $\bar{z}_{CS}$ increases in H^* / L^*. Hence, the probability that $\bar{z}_{CS} > \bar{z}_{IS}$ increases with H^* / L^*. Actually, it is fairly easy to show by using equations (12.17) that the 'locus' of labour market parameters and technology parameters which would result in $\bar{z}_{CS} = \bar{z}_{IS}$ is given by:

$$\frac{\mu_0}{\xi} = \lambda = \varepsilon^{(\alpha\theta+1)/(1-\alpha)} (H^* / L^*), \qquad (12.19)$$

where the ratio $\lambda = \mu_0 / \xi$ is actually equal to the steady-state ratio of 'new varieties' versus 'old varieties' in the Krugman model, that is, the ratio to which λ would converge when either profits would be the same in both cases, or $\theta = 0$. It can readily be verified that (12.19) also describes the 'locus' for which $\bar{z}_{IS} = \bar{z}^*$. Note that equation (12.19) describes a line through the origin in the λ, H^* / L^* plane, which divides that plane into two parts. Given the nature of (12.19), the part of the plane above this line is associated with the situation where $\bar{z}_{IS} > \bar{z}_{CS}$, $\bar{z}_{IS} > \bar{z}^*$ and $\bar{z}_{CS} = \bar{z}^*$, that is, case B from Figure 12.1. The part of the plane below the line is associated with the situation where $\bar{z}_{IS} < \bar{z}_{CS}$, $\bar{z}_{IS} < \bar{z}^*$ and $\bar{z}_{CS} < \bar{z}^*$, that is, case A. Case B has complete specialization as its steady state, case A has incomplete specialization as its attractor. Note that the slope of the 'locus' depends positively on the profit-elasticity of technology adoption θ. A relatively high value of this elasticity would decrease the probability of finding the country in question in a situation of complete specialization (that is, after all, what adoption is supposed to do), *ceteris paribus*.[8] Countries with a high value of λ will have a relatively high probability of being in or moving towards a situation of complete specialization. This causes high-skilled labour to become more scarce. Likewise, countries with relatively low values of the ratio H^* / L^*, which is an indicator of the participation/availability of high-skilled labour *vis-à-vis* low-skilled labour, will also have a relatively high probability of being in or moving towards a situation of complete specialization for any value of their λ. The above is summarized in Figure 12.2.

The problem now is to try to define the differences between Europe and the USA in terms of these parameters, and to see what a difference in characterization of Europe and the USA in terms of these parameters would imply for wage divergence. The idea is that the USA has a 'comparative advantage' in innovative activity, that is, a relatively high value of λ, while Europe has a relatively high value of H^* / L^*, for the same actual skill composition of the labour force. A relatively high value of λ would increase the probability of finding a country in or moving towards the complete specialization labour

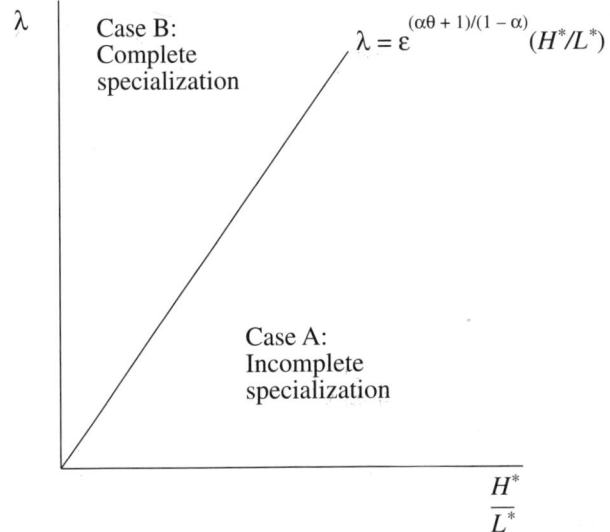

Figure 12.2   Parameter constellations and labour market regimes

market regime, for any value of H^* / L^*. Likewise, a relatively high value of H^* / L^* would increase the probability of finding a country in or moving towards the incomplete specialization regime. With the characterization of the USA and Europe suggested above, we would expect the USA to be more likely in a situation of complete specialization, and hence to experience relatively large movements in relative wages by skill, whereas the opposite would be more likely to hold for Europe. We see then that, even though production technologies are the same by assumption, differences in technological capabilities and/or labour market participation rates between countries may actually cause wage developments to be different. For countries with fairly low participation rates of the low-skilled, one would not expect to find wage divergence as often as in countries with high 'forced' participation of low-skilled workers. In relatively innovative countries, one would expect to find wage divergence more often than in less innovative countries.

An interesting possibility occurs for constellations of parameters close to the dividing line given by (12.19). Small random fluctuations in the profit-elasticity of the rate of adoption, theta, the value of epsilon, the relative efficiency of high-skilled workers on low-tech jobs, the elasticity of demand, alpha, or the relative labour participation (H^*/L^*) can cause regime shifts. However, employment by skill would not be affected by construction. Thus we could observe fluctuations in relative wages by skill, without any noticeable fluctuations in relative employment by skill.

5.   SUMMARY AND CONCLUSION

In this chapter we have addressed the issue of divergence in wages and employment opportunities by skill between Europe and the USA. We have shown that a simple stylized fact regarding asymmetries in employment opportunities between high-skilled and low-skilled workers, in combination with a high-skill bias in innovative activity, may be used to construct a model in which the US case of wage divergence and the European case of stable wages with crowding-out are two 'natural' states of that model. Which state will actually materialize depends on both technological and labour market characteristics of a particular country.

The model is a simple 'love of variety' product innovation model, based on Krugman's North–South model of international trade. In the model we have two types of varieties: 'new, high-tech' varieties which can be produced using high-skilled labour only, and 'old, low-tech' varieties for which the production process has been standardized. This enables producers to use both low-skilled workers and high-skilled workers as perfect substitutes in the production of low-tech varieties. We assume that high-skilled workers enjoy the same wage rate in both sectors, whenever they are employed in both sectors. However, a relatively high rate of innovation biases the demand for labour in favour of high-skilled workers, who then leave the 'low-tech' sector, until employment of the high-skilled is fully concentrated in the high-tech sector, and employment of the low-skilled is concentrated in the low-tech sector. In that case, the labour market is completely specialized, and wages by skill will diverge as a consequence of innovative activity. We have shown that the relative profitability of using high-skilled workers in a high-tech production process, and using low-skilled workers after switching to low-tech production techniques, may be used to define two possible steady states in the model: one in which the labour market will be specialized completely and where relative wages will continue to diverge, and one in which relative wages are stable and where low-skilled workers and high-skilled workers compete for the same jobs. We show that these possible steady states depend on certain technology and labour market parameter constellations, such that the USA is more likely than Europe to experience wage divergence.

The results which the model generates depend on asymmetries in employment opportunities by skill, which become more pronounced in situations where innovative activity is high. The latter makes high-skilled workers more scarce than low-skilled ones, almost as a matter of principle. Relative income prospects as well as employment prospects for the low-skilled may be bleak in these circumstances, since they would have to face a permanent fall in their relative wages for each increase in innovative activity. Nonetheless, the latter would also raise total utility and total real income by increasing the number of varieties.

NOTES

1. This section closely follows Krugman (1979).
2. By assumption, the South has the same production technology as the North.
3. This interpretation resembles Vernon's life-cycle hypothesis (Vernon, 1966). According to the life-cycle hypothesis, new goods tend to be produced in developed countries first, and then, after the production technology has been standardized, by less developed countries making use of their comparative advantage with regard to low-skilled workers.
4. For a similar approach in a somewhat different setting, see van Zon *et al.* (1997).
5. The latter reflects the assumption that wage differentials by skill reflect intrinsic productivity differences.
6. The term 'low-tech' may be somewhat misleading, in that it refers to the possibility of using low-skilled workers instead of high-skilled workers to produce a certain variety. This might well entail sophisticated standardization activities, which renders a production process only seemingly low-tech.
7. We do not specify how such a low-tech production technology actually comes about. This is left for future extensions of the model.
8. This is because this increase in the profit-elasticity would rotate the dividing line in a counter-clockwise direction, thus increasing the area of the plane associated with incomplete specialization at the expense of the area associated with complete specialization.

REFERENCES

Agenor, P.R. and J. Aizenman (1996), 'Wage Dispersion and Technical Progress', NBER Working Papers Series, no. 5417.

Beer, P. de (1996), 'Laag opgeleiden: minder kans op een baan, meer kans op ontslag', *Economisch Statistische Berichten*, 908–12.

Brauer, D. and S. Hickok (1995), 'Explaining the Growing Inequality in Wages across Skill Levels', *Economic Policy Review*, Federal Reserve Bank of New York, 1 (January), 61–72.

Burtless, G. (1994), 'International Trade and the Rise in Earnings Inequality', *Journal of Economic Literature*, 33 (June), 800–816.

Drèze, J.H. and E. Malinvaud (1993), 'Growth and Employment, the Scope of European Initiative', mimeo, CORE/Collège de France, Louvain-la-Neuve/Paris.

Groot, W. and H. Maassen van den Brink (1996), 'Overscholing en verdringing op de arbeidsmarkt', *Economisch Statistische Berichten*, 74–7.

Howell, D. (1995), 'Collapsing Wages and Rising Inequality: Has Computerisation Shifted the Demand for Skills?', *Challenge*, 38 (1), 27–35.

Jackman, R. (1995), *Unemployment and Wage Inequality in OECD Countries*, London: London School of Economics/Oxford: Oxford University Press.

Krugman, P. (1979), 'A Model of Innovation, Technology Transfer and the World Distribution of Income', *Journal of Political Economy*, 87 (2), 253–65.

Krugman, P. (1995), 'Technology, Trade and Factor Prices', NBER Working Paper Series, no. 5355.

Laan, L. van der (1996), 'Substitution in the regional labour markets in the Nether-lands', KNAG/Faculty of Geographical Sciences Utrecht/Economic Geographical Institute Rotterdam.

Lawrence, R.Z. and M.J. Slaughter (1993), 'Trade and Wages: Giant Sucking Sound or Small Hiccup?', *Brookings Papers on Economic Activity*; Microeconomics, 2, 161–210.

Leamer, E.E. (1994), 'Trade, Wages and Revolving Doors Ideas', NBER Working Papers Series, no. 4716.

Leamer, E.E. (1995), 'A Trade Economist's View of US Wages and Globalization', in S. Collins (ed.), *Imports, Exports and the American Worker*, Washington: Brookings.

OECD (1994), *The OECD Jobs Study; Evidence and Explanations. Part I: Labour Market Trends and Underlying Forces of Change*, Paris: OECD.

Vernon, R.G. (1966), 'International Investment and International Trade in the Product Cycle', *Quarterly Journal of Economics*, 80 (2), May, 190–207.

Zon, A.H. van, H. Meijers and J. Muysken (1997), 'On Trickling Chimneys and Other Unemployment Misery', MERIT Research Memorandum, no. 97-006.

PART IV

Institutional Change

13. Changing working time patterns

Vincenzo Spiezia and Marco Vivarelli

1. INTRODUCTION

The reorganization and reduction of working time is currently on the policy agenda of most European countries and of the European Union as a whole. Working time, indeed, is an essential indicator of welfare. Not only is the total amount of working time the appropriate measure of the employment possibilities existing in a country, but per capita working time is a measure of 'quality of life' of its workers.

The aim of this chapter is to analyse the main changes in the patterns of working time over the last 35 years with respect to three important aspects. The first one is the interrelation between the dynamics of total working time and per capita working time in determining the dynamics of employment. On the one hand, for a given growth rate in total working time, shorter working hours are a necessary condition for an increase in employment. On the other, the reduction in the average working time may exert an upward pressure on labour costs and may result in a decrease in labour demand.

The second issue is the analysis of the factors that affect the evolution of per capita working time in each country and that may explain the observed cross-country differences in working time patterns. Country-specific factors are likely to play a major role in shaping working time patterns and the identification of these factors represents an important benchmark for policy makers.

Finally, we will examine how the changes in average working time have been translated into changes in the distribution of working time across individuals. Indeed, the economic and social effects of a given reduction in average working time are significantly different according to the way such reduction is redistributed across different individuals in the labour market.

The chapter is organized as follows. Section 2 shows the major trends in working time and employment in some industrialized countries and discusses the main predictions of the economic theory about the effects of a reduction in average working time. Section 3 attempts to provide an explanation of the observed differences in working time patterns across countries and through

time. In section 4 we analyse the recent shift which occurred in the distribution of usual working hours across individuals within countries. Finally,
section 5 summarizes the main findings of the chapter and discusses their
implications for working time policy in the European Union.

2.   EMPLOYMENT AND WORKING TIME: THE EMPIRICAL EVIDENCE

Although most economic studies are mainly focused on employment and
unemployment trends, working time has always represented an essential
dimension of economic development. Total working time, in fact, represents
the only correct measure of the demand for labour, while average working
time is the means through which a given level of labour demand is shared
across individuals in the labour market. The dynamics of employment is then
the result of the interplay between the dynamics of total and average working
time: for a given level of labour demand, the level of employment is inversely
related to the length of working time per employed person.

Figure 13.1 presents the annual changes in total working time and in
average working time in ten industrialized countries (six European plus Australia, Canada, Japan and the United States) over the period 1970–95 (actual
figures are reported in Table 13.1). As is apparent, there has been a significant
variation in the combinations of the two variables: in some countries (France,
West Germany and the UK) both total and average working time have decreased; in others (Australia, Canada, Japan, Norway and Sweden) total
working time has increased but working time per employed person has decreased; only in the USA have total and average working time increased at
the same time.

To evaluate the impact of these different working time patterns on the
dynamics of employment, we have drawn a straight line through the origin
that indicates these combinations of total and average working time that
would be consistent with a constant level of employment. The points above
the line correspond to an increase in employment, while the opposite occurs
for the points below the line.

Over the period considered, employment has increased in all countries,
but employment growth in each of them has been due to a different combination of changes in total and average working time. In a first group of
countries, total working time has decreased (France and the UK) or has
remained almost constant (West Germany, Italy and Norway), so that the
reduction in average working time has been the crucial factor in determining the growth in employment. In the remaining countries, the demand for
labour (measured in total working time) has increased, but the way this

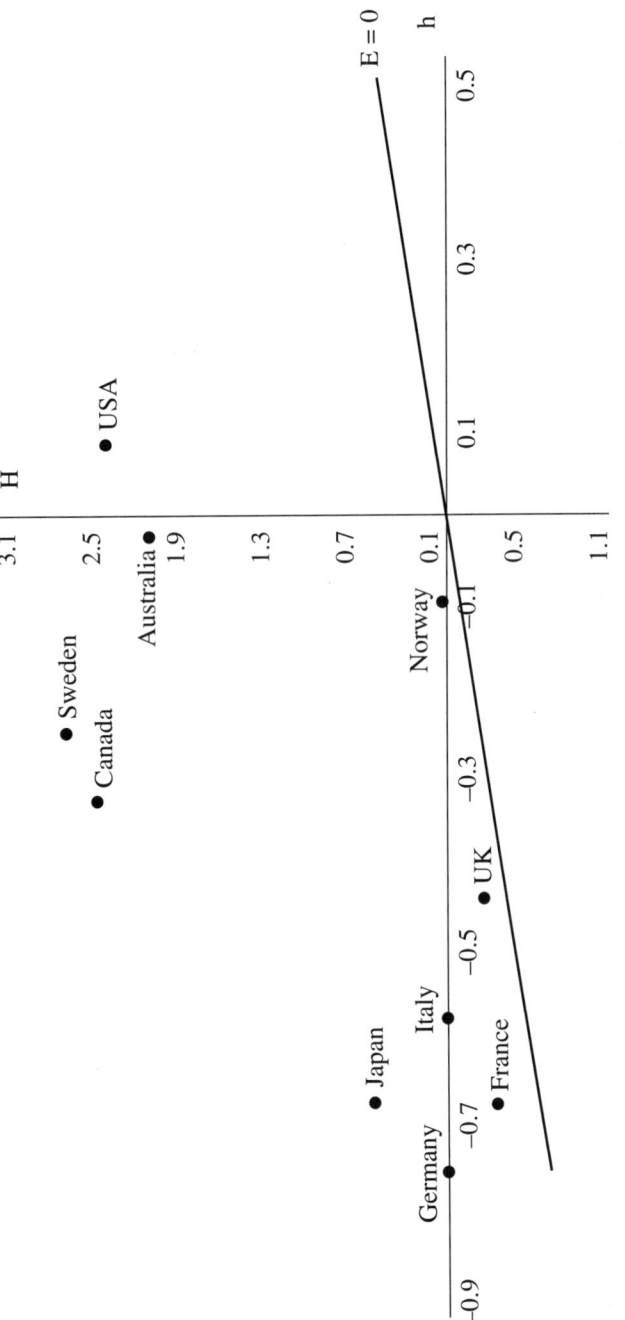

Source: Secretariat database on annual working hours of work; OECD Analytical Database.

Figure 13.1 Average annual changes in total working time (H), average working time (h) and employment (E)

Table 13.1 Long-run changes in total working time, average working time and employment (average annual percentage rate)

	Japan	Australia	USA	Canada	UK	Western Germany	France	Italy	Sweden	Norway
Average working time*	-0.67	-0.02	0.08	-0.33	-0.44	-0.75	-0.68	-0.58	-0.25	-0.10
Employment	1.18	2.14	2.35	2.82	0.16	0.74	0.30	0.58	2.96	0.12
Total working time	0.50	2.12	2.43	2.50	-0.28	-0.01	-0.37	0.00	2.71	0.02
Period	1972–94	1978–95	1970–95	1970–95	1973–95	1973–95	1970–95	1970–94	1970–95	1972–95

Note: * Total hours worked per year/total employment.

Source: Secretariat database on annual hours of work; OECD Analytical Database.

increase has been shared among the participants in the labour force has been quite different. In Japan, the reduction in working time per employed person has been substantial and has significantly contributed to the growth in employment. In Canada, Sweden and Australia, the reduction in average working time has been more limited and employment patterns have been mainly determined by the trend of total working time. The case of the USA is somewhat peculiar: labour demand has significantly increased, but this increase has been partially translated into a longer working time per employed person.

Although this exercise does not control for the direction of causality between variables, average working time does not appear to be the main determinant of the observed cross-country differences in the rate of employment growth. For instance, the increase in employment in Sweden and Canada (measured by the distance from the straight line) has been higher than in the USA, despite the fact that the former countries have reduced average working time. Conversely, the reduction in average working time has been of the same magnitude in France and Japan, whereas employment growth has been much higher in the latter country.

The lack of a clear-cut relationship between employment and average working time is somewhat at odds with the predictions of the economic theory. Although shorter working time should redistribute a given level of labour requirement over an increased number of workers, it is commonly argued that such a measure would tend to generate several counter-effects that may lead to an overall reduction in employment. The main counter-effects identified in the economic literature are as follows.

1. A reduction in working time would induce workers and trade unions to demand a higher hourly wage in order to preserve the total income of workers (Booth and Schiantarelli, 1987). For a given level of labour productivity, an increase in the hourly wage would determine a reduction in employment.

2. Wages are not the only component of labour costs, which also include fixed costs. Therefore, even at a constant hourly wage, a reduction in working time and a corresponding increase in the number of employed would imply an increase in fixed labour costs (Hart, 1987).

3. Firms can respond to the rise in labour costs induced by shorter per capita working time by increasing overtime, without hiring additional workers (Calmfors and Hoel, 1988).

4. Firms can organize production on a higher number of shifts in order to maintain an efficient use of capital. More shifts imply more efforts by workers, who will demand a higher wage which, in turn, would reduce the demand for labour (Calmfors and Hoel, 1989).

5. The skill profile of the new workers hired as a result of shorter working time may be inadequate for the firms' requirements and this would involve higher training cost (Hoel and Vale, 1986).
6. These negative effects are amplified by the loss of international competitiveness determined by the increase in costs and/or reduction in productivity.
7. A shorter working time may cause a fall in productivity either because the incidence of non-productive times (such as breaks) increases or because of the costs involved in reorganizing production with a larger number of workers (Andersen, 1987).

In short, the economic theory points out that a reduction in average working time does not ensure an increase in employment but, on the contrary, is likely to cause a fall in it. Yet this prediction appears far from general (see, for instance, Withley and Wilson, 1988; Corneo, 1994; Malinvaud, 1995). The technological and institutional framework specific to each country is likely to be a major factor in determining the impact of a change in average working time on employment (Bosch, 1986, and Seifert, 1991, report on the interesting policy debate in the former Federal Republic of Germany). Furthermore, any change in working time patterns does not occur in a vacuum but is normally supported by a set of complementary policy measures designed to minimize the counter-effects listed above (OECD, 1995; ETUI, 1996).

To derive a more comprehensive assessment, one should therefore consider a wider set of country-specific conditions, including the institutional settings, the state of industrial relationships, the preferences of workers and the characteristics of technological progress. In brief, one can list the following main points.

1. Employment decreases to the extent that the increase in wage necessary to compensate the reduction in the income of workers due to shorter working time exceeds the growth in productivity. The support of trade unions and the agreement of workers is therefore important in order to ensure that the increase in wages is consistent with the dynamics of productivity.
2. The increase in labour costs due to a higher incidence of fixed costs is small in those countries where the fixed costs arising from fiscal and social insurance obligations are low.
3. Overtime is discouraged in those countries where the overtime premium increases progressively with the length of overtime worked by each employee (Santamäki-Vouri, 1984; Toedter, 1988).
4. An increase in the number of shifts does not necessarily imply an increase in wages if workers are willing to accept a more flexible work

schedule in return for shorter working time (Calmfors and Hoel, 1989). Also in this case, the negotiation of these policies between firms and trade unions represents a central issue.

5. In those countries with an effective public training system, the skills of the unemployed can be less costly to update to match the requirements for the vacancies created by a reduction in working time.

6. The reduction of working time in a given country is unlikely to cause a loss of its international competitiveness if it is agreed with or if it is made at the same time as its trading partners'. In this respect, the European Union may represent an institutional and economic framework favourable to the negotiation of a reduction in average working time.

3. THE DRIVING FORCES OF CHANGING WORKING TIME PATTERNS

Historically, the reduction in average working time has been the major channel through which industrialized economies have redistributed the gains from technical progress between leisure and income. In fact, as productivity increases, each individual and the society as a whole are called to choose whether to convert this increase into a higher level of income or into more leisure time. Individual and social preferences are then a crucial factor in explaining the dynamics of per capita working time and income.

Since the early stages of industrialization, the growth in per capita income has gone along with the reduction in average working time. Such reduction has taken the form of shorter working days, shorter working weeks, shorter working lives, longer vacations, paid leave and so on. According to some recent estimates for the OECD countries (Maddison, 1979, 1991), over the period 1870–1970 the number of annual hours worked per person in employment has fallen by over 37 per cent. In recent years, however, this long-run tendency to the reduction in per capita working time appears to have slowed down significantly. Table 13.2 reports the growth rates of hourly productivity and average working time per person in employment in 11 industrialized countries (eight European plus Canada, Japan and the USA) from 1960 to 1995. Looking at figures in the table, there seems to be an inverse relationship between hourly productivity and average working time either across countries or through time. On average, those countries where hourly productivity has grown the most tend also to have the highest rate of reduction in working time per employed person. Correspondingly, the decrease in the average working time has been faster during a period (1960–79) of high productivity growth than in the following phase (1980–95) of productivity slowdown.

Table 13.2   Average annual changes in hourly productivity and in average working time, 1960–95

Country	1960–79		1980–95	
	Hourly productivity[1]	Average working time[2]	Hourly productivity[1]	Average working time[2]
Canada	3.4	–0.7	2.3	–0.2
Finland	5.2	–0.6	2.7	–0.3
France	5.8	–1.0	2.0	–0.7
West Germany	5.7	–1.0	2.2	–0.4
Italy	6.5	–0.6	2.7	–0.9
Japan[3]	3.7	–0.7	2.4	–0.4
Netherlands[3]	4.4	–1.6	2.9	–0.6
Norway	4.2	–1.2	2.3	–0.1
Sweden	4.5	–0.8	1.8	0.0
United Kingdom[3]	2.8	–0.7	1.7	0.1
United States	1.5	–0.2	0.5	0.3

Notes:
1.   Real gross domestic product/total hours worked per year.
2.   Total hours worked per year/total employment.
3.   1970–79.

Source: Secretariat database on annual hours of work; OECD Analytical Database.

Notwithstanding these common trends, the existence of country-specific patterns observed in section 2 appears to be confirmed. For instance, in the sub-period 1960–79, hourly productivity in the Netherlands and Sweden grew at about the same rate, but the reduction in average working time in the latter country has been only half that in the former one. Conversely, the decrease in the working time per employed person has been larger in Canada than in Italy, despite the evidence that productivity growth in the latter country has been about double that in Canada. In addition, productivity slowdown in Italy has been associated with an increase, not a decrease, in the rate of reduction of average working time.

In attempting to explain the observed cross-country differences in average working time, it is important to consider the alternative ways in which the growth in hourly productivity can be redistributed within each country. On the one side, productivity increase can result in an increase in hourly wage or in a decrease in the labour share of income; on the other side, for a given increase in the hourly wage, workers can choose to have a higher individual wage or a shorter working time.

The first change depends on the national determinants of income distribution between productive factors; the second choice involves the individual and social preferences regarding work and leisure. To the extent that such preferences differ across countries, the dynamics of average working time induced by a common increase in hourly wage will also tend to follow a country-specific pattern.

To control for the role of cross-country differences in preferences, Table 13.3 reports the growth rates of the hourly wage in the same sample of countries. The increase in hourly remuneration has been split between the increase in the wage rate and the reduction in the number of hours worked per person in employment.

Comparing the two sub-periods 1960–79 and 1980–95, there appears to have been a significant widening of the cross-country differences in their preferences about work and leisure time. At the one extreme, there are some countries (France, West Germany, Italy and the Netherlands) that have chosen to transfer

Table 13.3   *Distribution of the average annual growth rate in hourly wage between wage per employed person and average working time, 1960–95*

	1960–79			1980–95		
Country	Hourly wage*	Wage per employed person (% share)	Working time per employed person (% share)	Hourly wage	Wage per employed person (% share)	Working time per employed person (% share)
Canada	4.5	85	15	1.9	92	8
Finland	6.7	91	9	2.6	90	10
France	6.8	86	14	1.3	44	56
Western Germany	6.7	85	15	1.1	59	41
Italy	9.0	93	7	2.6	67	33
Japan	5.6	88	13	2.2	83	17
Netherlands	5.2	69	31	1.3	54	46
Norway	5.5	78	22	1.9	97	3
Sweden	5.1	84	16	0.9	104	–4
United Kingdom	2.6	73	27	1.0	110	–10
United States	1.7	91	9	0.3	180	–80

Note: * Total compensation of employees/total hours worked per year.

Source: Secretariat database on annual hours of work; OECD Analytical Database.

a larger percentage of the increase in the hourly wage into a reduction in average working time. This percentage ranges between the 56 per cent of France and the 33 per cent of Italy. At the other extreme, we find the UK, the USA and Sweden, where the growth in the hourly wage has resulted in a more than proportional increase in wages so that working time per employed person has also increased. In between these two groups, there are Finland and Japan (where the working time share of the hourly wage growth has moderately increased), Canada and Norway (where, instead, it has decreased).

To sum up, the historical decline in average working time has slowed significantly in the last 15 years and this seems to be associated with a generalized slowdown in the hourly productivity growth. Smaller increases in productivity have reduced the room for increasing wages per employed person and reducing the length of working time, so that these two objectives have come increasingly into conflict. However, within the constraints imposed by a lower growth in productivity, in some countries the increased preference for shorter working hours has partially counter-balanced this tendency. This seems to have been the case of most European countries (France, West Germany, Italy and the Netherlands) that have decidedly taken the road of translating higher hourly wages into a lower working time per employed person. At the other extreme, in Sweden, the UK and the USA the effects of the productivity slowdown have been reinforced by a shift in preferences towards higher wages. As a consequence of this change in preferences, in these three countries the number of hours worked per employed person have significantly increased in recent years.

4.   FLEXIBLE WORKING TIME

Although the dynamics of total and average working time represent two fundamental aspects of economic growth, there is a third important dimension, namely the distribution of working hours across employed persons. A reduction in average working time, in fact, may be the result either of a uniform decrease in average working time or of a polarization in the number of hours worked by different groups of the labour force. Clearly, the economic and the welfare implications of a given reduction in working time per employed person are remarkably different, depending on which of these two causes applies. In the case of a uniform shift in the distribution of working hours, all employed persons benefit equally from the reduction in average working time. In contrast, a reduction in average working time due to a widening of the working hours distribution may cause a decrease in welfare if some employed persons are rationed in their working time while some others are overworked (Schor, 1991).

In recent times, there has been an increasing concern that the radical transformations that are taking place in the economic structure of all industrialized countries may push them in the direction of a higher dispersion in working time. The so-called 'Fordist paradigm', which dominated the development of Western countries from the end of World War II to the early 1970s, has progressively given way to a new technoeconomic paradigm centred on information and communication technologies (Freeman and Soete, 1994; Boyer, 1988a).

The main feature of ICTs is their capacity to store, process and disseminate information in a short time and at a minimum cost (Soete, 1987). The information-intensive nature of the new paradigm has generated at least three major changes in the economic systems: it has increased flexibility in manufacturing production; it has widened the supply of non-material commodities, such as services; and it has raised the importance of knowledge as a productive factor (Boyer, 1989). These changes seem to have significantly increased the demand for more flexible working time arrangements, both from firms and from workers.

First, production has become more closely related to changes in demand, and this has expanded the use of overtime and part-time work to cope with fluctuations in orders (Bosch, 1986). This effect has been reinforced by the increased weight of the service sector, where working time is typically more flexible than in manufacturing (European Commission, 1997). Second, the spread of new services has raised the time intensity of consumption and this is likely to have modified individuals' preferences in favour of leisure time (Becker, 1965). Third, the centrality of knowledge as a productive factor has increased the importance of lifelong learning and it has modified the efficient mix between working time and learning time in favour of the latter.

Among the different types of flexible working arrangements, part-time seems to have been by far the most important. Figure 13.2 shows the estimated contributions of part-time working (that is, the number of hours worked as part-time) to the changes in average annual hours of employees in 15 industrialized countries over the decade 1983–93 (estimates are drawn from OECD, 1997).

In the large majority of countries, part-time working has given a determinant contribution to the reduction in working time per employee. In Ireland and the Netherlands, about 80 per cent of the reduction in average working time is explained by the increase in part-time. In Canada, France and the UK the spread of part-time work has been so large as to offset the increase in average working time of full-time employees.

Only in a few cases have the observed changes in average working time been led by a decrease in the number of hours worked by full-time employees. In Italy, the increase in part-time work accounts for only 13 per cent of

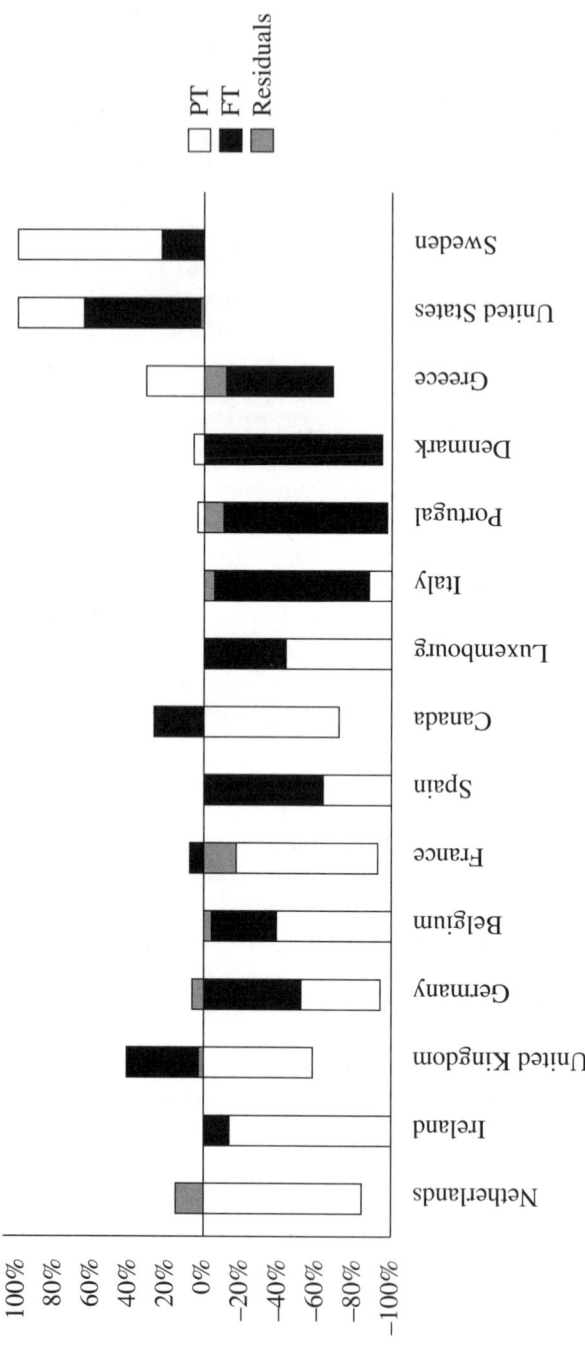

Notes: Initial year is 1985 for Denmark, 1986 for Portugal, 1987 for the Netherlands, Spain and Sweden. Estimates are based on a shift-share analysis, and residuals are due to the interaction term between FT and PT.

Source: OECD (1997).

Figure 13.2 Estimated contribution of part-time (PT) and full-time(FT) working to changes in average annual hours of employees, 1983–93

the total decrease in average working time; in Denmark, Portugal and Greece, part-time work has decreased but such variation has been more than compensated by a reduction in the average working time of full-time workers.

Demand for higher flexibility in working time seems to have resulted not only in the diffusion of part-time working but also in an increase of overtime. Table 13.4 reports the number of hours worked per employee as overtime and the number of part-time workers as a percentage of total employees in five countries over the period 1983–94.

Table 13.4   Overtime hours and part-time working, 1983–94

	Weekly overtime hours per employee		Part-time workers	
Country	Ratio (%) 1994	Average yearly change (%) 1983–94	Share (%) 1994	Average yearly change (%) 1983–94
Canada	0.5	2.3	18.8	1.08
West Germany	1.3	0.2	15.8	2.31
Italy	1.7	5.4	6.2	3.16
UK	3.1	0.3	23.8	2.36
USA	4.7	5.2	18.9	0.25

Source: OECD (1997), European Commission (1997).

The UK and the USA are the countries where both the percentage of overtime hours and of part-time workers is the highest. In Canada and Germany also part-time workers represent a significant share of total employees, but the use of overtime is more limited. In comparison with these two countries, Italy has a higher average number of hours worked as overtime but its share of part-time workers is definitely lower. Quite interestingly, in all five countries there has been an increase in both overtime and part-time and this suggests that the dispersion in usual working hours per employee may have risen in recent years.

For the European countries, the Eurostat Labour Force Survey[1] provides figures on the distribution of weekly working hours (including overtime) for full-time dependent workers over five time intervals: 1–35 hours, 36–9 hours, 40 hours, 41–5 hours and 46 hours and more. Table 13.5 shows the ranking of 10 European countries based on the mode of the working time distribution in 1995, that is, on the working time interval with the highest frequency. The frequency associated with the mode represents an inverse measure of the

Table 13.5   *Percentage of full-time dependent employees working most frequent usual weekly hours (mode), 1985 and 1995*

Country	Mode (weekly hours) 1985	Mode (weekly hours) 1995	Percentage 1985	Percentage 1995	Variation 1985–95
UK	36–9	46+	29.3	30.9	1.6
Ireland	40	36–9	62.9	34.1	−28.8
Italy	40	40	56.8	47.6	−9.2
Greece	40	40	49.3	51.1	1.8
West Germany	40	36–9	74.9	54.8	−20.1
Netherlands	40	40	65.0	62.1	−2.9
Belgium	36–9	36–9	61.9	65.3	3.4
France	36–9	36–9	63.6	65.9	2.3
Denmark	40	36–9	83.1	75.6	−7.5
Luxembourg	40	40	90.0	82.3	−7.7

Source: Eurostat (1997).

dispersion of the working time distribution: in fact, the lower its value, the higher the proportion of full-time employees who work an 'atypical' number of hours.

The UK and Ireland are by far the countries with the highest degree of dispersion in working time, with only about one-third of their employees working the typical number of hours. Dispersion appears significantly high also in Italy, Greece and Germany, where the percentage of employees with an atypical number of working hours represent about half of total dependent employment. In the remaining countries (Netherlands, Belgium, France, Denmark and Luxembourg), instead, workers with atypical working time tend to be a minority.

Over the decade 1985–95, the mode of the working time distribution has remained constant or it has decreased (from 40 weekly hours to 36–9 hours), with the only exception being the UK, where it has increased to more than 46 hours a week. In the same period, dispersion has significantly risen in many countries, and particularly in Ireland, Germany and Italy. Whether this increase reflects a shift in workers' preferences supported by a more favourable legislation or is instead driven by a change in the characteristics of labour demand it is not easy to assess. However, it is worthwhile noticing that the rise in the dispersion in working time per employed person has been associated with a remarkable increase in income inequality between workers and that such increase has been due not only to a widening of hourly wage differentials but also to a decrease in the number of hours worked by certain categories of employees (ILO, 1997). Although this tendency has been particularly intense in the Anglo-Saxon countries (the USA and the UK), in more recent times it has emerged significantly also in several European economies, including Germany and Italy. To a certain extent, the widening in working time distribution seems therefore to reflect an increasing inequality in work opportunities between different groups of the labour force and this tendency is likely to represent a major challenge to social cohesion within the European Union.

5. POLICY ISSUES AND CONCLUSIONS

This chapter has reviewed the major changes in working time patterns in several industrialized countries over the last 35 years. Attention has been focused on three important dimensions of working time: the interplay between total and average working time in determining the dynamics of total employment; the central role of individual and social preferences in the division of productivity gains between higher wages and shorter working hours; and the importance of working time distribution in assuring a satisfactory degree of equality of work opportunities and income for all individuals in the labour market.

Although the evolution of working time is greatly affected by country-specific factors, it seems possible to identify two major patterns, as summarized in Table 13.6. The Anglo-Saxon pattern (Canada, the UK and the USA) is characterized by a modest decrease in average working time (or even an increase in the case of the USA), by the tendency to translate productivity growth into higher wages and by higher inequality in the distribution of working time across workers. In the European pattern (particularly in France, Western Germany, the Netherlands and Italy), the reduction in average working time has been more substantial, shorter working hours have been the most frequent result of productivity growth and working time distribution has remained less dispersed.

Which of these two patterns should be preferred from a policy perspective is unclear. Labour demand (measured as either employment or total working hours) has sharply increased in Canada and the USA, but this has been associated with a higher dualism in labour income and working time. In the European Union, in contrast, total working time has decreased but the reduction in average working time has represented the main instrument for ensuring employment growth. Furthermore, a more concentrated distribution of working time has implied that the cost of the decline in labour demand has been

Table 13.6   Patterns of changes in working time

Pattern	Anglo-Saxon	European
Typical countries	Canada, USA, UK	France, Italy, Netherlands, West Germany
Evolution of average working time (AWT)	Modest decrease (increase in the USA)	Substantial decrease
Productivity growth is translated mainly into	Higher wages	Shorter AWT
Distribution of working time	Uneven	Even

shared more equally between workers. This is undoubtedly a positive outcome.

In recent years, however, a tendency to increasing dispersion in working time per employed person has emerged also in some European countries. To the extent that this tendency is translated into a higher inequality in work opportunities between different groups of the labour force, it may undermine social cohesion within the European Union. In addition, it has been stressed (see section 1) that the effects of a reduction of average working time on employment depend on the particular economic and institutional framework in which it is carried out.

These considerations seem to call for a policy on working time in the European Union and within its member states. A reduction and reorganization of working time would, in fact, require a wide range of policies aimed at different levels (Cette and Taddei, 1997). At the firm level, production activity and the system of shifts would be restructured in order to ensure consistency between firms' profitability and reduced working time (Taddei, 1995). In addition, the reorganization of working time would require a change in the modality of bargaining between trade unions and management, in order to include working time more explicitly as a factor of negotiation. Correspondingly, length of working time may not be defined on a weekly basis but over a longer period, such as a year: within a certain number of hours to be worked in a year, workers and firms might negotiate a flexible working schedule which is consistent with both firms' economic imperatives and workers' needs (Corneo, 1994; Gorz, 1989, 1993).

The reorganization of working time also requires appropriate policy measures aimed at the local socioeconomic system. On the one hand, a different time structure of all social, institutional and economic activities is required in order to support a different organization of working time. This includes the opening times of shops, banks and public offices, as well as schooldays, social and personal services and so on. On the other hand, local communities must be able to exploit the opportunities presented by a shorter working time: these include a new demand for cultural, tourist and leisure activities.

At the national level, the reform of labour legislation is clearly a crucial factor in the reorganization of working time. Governments and policy makers should define a new legal framework within which social and economic agents may negotiate flexibility and reduction in working time at different levels. At the European level, the coordination of national policies is essential to ensure consistency between the reorganization of working time and the competitiveness of the member states.

Clearly, the feasibility of such policies depends on the existence of a sound political consensus in that direction. In particular, the role of unions and employers' associations appears to be crucial for the practical implementation of a reduction and reorganization of working time. Given a national generally oriented law, working time schedules have to be defined at the micro level, taking into account the local communities' standards and specific needs, the local, national and international competitive arenas, the specific needs expressed by the firms and the corresponding workforces.

NOTE

1. The Labour Force Survey (LFS) database contains information relating to about 700 000 households annually (617 809 in 1993). The labour force characteristics of each person interviewed refer to their situation in a particular reference week. As a general rule, the reference week is a normal week in the spring excluding bank holidays. Some countries, such as Italy, use a fixed week, but others use the week preceding the interview, so it cannot be guaranteed that the reference week represents a 'normal' week. The survey is intended to cover the whole of the resident population, that is, all persons aged 15 years and above whose usual place of residence is in the territory of the member states of the European Union (EU). For technical and methodological reasons, however, it is not possible for every country to include the population living in collective households. Consequently, to harmonize the results of the study, results are compiled for the population of private households only. This comprises all persons living in the households surveyed during the reference week. This definition also includes persons absent from the household for short periods owing to studies, holidays, illness, business trips, and so on.

    As with any sample survey, the results of the European LFS are subject to sampling error. A significant source of sampling error in the LFS is the effect of non-response on the results. Non-response rate is normally less than 10 per cent in the countries where the survey is compulsory, but can reach 40 per cent in countries where it is voluntary. Adjustment for non-response in the majority of countries is made by reweighting the results, but in some countries the method used is 'duplication' or 'substitution' of units.

    Perfect comparability among 12 (or 15) countries is difficult to achieve, even were it to be by means of a single direct survey, that is, a survey carried out at the same time, using the same questionnaire and a single method of recording. Nevertheless, the degree of comparability of the Community Labour Force Survey results is considerably higher than that of any other existing set of statistics on employment or unemployment available for member states.

Since 1983 improved comparability between results of successive surveys has been achieved, mainly owing to the greater stability of content and the higher frequency of surveys, and this continuity will be maintained in the latest series of surveys (from 1992).

BIBLIOGRAPHY

Andersen, T. (1987), 'Short and Long-run Consequences of Shorter Working Hours', in P. Petersen and R. Lund (eds), *Unemployment: Theory, Policy and Structure*, Berlin: Walter de Gruyter.

Aznar, G. (ed.) (1993), *Travailler moins pour travailler tous. 20 Propositions*, Paris: Syros.

Becker, G.S. (1965), 'A Theory of the Allocation of Time', *Economic Journal*, 75 (299).

Booth A. and F. Schiantarelli (1987), 'The Employment Effects of a Shorter Working Week', *Economica*, 54, 234–48.

Bosch, G. (1986), 'The Dispute over the Reduction of the Working Week in West Germany', *Cambridge Journal of Economics*, 10, 271–90.

Boyer, R. (1988a), 'Technical Change and the Theory of Regulation', in G. Dosi *et al.* (eds), *Technical Change and Economic Theory*, London: Pinter.

Boyer, R. (1988b), 'New Technology and Employment in the 1980s: from Science and Technology to Macroeconomic Modelling', in J.A. Kregel, E. Matzner and A. Roncaglia (eds), *Barriers to Full Employment*, London: Macmillan.

Boyer, R. (1989), 'Synthesis Report: New Directions in Management', OECD Helsinki Conference on Education Practices and Work Organisation: Technological Change as a Social Process, 11–13 December, Paris.

Boyer, R. and E. Caroli (1993), 'Production Regimes, Education and Training System', RAND Conference on Human Capital, 17 November, Santa Barbara.

Calmfors, L. and M. Hoel (1988), 'Work Sharing and Overtime', *Scandinavian Journal of Economics*, 90, 45–62.

Calmfors, L. and M. Hoel (1989), 'Work Sharing, Employment and Shiftwork', *Oxford Economic Papers*, 41, 758–73.

CEC (1993), 'Growth, Competitiveness, Employment', White Paper (Delors Report), December, Brussels.

Cette, G. and D. Taddei (1997), *Réduire la durée du travail: De la théorie à la pratique*, Paris: Le Livre de Poche.

Corneo, G. (1994), 'On the Welfare Effects of Work Sharing', mimeo, Direction de la prévision.

Drèze, J.H. (1986), 'Work Sharing: Some Theory and Recent European Experience', *Economic Policy*, 3, 561–619.

ETUI (1996), *Collective Bargaining in Western Europe 1995–1996*, Brussels: ETUI.

European Commission (1997), *Employment in Europe*, Luxembourg.

Eurostat (1997), *Labour Force Survey: Results 1995*, Luxembourg.

Freeman, C. and L. Soete (1994), *Work for All or Mass Unemployment? Computerised Technical Change into the Twenty-first Century*, London/New York: Pinter.

Freeman, C., J. Clark and L. Soete (1982), *Unemployment and Technical Innovation*, London: Pinter.

Gershuny, J. (1994), *Changing Time*, Oxford: Oxford University Press.

Gorz, A. (1989), *Critique of Economic Reason*, London: Verso.

Gorz, A. (1993), 'Préfacions', in G. Aznar (ed.), *Travailler moins pour travailler tous. 20 Propositions*, Paris: Syros.

Hart, R. (1987), *Working Time and Employment*, London: Allen & Unwin.

Hewitt, P. (1993), *About Time: The Revolution in Work and Family Life*, London, IPPR/Rivers Oram Press.

Hoel, M. (1986), 'Employment and Allocation Effects of Reducing the Length of the Workday', *Economica*, 53, 75–85.

Hoel, M. and B. Vale (1986), 'Effects of Working Time in an Economy where Firms Set Wages', Memorandum from Department of Economics, n.2/85, University of Oslo.

Holland, S. (1993), *The European Imperative: Economic and Social Cohesion in the 1990s: A Report to the CEC*, Nottingham: Spokesman.

ILO (1989), *Conditions of Work Digest: Part-time Work*, Geneva: International Labour Office.

ILO (1997), *Youth, Older Workers and Social Exclusion: Some Aspects of the Problem in G-7 Countries*, Geneva: International Labour Office.

Lundvall, B.Å. and B. Johnson (1994), 'The Learning Economy', *Journal of Industrial Studies* 1 (2), December, 23–42.

Maddison, A. (1979), *Phases of Capitalist Development*, New York: Oxford University Press.

Maddison, A. (1991), *Dynamic Forces in Capitalist Development: A Long Run Comparative View*, Oxford: Oxford University Press.

Malinvaud, E. (1995), 'Preface', in D. Anxo *et al.* (eds), *Work Patterns and Capital Utilisation – An International Comparative Study*, Dordrecht: Kluwer Academic Publisher.

OECD (1995), *Flexible Working Time – Collective Bargaining and Government Intervention*, Paris: OECD.

OECD (1997), *Working Time: Trends and Policy Issues*, Paris: Château de la Muette.

Perez, C. (1981), 'Structural Change and Assimilation of New Technologies in the Economic and Social System', *Futures*, 15, 357–75.

Petit, P. and L. Soete (1997), 'Is a biased technological change fuelling dualism?', paper presented at the TSER conference on 'Technology, Economic Integration and Social Exclusion', 17–18 October, Paris.

Rothwell, R., C. Freeman, A. Horlsey, V.T.P. Jervis, A.B. Robertson and J. Townsend (1974), 'SAPPHO Updated: Project SAPPHO Phase II', *Research Policy*, 3 (3), 258–91.

Santamäki-Vouri, T. (1984), 'The overtime pay premium, hours of work and employment', working paper, Helsinki School of Economics, vol. 75.

Schor, J. (1991), *The Overworked America: The Unexpected Decline of Leisure*, New York: Basic Books.

Seifert, H. (1991), 'Employment Effects of Working Time Reductions in the Former Federal Republic of Germany', *International Labour Review*, 130 (4).

Soete, L. (1987), 'The Newly Emerging Information Technology Sector', in C. Freeman and L. Soete (eds), *Technical Change and Full Employment*, Oxford: Basil Blackwell.

Taddei, D. (1991), 'Temps de travail, emploi et capacités de production, la réorganisation–réduction du temps de travail', *Europe Sociale*, Supplément 4/91.

Taddei, D. (1995), 'Capital Operating Time (COT) in Macroeconomic Modelling', in D. Anxo *et al.* (eds), *Work Patterns and Capital Utilisation – An International Comparative Study*, Dordrecht: Kluwer Academic Publisher.

Toedter, K.H. (1988), 'Effects of Shorter Hours on Employment in Disequilibrium Models', *European Economic Review*, 32, 1319–33.
Whitley, J.D. and R.A. Wilson (1988), 'The Impact on Employment of a Reduction in the Length of the Working Week', *Cambridge Journal of Economics*, 10, 43–59.

14. Europe's system(s) of innovation

Bruno Amable and Robert Boyer

1. INTRODUCTION

The debate concerning the sources of growth was renewed with the literature on endogenous growth theory. Technology as an exogenous trend was a common assumption in most growth models until the mid-1980s, at least as far as mainstream economics is concerned, since both the Kaldorian[1] and neo-Schumpeterian[2] schools used an endogenous determination of productivity improvements in their models. However, in the most widely accepted perspective concerning growth theory, if long-run development was seen as a consequence of innovation and the progress in knowledge, it was nevertheless implied that economists had not much to say about such phenomena and should hence take long-run technical progress as exogenous.

The recognition that innovation and technical progress did not fall from heaven and the fact that a theory of long-run growth should explain technical change endogenously as well as, maybe, the confidence placed by economists in the explanatory power of the model of economic behaviour led to a questioning of the very cautious attitude held by traditional neoclassical modelling. Innovation and technical progress can be seen as the result of purposeful activities such as investment, either in physical or human capital, R&D expenditures and so on. Such investments are motivated by profit expectations and thus can be explained by standard economic models.[3] Therefore, if one knows how to relate growth to innovation, knowledge accumulation or increases in the level of human capital of the labour force, one has a theory of long-run growth. All that is necessary is a macroeconomic production function incorporating the advancement of knowledge or the consequences of innovation.

But since the potential sources of long-run growth are plentiful, one actually has not one but several theories of growth, some of them emphasizing innovation and R&D, others human capital accumulation or, more generally, increases in knowledge and competence through education and training, public infrastructures or private capital accumulation, and so on.[4] This diversity echoes the diversity of factors taken into account in growth accounting explanations of the postwar growth of developed economies.[5] Decomposing the Solow residual with the help of various statistical indicators documenting structural change, the age of capital equipment or the structure of the labour force never provided a theory of long-run growth but was nevertheless helpful in pointing at the many sources behind productivity improvements.

But the theory of long-run growth does not amount to just identifying the various 'technical' sources of growth and incorporating them into macroeconomic models. There is a tradition, to which, for instance, Abramovitz has contributed, which enriches growth accounting exercises with institutional considerations on the prospect of growth and catch-up from less advanced countries. In this tradition, growth and technological development are not just a matter of getting the right production function and the right combination of factors; they also involve some degree of compatibility between a country's technological capability and the current technological best practices. The fact that some countries are relatively backward from an economic and technological point of view leads to asking oneself at least two questions: can they catch up with the technologically leading countries, and why are they backward in the first place? Both questions deal more or less with the structure of institutions and types of organization of a country.

In Abramovitz (1979), relative backwardness implies a potential for catch-up-driven growth by imitation of technologies from the frontier countries, conditional on the 'social capability' of the economically backward country. However, the concept of social capability, borrowed from Okhawa and Rosovski (1973), is very broad and somewhat ill-defined. It refers to the idea that, for successful catching up, a country should actively engage in activities aimed at assimilating knowledge spillovers from abroad. These activities are, to a large extent, dependent on institutional factors such as the educational system (which supplies human capital and more specific product- or sector-related competencies necessary for assimilating knowledge spillovers), the financial intermediation system (which supplies financial capital for catch-up-related investment), the political system, and so on. In many cases, social capability and technological congruence are interrelated, for example when government stimulates foreign trade in the above case of scale economies and a small domestic market. But little in the way of a theory is proposed to incorporate institutional arrangements and modes of organization into a long-term view of economic and technological development.

North (1990) too emphasizes the importance of institutions – defined as 'humanly devised constraints imposed on human interaction' – in the growth process, but is a little more precise and analytical in his treatment of institutions than the social capability approach in (a) linking the role of institutions[6] to individual behaviour since institutional arrangements define the incentive

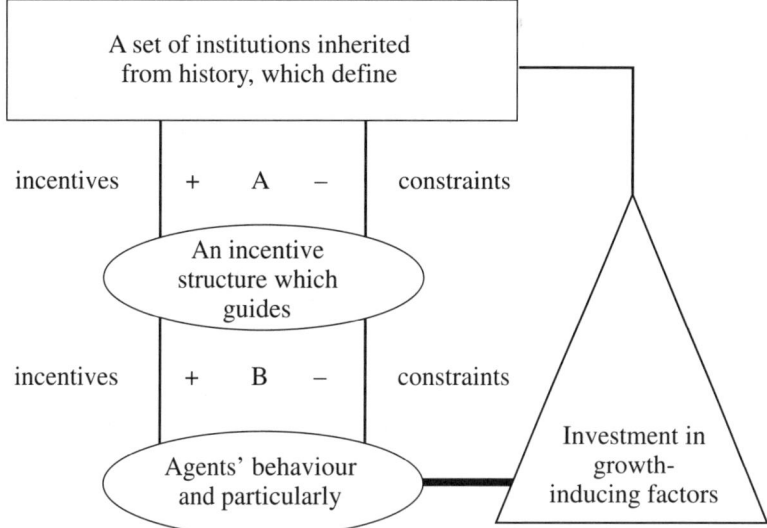

Figure 14.1   Why institutions shape the growth regime

framework which agents face, (b) insisting on the fact that institutions are generally not optimal and (c) providing an explanation why non-optimal institutions can still endure. An institutional basis for an explanation of non-converging growth trajectories would then imply the elements identified in Figure 14.1.

Therefore institutions can be 'inefficient' from a development point of view if they lead agents to adopt socially sub-optimal strategies, as for instance in a prisoner's dilemma type of equilibrium. But agents can modify institutions, provided they act collectively. Therefore explaining why inefficient institutions can persist in backward or declining countries necessitates mobilizing arguments involving the 'fixed cost' aspect of an institutional architecture as well as its 'increasing returns to adoption' aspect. Institutions are subject to the same phenomena as techniques: increasing returns to adoption and hence the possibility of lock-in to inferior solutions.[7] Inefficient institutions can continue to exist because a sufficiently powerful coalition has an interest in their existence, or because individual agents cannot coordinate their actions to change them, for instance.

All these approaches point to the importance of institutions in the process of economic development. But stating that 'institutions matter' is by no means enough. The treatment of institutions in economics is often partial, limited to specific aspects of the economy, and institutions are thought of as designed with respect to the function they have *ex post*. It is necessary to go beyond the consideration of the origins, the functions or consequences of specific institutions taken separately, and to raise questions related to the complementarity[8] and, possibly, hierarchy of institutions.

Some authors have investigated in this direction of research, relating the whole institutional architecture of an economy to its main economic characteristics: the regulation school (Aglietta, Boyer, Lipietz),[9] the Social Structures of Accumulation (SSA) school (Bowles, Gordon),[10] the Comparative Institutional Analysis (CIA) approach (Aoki),[11] macroeconomists concerned by the consequences of various coordinating mechanisms (Soskice),[12] economic historians concerned with the importance of institutions (David) and sociologists and political scientists who have studied the de facto complementarity of the monetary regime, unions' strategies, firm organization and economic specialization (Streeck).[13] These approaches ask the question, how do different institutions combine into a more or less coherent system? The work concerning systems of innovation[14] comes close to the concerns of these scholars, with a more pronounced emphasis on science and technology-related institutions and organizations.

This is the more relevant the more ancient and diverse institutions governing the contemporary economies are. This is especially so for Europe and Japan. The previous analytical framework suggests that institutions do not necessarily promote overall efficiency, hence their plurality. If this is so, convergence neither in institutions nor in productivity levels is guaranteed. Therefore this chapter starts with a reinterpretation of the now trite statement that 'institutions matter' (section 2). They do so not only when it comes to catching up and, conversely, clamping on or falling behind but also for technologically and economically leading countries. In order to check this influence and diagnose the existence of diverse institutional settings, it is important to collect a series of data about the various sub-systems related to scientific advances, technological innovation, industrial specialization, human resources and financial systems, and relate them to various indexes of economic and social performances (section 3). It turns out that at least four coherent innovation and production models tend to coexist among OECD countries and they derive from different coordinating and adjusting principles. These configurations explain also contrasted direction and intensity of innovations, even in the context of more interdependent national economies (section 4). Thus the institutional setting is quite important in shaping the various growth trajectories and may play a role even in the long run, contrary to a maybe no longer so widely held belief that institutions induce only minor short-run frictions around a long-run Walrasian or classical equilibrium growth path. This finding has definite consequences for the present debate about the future of European integration (section 5).

2.   A DIVERSITY OF INSTITUTIONAL ARRANGEMENTS

The Importance of Institutions for Growth Trajectories

One should expect that a large variety of institutions and organizations can influence the growth trajectories followed by nations. If the accumulation of individual competence drives long-run growth, then the pattern of industrial relations may be as important as the organization of the scientific system for economic development. Indeed, this is compatible with the new theories of endogenous growth. Table 14.1 lists the various sources of growth on which endogenous growth models can be built. The nature of the growth factors considered implies the presence of (mostly positive) externalities in most cases, the existence of which gives a role to correcting policies as well as non-market arrangements such as organizations or institutions. Some modes of organization and some institutional arrangements may be more or less effective in having agents internalize external effects, promoting cooperative arrangements or facilitating coordination than others. Therefore the characteristics of national institutions, along with more technological characteristics, will determine the accumulation of capital, investment in R&D, the type of education of the labour force and so on, and hence the growth path.

Do Institutions Matter?

As a first step, one can discriminate countries according to their institutional characteristics, as in the 'social capability' approach of Abramovitz. However, if it is helpful in distinguishing Lesotho and Mozambique from the USA and Japan, such an approach remains at too broad a level to be able to discriminate within the group of leading countries and gives only partial information on the differences between the leaders and the catching-up group. If one concentrates on developed countries (the USA, Japan and the EU), it is necessary to delve further into institutional analysis.

The importance of institutions and modes of organization is increasingly widely acknowledged in the economics literature. The most fundamental question regarding the institutional design of a developed economy is probably: why does economic competition not end up in chaos? Adam Smith's answer was that the propensity to exchange lies behind the invisible hand, making decentralized decisions compatible. More modern expositions of this argument can be found in the different versions of general equilibrium theory. However, the development of general equilibrium-based economics, neoclassical theory broadly defined, has led to incorporating specific features and taking account of particular institutions whose very absence was at the heart of the Walrasian project. If new classical economics remains true to the

Table 14.1 Institutions and organizations influencing endogenous growth

Source of growth	Externalities	Empirical support	Concerned institutions and organization forms	Economic policy instruments
Physical capital	Positive and strong (except in the simplest 'AK' model)	Relatively high, investment and growth are strongly related in most empirical studies	• financial intermediation system • firms (mode of organization, strategies, time horizon …) • the state	• taxes and subsidies to promote investment • adequate policy mix and macroeconomic stability
Human capital	Positive and strong	Supposedly high, but not very accurately measured Empirical tests may turn out to be disappointing	• general education and training system (public or private) • technical education system • firms' internal training systems • wage–labour nexus	• education and training systems reform • subsidies to education and training • regulation on diplomas
Innovation	Positive and strong	Strong at the micro level, generally positive at the macro level but difficulties in measuring the effects Direct tests of the endogenous growth innovation-driven models are somewhat disappointing, however	• scientific system • higher education system • firms and research labs • the state • financial intermediation system	• laws on patents and intellectual property • improving the relationships between science, technology, higher education and the firms • promoting the availability of 'patient' capital • promoting the availability of capital for risky investments • research and technology policy

Knowledge and ideas	Supposedly very strong	Difficult to measure but indirect support (international openness and location externalities)	• networks (firms, scientific community) • education and science systems • international regime	• human mobility • capital mobility • education subsidies • openness policies • location policies
Public infrastructures	Positive in general	Most of the estimations tend to support a positive effect of public infrastructures, but the magnitude of the effects is subject to variations according to the sample considered and the testing method used	• the state • local public authorities • firms	• public investment programmes • coordination between central and local authorities • taxes and subsidies
Learning effects	Strong but declining with time if applied to the same technique	Weak support at the macro level because of the difficulties in measuring learning Stronger support at the micro level on certain techniques	• firms' mode of internal organization • education and training system • wage–labour nexus	• labour force stability • technical education of the labour force • promoting cooperation on the shop floor • trade policy (specialization in goods with different learning potentials)
Competition	Ambiguous (positive or negative)	Supposedly positive, but too few estimations to obtain a definite answer	• the state • firms • networks	• competition policy (regulation) • industrial policy • taxes and subsidies

Source: Modified from Boyer and Didier (1998).

431

representative agent endowed with rational expectations and full rationality, contemporary research is not hostile to the consideration of context-specific hypotheses: the rationality hypothesis may be relaxed, asymmetric information is a common feature in the economics of industrial organization, modern macroeconomic theory takes into account imperfect competition and increasing returns to scale, and so on. Last, new institutional economics is concerned with the impact of institutions on individual behaviour.

From a theoretical point of view, two aspects of institutions must be distinguished, namely complementarity and hierarchy. On the one hand, complementarity is observed *ex post* and could be interpreted as (a) the emerging property of the process of coevolution of several institutions, and (b) the stable coexistence of several institutions associated with self-reinforcing mechanisms. On the other hand, a hierarchy can be associated with an institutional architecture if the inner design of one institution takes into account the constraints and incentives associated with another one.

Regulation theory[15] has incorporated institutional determinants in its analysis of modern capitalist economies. The basic idea was to attribute the properties of growth regimes to a special mix of institutional forms.[16] For instance, the 'Golden Age', that is, the postwar high-growth period, was derived from the central role of the capital–labour accord (also known as the wage–labour nexus) and this accord permeated the whole economic system via a new style for state intervention, oligopolistic competition, the credit regime and a stable international regime. In this respect, during this period, the wage–labour nexus accord was the driving institutional form of the growth regime. Nowadays, this hierarchy seems to be challenged, since the forms of competition and forces coming from the international regime directly affect the inner organization of the post-World War II capital–labour accord, which gives a second definition of the hierarchy of institutional forms: one form is driving the transformation of others.

An implicit hierarchy of institutions seems to inhabit the line of research on national systems of innovation (NSI). Since technical progress is at the root of long-run growth in the neo-Schumpeterian representations, it is somehow logical that the institutions directly involved with the advancement of science and technology should come to the forefront of most of the analyses on NSI:[17] R&D laboratories, universities and so on. It is indeed true that economic development and R&D expenditures are related, but this does not provide a causal link between R&D and growth, nor does it imply that technology drives all economic institutions.

The analysis of social systems of innovation found in Amable *et al.* (1997a, 1997b) not only intends to go beyond the strictly national aspect of systems of innovation (and production), but considers a large range of institutions and organizations belonging to different sub-systems in order to have a broader view of the institutional architecture of the economy and the possible interrelations between the sub-systems, without necessarily giving a prominent role to science and technology, but more generally by putting the technological aspects of the production systems in a broader perspective. The technology-related institutions and modes of organization may or may not be driving the transformations of production systems, the hierarchy of institutions is not a priori defined (see Table 14.2).

3. SEVERAL CONFIGURATIONS OF SYSTEMS OF INNOVATION

Research on national systems of innovation (NSI) has stressed the necessarily local character of learning and hence of the set of national institutions that govern the generation of knowledge. These studies have shown the extreme diversity of national organizations in the field of innovation, which are far from being simply variants of the same model. However, these studies have two major weaknesses.

1. When one concentrates on national case studies, it is tempting to have as many configurations as countries. What one can gain in precision is lost in the generality of the principles.
2. Second, at a time when the internationalization of research is becoming more widespread, when diffusion of new techniques and modes of organization is becoming more rapid, it may be an exaggeration to maintain the strictly national nature of learning mechanisms. It seems preferable to consider the notion of social systems of innovation (SSI), which leaves open the question of the territory over which the cumulativeness of learning operates.

There is no need to postulate an optimal SSI and rank the observed configurations according to their relative distance *vis-à-vis* this optimal reference. An implicit classification can be found, for instance, in Ergas (1984), where innovation is predicted to be all the more dynamic when demand is strong in volume and variety, and when many scientists and industrial structures favour both competition and cooperation on new-product development. Nevertheless, such a combination of favourable factors is far from being guaranteed. It may be that innovation results from low entry-barriers and a high level of competition, allowing for a rapid renewal of products, firms and processes. Or, on the contrary, one may have a production model where innovation takes place in the large firm, based on learning effects from one product line to the next. In fact, forms of cooperation and competition constitute two poles and

Table 14.2   Institutional analyses of innovation and production systems: a long way from the 'social capability' approach

Notion	Content	Implication	Analytical tools	Author(s)
Localization of technological trajectories	Ability to develop technological paths, because of the local character and learning effects	Possible coexistence of different productive systems in contexts of different prices and demand patterns	Economic analysis of localized change and comparative studies of growth	David (1975)
Social capability	Ability to use the technological potential, taking account of the local development	No guaranteed catch-up unless a country is endowed with 'adequate' characteristics	Analysis of Japanese growth Long-period study of comparative growth performances	Ohkawa and Rosovsky (1973) Abramovitz (1979)
Societal models of innovation	Organization of the supply of local public goods, experimentation and redistribution of the income associated with technical change	Possibility of divergence in terms of growth, productivity and employment	Mobilization of 'exit' and 'voice'	Ergas (1984)
National system of innovation	Network of institutions that contributes to creating and diffusing technological change	Some social and institutional configurations favour rapid growth and catching up Role of public policy, importance of the education and training system	Study of Japan	Freeman (1987)

Concept	Description	Characteristics	Methodology	Reference
Social technology absorption capability	The type of regulation, competition and financial system affects the process of specialization and local learning	The heterogeneity of institutions can account for the coexistence of diversely innovative economies	Theory of growth and innovation	Stiglitz (1991)
Innovation performance	Performance does not depend only on scientific and technological policy but also on macroeconomic, financial and international policies	Initiatives in terms of industrial policy may have minor or negative effects if the point of view is too local	Historical studies and international comparisons	Rosenberg, Landau and Mowery (1992)
Dynamics of constructed advantages	Innovation is a process that involves R&D, workforce, capital and public markets and policies	Competitiveness comes from the synchronization of several components. Economic policy can influence this process	Comparative studies, statistics, models	TEP Programme OECD (1992)
National systems of innovation (NSI)	Learning and innovation process based on certain norms and types of coordination at the national level	Vicious or virtuous circles are a consequence of interdependence between the components of NSI	Detailed studies of the components of NSI	Lundvall (1992)
National systems of technological innovation	Set of institutions whose interactions determine the innovative performance of a country	Systems differ because of national history, including the date of the industrial revolution	National studies put into perspective	Nelson (1993)

Table 14.2 continued

Notion	Content	Implication	Analytical tools	Author(s)
Worlds of production	There exists a multiplicity of coordination principles for social and economic activities, giving as many national or sectoral configurations	There are at least four worlds of production – interpersonal, market-based, industrial and non-material – that have no reason to converge	Socioeconomic analyses, data analyses on sectoral statistics, history of industrialization paths, international comparisons	Salais and Storper (1994)
Social systems of production (SSPs)	Complex of scientific and technological organizations, education systems, industrial relations, financial systems, governing production	The different SSPs are more or less efficient according to the international context and the nature of the dominant production paradigm Because of institutional inertia, there is no convergence among SSPs	Comparative study of the current configuration of SSPs for the large industrialized countries Long-run historical studies	Hollingsworth (1997)
National systems of production	'Technoeconomic' environment of a given country; context in which firms follow their particular trajectories	The environment defines a set of constraints and opportunities for firms' development	Historical and institutional comparisons of technical change	Von Tunzelmann (1995)

not one unique SSI. Furthermore, good scientific and technical capacities may be at odds with the inadequacy of an industrial structure, resulting in a disappointing innovative performance. Complementarities between the different components of an innovation system may make it more difficult to give an average grade.

Therefore the issue of the possible diversity of production systems is to a large extent an empirical matter. An empirical analysis of innovation and production systems was made in Amable *et al.* (1997a), concerning 12 developed countries: the USA, Japan, Germany, France, the United Kingdom, Canada, Australia, Italy, the Netherlands, Finland, Sweden and Norway. These countries are defined by a large set of indicators[18] that are classified in the six following sub-systems :

- Science,
- Technology and innovation,
- Industry,
- Labour force and human resources,
- Education and training,
- Financial system.

The macroeconomic *performance* (for the 1980s) indicators constitute a seventh set of parameters. These include indicators on productivity, unemployment, participation of the labour force, industrial self-sufficiency, income distribution inequality and growth.

It is necessary to set apart the indicators of economic performance from the other indicators, for a number of reasons. First, structural indicators in the six sub-systems referred to above change slowly (pattern of scientific specialization, industrial structure, and so on) since they reflect deep structural characteristics of economies. Second, performance indicators can change very rapidly (Sweden turning into a high unemployment country over a couple of years) and only reflect a country's production structure when taken in the long run. Third, and as a consequence of the previous two points, there is a difficulty in linking structural indicators to performance indicators since the systems (of innovation, production and so on) approach is a long-run perspective and assumes a relationship between structure and performance. It is thus necessary to correct for the short-run effects of, for instance, macroeconomic policy mix. Slow growth in Europe in the 1990s had more to do with monetary policy and the nominal anchor to the Deutschmark (DM) than with inadequate technological development.

For the study of innovation and production systems, a factor analysis was performed, followed by an analysis of distances (hierarchical classification) which led to the separation of the 12 countries of the sample into four groups that correspond to four different productive models: the social-democrat model (Sweden, Norway, Finland); the meso-corporatist model (Japan); the market-based model (USA, UK, Canada and Australia); and the public institutions-based model (France, Germany, Italy and the Netherlands).

The first dimension to take into account is defined mainly by the indicators of industrial competitiveness, income distribution, scientific and technological specialization (physics and chemistry versus biology and natural sciences) and the education and training system. This sets the Anglo-Saxon countries apart from continental Europe and Japan. The second dimension is defined by the level of unemployment, income distribution and characteristics of the education and training system (skilled labour force). The results can be summarized with the main characteristics of the four types of productive models.

The Market-based Model

The US, UK, Canada and Australia form this group. Industrial competitiveness, expressed in terms of relative trade balance, is low, unemployment and growth are at average levels. Other characteristics include very little adaptation to the new productive model; strong sectoral specialization in aerospace and chemicals; important military R&D; a high level of publications/GDP but low levels of patents/GDP; flexibility of the labour market; and high levels of expenditure in education but weak training. The financial system is sophisticated, and venture capital is available.

The Social-democrat Model

Sweden, Norway and Finland constitute this group. Competitiveness is average, unemployment (during the 1980s) low and growth moderate. Foreign-based subsidiaries of domestic firms are important for their laboratories. Firms have adopted most of the principles of the 'new production model'. Sectoral specialization is strong in resource-intensive activities. Public research plays an important role, scientific specialization is clear in medicine and biomedical research. Mobility and skills of the labour force are rather high. Education expenditure is high. The financial system is not very sophisticated, no venture capital is available and the cost of capital is rather high, but the ability to raise funds on international markets is good.

The Meso-corporatist Model

Japan belongs to a group of its own. In terms of performance, Japan is characterized by a high level of competitiveness, a low level of unemployment and rather high growth. Sectoral specialization is strong in electronics, equipment goods and transport goods. Firms have adopted many attributes of the new production model (long-term management, quality management, time to market, time to innovate and so on). The ratios R&D/GDP and patents/GDP are high, but the scientific publications/GDP ratio is low. Scientific specialization is in physics, engineering sciences (publications) and electronics (patents). Education expenditure is rather low except for public expenditure for primary and secondary education. Training is very important and the schooling rate is high. The financial system is characterized by a low cost of capital.

The Public Institutions-based Model

France, Italy, Germany and the Netherlands form this group. Competitiveness is rather high, unemployment is high and growth is slow. Income distribution is favourable to the poorest 20 per cent of the population. Adaptation to the new productive model is average. Otherwise, this group is characterized by a certain number of specialization characteristics. Scientific and industrial specialization is marked in pharmaceuticals, aerospace and chemicals. There is a low level of scientific publications/GDP. External flexibility and mobility of the labour force is low. The financial system is relatively unsophisticated. Public expenditure in education is relatively high, but total education expenditure is moderate.

As we see here, the scientific profile does not mechanically determine technological specialization, any more than technological specialization mechanically determines macroeconomic performance. It is neither necessary nor sufficient to have a dominant position in high technology to enjoy low levels of unemployment and high growth. Institutions that organize the innovation and production process are determinant: there is no ideal configuration to organize the relationships between science, technology and the economy.

4.   THE DYNAMICS AND TRANSFORMATION OF THE PRODUCTIVE MODELS

The Coherence of the Productive Models

Theoretical developments stress the historically determined, localized and partial character of innovation and technological learning processes. The empirical analysis in Amable *et al.* (1997a) confirms the multiplicity of 'coherent' systems. In a sense, this coherence appears *ex post*, through the ability of a network of institutions to accommodate internal evolution and external shocks. In an evolutionary perspective, one could consider that the four configurations are the results of a process associating selection and learning. But it would be a naive vision of evolutionism – whereby all that exists has been and is efficient – to limit oneself to this sole argument. What matters is that the fundamental principles of each configuration, structuring social systems of innovation or modes of regulation, are coherent with each other in the sense that, for instance, the mode of operation of the wage–labour nexus does not contradict the efficiency principles of other institutional forms.

A summary of the logic of the different models can be expressed as follows.

The market-based model

One finds at one end of the spectrum a market-based logic, held to govern almost all social and economic activities, but which is not contradictory to a sometimes heavy involvement of the state in some activities (the USA and defence being a blatant example). Advances in fundamental science are the result of competition between laboratories, and the patent race determines the R&D strategies of private firms. A strict definition of property rights and their enforcement by a very developed legal system form two of the pillars of such an SSI. Sophisticated financial markets make it possible to mobilize venture capital and give new sectors an opportunity to emerge. Very flexible labour markets do not in general allow skill accumulation within the firm, in a context where vocational training does not develop for lack of market-driven incentives. On the other hand, labour movements between firms favour the dynamism of new firms.

The social-democrat model

At the other end of the spectrum, one may find a model based on compromise and negotiation between the involved parties. Most of the forms of organization have their origins in the confrontation of the respective objectives of firms, salaried workers and the state. These parties seek to come to an agreement on principles liable to guarantee simultaneously firms' competitiveness and the reduction of inequality. In this sense, this model may be called 'social-democratic'. The egalitarian objectives in terms of income and education prevent the polarization of skills and induce pressures in favour of high value added sectors. This permanent adaptation process presumes the existence of public institutions whose purpose is to retrain the labour force rejected by non-competitive sectors. Relatively underdeveloped financial markets do not prevent the pursuit of long-term objectives in the field of innovation. In this model, one can see most clearly the construction of an SSI that has little to do with a market-based logic.

The meso-corporatist model

A third configuration can be called meso-corporatist, producing to some extent the same results as the social-democrat model, but according to a different logic and with the help of a different set of institutions. Large firms, linked through financial, personal or technological networks, are at the centre of the process of innovation. Public schooling institutions limit themselves to the general education of a large number of students, while universities do a certain amount of basic research. However, most of the innovations derive from the acquisition of competence within the firm, according to a learning process that enables a growing sophistication of products, processes and forms of organization. The control that the main bank has on an industrial firm permits long-term horizons. In this model, innovation is mostly governed by the search for products liable to satisfy the private demand, all the more so as the small size of military expenditures and public programmes based on incentives rather than rulings make competitiveness a major imperative of this SSI.

The public institutions-based model, or the model of European integration

A public institutions-based model defines the last configuration in which public institutions play a determining role in the impulse, codification and direction of innovation and are at the centre of the processes of economic adjustment. In opposition to the social-democratic configuration, tripartite bargaining is more difficult and a large part of the specialization derives from demand originating in public expenditures: transport and communication infrastructures, health and so on, as well as large public programmes, either civilian or military. Basic research is mainly public, as is the education system, so that the transferability to the private sector of research advances and the closeness of fit between training and the needs of private firms may be a problem. Traditionally, public regulations have a role in the allocation of credit by banks. This bank-based credit system tends to favour large firms over small businesses and new firms. In this model, public authorities are an important actor in the dynamic adaptation between supply and demand in the fields of innovation, production and credit.

Contrasted Performance and Evolution

The above considerations are necessarily schematic and sum up very briefly a very complex set of interrelations between economic structures and institutional forms, but what they suggest is that the four models are not variants of one unique pattern: each has an internal logic and synergetic effects that are entirely specific. One may have reservations when it comes to assigning a particular country to one model. There is no doubt that the USA and the UK belong to the market-based model, and the size of public (defence) spending on R&D in the USA is not relevant here as it has been shown that the particular US market-based model actually involves a peculiar involvement of public authorities.[19] France is the figurehead of the public institutions-based model, while Japan represents the meso-corporatist pattern. However, one may express some doubts on the classification of Germany: the *Sozialmarktwirtschaft* is, in a certain way, close to the social-democratic model, although the empirical classification puts Germany alongside France in the public institutions-based model. This is notably because they have very comparable macroeconomic performances and the public aspect in science and technology is dominant.

Nevertheless, broadly speaking, each institutional setting expresses a set of objectives which are implicitly or explicitly favoured.

- The market-led model puts a strong emphasis upon the need for quick adjustments to an uncertain environment, at the possible cost of dynamic efficiency. Similarly, the enforcement of property rights is perceived as much more important than social justice and the implementation of a complete welfare system.
- By contrast, the social-democrat model puts forward the objectives of social justice and universal welfare and tries to organize collectively all the adjustments required by innovation and technical change.
- The meso-corporatist model is another alternative to the market-led system: the common interest of permanent workers of the large company is perceived as more important than the enforcement of property rights by shareholders. Therefore dynamic efficiency is privileged against static efficiency, whereas most of the welfare is linked to each large company.
- The public institutions-based model is also different, since state interventions are highly developed within any sector of economic activity: the implicit objective is to alleviate recurring social conflicts between groups and simultaneously the legitimacy of public authorities is closely related to the monitoring of a successful growth regime.

Thus it is not a surprise to observe that each model exhibits different patterns of evolution in terms of scientific specialization, innovation, factors of competitiveness and therefore growth and employment patterns. Even though they belong to the same economic zone, the countries which follow different models have different trajectories, at least until the early 1990s.[20] Since then, some structural transformations have taken place. For instance, the Scandinavian countries belong clearly to the social-democratic trajectory, but their recent evolution has made their macroeconomic performances more similar to the rest of Europe. Similarly, the size of the country and the precise policy followed may explain differences in performances among countries which belong to the same broad model of production and innovation. The trajectory followed by the Netherlands is quite original, since the reforms in labour market institutions as well as new trends for innovation policy have promoted a clear decline in unemployment, contrary to what is observed elsewhere in continental Europe. Conversely, France and Germany, which belong to the same broad model but experienced significant institutional differences, tend to converge in terms of the reforms implied by the adhesion to the Euro.

The Diversity of National Innovation Systems: An Advantage and a Challenge

In any case, economic institutions remain significantly different and clearly Europe comprises different countries, on several counts.

- The technological level of countries differs substantially. There are a number of countries which can be considered technological leaders, and a number of countries which rely more on technology diffusion for growth.
- Second, especially for technology (business R&D), investment of re- sources in technology, as well as technological activities, vary between countries.
- European countries seem to belong to several different social systems of innovation, with markedly different organizing principles in terms of institutional architecture.

This raises the question to what extent these differences between European nations have an impact on economic performance, for example in terms of growth. This is an important question in light of the recent discussion on convergence and economic and monetary union (EMU). This discussion looks mostly at convergence of monetary phenomena such as inflation and interest rates, or limits the realm of 'sustainability' to the size of government debt. 'Real' factors, with the recent exception of unemployment, have largely remained unconsidered. It is therefore relevant to ask whether differences between European countries in such real factors, for instance technology, or the institutional arrangements may have an impact on economic performance and cohesion of the Union.

5.   CONCLUSION AND POLICY CONSIDERATIONS

The findings of this chapter have both theoretical and methodological consequences for the research on innovation and growth and, of course, they have
implications for technological and economic policy.

Institutions at the Core of Innovation and Growth

One or two decades ago, economic theory was observing a clear dichotomy:
on one side, innovation was supposed to derive from advances in basic
research; on the other side, the related technical change was added to the
neoclassical growth model. Of course, some neo-Schumpeterian economists
had a more refined analysis: the pattern of growth and employment was
related to the compatibility between the leading direction of innovation and
the current economic institutions. But in both cases, technical change was
assumed devoid of any institutional content.

Fortunately, this dichotomy has broken down under the pressure of recurring anomalies between the prediction of the theory and the observation.
Nowadays, after a long process of trial and error, a wide spectrum of economists agree upon the crucial importance of the institutions governing
innovation. This research agenda started from the concepts of a localized
technological change and of social capability, and progressively unfolded the
full institutional content of basic research, R&D strategies by the firms and
the general context which allows innovation to diffuse to the whole economic
system.

In a sense, this chapter has provided a systematic analysis of the social or
national system of innovation which had been pioneered and promoted by
Freeman (1987), Lundvall (1992) and Nelson (1993). Its major contribution
is to show the clustering of scientific, technological, social and financial
institutions around a definite set of configurations, which can be interpreted
by reference to key principles of adjustment: the market, the negotiation, the
leading role of the large company or a complete set of public interventions.

Need for Coherence, but Not Necessarily Uniformity of Systems of Innovation and Production

The diversity of systems of innovation within Europe is a potential problem
for the cohesion of the Union and the efficiency of European policy, whether
in the area of science and technology or not. At the microeconomic level, the
institutional arrangements present in the different countries provide different
sets of incentives to agents for opportunism or trust, short-termism or long-
termism, flexibility and so on. Therefore one may expect not only very

different patterns of investment, types of specialization and forms of innovation, which is what is found, but also very different patterns of response to structural policy impulses. To give an example, a set of European research policy measures based on market arrangements is more likely to have strong responses from the market-based countries than from others. Or, to go one step further, a competition based on price competitiveness is likely to be more detrimental to countries which have highly elaborated sets of non-market mechanisms. Conversely, large, old-style European technological programmes may benefit mainly the public institutions-based model, to the detriment of other systems, with different features, that is, strengths and weaknesses. The challenge to any European policy is to promote this institutional diversity, while creating more and more complementarity among economic specializations.

Catching up is Not an Inevitability but a Matter of Institutional Setting

This analytical framework has a broader consequence concerning contemporary controversy among growth theoreticians. On one side, some authors state that economic convergence of productivity will take place as soon as the relevant factors such as education and research are considered. On the other side, new growth theorists maintain that, according to their initial endowment and strategic choices, countries may experience diverging growth trajectories. This chapter provides a different answer and points out how the inherited institutional setting is framing the set of choices. The congruence of these institutions and their adequacy to the emerging productive principles are key factors in explaining why a given country will fully catch up, clamp onto or finally fall behind the technological leaders. This general result is valid for the group of follower countries, but for the leading ones as well: they belong to four configurations with distinctive features, and this explains why their specializations are different, none of the systems of innovation being superior in any sector.

The Euro puts a Strong Emphasis upon the Enhancement of the Structural Competitiveness of each Innovation System

One should not underestimate the impact of European monetary integration upon the dynamism of innovation, as a method for reconciling dynamic efficiency and the preservation of an extended welfare system. Back in the 1970s, an economy which experienced an erosion of its competitiveness could always adjust the exchange rate. Over the last two decades, the objective of stable exchange rates among European countries has prompted various strategies, the most frequent being a movement towards an increase in labour market 'flexibility'. But an unexpected currency crisis in one country could always erode within one day the structural competitiveness built up by another country over a decade or more. The Amsterdam Treaty is clearly putting an end to such possibilities. If a purely defensive strategy in terms of wage reduction and welfare slimming down is to be prevented, a strong and dynamic innovation policy has to be promoted, at both the national and the European level.

The discussion on the importance of the coherence and dynamic efficiency of institutional arrangements made above points to a set of policy recommendations other than just a matter of industrial specialization. This is why a differentiated set of policy measures adapted to the particularities of the different systems of innovation may be more suited to the present situation of Europe. One may object that such a differentiation would neglect the process of integration within Europe.[21] It is true that EU scientific programmes and measures favouring human capital mobility have made steps in the direction of a unified Europe, but the integration is stronger in the goods market and in the area of monetary policy, and this has had, for the moment, few echos in many institutional areas which are relevant for the systems of innovation.

- There is no unified financial system for Europe, which incidentally may affect the pattern of response of European countries to a common monetary policy, in spite of a very strong movement of financial liberalization. Besides, this pattern of unification towards a more financial markets-based system is as much imposed as chosen, and may not be the improvement it is meant to be, particularly in terms of financial fragility and investment projects monitoring abilities.
- There is no unified wage–labour nexus within Europe, and differentiated patterns in terms of labour market flexibility, social protection or modes of competence acquisition may render the building of a European model of production more problematic and may lead to a destabilizing social competition.

Besides, large European industrial and technological programmes have been more successful when they corresponded to a public action organized around some large projects than in the promotion of networking and more decentralized industrial and technological integration. But this problem is not specific to the Union and is present at the national level too. This is more a matter of adapting public policies to changes in the forms of innovation and technological competition at whatever level than a strictly European problem. The lack of coordination between member states and their respective policies may exacerbate the problem, but it did not create it.

NOTES

1. See Boyer and Petit (1991).
2. See Dosi *et al.* (1988).
3. If marriage and the number of children can be explained (?) with the help of economic models, why not technological innovation and scientific discovery?
4. See Barro and Sala-i-Martin (1995) or Aghion and Howitt (1998) for textbooks on endogenous growth.
5. Maddison (1987).
6. Institutions are 'carriers of history' in the words of Paul David.
7. Arthur (1989).
8. For further developments on this issue, see Amable (2000).
9. Aglietta (1979), Boyer and Mistral (1982), Boyer (1989).
10. Kotz *et al.* (1994).
11. Aoki (1995).
12. The 'variety of capitalism' approach. See Soskice (1996).
13. Streeck (1992).
14. Freeman (1987), Lundvall (1992), Nelson (1993).
15. See Boyer (1989) for an exposition.
16. Defined as a special configuration of a basic social nexus. A configuration is to be distinguished from a system in the sense that in the former case, the compatibility of various institutions is partial and not necessarily long-lasting.
17. Freeman (1995).
18. See appendix to this chapter.
19. See Bellon (1986) for instance.
20. For a full analysis of the 1990s, see Amable *et al.* (1997, pp. 265–312).
21. Caracostas and Muldur (1997).

BIBLIOGRAPHY

Abramovitz, M.A. (1979), 'Rapid Growth Potential and Its Realization: The Experience of the Capitalist Countries in the Post-war Period', in E. Malinvaud (ed.), *Economic Growth and Resources*, vol. I, London/New York: Macmillan.

Abramovitz, M.A. (1994), 'The Origins of the Post-war Catch-up and Convergence Boom', in J. Fagerberg, B. Verspagen and N. von Tunzelmann (eds), *The Dynamics of Technology, Trade and Growth*, Aldershot, UK and Brookfield, USA: Edward Elgar.

Aghion, P. and P. Howitt (1998), *Endogenous Growth Theory*, Cambridge: MIT Press.

Aglietta, M. (1979), *A Theory of Capitalist Regulation: The American Experience*, London: New Left Books. (French edn 1976.)

Amable, B. (2000), 'Institutional Complementarity and Variety of Social Systems of Innovation and Production', *Review of International Political Economy*, 7(4) 645–87.

Amable, B. and M. Juillard (1997a), 'The historical process of convergence (1)', paper presented at the 'Théories et Modèles de la Macro-économie' Conference, Louvain-la-Neuve, Belgium.

Amable, B. and M. Juillard (1997b), 'The historical process of convergence (2)', paper presented at the conference 'Economic growth in closed and open economies', 20–24 September, Castelvecchio pascoli, Italy.

Amable, B., R. Barré and R. Boyer (1997a), *Les systèmes d'innovation à l'ère de la globalisation*, Paris: Economica.

Amable, B., R. Barré and R. Boyer (1997b), 'Diversity, coherence and transformation of innovation systems', in R. Barré, M. Gibbons, Sir J. Maddox, B. Martin and P. Papon (eds), *Science in Tomorrow's Europe*, Paris: Economica International.

Aoki, M. (1995), 'Towards a Comparative Institutional Analysis', *Japanese Economic Review*, 47(1).

Arthur, B. (1989), 'Competing technologies, increasing returns and lock-in by historical events', *Economic Journal*, 99, 116–31.

Barro, R. and X. Sala-i-Martin (1995), *Economic Growth*, Cambridge: MIT Press.

Bellon, B. (1986), *L'interventionnisme libéral*, Paris: Economica.

Boyer, R. (1989), *The Regulation School, a Critical Introduction*, New York: Columbia University Press.

Boyer, R. and M. Didier (1998), 'Innovation et Croissance', *Rapport pour le Conseil d'Analyse Economique*, Paris: La Documentation Française.

Boyer, R. and J. Mistral (1982), *Accumulation, Inflation, Crises*, Paris: Presses Universitaires de France.

Boyer, R. and P. Petit (1991), 'Kaldor's Growth Theories: Past, Present and Prospects for the Future', in E. Nell and W. Semmler (eds), *Nicholas Kaldor and Mainstream Economics*, London: Macmillan.

Caracostas, P. and U. Muldur (1997), *La Société, Ultime Frontière. Une vision européenne des politiques de recherche et d'innovation pour le XXIème siècle*, Brussels: Commission Européenne, Office de Publications Officielles des Communautés Européennes.

David, P. (1975), 'Technical choice, innovation and economic growth', *Essays in American and British Experience in the Nineteenth Century*, Cambridge: Cambridge University Press.

Dosi, G., C. Freeman, R. Nelson, G. Silverberg and L. Soete (eds) (1988), *Technical Change and Economic Theory*, London: Pinter.

Ergas, P. (1984), *Why do some countries innovate more than others?*, Centre for European Policy Studies, Brussels.

Freeman, C. (1987), *Technology Policy and Economic Performance: Lessons from Japan*, London: Pinter.

Freeman, C. (1995), 'The National System of Innovation in Historical Perspective', *Cambridge Journal of Economics*, 19(1), 5–24.

Gerschenkron, A. (1962), *Economic Backwardness in Historical Perspective*, Cambridge, MA: Harvard University Press.

Hollingsworth, R. (1997), 'Continuities and changes in social systems of production', in R. Hollingsworth and R. Boyer (eds), *Contemporary Capitalism: The Embeddedness of Institutions*, Cambridge: Cambridge University Press.

Kotz, D., T. McDonough and M. Reich (eds) (1994), *Social Structures of Accumulation: The political economy of growth and crisis*, Cambridge: Cambridge University Press.

Lundvall, B.-A. (ed.) (1992), *National Systems of Innovation. Towards a Theory of Innovation and Interactive Learning*, London: Pinter.

Maddison, A. (1987), 'Growth and Slowdown in Advanced Capitalist Economies: Techniques of Quantitative Assessment', *Journal of Economic Literature*, 25(2) 649–98.

Maddison, A. (1995), *Monitoring the World Economy 1820–1992*, Paris: OECD Development Centre.

Nelson, R.R. (ed.) (1993), *National Innovation Systems. A Comparative Analysis*, Oxford: Oxford University Press.

OECD (1992), *Technology and the Economy: The Key Relationships*, Paris: OECD.

Ohkawa, K. and H. Rosovsky (1973), *Japanese Economic Growth: trend acceleration in the twentieth century*, Stanford: Stanford University Press.

Rosenberg, N., R. Landau and D. Mowery (eds) (1992), *Technology and the Wealth of Nations*, Stanford: Stanford University Press.

Sala-i-Martin, X. (1997), 'I just ran 2 million regressions', *American Economic Review*, 87(2), May, 178–83.

Salais, R. and M. Storper (1994), *Mondes de Production*, Paris: Editions de l'EHESS.

Soskice, D. (1996), 'German Technology Policy, Innovation and National Institutional Frameworks', WZB Discussion Paper FS I 96–319.

Stiglitz, J. (1991), *Governments, Financial Markets and Economic Development*, NBER Working Paper 3669.

Streeck, W. (1992), *Social Institutions and Economic Performance: Studies of Industrial Relations in Advanced Industrialised Countries*, London/Beverly Hills: Sage.

Von Tunzelmann, G.N. (1995), *Technology and Industrial Progress*, Aldershot: Edward Elgar.

APPENDIX: INDICATORS USED IN THE ANALYSIS OF SOCIAL SYSTEMS OF INNOVATION

Six sub-systems are considered in section 3 of the main text.

1. *Science*: indicators on scientific specialization on two dates, 1983 and 1993, according to a classification in eight scientific disciplines. Data on scientific publications are used.

2. *Technology and innovation*: indicators of technological specialization for seven technological sectors using patents data; R&D/GDP, share of public R&D, military R&D. Some indicators on the 'new production model': total quality management, long-term management, time to innovate, and so on. These last indicators concern expert advice and indicate how far from the traditional Fordist model a country has gone.

3. *Industry*: indicators on international specialization, foreign trade performance and industrial structure for eight manufacturing sectors; share of military expenditures and relative share of industry, agriculture and services sectors.

4. *Labour force and human resources*: indicators on labour market flexibility, mobility and skill level of the labour force and unemployment rates according to the skill level of the labour force.

5. *Education and training*: indicators on education expenditures, schooling rates and indicators of academic achievement.

6. *Financial*: indicators on cost of capital, level of sophistication of financial markets, internationalization of the financial system.

The macroeconomic performance indicators constitute a seventh set of parameters. These include indicators on productivity, unemployment, participation of the labour force, industrial self-sufficiency, income distribution inequality and growth. The parameters have been measured for the 12 countries considered, utilizing a variety of sources. Subsequently, a factor analysis has been performed, followed by a hierarchical classification, leading the typology (cluster analysis).

15. Employment, unemployment and ageing in the West European welfare states

Lars Mjøset[1]

The postwar West European welfare state, which matured from the 1960s, represents a major historical achievement: it has granted the broad masses of citizens social rights and security at an unprecedented level. But not long after these welfare states matured, since the early 1970s, most West European countries have experienced a movement to higher levels of unemployment (Table 15.1). A number of other worrying features of the West European situation have been noted: higher rates of long-term unemployment, an increasing volume of part-time work with little job protection, and an increasing number of people in precarious labour market positions.

The trend is striking: with the exception of the UK (with its very restrictive statistical counting of unemployed persons), all the larger EU countries were in the high unemployment group (more than 10 per cent). Exceptions are getting more scarce: by the late 1990s, among the countries listed in Table 15.1, only the UK, Denmark and Ireland – once in the high unemployment group – have been able to escape back to a medium level (between 5 and 10 per cent), and only Norway and the Netherlands have been able to jump back from medium to low unemployment (below 5 per cent).

What part has the welfare state played in this development? In the following analysis, we explore the relationship between the welfare state and labour market developments, unemployment in particular. Our analysis relates potential welfare state/unemployment links to major processes of socioeconomic transformation since the late 1960s. A brief sketch is given of the way the problem has been treated within comparative analyses by economists and sociologists. In particular, the typology – most thoroughly elaborated in the work of Esping-Andersen – distinguishing a Nordic, a continental European and a US pattern of welfare state–family–labour market interaction is discussed. The typology is shown to be too crude. This is shown in some detail for developments in the 1990s, especially for the case of the large continental European countries. The typology gives no clear understanding of present

Table 15.1   The West European unemployment experience

	Low unemployment <5%	Medium unemployment >5%, <10%	High unemployment >10%
Germany	1973–81	1982–95	1996–99
France	1973–77	1978–84	1985–99
UK	1973–75	1976–81, 1988–92, 1994–99	1982–87, 1993
Ireland		1973–81, 1996–99	1982–95
Netherlands	1973–74, 1999	1975–81, 1987–98	1982–86
Italy		1973–84	1985–99
Spain	1973–76	1977–79	1980–99
Portugal	1990–92	1983–89, 1993–99	
Greece	1973–81	1982–94	1995–99
Denmark	1973–76, 1979	1977–78, 1980, 1984–92, 1994–99	1981, 1983, 1993
Finland	1973–76, 1980–81, 1988–90	1977–79, 1982–87, 1991	1992–99
Norway	1973–89, 1995–99	1990–94	
Sweden	1973–91	1992–99	

Note: Projections for 1998 and 1999; Germany before 1991 refers to West Germany.

Sources: OECD, *Historical Statistics*, various editions; OECD (1998b).

short-term problems of unemployment. For such research purposes the continental European countries should be treated separately. The typology may, however, be relevant in the discussion of longer-term challenges such as ageing, increasing female preferences for educational attainment, and changes in the family structure. The policy conclusions sketched at the end of our analysis concern the relation between these longer-term problems and the present, immediate problems of high unemployment.

1. THE THREEFOLD TYPOLOGY OF WESTERN WELFARE STATES

Owing to the development of the welfare state, the 'Golden Age' period of the 1950s and 1960s may be characterized as the peak era of the nation state, in which the economies of Western Europe became national economies (Mjøset, 1997). The slightly strange term 'nation-welfare state' fits the postwar period well.

There were external influences on all the nation-welfare states of West Europe in the postwar period: increasing volumes of international trade, the role of international institutions such as the UN and the IMF, there was US cultural industry, a Cold War ideological discourse about the free world, and influence from the continuing process of West European integration. But the various states entered the postwar era with different structural characteristics and institutional preconditions. Thus different welfare states developed and, furthermore, each 'national' economy had peculiarities in terms of economic structure, patterns of organization and economic policies. Despite the growing importance of both the EU and Western economic interdependence, the 1950s and the 1960s saw the consolidation of distinct economic–political models.

The old European state system became transformed into a state system of welfare states. Differences between the units of this state system of welfare states should thus be traced both at the economic–structural level, in nationally specific relationships between state and civil society (including the welfare state) and in economic policies.[2] Various terms have been employed as a general label for the constellations that evolved: corporatism, class compromises, Keynesianism, and so on.

The postwar welfare states were all based on older legacies, but they were extended with a more comprehensive aim than earlier: securing social citizenship for all inhabitants of the nation state. Based on the work of Titmuss and Korpi, Esping-Andersen developed an influential typology of postwar West European welfare states (Esping-Andersen, 1990; Korpi, 1983; Titmuss, 1974). With special reference to the composition of social policy spending, the typology distinguished the continental European insurance model, the Nordic citizens' rights universal model and the British residual model. In the Fordist Golden Age period, all three of these models corresponded to the prevailing type of work organization (lifelong, uninterrupted male employment) and a particular structure of the family (organized around the incomes of this one male breadwinner).[3] The welfare states catered for pre- and post-work phases of the life cycle. The pension systems were biased towards the male breadwinner.

Esping-Andersen starts from the degree to which social policy transfer payments are 'decommodified', that is, made independent of the market mechanism. Roughly speaking, in the Nordic case, most of these payments have developed into universal citizens' rights; in the continental case, they are based on occupation-specific schemes: pension rights exist as claims on a myriad of associations (specified according to occupation or otherwise) or on large firms;[4] in the British case, decommodification is weakest, amounting only to a quite low flat-rate payment, while most additional pensions must be covered by private insurance. The US system is even more reliant on market mechanisms, and often Britain and the USA are quoted together as an 'Anglo-American model'.

In various comparative analyses, Esping-Andersen has shown that this typology provides a powerful first cut to a number of contemporary problems relating to the welfare state, including the problem of unemployment, which is of specific interest in this chapter. Importantly, he challenges the idea of a convergence between all West European welfare states, by pointing to non-convergence between his three groups of welfare states since the mid-1970s. But this leads to two interconnected problems. *First*, he is led to exaggerate convergence *within* his groups, and this is particularly a problem with the continental type, which is also the largest one (this group of welfare states caters for a population of more than 200 million, while the Nordic population is about 33 million and the UK about 57 million). *Second*, his analysis of structural processes of transition (from Fordism to post-industrialism; cf. Table 15.2, I.5 to III.5) implies that these are seen as convergent across all three models. But, as just emphasized, the differences between the welfare states must also be related to differences at other levels: economic structures, economic policies and a broader range of state–civil society relations. Relating to these dimensions, however, Esping-Andersen does not proceed in a thoroughly comparative fashion. Besides the rather global conceptualization of changing patterns of economic structure and economic policies, differences in the family structure of different countries are introduced only unsystematically.

As a consequence of these two features, Esping-Andersen fails to relate his analysis to the tradition of comparing different varieties of Fordism in postwar Western Europe. A link to the important institutionalist traditions that focus on 'varieties of capitalism' is missing. In addition, there is also a third problem, namely that Esping-Andersen's framework has no way of incorporating the broader institutional and regional setting, and particularly the growing importance of positive integration in the EU (the regulatory framework of the single market, economic and monetary union). In sum, Esping-Andersen fails to appreciate the historical individuality of both the postwar continental welfare states and the historical individuality of their integration in the EU.[5]

In the following, we shall show that specifications are needed on all these accounts.

2.   POSTWAR TRENDS AND TRANSFORMATIONS

Esping-Andersen regards the transition from the Fordist pattern of industrialization to post-industrialism as a major background against which processes of adjustment in the three models are analysed. Our plea for a link to studies of the 'varieties of capitalism' does not mean that we can escape a similar focus. We need some general notion of the challenges facing the postwar welfare states after the Golden Age.

Studies in the modernization tradition of the 1950s and 1960s (Kerr *et al.* 1962; Bell, 1973) contain many important findings on trends such as tertiary sector employment, the growth of education, the rising level of living, but, even though they show some awareness that these trends were basically limited to the Western world, the temptation to generalize mostly gained the upper hand, and a fully modernized future was envisaged. Sometimes it seems that contemporary prophets of globalization commit the same fallacy. We shall offer no extensive criticism of these views, but we shall pursue a more modest approach: we replace the logic of industrialism with a neo-Schumpeterian emphasis on shifting technoeconomic paradigms (that is, the shift from a Fordist mass production/mass consumption to an information and communication technology (ICT) paradigm), and we replace the logic of modernization with a comparison of *varieties of capitalism*.

On this basis, we can rework some of the aggregate structural changes mentioned by Esping-Andersen (1999a), and add certain others, providing a historical sketch (see Table 15.2) of the postwar period according to dimensions important to our study. In the Golden Age (I), the West European states became rich, and quite a lot of this wealth was devoted to peacetime security for the people, who looked back at the risky inter-war and wartime periods. The following dimensions are important: the welfare state (I.1), the expansion of institutions for (higher) education (I.2), the labour market (I.3),

Table 15.2 Historical background to Western Europe's present challenges

I The Cold War, Golden Age period	II The 1965–75 watershed period	III The troubled period starting around 1975	IV Turn-of-the-century challenges Immediate problems	Longer-term problems
The nation-welfare state 1. Legitimation: the promise of security (economic containment)	*Generalization of welfare state promises	Labour market policies against unemployment. Fiscal crisis of the state. Dilemmas of economic policy management, pushing towards neoliberal consensus	†Rigid welfare state institutions (*Welfare state crisis*)	Preparedness for the 'ageing society'
2. Technocracy/knowledge: higher education	Student revolt	Challenge of industrial and R&D policies in knowledge society (increasing levels of educational attainment)		
Labour market 3. Successful deruralization	Tight labour markets – generalized worker protection	Unemployment (context of deindustrialization different from context of deruralization)	†*Failure to restore full employment (*Labour market crisis*)	
Industrial relations 4. Nationally based adoptions of Fordism	*Labour unrest. Growth of labour protection	Flexibilization. Destandardization of the life cycle. End of lifelong careers and of the homogeneous manual labourer	Crisis of industrial relations	

Technoeconomic paradigm 5. Fordist mass production	Restructuring. Emergence of the new ICT paradigm	De-industrialization. Tertiarization. Challenge of reforming national innovation systems	
Family 6. Mass consumption One-income family. Baby boom	Increasing female labour force participation. End of the extended family. Second sexual revolution	Increasing economic independence of women – more single mothers, single household. Lower birth rates	†Crisis of the family
International dimension 7. 'Embedded liberalism'. Keynesian interventionism *and* expansion of international trade and investments	Monetary instability Breakdown of Bretton Woods	Interaction between financial tensions and burden-sharing. International net-working and mergers between firms. Decontrol of international finance	
West European dimension 8. Founding of the European Union	A first integration offensive: 1972 widening, Werner plan for economic and political union	Integration setback, but leading to more positive integration in the 1990s: single market, 1992; EMU, 1999	

457

Note: The interpretation favoured by neoclassical economists links the two asterisks in column II to the asterisk in column IV, which is the outcome. Esping-Andersen's (1999) analysis, on the other hand, focuses on the interaction between the three 'crises' marked with obelisses.

industrial relations (I.4), the technoeconomic paradigm (I.5) and the family (I.6). Most of the trends that we single out are based on processes at the national level, but some are also traceable to the international or European level, that is economic integration in trade, investments and money (I.7) and the founding of the EU (I.8). International factors have no special priority in this scheme.

Throughout the rest of this chapter, we discuss how various welfare states adapted to these trends and transformations. There *is* a dilemma involved in deriving the 'challenges' in isolation from the development of the various countries, but the general overview of trends should be regarded as preliminary, with no autonomous value except when related to the specific national economic and political constellations. Just as an example, the 'Golden Age' labels cover marked differences such as Germany's retained craft-based manufacturing system and corporatist industrial relations, versus France's more mass production-oriented production regime and fragmented industrial relations. Similarly, for education, there is a major difference between Germany's dual-track system and the less differentiated French system. Despite the fact that our methodological approach in this chapter requires such specifications, we still need such a general scheme to be able to analyse the macrohistorical conjunctures of the last decades. Construction of such trends and transformations actually has a dual character: they are aggregates of developments within distinct models, but they are also constructions that matter in the discourses of policy makers.

Column (II) in Table 15.2 covers the culmination and/or transformation of our Golden Age trends in the watershed 1965–75 period. This was a period of worker militancy, tight labour markets and strong union bargaining power (Buechteman, 1993). At that time, extensive systems of employment protection were institutionalized. They were either made into law or consolidated in collective agreements (II.3). At the same time the income maintenance systems of the welfare state systems were extended (Esping-Andersen, 1999a, pp. 1, 2, 15, 54) and it was only at that time that the welfare arrangements promised earlier (benefit adequacy and universal coverage) began to yield substantial results for the earlier risk groups, such as the aged (II.1). These twin developments – welfare state generalization and labour market regulations – have been the focal point of a series of economic analyses of the relationship between the welfare state and recent labour market problems, as we shall see in the next section.

As for other trends and transformations covered in the historical part of Table 15.2, we shall be briefer. An important part of state policies in the Cold War period was the ambitious programme for upgrading the human capital of Western populations after the late 1950s 'Sputnik shock' (I.2 to II.2). Technocratic views dominated at the time. In the 1965–75 period, complex historical processes gave rise to student revolt among huge strata of the recently en-rolled students. There are interesting lines to be drawn from these developments in the field of knowledge (II.2) to the new technoeconomic ICT paradigm (II.5 to III.5) and to tertiarization, which is again related to the growth of the 'knowledge class' and the new middle classes, but we cannot deal with these here.

As for industrial relations and the technoeconomic paradigms, we have already noted the rich traditions studying the Fordist mass production regime. The tight labour market of the 1965–75 period (II.4) influenced the early restructuring of the technoeconomic paradigm (II.5).

We shall cover the two features of the labour market and the family together (II.3 and II.6). Following Esping-Andersen, we shall analyse the 1950s–60s process of deruralization in order to compare it to the process of deindustrialization since the mid-1970s. In the Golden Age, the labour redun-dancies created by deruralization (I.3) were absorbed by Fordist industrialization. Successful deruralization was due to a large flow of low-skill workers absorbed into well-paid, secure jobs. Polarization and downward wage pressure were avoided.

Let us consider the family (I.6) more closely. Especially in the low-wage strata, many women were moving from agriculture in the countryside to a position as the housewife of a worker in the city. It has been noted that 'working-class families emulated the middle-class norm of housewifery at the very moment that middle-class women began to distance themselves from it' (Esping-Andersen 1999a, p. 28). Since mass production boomed, postwar labour market dynamics fuelled demand for goods and services, and a female labour market opened up in sales and clerical jobs. As an increasing number of women (many middle-class) entered the labour market, demand for per-sonal and social services grew further: 'the one-earner-plus-housewife family appears to have been an historical exception rather than an institution, a fleeting, mid-twentieth-century interlude' (ibid., p. 53). In addition, historical variations in birth rates were important. Birth rates had been low in the 1930s, so the cohorts entering the labour market were not too large to be absorbed. These trends culminated in the tight labour markets of the 1965–75 period (II.3), despite more working women and despite substantial immigra-tion into many West European countries.

We note in Table 15.2 that the Golden Age was a baby boom period (I.6). We also note that the watershed period 1965–75 is marked by the second sexual revolution. This periodization refers to sex outside marriage. The first sexual revolution (mid-18th century) implied the acceptance of intimate rela-tions and premarital sex within certain social groups. The second sexual revolution was the generalization of this phenomenon, beginning in the 1960s (Shorter, 1975, pp. 123f, pp. 168f). But in our context it is more relevant to quote another aspect of that social change, namely that sophisticated contraceptive means (especially the pill since 1965) and diffusion of knowledge about this (again related to the broadening of the educational system) allowed people to engage in sexual intimacy with few worries about the female partner getting pregnant (ibid., pp. 116–18). Thus, with the second sexual revolution, marriage was no longer a guarantee of many children. This topic is linked to the completion of the 'demographic transition' to quite low birth rates. The founding fathers of the postwar welfare state did not think about its future with reference to demography. But, as we shall see, very low birth rates (linked to an increasing age at first pregnancy) emerged in certain countries during the troubled period (III.6). Both the growing number of pensioners and low fertility can be seen as unintended consequences of the welfare state.

Following the watershed period came the 'troubled era' since the early 1970s (III.1). At the basic macroeconomic level, this period is defined by the productivity slowdown and unexpected low rates of economic growth in the Western world (Maddison, 1991; Freeman and Soete, 1994). Furthermore, inflation shot up during this period, an important difference from the preceding deflationary crisis of the 1930s. In the neo-Schumpeterian account, the troubled period is analysed as a period of mismatch between a socioeconomic complex inherited from the Fordist period and the new technoeconomic paradigm (III.5 versus III.1–4, 6–7) based on information and communication technologies (ICT) (Perez, 1983; Freeman and Perez, 1988). The regulationist diagnosis is similar, but with more emphasis on how various institutions and structural economic features of the Fordist model become incompatible (that is, a tension between the various factors III.1–7) (cf. Boyer, 1990; Jessop, 1990).

In this troubled period, irreversibly rising unemployment emerged as a major problem in most of Western Europe (cf. Table 15.1). Analysing the turn into the troubled period, we should avoid reductionist temptations. The situation became substantially more difficult for policy makers, not just because of sluggish growth and inflation, but also because of a more difficult labour market constellation in the period of 1980s and 1990s deindustrialization. That whole context (III.3) was much less favourable to a balance between labour supply and employment growth than the 1950s and 1960s context of deruralization. There was now a flow of workers out of industry (III.5), and female employment was increasing (II.6, III.6). Furthermore, cohorts from the baby boom period (I.6) of the 1950s and 1960s now entered the labour market. In the 1980s and 1990s, the service sector has been able to absorb the employment surplus created by deindustrialization, but, since the growth in female employment and in the supply of new workers is so much larger than in the deruralization period, the scale of labour market problems has become much more severe than in the

Golden Age (Esping-Andersen 1999a, p. 28). The deruralization process is largely finished and nobody expects a revival of primary sector employment, so the major question is the extent to which the service sector can absorb the labour redundancies created by deindustrialization.

The phenomenon of youth unemployment illustrates how important it is to study the new labour market problems in the context of preceding processes of social change. In the 1930s, groups of unemployed youngsters could return to the rural sector. Today, many West European countries have a huge outsider population of young people. Only a minority can go back to the farms. These young outsiders are the children of youth revolution, popular culture and of Golden Age living standards. This has affected their motivation. Young people today have few barriers to seeking help. In a situation of high youth unemployment they may not hesitate to switch from family dependence to state dependence. They have never experienced the need for self-help. Moreover, many of their peers also seek assistance, so there are few social barriers. Finally they own little, so, unlike for an old person, 'welfare' for them does not mean they lose accumulated savings even if they accept social assistance.

Western Europe's nation-welfare states relied on economic restructuring processes that maintained and/or created strong innovation systems (III.5). In the troubled period, all the West European nation-welfare states were exposed to the challenge of industrial policies, competitiveness and the pursuit of conscious policies to strengthen national or regional innovation systems (III.2). There were worries about the emerging competition in manufactured goods, produced by the East Asian NICs (III.7). Coordination of policies at the EU level largely failed in the early 1970s (II.8). Since the mid-1970s, the strategies chosen were in line with the biases established by the national systems of production and innovation. In the late 1970s, several countries attempted to counter deindustrialization with defensive and expansive industrial policies (saving jobs in exposed sectors such as textiles, steel and shipyards). Thereafter the general response was labour supply management. We shall see later how the national responses differed here. There was Britain's labour-cheapening approach, the Nordic public employment-led strategy, and the continental strategy of supply reductions. Let us first return to some general trends.

The new constellation put the welfare states under pressure. We shall see below how welfare state transfers grew (broader coverage, higher benefit levels and more unemployment relief), while the tax base was weakened owing to the situation of slow growth. Economic policy making encountered severe dilemmas as inflation rates were high and tended to destabilize existing institutional arrangements (Notermans, 1993). The inflationary surge broke down some of the state interventionist aspects of the postwar economic

policy management (III.1). The strengthening of neoliberal views on economic policies in most countries must be seen in this perspective. This led
further on to deregulation, dismantling of capital controls and higher interest
rates, which set off a flourishing new sector of financial dealings, including a
greater variety of individualized offers of pension insurance. The EU
deregulated capital flows in 1988. Since the early 1990s, the deregulation of
internal and external capital movements has grown into a major challenge.
Again note that we shall encounter specific varieties of these trends later.

Certain features relating to the development of the family (II.6) in the
1965–75 period parallels the 'student revolt': this is the start (first in Scandinavia) of a middle class-based women's movement, especially based on those
gaining a higher education (II.2, II.6 and III.2 leading to III.6). There are
links here to sociostructural changes such as those termed the 'second sexual
revolution', but we shall only note the significant impact of women's increasing educational attainment, work preference and the higher divorce rates in
family relations. These processes of women's liberation can largely be seen
as a consequence of the evolution of the welfare state (broadly conceived:
including education and health systems, not just social security), especially
its provision of better social security and educational opportunity.

In the 1970s and 1980s, the international system was also marked by
considerable turmoil (III.7). The Atlantic scene was haunted by Euro-Ameri-
can conflicts in questions of trade integration and monetary management,
interacting with contemporary problems of burden sharing within Nato (Calleo,
1987). But we shall not study these more closely here.

In the EU, institutional destabilization and international tensions interacted
with growing positive integration (III.8). In Western Europe, EU efforts at
integration were strengthened after the mid-1980s, partly in response to the
international tensions (III.7). The first major step was the single market
(1992). With the implementation of EMU from 1999 onwards, the EU is
developing towards a system which 'nationalizes' the intra-EU capital flows.

The trends we have sketched are more complex than in standard globalization diagnoses. Some processes (such as the East Asian challenge in
manufacturing products) take place at the level of the Western Bloc (the
OECD area), while other processes stem from the declining importance of
borders within the EU. Note also that globalization often works via the state
(regulatory competition or EU-convergence criteria worsen fiscal restraint) or
via structural change (firms constrained by low-cost competition). Other
trends are directly related to the evolution of welfare states. Ageing populations,
for instance, are not an effect of globalization. They are a demographic
dynamic that can be linked to the welfare state (health systems, living standards). Even termination of capital controls need not be seen as international
forces imposing themselves on the state: they have been interpreted as the

lagged response by economic policy-making elites to the 1970s surge of inflation (Notermans, 1993).

If a general formula is needed, the 'troubled period' may be summed up as a crisis of the Fordist model of economic growth. But, under this general label, it is possible to trace at least four crises, and their main bases can be indicated with reference to our scheme. First, dilemmas relating to the future of the social security commitments generalized in the 1970s–80s (III.1 to IV.1), as well as the more complex question of the impact on work incentives, can be labelled *a crisis of the welfare state*. The unemployment problem emerges as the main indicator of a *crisis of the labour market* (III.3 to III.4). Third, flexibilization, growth of service employment and the end of lifelong careers are indicators of a *crisis of industrial relations* (III.4 to IV.4). In the following, our focus is mainly on the interaction between the two first crises, but, for some of our arguments, we also pay attention to industrial relations. We shall, however, deal more extensively with what loosely – and with no moralistic implications – has been called a *crisis of the family* (III.6 to IV.6).

The far right column of Table 15.2 sketches current challenges. Here we distinguish between immediate and medium/long-term problems. The immediate problem is the challenge to the legitimacy of politicians and the state as they continuously fail to restore full employment. Although the legitimacy of the welfare states is high, there is also concern that the state may not be able to keep its social security promises. Concerning these immediate challenges, one very influential interpretation has throughout the whole of the 1980s and 1990s supported the neoliberal turn in economic policy making: the claim is made, often within a theoretical framework based on neoclassical micro-economics, that the increased labour protection and generalized welfare of the watershed period are the factors responsible for the chronic unemployment (the links between II.1 and II.4, leading to IV.3 are marked with asterisks in Table 15.2). We shall deal with this in some detail in a later section.

But the longer-term challenge is to some extent the opposite. Towards the middle of the 21st century, 2030–40, bureaucrats and insurance statisticians see another crisis, which is linked, not to unemployment, but to ageing. Low birth rates and increasing life expectancy interact to create a future situation in which a too low supply of labour may be a more severe problem than unemployment. Let us now deal in more detail with this trend of rising life expectancy, which is at the core of this longer-term challenge. This feature is related to rising living standards and risk reductions typical of the welfare state, such as health system improvements. Let us consider briefly the crude, but synthetic, indicator of average life expectancy (Table 15.3).

We see that the rise in life expectancy at birth was greater between the 1930s and early 1970s than in the period from the early to the late 19th

Table 15.3   *Growing life expectancy (males, at birth) in selected European*
countries, 19th and 20th centuries

	Early 19th century	Late 19th century	1930s	Early 1970s	1997
England/Wales	40.3	44.0	58.7	68.9	74.0
Germany	—	35.6	60.0	68.0	73.0
France	38.3	40.9	55.9	69.0	75.0
Sweden	39.5	49.0	63.2	72.1	75.7
Italy	—	35.1	53.7	68.9	75.0

Source: Flora *et al.* (1987); *The World Almanac and Book of Facts* (1998).

century. Going by these rough numbers, we see that in the period (1950s, 1960s) throughout which the early postwar welfare programmes (old age pension system, health system) were established, it did not seem that pensioners would live very long after retirement. In a longer historical perspective we realize that only at a certain point in historical time, did the notion of the retirement age become meaningful.

We shall give no comprehensive interpretation of these demographic data. The trend seems secular over the last two centuries, but there is a spurt since the 1930s. Developments in the 19th century are surely linked to the 'hygienic revolution', and the struggle to combat infant mortality. These efforts started as an attempt by the upper classes (the medical profession particularly) to counteract the 19th century's 'risk society', which had been created by rapid urbanization. Later, early social policies, as well as a general rise in living standards linked to phases of economic growth, are important factors too. From the 1930s, there is a long phase of extensive state intervention. Improvement of public health systems was a major feature of the rise of the postwar welfare state.

We now consider data on the distribution of age groups as a percentage of total population.[6] In the second half of the 19th century, the distribution was relatively stable. In the case of Germany, the share of 0–9-year-olds was 25 per cent, 10–19-year-olds about 21 per cent, and then there are declining percentages to about 5 per cent for the 60–69 group and 3 per cent above 70 years. By 1970, this had changed: now 7 per cent of the German population was above 70, and as many as 11 per cent were between 60 and 69. The 0–9 and 10–19 age groups had shrunk to 16 and 14 per cent, respectively. A similar ageing of the population can be traced for the other European countries. More sophisticated indicators are available: an OECD estimate of average years in retirement for men (aggregate for 15 industrialized countries) shows that, in 1960, it was only about two years, increasing gradually to more than ten years by 1995 (OECD, 1998a, Figure A1).[7]

We have provided a messy general sketch of a number of trends, of which we shall later encounter specific national varieties. The sketch is as general as we can possibly allow it to be. It certainly gives no simple and straightforward transition from Fordism into post-industrialism, and even less from modernity to post-modernity. But it provides a point of departure – a map of some kind – from which we can later trace the dynamics of specific cases in specific periods.

3. WELFARE STATES AND THE CHANGING RISK STRUCTURE

A historical periodization of risk structures may be useful in linking and broadening the analysis of the short- and long-term crises. When social security systems first emerged with the spread of the industrial society in the late 19th century, the problems of age and invalidity were the most prominent risks, and they related above all to workers. In all other areas of society, the extended family was effective: family members capable of working took care of both the young and the old, and among artisans and farmers there was extensive family labour.

But given the dismal housing standards and the absence of savings among workers, an ageing worker who lost his job or was made invalid by a work accident would have nothing to rely on to support his family, unless he had a working wife. And if the work accident was fatal, a worker widow with many children had few possibilities of working, and ran a high risk of falling into poverty.[8] As in the most recent decades, there were many children growing up with single parents, but, in the 19th century, this was due to the risks of one parent dying at a young age, not to the (present) risk of divorce. Typically, in Germany, it was expected that, if poor widows or retired workers got support, they should move to the countryside, where living costs were lower (Döring, 1997, pp. 21f).

Risks at a young age were infant death and incomplete schooling, barring the access of young people to basic skills (rates of illiteracy would be the main indicator). For instance, if work was available for children, schooling would easily be displaced. Already in the 19th century, the emerging nation states were starting to address the challenge of schooling, and state intervention in the health sector became an important factor behind the proto-welfare state developments of the late 19th century.

Through the extension of the welfare state in the postwar era, the welfare regime has come to encompass ever-larger parts of the population. This reflects the decline of the extended family: the states of Western Europe now largely manage the education and health systems, supporting the various pension schemes. Thus the welfare states now cover the classic risks of childhood, invalidity and old age. Present pension systems 'replace' the extended family: working age citizens support – through the state as a large and very trustworthy insurance company (either as such or as an 'insurer of last resort' in a system of fragmented pension funds) – children and youth in their pre-work years and the old and non-employable citizens who do not earn wages. 'This three-generation relationship, now shifted to society as a whole, looks like the old relationship between generations in the pre-industrial extended family' (ibid., p. 51).

In the troubled period since the mid-1970s, as we have seen, these commitments to insure against old risks have imposed a heavy burden on the welfare states. But, even worse, the troubled period implies new risks: life course risks now increasingly emerge in youth and prime age adult life (Esping-Andersen, 1999a, p. 42). Table 15.4 relates to the various features of the troubled period, specifying the challenges to the welfare state and the new risks facing employees and citizens. Concerning consequences for citizens, the main point about the new risks is that they strike people in the active, adult part of their life cycle in a situation where escape into a low-cost existence in the rural economy is historically blocked.

The range of new risks stretches from immediate ones to the long-term challenge of ageing. Unemployment may hit young and old. For young people it may block the career they had prepared for through education. For older people it forces unexpected requalification on them, with no guarantee of new employment. There are also new risks facing those who are working: divorce rates have increased chronically, a major sign of the crisis of the family. A divorce process means, not just psychological strain, but (often) the costs of dividing up one household into two.

As more people are hit by the new risks, demands on the welfare state increase, while its tax base is threatened by erosion. The long-term implication, finally, is the already mentioned challenge of ageing. For future employees, it means that they may have to support a much larger share of dependent citizens than earlier. For future pensioners it means that, when they retire, the amount and quality of welfare services and pensions may not be in tune with their expectations.

The various features interact, and the grand challenge, obviously, is to avoid these elements turning into a vicious circle, spurred by a welfare state that responds inadequately to the new risk structure.

Esping-Andersen notes that the postwar welfare regimes were related to egalitarian ideals and risk profiles of the earliest postwar period (that is, childhood and old age). They now face problems because post-industrial

Table 15.4   The new risk structure

Challenge	Consequences for the state	Consequences for employees/citizens
Unemployment (III.3) Social exclusion	Supports/finances unemployment schemes: higher payments strain fiscal potentials. Also pressure on social assistance systems	If long term: severe decline in incomes Blocking of career strategies, especially for youth
Weakened tax base (III.1)	Fiscal deficits, pressure to scale down social benefits	Benefits reduced, lower-quality welfare services
Structural change/ flexibilization (III.4–5, III.2) Internationalization (III.7, III.8)	Challenge to educational, industrial and R&D policies	Destandardization of the life cycle, end of lifelong careers: more insecure careers, danger of skill devaluation, early retirement
Crisis of the family (divorce, single households, cohabitation) (III.6)	Challenge to adjustments in family policies	Problems of adequate incomes. Higher costs of child care. More single households
Low birth rates (III.6)	Long-term political challenge of balancing pronatalist family policies and immigration policies	Longer term: problems arising from an increasing ratio of pensioners per employed

Note: A number of these trends are discussed in Esping-Andersen (1999a, *passim*).

society alters the structure of social risks, while welfare state institutions remain static. Preoccupied with the old risk structure, the welfare state may end up supporting only insiders, those already covered, creating ever more outsiders. There is a 'growing disjuncture between existing institutional arrangements and emerging risk profiles' (Esping-Andersen, 1999a, p. 146).[9]

Esping-Andersen talks about the forces behind the new risks in terms of an 'exogenous shock', like a Trojan horse offered as a 'gift' to the welfare state: inside the horse is 'globalization, ageing and family instability; a simultaneous market and family failure'. This leads to 'a severe trade-off between welfare and jobs, equality and family' (ibid., pp. 148, 173). The Trojan horse metaphor is perhaps not the best choice, given its image of problems emerging as a consequence of a gift from outside. We have noted already that many of the structural changes we are discussing here are actually more or less unintended consequences of postwar developments, partly of the welfare state programmes themselves.

The risk structure, it seems, changes more because of changes internal to various nation-welfare states, than because of globalization. It should also be taken into account that the security produced by the postwar welfare state creates its own internal dangers. For instance, a pressure on the health system is not just due to ageing.[10] Generally, the health system increasingly mends people stricken by the risks linked to a high living standard (way of life, cars, drinking, drugs and so on). Life-style features such as diets, adventurous leisure activities and so on expose people to risks. This relates to the evolution of people's constructions of what quality of life really is. All this can be seen as a specific part of the new risk structure.

An influential general diagnosis of the current situation says that Westerners are now living in the post-modern age. The working class is heterogenized, it plays less of a role as a collective actor than previously. Social movements are today mainly defensive, defending life styles and identities. Employment no longer fosters collectivist attitudes. Instead people with relatively secure employment are polarized against a heterogeneous group of disadvantaged persons, such as part-time workers, single mothers, long-term unemployed and welfare dependants. The role of the family is being reduced; urban society can increasingly be seen as consisting of single households. The growth of smaller, more flexible units (especially in services, practising less hierarchical patterns of organization; the 'organic' firm), rising educational attainment (continuous reschooling) and a greater impact of professions – all these trends point towards individualization rather than collective action. But there is considerable national variation in terms of how important collective actors are and, in the field of economic policies and industrial relations, they are crucial in many countries.

In the postwar period and through the watershed period, the state had coordinated (in varying ways) social security and emerged as the guarantor of the social security and (more specifically) pension systems. The state had been trusted as the 'insurer of last resort', a trust spurred by memory of the inter-war and wartime risks. But in the 1980s and 1990s, the state itself was exposed as overcommitted to social security and not able to counteract unemployment and increasing inequality. In the same period, the memories of the wartime risks faded.

This is an important condition for the launching of private, individualized insurance schemes, indicating lower enthusiasm about general risk pools (Øverbye 1998, pp. 257–9). However, again, we must note that the legitimation of standard social security schemes varies across countries, and it is frequently found that the general risk pools still enjoy very solid popular support. Some scholars envisage a peaceful coexistence of public and private pension schemes at this high level of income (Kuhnle, 1999).

Our sketch of challenges to the welfare state gives no priority to processes of internationalization or globalization; such processes must be analysed together with processes mainly operating at the level of the nation-welfare state. With this in mind, we turn to the period during which West European unemployment exploded.

4. WELFARE STATES AND THE LABOUR MARKET AFTER THE GOLDEN AGE

We have distinguished short-term and long-term challenges. Our scheme (Table 15.2) uses historical periodization to sketch the interaction between the various relevant areas. We shall now see that both the short- and long-term crises take quite different forms when we study the national level. We shall refer to recent discussions of the relationship between welfare state, labour market and family. First, we consider briefly comparisons in terms of two clusters (the USA versus Europe), then we turn to Esping-Andersen's clusters of European countries, and finally we study even more disaggregated groupings.

Comparing the USA and Europe

A number of mainstream economists studying job creation have concluded on a contrast between 'flexible' US and 'sclerotic' Europe. Only in the 1990s was the fact that the US jobs miracle was associated with a dramatic rise in earnings inequalities taken fully into account.[11] In particular, the earnings of low-skilled US workers dropped dramatically. It was thus emphasized that the competitive US labour market has flexible wages, so a relative reduction in the demand for unskilled workers just led to sharp declines in their wages, and a huge supply of low-skill labour was put to work at very low wages. The strata of working poor, with incomes below the poverty line, exploded in the USA. In Western Europe, wage formation is more institutionalized and there are barriers to wage inequality, so that low-skill labour is priced out of the market, becoming long-term unemployed. Thus, while the risk of becoming unemployed is not as great in Western Europe as in the USA, the Europeans who become unemployed, have a higher risk of staying unemployed over the long term (Røed 1998, p. 171).[12] Put another way, in the USA, a large number of people are unemployed over short spells and unemployment compensation is anyway only provided for a very short time. In Western Europe, a smaller number of people are unemployed, but over longer periods.

As long as the evaluation of the US performance was on the whole favourable, the US/West European comparison supported a broadly neoliberal conclusion, suggesting that Western Europe needed more wage flexibility, interpreted as more income differentials; that is, a US model.[13] But the emphasis on polarization in terms of incomes led to doubts about the virtues of the US model. Richard Freeman (1995) pointed to the following problems: first, the explosive rise in inequality leads to dramatic poverty among unskilled workers; second, the lowering of wages for the less skilled did not lead to more employment for these groups; third, although the USA avoided long-term unemployment, that contrast with Europe can be counted as a triumph only as long as US crime rates and prison populations are kept out of the analysis. With this last point, Freeman introduces inter-disciplinarity into the economists' USA/Europe comparison. These differences regarding the rates of imprisonment and incarceration have long been known to criminologists (see, for instance, Christie 1993, 1998).

In 1993, the US prison population was 1.9 per cent of the US male workforce, and 4.7 per cent were on probation or parole. Thus Freeman estimates that roughly 7 per cent of the male workforce (nearly five million people) were supervised by the criminal justice system. He reminds us that the number of people in US prisons and jails (1.35 million) is about equal to the share of long-term unemployed men in Western Europe (Freeman, 1996).[14] This trend was established by the US neoliberal offensive from the early 1980s onwards. If it continues, the USA will by the year 2000 have more of its workforce imprisoned than Europe will have long-term unemployed backed by welfare transfers (Freeman, 1995, p. 70). These groups are in some respects similar: they mainly consist of the less skilled, and they are costly for the state. US criminals certainly have done more harm to others than Europe's unemployed. On the other hand, some of the people on parole are working, and very few are women, while in Europe, both sexes are in the group of long-term unemployed.

In the USA, the liberal state makes unemployment so unbearable that people have to work even below the subsistence level. Unemployment is seen as a pathology; welfare is constructed as disabling. The state's task is seen in moralistic terms, it must 'enable' people to work: entitlements may be useful for winners, but the losers cannot handle them. Thus, besides the unemployed (receiving very little social assistance), there are the many working poor. This US approach above all implies moralism towards young, black single mothers, the main group on welfare in the USA. Many young black males see another option. They try to escape poverty by pursuing a criminal career, often dope-related. There is thus a connection between the massive wage inequalities (the number of working poor) and the explosion of crime: the criminals are mostly less educated men (disproportionately black), and the low wages earned by those less educated who get work is below the perceived incomes from dope dealing and other law-breaking activities in the inner cities (Freeman, 1996, p. 71). But when they do not comply with the laws, they clash with the state monopoly on legitimate violence.

While Western Europe has its insider/outsider problems, the USA transforms quite a share of their outsiders into *insiders,* in the *prison* sense of the word. Europe's outsiders are counted as part of the labour force, while US prisoners are not! The ironic message to the US poor is: work for a very low wage, or go to jail. Continental European outsiders are taken care of by the welfare state, via unemployment relief, early retirement or social assistance. US outsiders are either inside a prison, on parole (that is, supervised by the penal system) or working poor – less often unemployed.

Many ardent neoliberals are promoting the US model, lamenting how the European welfare state undermines the work incentives of long-term unemployed. Such an interpretation is in itself questionable, but even more striking is the fact that the mechanisms whereby the absence of worker protection and the high income differentials stimulate criminal careers and fill up prisons are not mentioned.

Freeman's alternative analysis, however, remains wedded to a dualist comparison of the USA and Europe. Its interdisciplinary orientation is fascinating, and it says crucial things on the historical specificity of US developments. Figure 15.1 provides a stylized view of the long-term implications of the US

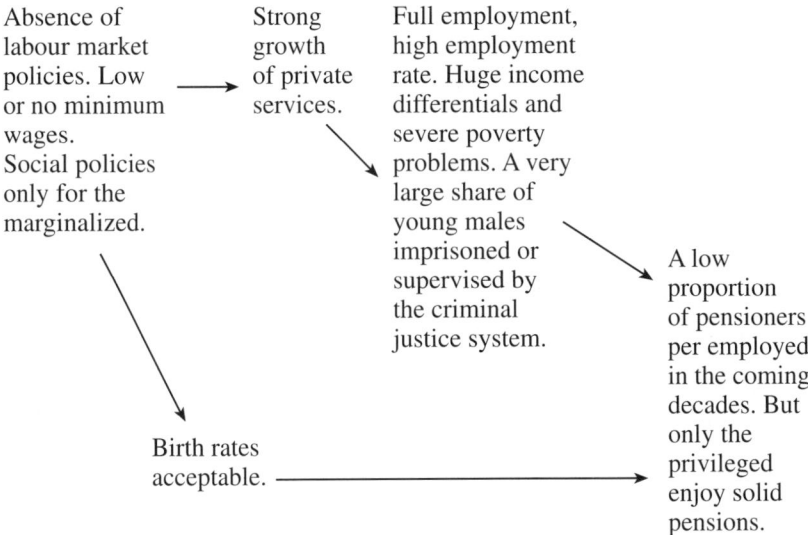

Figure 15.1   Long-term implications of the US strategy

relationship between social policies, employment, birth rates and the number of pensioners. But the analysis tells us little about Western Europe.

It is not difficult to point out the weaknesses of the economists' USA/ Europe contrast as a comparative venture. Their conception of institutional inertia equates continental Europe and Scandinavia, areas with very different development of employment and unemployment (cf. our next two sub-sections). Esping-Andersen's typology (see Table 15.5) provides important perspectives on this difference, and also indicates how some of the US patterns mark the British case in Europe, but with very different outcomes in terms of labour market dynamics.

The UK and Ireland do have poverty problems extensive enough to prompt discussions about 'a new underclass'. While the continental and Nordic welfare states succeed in maintaining some kind of equality, the British one failed on this count. This is similar to the US case, but, unlike that country, Britain had no job-creation miracle.[15] Conservative governments since the late 1970s successfully attacked the union movement,[16] pursued extensive deregulation and privatization of public activities and scaled back the social wage (Esping-Andersen, 1999b, p. 298). Thus in Britain there are increasing earnings inequality, lower social benefit levels and weaker unions. Relatively low labour costs (in comparison with continental and Nordic countries) and an employment structure marked by a quite large share of private service activities, however, did not prevent the level of unemployment from staying at continental European levels (cf. also Esping-Andersen, 1997, pp. 67f).

As in the USA, the problem with the British pattern of development is the proliferation of low-status jobs and the explosive growth of poverty.[17] In this perspective it is not surprising that the British conservative government in the early 1990s opted out of Maastricht's social clause, and that it suggested a full integration of Eastern Europe into the EU with no defence against the low labour costs of these areas.

Given these weaknesses of the economists' USA/Europe dualism, let us see what Esping-Andersen's threefold typology has to offer: we discuss the Nordic model first, then the continental one.

Adjustments in the Nordic Model

In both the Nordic and the continental models, the welfare state still has a concern for equality, and they prevent an explosion of poverty of British proportions. But they do this in different ways. The continental welfare states pursued income maintenance, while the Nordic ones went for work promotion. The goal of full employment had different meanings in the continental and Nordic contexts. On the European continent, full employment for male breadwinners is the crucial goal, while in the Nordic area both males and

Table 15.5   The three welfare models in the post-Golden Age period

Labour market policies	Effect on the creation of service-sector jobs	Relation to prevailing family structure	Main problems as of the 1990s
Universalist Nordic 'social-democratic' model (welfare state-led social services model): priority to welfare state-based solutions (decommodification and defamilialization)			
Supporting full employment by means of active manpower policies.	Creating jobs for women in the public social services. Relatively flexible labour markets.	Two-income, dual-career families dominate. Household security even with labour market turmoil.	High wages in public service sector due to general ethos of income equality. Recent surge of unemployment puts pressure on the compensation system.
Corporatist 'conservative' continental model ('jobless growth'): priority to family-based solutions (decommodification and familialization)			
Early retirement and other defensive strategies.	Job expansion in either public or private services largely blocked. Quite rigid labour markets, minimizing turmoil.	One-income, male breadwinner family persists. The households secure the welfare guarantees.	Insider–outsider cleavage. High youth unemployment.
Marginal 'liberal' British model: priority to market-based solutions (commodification and defamilialization)			
Non-intervention (regarding both the high and the low end of the labour market).	Job creation in private services. Deregulation of labour markets.	Female employment in between the two other models, more part-time work, fewer entitlements linked to part-time work. Cannot fall back on familial welfare, thus more precariousness, inequality, poverty.	Proliferation of low-status jobs, high-income inequalities, high growth of poverty.

females are implied in the understanding of what full employment entails. Esping-Andersen's analysis of the transformation typical of his three welfare regimes in the 1970s and 1980s is summarized in Table 15.5.

In contrast to the continental countries, the Nordic welfare states resorted to active manpower policies, and to extensions of welfare state social services. Thereby they (except for Denmark) sustained full employment and – this goes for Denmark too – promoted strong growth of female employment well into the 1980s. The state acted both on the supply side and on the demand side. There is also an egalitarian tradition: these countries have the lowest income inequalities and wage differentials in the OECD. Egalitarian and consensual incomes policies were part of the strategy.

Labour market policies secured relocation and retraining of workers. Compared to the continent, there was less of a decline in the employment of older males. More important in employment terms, however, was the growth of welfare state services, which sustained full employment and increased female employment, leading to a convergence of male and female employment rates: the dual-career/dual-earner family became the norm. While the continental welfare state remained a transfer state, the Nordic became one that also produced services allowing citizens (especially female citizens) to take up paid work. Education and health are publicly financed: this model provides free social services for families with young children, the old, handicapped, sick, drug addicts, immigrants and so on.

In connection with Table 15.2, the historical specificity of the postwar one-income family was noted. These families, based on the incomes of the male breadwinner with one lifetime career, were at the core of the social insurance systems of the Fordist period. But, from the early 1970s, this male-biased 'Fordist' employment pattern was to some extent transcended in Norden. Many social services require female skills; they attract women, they are secure, well-paid and flexible in terms of working hours. The many women who entered the labour market went into education, child care, health services and so on. The Nordic welfare states increasingly also focused on the active, adult part of the life cycle, partly responding to the new risk structure. Growth of public social services was the only viable source of employment creation, since, in the Nordic area, fixed high labour costs and egalitarian wage policies barred the growth of personal consumer services.

The policy goal of equalizing women's economic status was influential. Female employment in the public sector helped to produce around an 80 per cent participation rate for women in the relevant age group (Table 15.6(b) for 1983). The result is the dual-income family, with the consequence that employment as a share of total population is 10 to 20 percentage points higher than in continental Europe (Table 15.6(a) for 1983). A mix of private and

Table 15.6   Basic labour market indicators, 1983 and 1996

(a)   Total labour force

	Employment/ Population rate		Labour force participation rate		Unemployment rate	
	1983	1996	1983	1996	1983	1996
Denmark	71.8	74.9	79.6	80.1	9.7	6.8
Finland	73.2	62.2	77.4	74.1	5.4	16.1
France	60.2	59.6	67.4	67.8	8.0	12.1
Germany	62.2	64.0	67.5	70.3	7.9	9.0
Norway	77.3	76.8	79.3	80.8	2.5	4.9
Sweden	80.2	72.7	83.0	79.0	3.5	8.0
UK	67.0	71.0	75.9	77.3	9.6	5.4
Italy	55.0	51.3	60.1	58.5	8.3	12.2
Spain	49.5	48.1	59.6	61.8	17.0	22.2
Portugal	69.7	67.2	75.7	72.6	8.0	7.5
Greece	57.5	56.4*	62.4	62.0*	7.8	9.1*

(b)   Women aged 25–54

	Employment/ Population rate			Labour force participation rate			Unemployment rate	
	1983	1996	Change	1983	1996	Change	1983	1996
Denmark	76.8	75.8	−1	84.0	82.1	−2	8.5	7.6
Finland	82.5	73.2	−9	85.8	85.4	0	3.9	14.3
France	61.9	67.6	+6	67.0	77.8	+11	7.7	13.0
Germany	53.7	65.1*	+11	58.3	72.1*	+14	8.0	9.3
Norway	71.1	78.5	+7	73.2	81.7	+8	2.9	3.9
Sweden	84.9	80.1	−5	87.0	85.8	−1	2.4	6.7
UK	60.2	70.3	+10	66.7	74.5	+8	9.7	5.6
Italy	41.8	47.0	+5	45.5	54.8	+9	8.1	12.9
Spain	29.4	41.9	+12	33.3	56.8	+23	11.6	26.3
Portugal	55.8	71.1	+16	61.4	76.8	+15	9.1	7.3
Greece	40.1	49.0*	+9	43.8	55.0*	+11	8.6	10.9*

Note: *1995 (note that the German figure will be blown up by addition of the 'new *länder*', the former East Germany, which had a higher share of women employed than former West Germany).

Source: OECD (1997, Appendix Tables B and C).

public arrangements made for good provision of child care, facilitating life in two-income families.

Referring back to our discussion of the new risk profile, this indicates that the Nordic strategy to some extent catered for the new risks better than the two others did. In contrast to the US strategy, the Nordic countries avoided poverty traps (above all, a large proportion of single mothers work[18]). In contrast to the continental strategy, as we shall see, contraction of employment and low birth rates was avoided, but it should be noted that the Nordic strategy creates a segmented employment structure: men in private business and women in the public sector. Many of the jobs in the female/public sector are part-time jobs, but then again the rights connected with part-time work (pension rights and so on) are more extensive than in most other countries (Ellingsæter, 1992).

Well into the 1980s, the Nordic response, implying government service growth, active labour market policies and consensual bargaining, was a relatively virtuous circle. Figure 15.2 provides a sketch of how the Nordic strategy had implications for the longer-term challenge of ageing societies.

Adjustments in the Continental Type

The continental strategy was to maintain full employment via labour force reduction, using pension insurance to reduce the industrial labour force by means of early retirement. In addition, repatriation of foreign workers, discouragement of female employment and reduction of hours for those working influenced the supply of labour. The model implies a concern for equality mainly via income maintenance and job protection, but not via work promotion (the Nordic variety). There was no attempt to scale back or reform social security (Esping-Andersen, 1996, p. 77). The strategy was not sufficient to prevent rising unemployment from the mid-1970s onwards.

This strategy was sometimes portrayed as a high-skill, high-qualifications strategy: a most efficient core of workers would be retained, their adjustments would secure strong competitiveness, and the resulting productivity dividend would restrain budget costs, so that the income guarantees offered to unemployed and early retirees would not lead to escalating budget deficits. But such an outcome did not follow. There was still unfavourable development of employment and unemployment.

Esping-Andersen links the continental European strategy closely to the social insurance system. According to the Bismarckian model, workers, employers and the state would pay contributions to funds, which would be managed on a 'strictly actuarial' basis. After World War II, the systems were changed to pay-as-you-go systems, and the influence of the state varied. In the watershed period 1965–75, strict labour market regulations were enforced

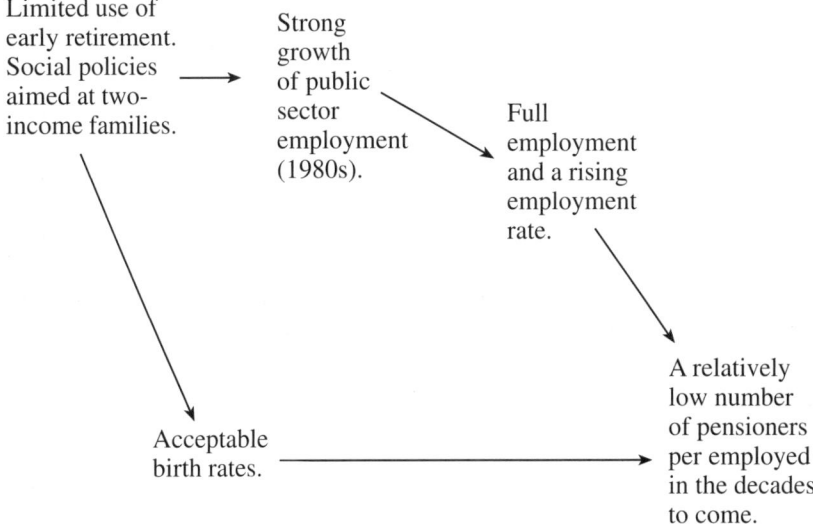

Figure 15.2 Long-term implications of the Nordic strategy

and social security systems were universalized. Already this added pressure on the state to make increasing transfers to the social security system. Then the post-industrial employment challenge made itself felt. The strategy of early retirement was the main response.

The historical roots of the early retirement response, however, are older than the mid-1970s economic setback. In France, a state commission dealt with the problem of old age poverty in 1960, suggesting measures to increase the employment of older people. In 1971, this line of argument was reintroduced, and the prospect of an ageing population was explicitly introduced into the debate. However, unions were strongly against such measures, and employers were ambivalent, too, since early retirement – given new techno-economic challenges – soon proved to be a way to lay off old and inadequately skilled parts of the workforce. The main point for the employers was that firms would themselves decide. Various policy measures in the early 1970s still retained 'insertion' of older workers as a goal, but the real and unexpected consequences were much more in line with the employers' preference for firm-driven early retirement. Early retirement in France developed from about 13 000 workers a year in 1969–71, then rising by about 20 000 more in 1972–5, passing 100 000 a year in 1977, 200 000 a year in 1982, and going to 400 000 in 1982 (Mares, 1998, ch. 6.2.1; Guillemard, 1991).

A similar story may be told for Germany. A 1969 federal social court decision opened the possibility that older workers unable to find a part-time job would be entitled to an occupational disability pension. Thus some of the risks catered for earlier by the system of unemployment insurance were shifted to the pension system (which also implied less of a stigma for the workers affected), with no matching increase in contributions. In addition, a 1969 Employment Promotion Act ruled that an unemployed worker could receive pension benefits starting at 60 (official retirement age 65). Many large firms exploited this act in the following way: they laid off workers at 59, supplementing their unemployment compensation for a year (only large firms had the means to do this). At 60, the worker would then be entitled to age pension. In 1967, about 4000 60-year-old workers received such a pension; in 1984, 76 000 did (Mares, 1998, ch. 6.2.2; Jacobs *et al.*, 1991). As in France, firms were allowed to have it their way. In the Nordic area, early retirement never attracted such numbers. The magnitude of early retirement seems to be a major difference between the Nordic and continental welfare regimes.[19]

In both France and Germany, the 1980s saw a response by the state and the unions to these privileges attained by firms. In both cases, explicit references were made to the danger that the practice of early retirement would fully destabilize the social insurance systems. But both the German Early Retirement Act of March 1984 and the French *Ordonnance* on these matters of March 1982 had been softened by the respective governments' need to compromise by means of side-payments and concessions to the unions (Mitterrand government) and the Christian Democratic social committees (Kohl government). The economic impact of this legislation was small (Mares, 1998, ch. 6.3).

Furthermore, many infrastructural costs, such as health system costs, carried solely by the state increased in line with the rising living standards. As for the social security contributions, the states now cover between 10 and 20 per cent of the financing, the employers 50–60 per cent, and the employees 30 per cent.[20] These fixed labour costs hover around 50 per cent (of the wage bill, married worker with two children) in the core continental countries (Belgium, France, Germany, Italy, the Netherlands) (Esping-Andersen, 1996, p. 72). Early retirement and unemployment worsened the fiscal problems: workers retire earlier, and, as we have seen, on average, they live longer (cf. Table 15.3), not least thanks to the improved health services. The state also had to increase transfers to unemployment funds.

Trying to spell out the differences between the Nordic and continental welfare states, researchers have focused on the role of non-wage labour costs. These studies avoid the problems of the economists' aggregate comparison of the USA and Western Europe. The role of the tax structure has been investigated by Scharpf (1999), who retains Esping-Andersen's threefold typology.

The standard measure of non-wage labour costs is the *tax wedge*, which measures the difference between the cost of employing an employee and the consumption that can be financed from the wages paid. The higher the wedge, the higher the difference between these costs and consumption. But such data bring out only the crude contrast between the USA and Western Europe, giving no striking differences between the Nordic and continental countries.[21] A comparison of tax structures is much more telling. Table 15.7 shows that a large share of continental taxes are levied as social security contributions on the wage bill, while both the Nordic and Anglo-American groups are marked by high personal and corporate income taxes. This means that in the Nordic case, a large share of the (large) tax volume is paid by all citizens and companies. In the Anglo-Saxon case, a similarly large share of the (small) tax

Table 15.7   Comparative tax structures (percentage of GDP, 1995)

	Impact on employment	Nordic	Continental	Anglo-Saxon
Total taxation	High share facilitates growth of public services	High	Medium	Low
Social security contributions (payroll taxes)	Little impact on manufacturing, transport, financial services, more on low productivity services where impact cannot be shifted onto workers or consumers (wage close to reservation wage)	Low (Dk) Average	High	Low
Consumption taxes (on goods & services)	Reduce demand for all products, but the most heavy impact is on low productivity services	High (Dk) Medium	Medium	Medium
Personal & corporate income taxes	Relatively low impact on employment, as these are progressive, and not collected on wages below the basic income.	High Very high (Dk)	Low	Medium to high
Impact of tax structure on employment growth (especially in private services)		Friendly (esp. Dk)	Unfriendly	Friendly

Notes: The cases analysed in Scharpf's project are: *Nordic*: Denmark (Dk), Sweden; *Continental*: Austria, Belgium, Germany, France, Italy, the Netherlands; *Anglo-Saxon*: Australia, New Zealand, UK. As for the impact on employment, Scharpf's judgement is reproduced.

Source: Scharpf (1999, pp. 10, 15).

volume has the same source. Scharpf argues that, in both cases, the tax structure is 'friendly' to employment creation, while the continental structure, a high share of payroll taxes in the (medium-sized) tax volume, is unfriendly to employment expansion (Scharpf, 1999). Esping-Andersen has also pointed to this role of the high payroll taxes.

To specify these employment effects, let us consider the expansions of the service sector. Data on France, Germany, Sweden and the USA over the 1960–87 period reported by Elfring (1992; cf. Table 6.1 in the present volume) show that there are no major differences for production and distribution services. Personal services are about constant for the continent, declining for Sweden, and at a high level in the USA. The social service share is increasing in all cases, but levelling out in the USA. Growth was quite large in Germany and Sweden, and Sweden reached an astonishing level in 1987 (more than half of the total service sector employment). Table 15.8 shows how health services are the crucial Swedish peculiarity.

Consider the difference between Denmark and Germany. The overall difference in service employment is 13.9 (52.6–38.7). Health and community services alone make up half of this difference, education adds another 2.4, commerce 1.3, transport, telecommunications 1.9, and business services 1.4. Even more disaggregation gives a similar impression. Considering (public)

Table 15.8   Service sector employment disaggregated, mid-1990s

	Germany	France	Denmark	Sweden
Commerce	8.9	8.2	10.2	8.6
Hotels & restaurants	2.0	2.0	2.0	1.8
Transport, telecommunications	3.4	3.8	5.3	4.7
Financial intermediation, insurance	2.3	1.9	2.4	1.3
Business services	4.0	5.1	5.4	6.7
Public administration	5.5	5.7	4.7	3.6
Education	3.3	4.5	5.7	5.2
Health and community services	5.8	6.2	12.7	14.1
Other services	3.3	2.5	3.8	3.8
Domestic services	0.2	1.3	0.2	0.0
Total services	38.7	41.4	52.6	49.8
Agriculture and industry	23.9	18.9	22.9	20.5

Note: NACE classification; share of total population between 15 and 64 years of age. French and Danish totals are not exact due to rounding and omission of 0.7 per cent employed in extra-territorial organizations.

Source: *Employment in Europe* (1997, p. 97), annual review.

spending on services for the elderly and disabled (per cent/GDP) in 1993, we find that it was 6.4 in Sweden, 4.4 in Denmark and 3.0 in Finland, compared to around 1.2 in Luxembourg, the Netherlands, France and the UK, 0.7 in West Germany and less than 0.5 in Ireland, Greece, Italy, Belgium, Spain and Portugal (Scharpf, 1997b).[22]

In the Anglo-Saxon cases, as we have seen in the case of the USA, there is strong expansion in private low-productivity, low-skill services. In the Nordic cases, public sector (female) service employment expanded rapidly until the late 1980s. Continental welfare states are in the worst of both worlds. One could have expected these to have medium employment growth in both private and public sectors, producing high overall service sector employment, but they have low employment growth in both these areas. The continental countries are left in a vicious circle. If unemployment increases, or early retirement spreads, contributions to social security funds have to be increased, possibly crowding out even more jobs in low-productivity services. But still the 'deficits' of the funds may increase, and the state (depending on the specific systems in function) will have to cover these deficits. Thus the fiscal position of the state also worsens.

Both Esping-Andersen and Scharpf see here a main mechanism behind the continental jobless growth and low employment rates. But there are certain differences between their accounts. Scharpf emphasizes that the contemporary unemployment problems are due, not to the size of the welfare state, but to its 'characteristic structure and mode of financing'.[23] The institutional reforms needed are not such that destroy social policy support for disadvantaged groups. The rigidities that matter 'are the rigidities of political systems that are incapable of efficient reforms, rather than alleged rigidities of the labour market' (Scharpf, 1997a, p. 22). The continental countries, claims Scharpf, destroy more employment opportunities in the private sector than even the Nordic welfare states.[24] As we have already seen, Scharpf explains this with reference to the financing of continental welfare benefits primarily through payroll taxes, which signifies a more employment-unfriendly tax structure. Denmark is held up as the purest Nordic counter-case to the continent's vicious circle, with 83 per cent of total social spending financed by general tax revenues (Scharpf, 1997b, p. 15).[25]

But Table 15.7 on tax structures indicates an internal Nordic difference. Denmark's financing of its social security system is nearly completely tax-based, whereas both Sweden and Norway finance welfare via payroll taxes. Denmark has a very high rate of VAT taxation, which should depress employment creation. Sweden has one of the highest rates of payroll taxation among the 13 OECD countries studied by Nickel and Bell (1995, Fig. 5, p. 60). This study showed no significant relationship between unit labour costs and payroll taxes. The reason is that payroll taxes in the long run are shifted onto employees (ibid., p. 59). The same goes for income and excise taxes. Nickel and Bell conclude from this that across-the-board cuts in taxes on employment are not a relevant policy proposal. The more relevant recommendation is tax reductions (or subsidies) on unskilled workers. But if payroll taxes have been shifted onto employees, the only expansive consequence of payroll tax cuts is a rise in take-home pay, which, at least in the British case, will reduce the rate of unemployment compensation (ibid., p. 60).

The Nordic experience creates problems for Scharpf's blaming of payroll taxes: Norway's and Sweden's labour market performance was satisfactory with high payroll taxes until the late 1980s, as was Denmark's in the 1990s, without such a system of taxation. Scharpf could defend his explanatory preference for payroll taxes by restricting the conclusions to the 1990s, claiming that Sweden's dismal performance in the 1990s must be linked to the financing of social insurance through payroll taxes. But Norway would still be a counter-case. Scharpf might explain that exception with reference to Norway's oil wealth, but then why should one not apply a multi-factor explanation also in the other cases? This is in fact what he does. Comparing Sweden and Denmark more closely, he notes that many of the social services provided to Danish families imply means-tested co-payments, while, in the Swedish case, these services are exclusively reliant on tax revenues, with no link to income differences (Scharpf, 1999, p. 21). Finally, Denmark has low job protection, but unemployment benefits are provided very generously (90 per cent wage replacement over five years), so unions accept adjustments. Sweden, in contrast, has a traditional kind of job protection: jobs are protected by strong unions.

But the reference to payroll taxes understood as a specific percentage (of total wage costs) may be misleading. In fact, the institutional framework surrounding social security contributions may matter more. In the continental cases, social security has a legal definition. They are regarded as property rights, and there are very strong legal entitlements to expected benefits. There are passing references to this from both scholars. Such systems, claims Scharpf (1999, p. 23), are not easily reformed. Similarly, Esping-Andersen (1996, p. 68) points to the 'unusually broad legitimacy' enjoyed by continental social security systems, since such contributory social insurance gives a 'sense of individually earned contractual rights'. Even the elites think the system is adequate, only the super-rich go for private solutions. Thus, unlike Scandinavia, Britain and the USA, the continental models have not seen political anti-welfare state revolts.

But, mostly, both Esping-Andersen and Scharpf refer only to the magnitude of payroll taxes, and Esping-Andersen also quite often refers to the more or less fragmented nature of social insurance funds in the continental case. The continental problems are primarily related to the fiscal crisis of the state (later reinforced by EU convergence criteria), interacting in a vicious circle with high non-wage labour costs, which results in a low level of service provision in both public and private sectors. In further research, their brief remarks on the organization of social insurance should be expanded on in a comparative framework. In most advanced welfare states, parliament determines pension age, benefits, coverage and so on, but in the continental systems, this determination depends on delicate balances with employers and unions, that are running – in various ways – the many funds of which the social insurance system consists. In Germany, for instance, parliament can only decide on these matters after formal consultations between unions and employers. In France, there are individual administration units (employees/employers organizations). While superior decisions on contribution rates and benefit amounts are made by the parliament/ government, the state has to administer the system with due attention to the existing contractual arrangements between social partners (Charbonell, 1994, p. 122).

The Nordic countries are in this respect statist contrasts. In Norway, Denmark and Sweden, employers and workers are not formally represented on any of the social insurance boards. The funds have less autonomy from the state, and the parliament is rather sovereign in adjusting the terms of the system (Denmark is an extreme case in point). Opposition to these decisions becomes part of the general political discourse, not part of complex corporatist tinkering *vis-à-vis* a multitude of funds. This point is important also because it allows us to link the analysis of the welfare state more closely to comparative industrial relations. The need to establish such a link is increasingly realized in welfare state research.

When Scharpf argues that systems which rely more on general taxation are less resistant to cutbacks and means testing, this decision-making feature seems the crucial one. In addition there is the economic argument that general taxes are levied on larger parts of the population than payroll taxes. Scharpf's analysis is above all focused on the decisions of employers and firms. Esping-Andersen's analysis provides much more detail on the state and the family. He emphasizes the differences between the continental transfer state and the Nordic service-providing state.

Among the three welfare state types, implies Esping-Andersen, the continental one is the most inert one, since it is still tightly wedded to the old risk structure. He holds that the continental difference to Norden is a consequence of the Bismarckian welfare state pattern: the focus is only on risks faced by a single (male) breadwinner unable to support his family through full-time work. The fact that developments in the recent decades have left the state with a huge transfer burden leaves it with little to finance public services. All its resources are absorbed in transfer payments (disability, retirement, unemployment) while social services that would complement/compete with the social functions performed by daughters, mothers and wives are disregarded. Thus the Nordic creation of well-paid, part-time public sector jobs (primarily for women) is not seen in the continental welfare regimes. The Nordic strategy implied jobs for women in branches of the public sector which were partly related to new risks: old age homes and hospitals related to ageing, kindergartens and after-school arrangements for children. We have seen that health, community services and education account for larger shares of Nordic employment. At the same time, this growth of female employment provides women with returns to their education, and serves as an insurance in case a family crisis (divorce) occurs. The dual-income family becomes the rule and employment rates stay high. The Nordic states have more generous maternity leave arrangements. These multiplier effects are much less marked in the continental case.[26] The fact that the continental model does not provide as much of this as the Nordic ones is fundamentally related to a familialist orientation, a resistance towards fully leaving the extended family behind. It resists the Nordic (and Anglo-American) implication that a larger proportion of females may do paid work: women should rather work at home, taking care of children and the old.

Thus, even if unemployment levels had been the same on the continent as in Norden (note that in the case of Denmark, unemployment had continental proportions in the 1980s), the continental welfare regime produced *jobless growth*, in contrast to the Nordic *growth with job creation* in the 1970s and 1980s. In the longer-term perspective, this leaves the continental welfare states in a more difficult position than the Nordic ones when it comes to tackling future problems of the ageing society. Figure 15.3 (to be compared with Figures 15.1 and 15.2) gives a stylized portrait of this dynamic.

The continental response to 'post-industrial' pressures is family-based solutions: the welfare state bureaucracies decide not to invest in the provision of public sector jobs, and families remain (to a larger extent) one-income families, so that they actually produce the necessary services (child care and tending for the elderly at home). They have also opted for the early retirement solution. It is here, in these policy packages, the way the tax incomes are spent, and not in the rates of payroll taxation, that the main difference *vis-à-vis* the Nordic area lies.

The fact that the Nordic countries do not spend that much on pensions is related to the lesser reliance on early retirement. The conjunctural link between the system of financing, structure of the system and decisions on early retirement separates the Nordic from the continental models. If an employee quits two years earlier, this costs the state a lot of money. This makes more of a difference than how much the state pays and how the system of financing is organized. Table 15.9 shows that, in terms of effective retirement age in 1990,

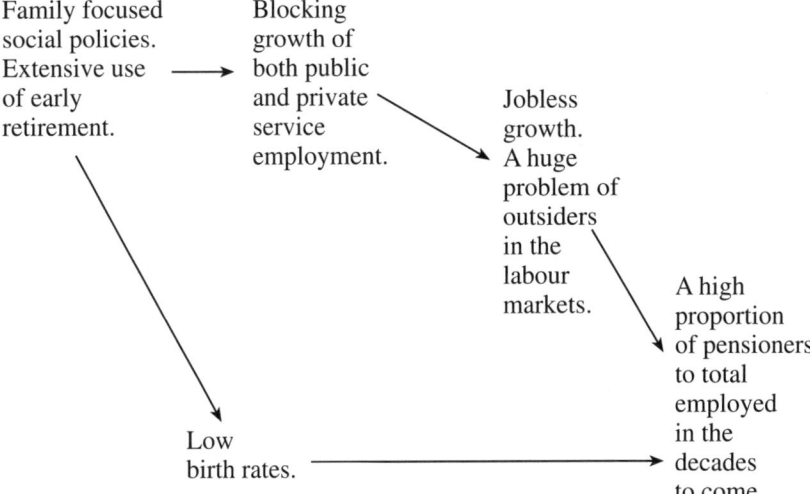

Figure 15.3   Long-term implications of the continental type

France and Germany are at the low end, although one Nordic country, Finland, is at about the same level.

Esping-Andersen emphasizes that, in the continental countries, Catholic political forces were always committed to traditional familialist social policies: young women might gain a brief spell of formal work experience, but, once married, they should turn to household work, taking care first of chil-

Table 15.9   Retirement age

| | Effective | | | |
	1950	1990	Change	Statutory
Denmark	64.2	62.6	−1.6	67
France	64.4	59.9	−4.5	60
Germany	63.7	60.8	−2.9	65
Italy	63.9	61.1	−2.8	65
Spain	65.5	62.5	−3.0	65
UK	64.4	62.4	−2.0	65
Finland	—	60.9	—	65
Sweden	—	63.1	—	65

Source: Bovenberg and van der Linden (1997, pp. 176f, Tables 6.6 and 6.7).

dren and then of the elderly. This factor interacted with the state's huge transfer burden, blocking extensive growth of public welfare services. Sluggish job creation in private personal services was counteracted by self-employment (family labour, women who work in family businesses) and informal (black market) employment, but this was not enough to solve the unemployment problem. Female labour supply was discouraged, not only via family orientation in social policies, but also by the tax system.

The result for continental Europe has been jobless growth, high unemployment and the emergence of an insider–outsider cleavage in the labour markets of the core EU states. The insider labour force of mainly males (with high wages, job security and entitlements) is shrinking, and the outsider population consists of young people, early retirees and women, who depend either on the family or on the welfare state (social assistance). This insider–outsider dualism reinforces the traditional Fordist employment pattern: insiders rely even more on the orderly career life-cycle model, shortened from a 16–65 to a 20–55 age span. Ever more people depend on the family wage. Unemployed youth depend on family help and/or on social assistance (if available, as eligibility depends on prior employment). Also older long-term employed depend on family help, and/or social assistance, if they are not granted early retirement.

These differences in policy orientation were particularly striking in the 1970s and 1980s. They were also important for longer-term problems. We have noted that, as a consequence of health service improvements created by the welfare state, people live longer. As a consequence of broader access to schooling, broader groups of young people reach higher levels of educational attainment. Given this trend, as well as a more general change in attitudes, young women wish to work more (see the increase in labour force participation from 1983 to 1996 in Table 15.6b). Demand for female careers increases. The demographic consequence is delayed family formation, and thus low birth rates (stylized in Figure 15.3). The continental model 'reproduces family responsibilities but at the expense of family formation' (Esping-Andersen, 1997, p. 10).[27] The women of the southern European countries, as Esping-Andersen has phrased it, are on a 'fertility strike'.[28]

With the ageing of the population, social security becomes increasingly expensive, blocking (as already noted) employment creation.[29] The Nordic countries were at least able to push female employment rates high up, thereby preparing for a ratio of pensioners to employed that will remain more favourable than the continental European countries when the 'ageing crisis' strikes in the decades from about 2020, the longer-term problem noted above. This is the message when we compare Figures 15.1–3. Discussing early retirement, we mentioned that state planners and employers in France and Germany were already well aware of the problem of ageing in the early 1970s. Quantitative projections are hard to come by, but if we consider the purely demographic dependency ratios together with the present employment rates in the continental countries, the picture clearly looks bleaker there than in the Nordic area. There are also internal differences. It seems, for instance, that Germany is worse off than France.[30]

For young outsiders in the continental model, 'their fathers' high earnings and job security have become prime obstacles to their own chances of achievement' (Esping-Andersen, 1999b, p. 315). These groups are provided with a minimum of protection, mostly means-tested social assistance. Ironically, the older citizens of the continental model save quite a large share of their pensions, and these savings may eventually be inherited by their children (cf. Attias-Donfut, 1996), but this will be a mixed blessing, since this inheritance must be regarded as one of the reasons these people had to experience long spells of inactivity in their youth. Thus the continental regime implies a much more pronounced equality/jobs trade-off than the Nordic regime. This is not necessarily due to rigid worker protection, since these rigidities may themselves be regarded as the result of the need to protect – given strong familialism and low levels of female employment – the main breadwinner against labour market risks (Esping-Andersen, 1999a, p. 131).

The dilemma is the absence of acceptable alternatives to early retirement: families are then crucial. As in the liberal British case, it is an aim not to increase state influence. But, unlike it, the continental strategy avoids throwing families into poverty: instead the main breadwinner is allowed to retire earlier than planned. This solution is backed by both unions and the median voter, generating a national coalition against change (ibid.). According to Esping-Andersen (ibid., p. 141), it is typical of 'familialistic political economies' that 'the common good matters less than family welfare'.

Our survey has shown many strong features of Esping-Andersen's three-fold typology. Should we not retain this tidy analysis of basic structural changes, which generates three major welfare state responses?

5.   WHY THE HOMOGENEOUS CONTINENTAL MODEL CANNOT BE RETAINED

The answer to the question just posed depends to some extent on what aspects of the welfare state we are studying. In the following, we shall show that Esping-Andersen's typology does not provide a good guide to the immediate problems of unemployment. There seems to be no way of escaping the more conjunctural, national model-oriented approach we suggested above.

Esping-Andersen's interpretation has weaknesses even regarding the 1970s and 1980s, some of which will be noted below. The main approach in this section, however, is to combine criticism of the typology with an account of trends in the 1990s, a decade which in macroeconomic terms was marked by an early economic crisis, followed by a jobless economic recovery with unprecedented unemployment in a postwar perspective.

A main contrast between Nordic and continental European models is the higher Nordic employment/participation rates, even in times of economic setbacks. Does this contrast persist in the 1990s? The changes between 1983 and 1996 are given in Table 15.6a. In France, there was no increase. In Germany, there was a slight increase, to the tune of 2–3 percentage points. Italy decreased somewhat. The Nordic contrasts, however, now do not fare much better: Norway is roughly unchanged; Denmark, being the success story of the 1990s, has some increase; Finland and Sweden have significant reductions (about 10 and 3 in Finland, about 8 and 4 in Sweden calculated from Table 15.6a). Esping-Andersen's contrast persists, but the differences are now smaller, given the deeper crisis, especially in Finland and Sweden. His stylized facts here still hold true.

Given its priority to family-oriented solutions, the continental model has problems with new risks. One of these is the danger that young people do not get jobs when they finish education. Such an underlying logic should produce high youth unemployment, while it should remain low in the Nordic cases. But even in his own publications, Esping-Andersen is well aware that the largest continental economy, Germany, does not fit into this picture (Esping-Andersen, 1997, p. 67; 1999a, p. 41). Since 1987, Germany has consistently recorded the lowest youth unemployment in the OECD, and in 1983–6 it was only beaten by Austria and Japan (two other German-like welfare states) (OECD, 1992, Table 2.19). Again it seems that we must look outside the narrow spectrum of welfare state arrangements to explain this. No comprehensive explanation can be given here, but Germany's dual-track educational system, with a well-integrated track for vocational training, as well as coordinated employer strategies, might be further investigated.

Germany's peculiar performance with respect to youth employment can be linked to the skill hypothesis that we presented above. Empirically, the hypothesis that low-skill labour is a larger share of West European than of US unemployment has not been borne out. If the skill argument was to be the key, we would expect that the UK, which has greater earnings inequality than the continental countries (in the following we refer to Germany only),[31] would have less unemployment among the unskilled than Germany. In an analysis restricted to the male part of the labour force, Nickel and Bell found this not to be the case: unemployment rates for male unskilled are similar in the USA and Germany, and highest in the UK (Nickel and Bell, 1996, p. 305). Their tentative answer to this paradox is that the German system of innovation (which includes the dual-track educational system) is stronger than the British one.

The distinct German pattern – low unemployment for younger people, but increasing for age groups above 40 – is one of three patterns of unemployment by age in the 1990s, distinguished in Reyneri's recent comparative study (Reyneri, 1999, pt 5).[32] The second is the southern European pattern, most marked in the case of Italy, but applying also to Greece, Portugal and (partly) Spain: high unemployment for the youngest people, declining up to the age of 40; from there on, rather low unemployment for all older age groups. Finally, there is a UK profile: high unemployment in both the youngest and the oldest age groups, with lowest unemployment at mid-age (prime age). France (with Belgium) can be located in-between the Italian and British models, so it has high unemployment for the youngest, but also quite high for the various older age groups. Thus the classification of Germany together with France and Italy is in this respect most unfortunate.

Apart from the German case, Reyneri's study supports the more general point – found in many studies (Therborn, 1986; Esping-Andersen, 1990) – that high employment rates counteract employment discrimination. The lower the employment rate, the more youth is discriminated against (Reyneri, 1999, pt 5). The same conclusion applies with respect to gender discrimination: it is lower the better a country is at providing jobs for its citizens. Where jobs are few, men have privileged access; where jobs are many, women have some parity of access. Women looking for a job are not really considered unemployed in southern Europe (ibid., pt 4).[33] This fits well with Esping-Andersen's general typology, especially the contrasting of high Nordic and low continental employment rates. But, as we shall now see, other aspects of gender differences do not fit.

In Esping-Andersen's view, the Nordic welfare states support the employment of women. Table 15.6b, however, shows a considerable catching up by the large continental countries in the period 1983–96: increases in employment (participation) rates in the core age group are 6 (11) in France and 11 (14) in Germany. If the participation ratio may be taken as a proxy for the motivation of these women to work, we see that this motivation is increasing even more than their factual employment (employment rate); thus rates of unemployment for these women are quite high. The increase is also strong in Britain. The two Nordic problem cases record significant decline – minus 9 (0) for Finland, minus 5 (minus 1) for Sweden – while Denmark has little change, and Norway obviously had still more jobs opening for women in that period, with an increase of 7 (8).

As we have seen, Esping-Andersen implies a mechanism whereby both the low birth rates and the low participation rates of women are connected to the weak capacity of continental welfare states to provide public service infrastructure (kindergartens, old age homes and so on). But a case-oriented comparison of France and Germany shows that there are considerable differences between these two countries. France seems to match Sweden where family policies are considered: day-care provision seems not underdeveloped. France has little of the 'Children, Kitchen and Church' pattern that prevail in Germany (Schultheis, 1996; Théret, 1996, p. 446). Again this is something that Esping-Andersen notes himself in passing (Esping-Andersen, 1997, p. 67; 1999a, pp. 82, 93).[34]

Furthermore, differences in birth rates turn out to fluctuate quite a bit. It may be that the differences emphasized by Esping-Andersen for the 1980s are less clear in the 1990s. Italy, Spain, Greece and Portugal are really a low fertility league, joined by Germany. But France rather clusters with the Nordic countries. In the late 1990s Swedish birth rates had declined to 1.5 (Statistics Norway, 1999, p. 76). Again Esping-Andersen's continental regime turns out to be heterogeneous. Of the two largest countries in the group, only Germany fits regarding the slow growth of service employment and a somewhat marginal role for women in the labour market, while France fits in terms of very high youth unemployment. It would seem wise to search for different mechanisms ('underlying logics') at work behind the development of unemployment in the two cases.

Studying more closely the composition of job seekers, Reyneri finds that, in Greece and Portugal, young women predominate, whereas in Spain and Italy, young people of both sexes form the largest share. Furthermore, in southern Europe, *first job seekers* dominate. In Italy and Greece, more than 40 per cent of the job seekers are women seeking their first job. Had it not been for the Spanish policy of providing temporary jobs for young people, the same would be the case in Spain.

This analysis indicates that youth unemployment is a particularly grave problem in southern Europe (Reyneri, 1999).[35] In the core continental countries of Belgium and France, prime-aged, female job seekers are the most important group. In a more 'Atlantic' group (UK, Ireland and the Netherlands), prime-aged men are the most important segment. In France and Germany, job seekers are mainly prime aged, and they are mainly seeking a new job after leaving or losing another. In France, women predominate in this respect, while in Germany there is no gender bias. It is thus common to Germany, the Netherlands, the UK and Ireland that job seekers are predominantly prime-aged men. This implies that they are, to different degrees, protected by the welfare state: they get either unemployment benefits or social assistance. In the southern European group, in contrast, unemployment benefits mainly go to a small group of male breadwinners,[36] while many of the job seekers are young and/or female family members. (We shall look further at this difference in the next section.) These nuances, which must be important for the fine-tuning of policy measures, disappear if welfare state/labour market links are studied only with reference to the mechanisms traced in Esping-Andersen's broad continental group.

Concerning long-term unemployment, Italy and Spain combine relatively high unemployment and relatively high shares of long-term unemployment (more than 12 months, predominantly affecting women). Only Ireland is worse off. In this respect, Germany and France (together with Portugal, Denmark and Belgium) have somewhat 'easier' types of unemployment; that is, lower shares of long-term unemployed, somewhat lower unemployment rates. Even in this case, Italy does not belong in the same group as Germany and France.

We noted above that the differences between certain labour market indicators in the Nordic and continental cases are no longer as clear-cut in the 1990s as in earlier decades. The main challenge here is to explain the explosive, but unequal, development of employment (and unemployment) in the Nordic area in the 1990s. We shall not go into details about this, but it seems unlikely that these effects can be explained by the nature of the Nordic welfare state. Denmark has as generous a welfare state as Sweden, but still that country was much less affected by the economic setback of the early 1990s. What seems to matter here is the much stronger impact of the late 1980s financial deregulation in Sweden (as in Norway and Finland) compared to Denmark. The timing of economic policies was also crucial; it was most unfortunate in Sweden and Finland, more fortunate in Norway and Denmark. As for the 1990s, restrictions on tax increases were imposed in various ways in the Nordic countries in the early part of the decade. Politically, public sector growth evened out, private sector service activities were expanding, as was even manufacturing in the odd case of Norway. Finland and Sweden in the early 1990s saw a decisive halt to public spending: the public debt is a major problem (there were signs of this in Sweden even in the early 1980s). In addition, factors relating to the economic structure must be taken into account: the structural inertia represented by the impact of the huge Swedish industrial combines, the greater flexibility in the Danish industrial structure, oil in the Norwegian case, and the implosion of eastern trade in the Finnish case. What explains Nordic diversity in the 1990s is not just the structural traits, not just oil, but also the timing of adjustments (Erixon and Mjøset, 2001).

In sum, we conclude that Esping-Andersen's continental regime is a hybrid. It sometimes looks like a construction composed of Italian (or southern European) social relations (as specified in the next section), coupled *not* with specific structural traits of the respective economies, but with a general model of the Fordism/post-industrialism transition, a transition which, unfortunately, fits only the most industrialized countries, such as France and Germany (and perhaps northern Italy). The Mediterranean countries (especially Spain, Portugal and Greece, and also southern parts of Italy) did not develop Fordist industrial structures comparable to those of France and Germany. We shall see below the importance of this, as well as of certain state structures, for southern European developments.

6.   THE SOUTHERN EUROPEAN WELFARE STATE MODEL

The debates and criticisms relating to Esping-Andersen's typology have shown that a southern European model needs to be distinguished.[37] The core countries of this type are Italy (the most industrialized case, the regional leader and paradigm for the others) and Spain; but Greece and Portugal should also be included.[38] This model implies a very different version of the insider–outsider relationship compared to the northern parts of the continent. Its dualistic, polarized provision of social protection implies very generous protection (pensions) but only to a (male) *core* of the labour force. This core of insiders consists mainly of male breadwinners, 'hyperprotected beneficiaries', covered by the 'garantismo' which implies higher replacement rates than anywhere else in Western Europe (Ferrera, 1996, p. 19). As we have seen, the main job seekers in these countries are women and youth seeking first jobs. In addition to these job seekers, the underprotected outsider group consists of people in the irregular, non-institutional labour markets of the informal economy. As in most other continental countries, together with Britain, there are no minimum income schemes. The southern European countries have the highest incidence of household poverty in Western Europe.[39] The continental welfare states to the north, in contrast, have a less extreme insider–outsider cleavage, and lower rates of poverty. Two features, however, modify the southern insider–outsider gap.

First, the southern family serves as a 'social clearing house'. It is crucial for each family to have at least one family member in an insider position. The extreme would be the private sector urban couple with no children, having two insiders and no caring obligations and expenses. First job seekers are the largest share of the unemployed; they live with their parents[40] and the unemployment rate of the heads of households is very low. Thus their fathers most likely have either work and/or a pension. Most of the youngsters seeking work are unemployed, but not socially excluded. At the political level, they show no eccentric new political inclinations (Esping-Andersen, 1999b, pp. 308, 314). The 'welfare state' is in a sense located in the families. Demographic dynamics are sluggish, and single-income families prevail. The level of education is low, and the social status of jobs is strongly emphasized. Given the dominant paradigmatic role of the male industrial worker or civil servant, there are considerable barriers to part-time work. The further south we go, the fewer part-time/temporary ('atypical') jobs we find, and the more black or grey economies utilizing family labour. In these southern countries, part-time workers in the white economy enjoy very little job protection and accumulate few, if any, pension rights.

There is a 'southern dilemma' relating to the recent rise in educational attainment and the still quite backward economic development (the share of managerial/professional/technical workers in total employment is still only about 27 per cent, compared to about 35 per cent in the more northern countries) (Reyneri, 1999). Studying job seekers and levels of educational attainment, Reyneri finds that, in Italy, Spain and Portugal, job seekers with high- or medium-level education often wait for a job while living with their parents. Educated youth do not accept unskilled jobs. There is the perceived risk that if, one starts in a low-level job, one will get stuck there. This vicious circle is peculiar to the Mediterranean–Southern European case: low upward occupational mobility and a social status approach to jobs.

The second feature relates to the state and its capacities. Especially in rural areas, public sector jobs are crucial. Such jobs serve as a substitute for regional policies. As already noted, late industrialization and a large proportion of agricultural jobs in the labour force have marked the economies of these countries. Esping-Andersen has in fact himself noted that these countries have particularly grave problems because deruralization and deindustrialization coincided from the mid-1970s while, further north, deruralisation was historically first (Esping-Andersen, 1999a, p. 25). Here Esping-Andersen's own remarks undermine the homogeneity of his continental model.

To understand the peculiarities of the southern welfare states, the nature of their states must be considered. As Ferrera (1996) points out, these states are 'soft', they have little capacity to manage social developments and they have little control over the welfare institutions. The state apparatus is embedded in a 'closed, particularistic culture', marked by patron–client relationships which may be regarded a 'historical constant' in the region. Political parties are major actors. They have considerable control over welfare benefits and use this control for purposes of electoral manipulation. The mechanism is not the indirect political business cycle, known in all OECD democracies, but a direct exchange of individual votes for individual benefits: *political clientelism*. The labour market outsiders (especially in rural, agricultural areas) constitute a market for clientelism: state transfers supplement inadequate work incomes and votes are given in return. Parties are mass patronage parties, and their clients get job opportunities in welfare bureaucracies or assistance in attaining subsistence from welfare bureaucracies (disability or unemployment benefits). These parties are highly sophisticated political machines, penetrating all layers of the welfare administration, or bypassing these administrations by means of special commissions with discretionary power to provide single benefits (either directly or through unions in an intermediary role) (cf. also Mouzelis, 1986). Although, for example, disability pensions have been used counter-cyclically with some discretion even in the north, in the south this systematically conditions the strategies of most outsiders (Ferrera, 1996, p. 29). In Italy, the number of invalidity pensions rose from just above one million in 1960 to 5.4 million (nearly one in every ten Italians) such pensions in 1982 (ibid., p. 26).

Clientelism played an important role in smoothing two transitions. The first was the transition in Spain, Portugal and Greece from an authoritarian to a democratic regime since the mid-1970s. In this case, clientelism secured support for the new (socialist) parties. Second, it facilitated the transition from agriculture to industry. In the case of the Spanish countryside, the number of really unemployed is probably only 10 per cent of those receiving benefits. In such cases, the occurrence of unemployment can in no way be seen as a loss of legitimacy for the ruling government, but rather as a power resource exploited by the government in office. In the early 1990s, the Spanish Socialist Party (PSOE) had almost full control of the Spanish countryside (ibid., p. 28). In Italy, this system led the country into a major political crisis in the mid-1990s.

Both the Nordic and the northern continental states had completed the 'Weberian' transformation of their states. Germany was in fact a pioneer and the model for Weber's theory of rational legal dominance. Thus, as Ferrera (ibid., pp. 30f) emphasizes, welfare rights are here 'embedded in an open, universalistic political culture and a solid, weberian state, impartial in the administration of its own rules'. In contrast, the south is marked by the absence of public sector modernization 'prior to the mass expansion of welfare programmes'. Thus clientelism did not whither away as the dictatorships fell (in Spain, Portugal and Greece); it was strengthened and pushed to new levels of sophistication.[41] Features interacting with this development were the survival of a family-based parochialism, economic dualism/backwardness and the strength of the Catholic church.

We have already claimed that, by including Italy with Germany and France, Esping-Andersen smuggled the transition from Fordism to post-industrialism into the southern European area. We have now seen that processes of structural change have been different in the north and the south. The relation of family structures to these processes is also clearly specific to the south. For the analysis of unemployment patterns, it is crucial to note the 'paternalist' aspects of the southern European situation: the life of young and female unemployed is highly integrated into their families. Furthermore, Esping-Andersen's continental model blurs the crucial difference between a 'Weberian' bureaucracy, indicating quite strong state capacities (typical both of northern continental and of the Nordic states) and the soft south European states, penetrated by party-based clientelism. For broader analyses, France should be analysed as a separate case, while Italy may be counted with the southern European group. Still, in more specialized studies, one may use the typology flexibly. In some respects (for example, as regards youth unemployment and poverty) we understand the French case by placing it in-between Germany and the southern European cases. But in other respects (family policies, fertility) it is more similar to the Nordic group.

In sum, Esping-Andersen's claim that there is a general underlying logic typical of the whole continental area from north to south is not convincing. At most, the southern European group seems to contain some of the 'familialist' features of the larger continental group, but in a setting marked by different processes of political and economic development. This means that future conjunctures of national specificities and international forces may lead to different challenges and 'logics' within the broad continental group.

7. WELFARE STATES AND EMPLOYMENT AT THE TURN OF THE CENTURY

If we focus on the above-mentioned immediate challenge of countering unemployment, it seems that Esping-Andersen's continental type is of little help. We have seen above how the composition of the unemployed and of job seekers differs in the various countries in the 1990s. Specifically, we could refer to the unemployment problems of the two largest of the continental countries. French unemployment has been a chronic problem over several decades, possibly linked to tight fiscal policies and a hard currency approach, while German unemployment has fluctuated more, exploding recently in connection with the reintegration of the citizens of the old GDR into the West German economic system and welfare state (Czada, 1998; Ganssmann 1997).

As for the longer-term problem, however (the complex of social changes related to ageing and the expansion of service employment) Esping-Andersen's analysis of the continental model does point to important mechanisms. Furthermore, his threefold typology is well suited to discussing the overall visions (emphasizing the state, the family or the market) in current discussions on the future of West European societies.

Rather than suggesting a reordering of OECD countries into four, five or more welfare regimes, we suggest that Esping-Andersen's 'welfare regime' focus should be more closely linked to analysis of the distinct varieties of West European capitalism. In the following concluding notes, we argue further along this line, with a particular view to the current and future problems of the continental welfare states.

In Esping-Andersen's view, the continental type welfare regimes are those most strongly wedded to the 'old' structures (the traditional reliance on the family to solve social problems) as well as to the Fordist industrial structure and its corresponding risk structure. Thus, in debates on reform options, different social groups promote features from the two other types against the continental models.

We have questioned the US ideal as an alternative for Europe. Freeman holds that increasing US earnings inequality was not a necessary condition for the US jobs miracle: low aggregate real wage growth was enough. Thus, contrasting the conclusions in the OECD *Jobs Study*, Freeman suggests that Europe should *not* go for 'wage flexibility, US style'. There are options that are more in harmony with the prevailing West European social structures. As for promising policy options, Freeman suggests overall wage moderation negotiated by unions/employers' federations (incomes policies) and subsidized payroll taxes, as well as such wage inequalities that spur low-skilled people to move up the skill ladder (Freeman 1995, pp. 64, 72).

Such reforms may influence job creation at the low wage end of the service occupations. The institutional patterns, which counteract the growth of working poor on the European continent and in the Nordic area, will remain intact. Squeezing the worst layer of low-wage unskilled jobs has been an intended effect of the Nordic and European welfare states. What, then, about the public parts of service employment? We have seen that 'underdevelopment' on this count is the distinguishing feature of continental labour market developments when compared to the Nordic cases. Esping-Andersen links this to the difference between the transfer-oriented and the service-producing welfare state. The Nordic models imply extensive defamilialization, so the state produces a number of services that allow mothers to work outside the household. Above we provided a comparison of the virtuous and vicious circles of the Nordic and continental cases (cf. Figures 15.2 and 15.3).

This contrast holds at least between 1975 and 1990. It is true that, since 1990, the virtuous circle no longer applies generally in the Nordic area. As we noted, this seems not related to a general tax ceiling, but to varieties of capitalism within the Nordic area, including the more lucky timing of economic policy decisions in the cases of Norway and Denmark. If Sweden experiences a fiscal squeeze parallel to that of Germany, the challenge is still different, since Sweden is at a much higher level of employment and has a much larger health/community services sector. It is not the case that Sweden's problems in the 1990s prove that the Nordic model is unsustainable. The continental welfare regime, in contrast, has counteracted high female employment and favoured early retirement, and is thus less prepared to face the new risks.

The absence of a Nordic-type growth of public services in the continental area is not just an effect of isolated features of the welfare regime (the type of financing, that is payroll taxes, or the structure of the system: fragmented, occupation-oriented or the like). Above, we have seen this in the light of policy choices and, given that choices have generally been compromises, unintended consequences of the chosen policies. The continental policy of early retirement seems to have been a crucial triggering factor behind the different paths of the continental and Nordic labour markets in the 1970s and 1980s. Interestingly, the starting point of these policy choices was not the unemployment problem, or the problems of globalization, but worries about lower labour force participation of older workers during the watershed period: what we have called above the long-term challenge. The overall result of the policy choices – interacting with other features of the models – has been lower employment/participation rates, lower shares of service employment in total employment, and worrying prospective pensioner/employment ratios. The different policy choices in the Nordic countries have made these better prepared for the longer-term challenge of ageing.

Should the continental countries try to learn from the Nordic example? It has been argued that, for the continental countries, the Nordic option is gone. There are fiscal pressures, and the Nordic model requires either substantial tax increases or reduced welfare transfers. There is in continental Europe no large organized demand for additional public services, while public opinion is clearly against tax increases or reduced welfare. Business and neoliberals go for the US model. Unions and other political parties defend the existing institutional framework. Nobody advocates a Nordic model.

This may well be right, and we shall not deny that the Nordic models are facing difficult internal challenges. But if it is the case that the Nordic models are better prepared for the longer-term challenges mentioned above, we may still try to specify the continental problems in the light of the Nordic experience. This is all the more relevant since there are indications that the increasing educational attainment of young women leads them increasingly to choose work outside the household. If the continental welfare states are not able to get out of the jobless growth trap by creating more employment in social services, problems related to 'caring failure' may escalate. Since the longer-term problems are important in our discussion, we shall focus on younger generations, rather than on those who are at present early retirees.

Thinking back to our discussion of contemporary trends linked to the new risk structure, the following developments seem to interact, creating a pressure towards defamilialization: first, the reliance of employment creation on the growth of service jobs; second, the higher educational attainment and increasing preference for work among younger women; third, the increasing share of single households out of all households.

At the high-skill end of the service occupations (often indistinguishable from parallel activities classified as industrial), the knowledge economy leads to an increase in the strata of young employees, male and female, who live alone, work long hours and postpone family formation. We can assume that these ideals may also influence the behaviour of young people working in lower-paid service occupations. The sociological label 'individualization' can be related to such (but also to other) developments.

If these trends prevail even in the continental and southern European area, birth rates will remain critically low[42] and the risk of 'care failure' by young and middle-aged towards their ageing family members will increase. Already in the watershed period, the most industrialized countries in the continental group chose to invite foreign workers rather than to employ women. Thus they already have a multicultural labour force. A continuation of this strategy can counteract many of the economic problems resulting from low birth rates, but since projections indicate continued high unemployment in a transition period before the full effects of ageing are felt, such a strategy for the moment has many problems.

The Nordic approach has been to extend public services – kindergartens, pre- and after-school arrangements, extensive health services and old age home services for the elderly – whereby the state prevents 'care failure'. It is often argued that such huge groups of public sector employees are a heavy toll on the state and on private business, too. This is seen in the Nordic area as a challenge for incomes policies. Furthermore, it has been emphasized that there may be certain economic benefits of having a huge service sector. New ICTs may have *some* labour-saving consequences in health and community services. Sophisticated technological and institutional innovations may develop as innovative entrepreneurs interact with large public health sectors. If successful, these innovations may later be the basis of viable export production. However, health services (especially for the elderly) are a sector in which there are major limits to rationalization. As is often pointed out, caring activities need time and commitment. Service personnel must have social skills, while the technical skills required are often not so important.

We considered above younger generations that are working. Let us now consider young unemployed people. The present situation is historically unprecedented: the young are a new risk group. Young people were also at risk in the 1930s, but we have already noted that, at that time, many simply migrated back to rural agriculture, and even the most motivated would not necessarily get the education they wanted. The present situation differs, not just in terms of a much higher level of living for the broad masses of citizens. The vast majority of young people in Western Europe today have scarcely experienced direct poverty; their social exclusion is relative, as the huge social science literature of relative poverty shows. The liberal cure, the threat of starvation if a low wage position is not accepted, does not work. Young people mostly get the education they want, but, after deruralization, a return to agriculture is out of the question. Not just the young aspiring professionals with crowded time budgets and no children, but also non-starving drop-outs, are typical of the present knowledge economy.

Young people are no longer traditional in their values. They grew up in a rich society. Many of them also have the prospects of enjoying generational transfers, and probably already do. For young people with a job, such transfers may increase inequalities, making wealth more cumulative. For young unemployed it may be a relief. In this sense, there is a pressure on the family owing to the fact that there are few public arrangements to assist the young unemployed.

We have noted that young people without work seem to have few reservations about receiving social assistance if they are eligible. In all countries that experience high youth unemployment, there is thus a turn from passive to active welfare policies concerning these groups (Lödemel, 1997). Young people are considered able to work, they are able to choose, but they get support based on their citizenship. They get entitlements, but have not earned their rights as nationals. This is made into a moral issue. In the UK, there is 'welfare to work' and in France 'activation of passive expenditures'. In the USA, there is a rhetoric about the enabling state, and in Norden, there is a traditional work orientation. These are packages of policies that further labour market participation and give more balance between rights and obligations.

This should be analysed with reference to the relationship between the short-term (unemployment) and the longer-term challenges (ageing). It is obvious that, the lower the youth unemployment, the better for the country. Low youth unemployment minimizes the risk of wasting generations. But, given some amount of youth unemployment, the challenge is to counteract social exclusion, since, in the longer term, projections indicate that not unemployment, but bottlenecks may be the problem. Much of this hinges on the educational system. The question of what skills to nurture is a difficult one. It is easy to say that the knowledge economy requires ICT-related skills, but it is also the case that many health and community services require many social, but not so many technical, skills.

Educational reforms should address this dual challenge. First, youth should be provided with many and interesting educational opportunities, in order to minimize the temptation to drop out of the system. Second, reforms of the educational system should take into account the dilemma of the present service economy: it requires both social skills for welfare services and technical, ICT-related skills for business services.

Let us return to the question of privately provided versus publicly provided services. If the Nordic public service option is irrelevant for the continental countries, what about the voluntary sector? In the voluntary sector, social work came out of idealism and ideological conviction, often related to religious motives. If this idealism is maintained as the basis of the system, the voluntary sector is vulnerable to secularization, which is clearly involved when young women get more education and expect to work outside the family. Furthermore, if the voluntary sector is given more important tasks by the state, the pressure of professionalization and for decent wages will be felt.

The classical alternative is found in the liberal regime. The general line is that private, not public, activities should compensate defamilialization. These will be profit-oriented. Such market provision will be linked to systems of (supplementary) private pensions. But such systems are liable to produce huge problems of inequality as large strata of the poor simply cannot afford to establish insurance against these risks.

For this brief diagnosis, we have used the Esping-Andersen types strictly as ideal-types. It may be that the most important future developments will be various hybrid solutions: civil society voluntary associations, or private providers, may cooperate with the state. As for production of social services, the state may establish the basic framework, while there is competition between public and private providers. Nordic authorities have experimented with such privatization in order to expose public services to competition. Here there may well be important lessons to be learned as to how clever organization of such private–public interfaces may increase service employment without producing masses of working poor. For the continental welfare regimes facing the ageing crisis, job creation in the social services is as important as job creation in the business services.

NOTES

1. I am grateful for comments on earlier versions by Ådne Cappelen, Jon Erik Dølvik, Chris Freeman, Andreas Føllesdal, Kåre Hagen, Ton Notermans, Pascal Petit, Luc Soete, Einar Øverbye, and to Trine Stavik for research assistance.
2. For analyses which focus on one or more of these dimensions, see, for example, Piore and Sabel (1984), Boyer (1988), Castles (1993), Mjøset (1987).
3. For the economic and institutional analysis of the Fordist age of mass production and mass consumption, cf. Boyer (1990), and for the analysis of the relationship between the Fordist employment pattern and the welfare state, cf. Esping-Andersen (1990).
4. Note that our labels are political–geographical ones: Denmark is in the Nordic group, not in the continental European, and we only deal with Western Europe, mostly only with the EU countries in that region. Later we use the terms 'Southern European', 'southern' or 'Mediterranean', referring only to the Southern European EU members. Note also that we use the terms 'type' and 'model' interchangeably.
5. In his most recent book, *Social Foundations of Postindustrial Economies* (1999a), Esping-Andersen defends his typology with some quite interesting qualifications. The critical points mentioned here, however, are not discussed.
6. Such graphs are available in Flora *et al.* (1987).

7. This is an average figure; it seems that, in the Nordic countries, expected years in retirement did not change much during 1960–95.

8. Döring (1997, p. 21) notes the high risks of suicide or criminal careers among worker widows.

9. In a sense, this is a mismatch, parallel (and related) to the one traced by neo-Schumpeterian economists (Perez, 1983; Freeman and Soete, 1994) between the new technoeconomic paradigm (one crucial force behind new risks) and the set of institutions that served so well the earlier growth phase.

10. British nutrition scientists, for instance, recently reported that obesity was increasing at a worrying rate: 20 per cent of British women and 17 per cent of British men are now classified as obese. In the USA, which pioneered the mass consumption/car culture constellation, the corresponding figures are 40 and 30 per cent. (Reported on *Sky News*, 27 May 1999.)

11. For instance, Walwei and Zika (1997) find no connection between looseness of redundancy regime and employment growth.

12. We shall return to the empirical evidence for the skill hypothesis below.

13. This was reflected in the policy conclusions to the OECD (1995), *Jobs Study*. For a thorough criticism, see Crouch (1998).

14. Considering the proportion of adults per million of total population in prison in 1990, the USA is at 5236, Australia and New Zealand are at around 1000, France at 743, the United Kingdom at 566, and Italy as low as 180 (Scherer, 1997, Table 1.11).

15. Total employment in the USA grew at an average of 2.5 per cent between 1973 and 1979 and at 1.6 per cent in 1979–90 (with Canada the highest in OECD). Corresponding numbers for Britain were 0.2 and 0.5 per cent, roughly similar to the continental members of the OECD (OECD, *Historical Statistics* (1995), Table 1.6).

16. Union density (trade union membership as a percentage of wage- and salary-earners, employed members only) declined from about 50 per cent in 1980 to 39 per cent by 1990. Although the change was dramatic, even in 1990 the UK was above Germany (33 per cent) and France (10 per cent). Considering collective bargaining coverage, however, the picture is very different: above 90 per cent in both France and Germany in the 1980s and 1990s, while the decline was from 64 per cent (1980) to 47 per cent (1990) in the UK (OECD, 1994, Tables 5.7 and 5.8).

17. For poverty rates, see note 39 below.

18. In Sweden, the participation rate of mothers with small children (below six years of age) is higher than 80 per cent (Esping-Andersen, 1997, p. 77). The activity rate of single mothers (according to data from the Luxembourg Income Study, LIS) is about 80 per cent in Denmark and Sweden, 50 per cent in Germany, 45 per cent in the USA, and 30 per cent in the UK (Esping-Andersen, 1999a, p. 152).

19. For an analysis in terms of the three models, see Esping-Andersen (1990, pp. 150–53).

20. There are, however, nuances: in the Netherlands, employers pay, not 50 per cent, but about 25 per cent of the social security contributions. (The Netherlands is sometimes classified with Denmark.)

21. In 1991–2, the US tax wedge was below 40, while the average for OECD/Europe was 63 (Lazar and Stoyko, 1998, Table 1).

22. Scharpf (1997b) argues that continental public sector spending on services for the elderly and families with children are so low (about 25 per cent of the Nordic countries') because health care is provided on a fee-for-service basis by private sector doctors/hospitals (some non-profit). We cannot in this chapter engage in detailed comparisons of the health systems of the various countries.

23. The percentage of GDP absorbed by welfare state (social) spending is Sweden, 37; Continent, 25–30; the USA, 15 (Scharpf, 1997b).

24. This is indicated in the comparison of Germany and Sweden in Table 15.8, but Scharpf supports this conclusion also from other sources.

25. Cf. Walwei and Zika (1997) for a similar perspective. Wage-related costs, they argue, are crucial for putting the long-term unemployed and other disadvantaged groups, that is, low-productivity, unskilled, back into work. Their conclusion is that the impact of social

protection on the labour market depends on how that protection is organized (defines its costs) and financed. Who picks up the bill for social insurance is decisive. Simulations show that, for Germany, financing via fuel tax would be beneficial. However, the authors note that reorganization could not solve all of the unemployment problem.

26. Again there are differences: on the spending side, Italy devotes 50 per cent of its welfare budget to old age pensions, compared to 30 per cent in Germany (Scharpf, 1997b).

27. Reference is to 'countries like Italy and Spain'. There are clear parallels to Japan's welfare state here; cf. Goodman and Peng (1996). In Japan, 65 per cent of the aged live with their children's family; in Italy and Spain, that share is 40 per cent, in Sweden and Denmark less than 5 per cent (Esping-Andersen, 1997, p. 77).

28. Considering changes in the total birth rate between 1965 and 1990–95, we find two groups: the Nordic group (Denmark 2.6 to 1.8, Finland 2.5 to 1.8, Norway 2.9 to 1.9, Sweden 2.4 to 2.0) and a German/southern group (West Germany 2.5 to 1.3, Italy 2.6 to 1.2, Spain 3.0 to 1.3, Greece 2.3 to 1.4, Portugal 3.0 to 1.5). But France is in-between at 2.8 to 1.7. (*Sources:* for 1965, ILO, 1989; for 1990–95, United Nations, 1997.)

29. Cf. the overall rise in life expectancy (Table 15.3 above). Referring to the mid-1980s situation, the World Bank estimated the life expectancy at retirement (numbers given are for males first, then for females): France 19/24, West Germany, 14/18, Spain 15/18, Sweden 15/19 (World Bank, 1994, Table A.10).

30. Demographic dependency ratios are based on the assumption that all citizens in the 15 to 64 age group are economically active and contributing to the support of dependent population segments. For such ratios, see OECD (1996, Table A3). Projections of the elderly dependency ratio have Germany as the highest of all OECD countries: 49.2. France is at 39.1, which is similar to the Nordic countries. As for the total dependency ratio by 2030, Germany is at 75.1, only surpassed by Switzerland, while France is at 67.9, again similar to the Nordic countries. But, as noted several times, present employment rates are higher in the Nordic area. In this long-term perspective, pushing up the employment rate is a challenge for the continental countries. The point that projections of dependent citizens on the working population were more favourable in France than in Germany was discussed in connection with the 1993 French social insurance reform (cf. André, 1996). A good indicator for the future development of the system is contribution rate projections. Such projections provide an aggregate estimate (for employees and employers together) of the contributions needed to finance the pension scheme in the future. In the mid-1980s, French authorities published a best-case scenario (defined by birth rates, labour force participation and unemployment) which predicted an increase from about 19 per cent of gross salary in 1990 to 24 per cent by 2010 to 31 per cent in 2030. The worst-case scenario had 26 per cent in 2010 and 40 per cent in 2030. Other projections also included the effect of ageing on pension expenditure, arguing that from about 1985 to 2025, contributions would have to increase by 170 per cent or benefits would have to be halved. Given the French 1993 pension reform, new estimates indicated that under a best-case scenario (1 per cent growth of salaried employees, 1.5 per cent average real wage growth) contribution rates between 1983 and 2010 would rise by 2.73 percentage points, against 8.23 under the old system. A less favourable scenario had 7.26 against 12.45 (Bonoli, 2000, ch. 5).

31. In mid-1993, male D9/D1 earnings ratios (highest compared to lowest decile) were as follows: Germany 2.3 and stable; UK 3.3 and still rising; USA 5.6 and still rising (Nickel and Bell, 1996, p. 305).

32. All Reyneri's estimates are based on Eurostat's labour force surveys covering the 1986–95 period.

33. Reyneri's indicator of gender discrimination is an index showing the dispersion of the difference between male and female unemployment.

34. He defends his position by claiming that it is only in this particular area that France (and Belgium) break ranks with the continental/conservative regime. The question, then, is whether this feature is important enough to alter the causal conjuncture; that is, whether the 'underlying logics' are different.

35. But note the link between clientelistic arrangements and unemployment relief; cf. the example from Spain in the next section.
36. In the early 1990s, the proportion of all job seekers that received unemployment benefits was above 85 per cent in Belgium, Denmark, Ireland (men); between 40 and 65 per cent in Germany, the UK, France, the Netherlands; between 15 and 30 per cent in Italy and Spain; and below 15 per cent (but increasing) in Greece and Portugal (Reyneri, 1999).
37. This was emphasized in Leibfried (1992). Cf. also Kosonen (1994) who, however, accepts Esping-Andersen's inclusion of Italy in the continental group.
38. At a more detailed level of specification, one may of course also list differences between these four, just as differences can be traced among the Nordic countries.
39. Not all Italy, but southern Italy, can be included. Using a definition of the poverty line as 50 per cent of national average equivalent expenditure, the shares of household/personal poverty in 1985 were Denmark, 8/8; Germany, 9.2/9.9; France, 14.8/15.7; Netherlands, 7.9/11.4; United Kingdom, 18.9/18.2; Greece, 17.4/18.4; Spain, 17.8/18.9; Italy 14.7/15.5; Portugal, 31.7/32.7 (Ferrera, 1996, pp. 21, 24).
40. The share of 16–24-year-olds living with parents is 89 per cent in Italy, 82 per cent in Spain, 61 per cent in France, 50 per cent in Germany and in the Netherlands, and 47 per cent in the UK. In Italy, even at the age of 27, 60 per cent of the men and 40 per cent of the women live with their parents (Reyneri, 1999).
41. Cf. Ferrera (1996, p. 31) on the impact of the split left wing: social democrats tried to go for full universality (inspired by the Nordic model), but the communist parties and other political radicals defended the special interests of the industrial working class. The result was thus a dual strategy: defending and even strengthening working-class interests, plus a 'policy of particularistic attraction of marginal workers via the patronage system'. So even this system may be interpreted as a sort of 'compromise'.
42. Record low birth rates will add to the vicious circle, making the dependency rate look even worse.

REFERENCES

André, Christine (1996), 'Die Finanzierung der sozialen Sicherung und die Gewährleistung der Alterssicherung: Ein französisch-deutschen Vergleich', in Franz X. Kaufmann (ed.), *Sozialpolitik im französisch-deutschen Vergleich*, Wiesbaden: Chmielorz, pp. 721–44.

Attias-Donfut, Claudine (1996), 'Renten und Gerechtigkeit zwischen den Generationen', in Franz X. Kaufmann (ed.), *Sozialpolitik im französisch-deutschen Vergleich*, Wiesbaden: Chmielorz, pp. 745–63.

Bell, Daniel (1973), *The Coming of Industrial Society*, New York: Basic Books.

Bonoli, Giuliano (2000), *The Politics of Pension Reform. Institutions and Policy Change in Western Europe*, Cambridge: Cambridge University Press.

Bovenberg, Lans and Anja van der Linden (1997), 'Can we afford to grow old?', in OECD, *Social Policy Studies*, no. 21, Paris: OECD, pp. 167–87.

Boyer, Robert (ed.) (1988), *The Search for Labour Market Flexibility. The European Economies in Transition*, Oxford: Clarendon.

Boyer, Robert (1990), *The Regulation School: A Critical Introduction*, New York: Columbia.

Buechteman, Christoph F. (1993), 'Employment Security and Deregulation: The West German Experience', in Christoph F. Buechteman (ed.), *Employment Security and Labour Market Behaviour: interdisciplinary approaches and international evidence*, Ithaca: ILR Press.

Calleo, David (1987), *Beyond American Hegemony*, New York: Basic Books.

Castles, Francis G. (ed.) (1993), *Families of Nations*, Aldershot: Dartmouth.

Charbonnel, Jean Michel (1994), 'Social Security in France: Institutions, Actors and Reforms', in Niels Ploug and Jon Kvist (eds), *Recent Trends in Cash Benefits in Europe*, Copenhagen: Socialforskningsinstitutet, pp. 117–22.

Christie, Nils (1993), *Crime Control as Industry: Towards Gulags, Western style?*, London: Routledge.

Christie, Nils (1998), 'For a penal geography', *Actes de la recherche en sciences sociales*, no. 124, September, 68–74.

Crouch, Colin (1998), 'Labor market regulations, social policy and job creation', in Jordi Gual (ed.), *Job Creation*, Cheltenham, UK and Lyme, USA: Edward Elgar, pp. 130–64.

Czada, Roland (1998), 'Vereinigungskrise und Standortdebatte. Der Beitrag der Wiedervereinigung zur Krise des westdeutschen Modells', *Leviathan*, 26(1), 24–59.

Döring, Dieter (1997), *Soziale Sicherheit im Alter*, Berlin: Aufbau.

Elfring, Tom (1992), 'An international comparison of service sector growth', Ch.2 of 'Personal and collective services: An international perspective', Economic Commission for Europe, Discussion Papers, vol. 2, no. 1, United Nations, New York.

Ellingsæter, Anne Lise (1992), *Part-time Work in European Welfare States: Denmark, Germany, Norway and the United Kingdom compared*, ISF-report 92(10), Oslo: Institute for Social Research.

Employment in Europe (1997), annual review.

Erixon, Lennart and Lars Mjøset (2001), 'The End of Interventionism – Economic Policy in the Nordic Countries in the 1980s and the 1990s', forthcoming.

Esping-Andersen, Gösta (1990), *The Three Worlds of Welfare Capitalism*, Cambridge: Polity.

Esping-Andersen, Gösta (1996), 'Welfare States without Work: the Impasse of Labour Shedding and Familialism in Continental European Social Policy', in Gösta Esping-Andersen (ed.), *Welfare States in Transition*, London: Sage, pp. 66–87.

Esping-Andersen, Gösta (1997), 'Welfare states at the end of the century: the impact of labour market, family and demographic change', in *Family, Market and Community: Equity and Efficiency in Social Policy, OECD Social Policy Studies*, no. 21, Paris: OECD, pp. 63–80.

Esping-Andersen, Gösta (1999a), *Social Foundations of Postindustrial Economies*, Oxford: Oxford University Press.

Esping-Andersen, Gösta (1999b), 'Politics without class? Postindustrial cleavages in Europe and America', in Herbert Kitschelt, Peter Lange, Gary Marks and John D. Stephens (eds), *Continuity and Change in Contemporary Capitalism*, Cambridge: Cambridge University Press, pp. 293–316.

Ferrera, Maurizio (1996), 'The Southern Model of Welfare in Europe', *Journal of European Social Policy*, 6(1), 17–37.

Flora, Peter, Franz Kraus and Winfried Pfenning (1987), *State, Economy and Society in Western Europe 1815–1975*, vol. II, Frankfurt a.M.: Campus.

Freeman, Chris and Carlota Perez (1988), 'Structural crises of adjustment: business cycles and investment behaviour', in Giovanni Dosi, Christopher Freeman, Richard Nelson, Gerald Silverberg and Luc Soete (eds), *Economic Theory and Technical Change*, London: Pinter, pp. 38–66.

Freeman, Chris and Luc Soete (1994), *Work for all or Mass Unemployment?*, London: Pinter.

Freeman, Richard (1995), 'The limits of wage flexibility to curing unemployment', *Oxford Review of Economic Policy*, 11(1), 63–72.

Freeman, Richard (1996), 'Why do so many young American men commit crimes and what might we do about it?', *Journal of Economic Perspectives*, 10(1), 25–42.

Ganssmann, Heiner (1997), 'Soziale Sicherheit als Standordproblem', *Prokla*, 27(1), March, 5–28.

Goodman, Roger and Ito Peng (1996), 'The East Asian Welfare States', in Gösta Esping-Andersen (ed.), *Welfare States in Transition*, London: Sage, pp. 192–224.

Guillemard, Anne-Marie (1991), 'France: Massive exit through unemployment compensation', in Martin Kohli, Martin Rein, Anne-Marie Guillemard and Herman van Gunsteven (eds), *Time of Retirement. Comparative Studies of Early Exit*, Cambridge: Cambridge University Press, pp. 127–80.

ILO (1989), *From Pyramid to Pillar*, Geneva: ILO.

Jacobs, Klaus, Martin Kohli and Martin Rein (1991), 'Germany: The Diversity of Pathways', in Martin Kohli, Martin Rein, Anne-Marie Guillemard and Herman van Gunsteven (eds), *Time of Retirement. Comparative Studies of Early Exit*, Cambridge: Cambridge University Press, pp. 181–221.

Jessop, Bob (1990), 'Regulation theories in retrospect and prospect', *Economy and Society*, 19(2), 153–216.

Kerr, Clark, John T. Dunlop, Frederick H. Harbison and Charles A. Myers (1962), *Industrialism and Industrial Man*, London: Heinemann.

Korpi, Walter (1983), *The Democratic Class Struggle*, London: Routledge.

Kosonen, Pekka (1994), *European Integration: A welfare state perspective*, Helsinki: Helsingin yliopisto.

Kuhnle, Stein (1999), 'Survival of the European Welfare State', Arena Working Paper no. 19, June 1999.

Lazar, Harvey and Peter Stoyko (1998), 'The Future of the Welfare State', *International Social Security Review*, 51(3), 3–36.

Leibfried, Stephan (1992), 'Towards a European Welfare State? On Integrating Poverty Regimes into the European Community', in Zsuza Ferge and Jon Eivind Kolberg (eds), *Social Policies in a Changing Europe*, Frankfurt a.M.: Campus, pp. 245–79.

Lödemel, Ivar (1997), *Pisken i arbeidslinja: om iverksetjinga av arbeid for sosialhjelp*, Fafo-report no. 226, Oslo: Fafo.

Maddison, Angus (1991), *Dynamic Forces in Capitalist Development*, Oxford: Oxford University Press.

Mares, Isabela (1998), 'Negotiated risks: Employers' role in social policy development', PhD dissertation, Harvard University, December.

Mjøset, Lars (1987), 'Nordic economic policies in the 1970s and 1980s', *International Organization*, 41(3), Summer, 403–56.

Mjøset, Lars (1997), 'Stat, sivilsamfunn og verdensøkonomi fra renessansen til i dag', in Å. Birkeland (ed.), *Den moderne staten*, Oslo: Pax, pp. 94–237.

Mouzelis, Nicos (1986), *Politics in the Semi-Periphery*, London: Macmillan.

Nickel, Stephen and Brian Bell (1995), 'The collapse in demand for the unskilled and unemployment across OECD', *Oxford Review of Economic Policy*, 11(1), 40–62.

Nickel, Stephen and Brian Bell (1996), 'Changes in the distribution of wages and unemployment in OECD countries', *American Economic Review*, 87(2), 302–8.

Notermans, Ton (1993), 'The Abdication from National Policy Autonomy', *Politics & Society*, 21(2), 133–67.

OECD, *Historical Statistics*, 1992 and 1995 edns, Paris: OECD.

OECD (1994), *Employment Outlook*, July, Paris: OECD.

OECD (1995), *Jobs Study. Taxation, Employment and Unemployment*, Paris: OECD.

OECD (1996), *Ageing in OECD countries. A critical policy challenge*, OECD, *Social Policy Studies*, no. 20, Paris: OECD.

OECD (1997), *Employment Outlook*, Paris: OECD.

OECD (1998a), *Maintaining Prosperity in an Ageing Society*, Paris: OECD.

OECD (1998b), *Employment Outlook*, Paris: OECD.

Øverbye, Einar (1998), *Risk and Welfare*, Oslo: NOVA.

Perez, Carlota (1983), 'Structural change and assimilation of new technologies in the economic and social systems', *Futures*, 15(5), 357–75.

Piore, Michael J. and Charles S. Sabel (1984), *The Second Industrial Divide*, New York: Basic Books.

Reyneri, Emilio (1999), 'Unemployment patterns in the European countries: A comparative view', *La rivista elettronica di DML on-line*, no. 1, March [www.lex.unict.it/dml-online].

Røed, Knut (1998), 'Det europeiske arbeidsledighetsproblemet', *Søkelys på arbeidsmarkedet*, 15(2), 161–74.

Scharpf, Fritz W. (1997a), 'Economic integration, democracy and the welfare state', *Journal of European Public Policy*, 4(1), 18–36.

Scharpf, Fritz W. (1997b), 'Combating unemployment in Continental Europe', EUI, Florence, Robert Schuman Centre, Policy Papers, RSC, no. 97/3.

Scharpf, Fritz W. (1999), 'Advanced welfare states in the international economy', Paper to the European Community Studies Conference, 3–5 June, Pittsburg.

Scherer, Peter (1997), 'Socio-economic change and social policy', in OECD, *Social Policy Studies*, no. 21, Paris: OECD, pp. 13–62.

Schultheis, Franz (1996), 'Die Familie: Eine Kategorie des Sozialrechts? Ein deutsch-französischer Vergleich', in Franz X. Kaufmann (ed.), *Sozialpolitik im französisch-deutschen Vergleich*, Wiesbaden: Chmielorz, pp. 764–79.

Shorter, Edvard (1975), *The Making of the Modern Family*, New York: Basic Books.

Statistics Norway (1999), *Økonomisk utsyn*, 1/99, Oslo: SSB.

Therborn, Göran (1986), *Why Some Peoples Are More Unemployed than Others*, London: Verso.

Théret, Bruno (1996), 'De la comparabilité des systèmes nationaux de protection sociale dans les sociétés salariales: essai d'analyse structurale', in MIRE (Mission Recherche du ministère du Travail et des Affaires sociales), *Comparer les systèmes de protection sociale en Europe,* Vol. 2, *Rencontres de Berlin*, Paris: MIRE, pp. 439–503.

Titmuss, Richard M. (1974), *Social Policy*, London: George Allen & Unwin.

United Nations (1997), *Statistical Yearbook*, 42nd edn, New York: UN.

Walwei, Ulrich and Gerd Zika (1997), 'Social protection: An obstacle to employment?', *International Social Security Review*, 50(4), 7–25.

World Bank (1994), *Averting the Old Age Crisis*, Oxford: Oxford University Press.

16. Policy conclusions: on the future of European employment

Pascal Petit and Luc Soete

The overview of the contemporary aspects of technical change and their implications for employment growth and displacement – what we described in our introduction as the employment dynamics – presented in this book reveals at first sight some grim features for Europe. On the surface, the diversity of institutions and experiences appears to have limited the chances for Europe to gather endogenously its own growth momentum and propel it along an autonomous expansion path. The same diversity appears also to have seriously hampered Europe's reaction to external changes, whether on product, skilled labour or financial markets, or in the development and commercialization of new technologies. Yet it is obvious that diversity may also have its advantages. It might lead to more creative responses to external challenges and help in fostering new, different, possibly promising, endogenous developments. And Europe has of course undergone changes in all its dimensions and components over the last two decades. Many of these have barely had their full impact yet on European society and on the many, very different economic, social and cultural nations and regions of which it is composed.

In this concluding chapter, we provide some further insights where the findings of the research brought together in this book might help one to strike a balance between the advantages and disadvantages of such an evolving Europe. From this perspective, great care must be taken in assessing the time horizons and time lags that are implicit in discussing any prospective view. A first, in our view, essential, consideration is that, behind the diversity of the various European countries' performance at both the technology and employment levels, one finds different evolution processes, implying different kinds of learning, of time horizons and time lags.

We start by looking briefly at some of the major institutional changes likely to influence the dynamics of employment in the European member countries in the following decades. This assessment illustrates how different the last two decades have been compared to the earlier postwar period and highlights the kinds of institutional changes which successively played a dominant role, whether one considers the field of labour markets and job creation or of financial markets. We then turn to an appraisal of the organizational and structural changes in Europe. Without trying to be exhaustive, we shall stress the main lines of development regarding changes in organization within firms, as well as between them. Our analysis will focus on the reactions of firms with respect to the development, diffusion and use of new technologies, but will also pay special attention to sectoral changes where the expansion of services has played and continues to play a major role, redefining the conditions under which economic transactions emerge and are becoming commercialized. Finally, we shall come back to the dynamics of growth and employment, taking into account the demographic and societal factors which are likely to shape Europe's future.

The discussion of these various issues is organized in reverse order to the one followed in the four parts of this book. In this way, we hope to re-enforce somehow the aspect of a 'balance sheet' of past developments at different levels: institutions, firms, sectors and, finally, countries which can lead us to draw some broader analytical policy conclusions which might further or counter some of the trends identified.

1.   INSTITUTIONAL CHANGES IN LABOUR MARKETS: FROM LABOUR MARKET FLEXIBILITY TO EDUCATION AND TRAINING ISSUES

Institutional changes are continuing processes which change pace, accelerate or bifurcate in times of crises, as when situations in important spheres of activities appear to be unsustainable. Crises may of course be of various magnitude and coverage. The strong inflation and mass unemployment which successively developed in the late 1970s and early 1980s were from this perspective symptoms signalling and ultimately characterizing the period as a phase of economic crisis. The institutional changes of the 1980s were motivated by the need to get out of this crisis. They led to structural adjustments – a key phrase in the 1980s – aimed at reforming labour market institutions. Increasing the flexibility of labour markets became a widespread policy objective. A second field of reform, more typical of the late 1980s, was brought about by a more general concern not only to control wage inflation but more generally to expand the sphere of 'deregulated' market activities in order to take advantage of internationalization and the potential of information and communication technologies which modified the time and space dimensions of economic activities.

The 1980s were for all developed countries still an early phase in the diffusion of the new technological system centred upon the new technologies of information and communication. Still some countries were in advance and some lagged behind (see, for example, the spread in the ratios of ICT investment expenditures to GDP over the 1980s, compared to the more recent period: OECD, 1999). The growing internationalization of most developed economies was on its way and international competitiveness was thought to depend strongly on the mastering of these new technologies. Countries, conscious of going through a period of transition, launched under different conditions and with different strength and emphasis these two waves of reforms and institutional changes concerning the large intermediation services and the adjustment of labour markets. While deregulation of 'regulated' services appeared to be a worldwide phenomenon, which started in the early 1980s in the USA and spread progressively through the decade to the whole developed world, the need for the deregulation of labour markets concerned more especially Europe.

The notion of European sclerosis developed in the early 1980s to take into account the continuous rise in unemployment in Europe. This characterization was a partial view, omitting to take into account the quality of the jobs created and the overall rates of employment (and in particular the shares of age groups in the workforce) which themselves reflected different life styles and traditions. Still the debate in itself echoed a widespread pressure to reform labour markets. And indeed, although blaming rigid labour markets was often exaggerated and critics did not take into account the effective capability of labour markets to adjust, most European countries reformed their rules and practices towards more flexibility to short-run changes in product markets and to new patterns of organization (both within firms and across firms). In all these cases the change in the sectoral structure of employment did facilitate such transformation. The expansion of service activities, much less institutionalized than manufacturing activities on average, did help to create more flexible labour markets. This major, continuing change in the structure of employment was all the more open to all kinds of new forms of work organization as the large service activities, where labour was more organized and where more rigid practices constrained the organization of work, were under the strain of deregulation and the ensuing rearrangement of their production processes. Clearly, by the end of the 1980s, flexibility of the labour markets was no longer a major issue on the policy agenda – despite the popularity of international organizations' studies, such as the OECD *Jobs Study* addressing national unemployment and job creation issues (OECD, 1995a) – and the debate shifted more towards participation rates, life styles and new forms of socialization.

The debates about work, non-work and the status of work really only started during this last decade. While many of the concepts used are reminiscent of the 1960s, these debates are new and in the end do not replicate those of the aftermath of World War II, as most rich OECD societies in which they are taking place are now much more often composed of double-income families, living mostly in urban areas and nearly three-quarters of them occupying jobs in services. The old debates and the choices made in the postwar period, especially regarding the type of welfare states, do influence today's options, but the societies are more affluent and more educated. Therefore the issues involved in redefining the status and content of work over each person's individual life cycle are now more centred upon the kinds of training, the types of family structures preferred, the borders between public and private, personal and collective, as well as on the division between work and non-work activities over one's life cycle.

Lars Mjøset in his contribution stressed how pregnant are the various types of welfare states in current debates and ultimately choices of the individual European countries. The fact is that evolutions of welfare states occur in rather marginal and incremental ways, generally speaking by means of small discrete adjustments of primarily national systems which have evolved around some central options made five decades ago (themselves more or less rooted in long pasts) regarding, first, the role of the family in securing welfare and the relative place of the young and old generations, and second, the ways by which to finance social welfare coverage.

In effect, the relative continuity of these evolutions is worth underlining. One may also observe that these issues are far less often debated than they deserve in the political spheres of the various countries at the level of large societal choices that they imply. We would agree with Heilbroner and Milberg (1995) that there is a certain lack of vision in the political debates of our societies when addressing these basic issues. Four decades of macroeconomic national policy have to some extent paid the price and restricted most policy debates to broad macroeconomic objectives such as the reduction of inflation, the fight against unemployment and the fostering of economic growth. When these objectives began to appear out of reach of national policy makers, at least in the short run, the political debates failed to reformulate new policy objectives in broad, yet still concrete, terms of societal choices. In smaller countries (the Scandinavian countries, the Benelux, Ireland) where the limits of national policy actions were recognized much earlier, such debates took place to a much greater extent.

The debate about the reduction in working time is very indicative of such reductionism. In societies where the status of wage earner has become strongly predominant, the debate tends to be mainly concerned with the reorganization of work and much less with the reorganization of non-working time. Again the evolution of working time across countries does show that, beyond a common trend of reduction in working time, the patterns remain quite different. Countries combine in various ways flat reductions in working time across the board of all occupations with expansions in part-time work, as stressed in the contribution by Spiezia and Vivarelli in Chapter 13. This leads to two polar cases, one experiencing modest decreases in average working time and an uneven distribution of part-time work (close to the UK and the US experiences) and another experiencing a substantial decrease in average working time with a more even distribution of part-time work (more in line with the experiences of continental Europe countries such as France and Germany). To go further in the reduction of working time, reducing either the average working time in all occupations or increasing the diffusion of part-time work in all activities, clearly depends on whether it is voluntary and seen as improving welfare or imposed and seen as constrained by what is perceived as a scarcity of available jobs. Again the answer depends on the overall social project in which such arrangements are taking place. To reconstruct an effective macroeconomic policy one is therefore forced to debate also some of the fundamental options which our societies are being confronted with concerning life styles, education and training, environment and social cohesion.

What should be the role of education and training is certainly part of these interrogations, but again one has to allow for the new context in which the supply of qualified labour has been substantially enlarged and the use of knowledge in economic activities extended, be it in codified or tacit forms. Again these debates about the role of education and training may be too much oriented by a vision centred upon production processes, and not taking into account personal needs and well-being. Thus, following the debates in the economics literature, the issue of education and training has to focus on both the increase in the demand for skilled labour and the abundance of the supply of qualified labour, both shifts inducing changes in the effective role of training and education. Both phenomena obviously have their share in explaining the increase in skill and, more generally, knowledge intensity in society. On one side, the debate about the presumed 'skill-biased' nature of technological change emphasized the fact that firms upgraded their skill requirements in order to face uncertainty in the evolution of productive processes in times of rapid technological change and tense competition (once the direct effect of the trade with low-wage countries has been scaled down to its relatively minor dimension: see Petit and Soete, 1997, for a short survey of this issue). On the other side, studies have stressed that the supply of qualified labour had largely increased all over the developed world, to such an extent that in some countries it has in itself reduced the number of jobs for the so-called 'non-qualified' (Goux and Maurin, 1996; OECD, 1996a, ch. 4).

By contrast, very little has been said about two other major dimensions of the new challenge to education and training. One has to do with the social value of education and training, not only for participation in traditional social networks (conditioning, amongst other things, family structuring and job opportunities) but also for gaining access to new life patterns and taking advantage of complex new service provisions (not only in health and education but also in finance, transport, communication and distribution). From this perspective education and training has also to respond to new needs in economies in which relationships between users and producers have become essential knowledge links, with 'smart' consumers as the final link. The other major dimension has to do with the laws of obsolescence of knowledge brought about by education and training. Recent work on literacy (OECD, 1997) shows that the rate of obsolescence, as measured in comparing the effective capabilities of various education levels after some years, varied strongly among countries.

These two issues raise questions about the content of education and training, the forms and channels for its provision and its timing along the life cycle. These questions are rather different from the ones raised in the postwar era, where educational systems provided initial training of different kinds, giving access to specific professional trajectories determined from the start, more or less rigidly, across countries. The advantages brought by education remain effective but more widespread and open to various upgrading and downgrading hazards encountered on the job. It follows that the meritocratic model, whereby positions were gained according to the years of education, may not work as systematically and steadily as before. Rates of return may vary strongly for the same degree or course and it is not that easy to determine which study curriculum to follow – especially as careers not only correspond to a wide range of wages trajectories but also include, increasingly, non-wage incomes (such as stock options or standard financial benefits).

As a matter of fact, in the new economic environment of the 1990s, the question of wages and incentives has gained dramatically in importance. Many practical ways of alleviating the fiscal levy on the highest incomes, developing new schemes of profit sharing and of wage funds, as well as reducing minimum wages and therefore reducing their impact, were developed in individual countries. Development of various forms of pension funds can be seen from the same perspective of restructuring distribution, not so much to ensure solidarity between wage earners (as has been the case in the past) or to increase flexibility of labour markets – an objective sufficiently attained by the late 1980s – but to stimulate incentives for individuals whose personal response and degree of competence seemed to condition the success of the new productive processes.

Both issues of training and incentives are from this perspective central in the actual processes of institutional and organizational change. It still is a phase of complex and substantial trials and errors, as economies are facing new situations where the challenges are no longer to fight shortages of collectively defined qualified labour, but to provide and put to use individual competences (see also Vendramin and Valenduc, 1999). It is important to stress the greater individualization implied in this process and the ensuing difficulty of promoting social norms of training for the provision of competences which in the end rely on combinations of formal training and experiences at work and out of work.

We would thus conclude by arguing that a new approach to education, training and incentives must be high on the policy agenda, while the debate about the 'desired' level of flexibility of labour markets, insofar as it had already been more or less reached by the end of the 1980s, might well be moved to a somewhat lower level. Such a shift in policy priority setting is actually in accordance with the trend towards a more knowledge-based economy. As we mentioned before, this trend is prone to favouring, at least in a first phase, unequal access to new forms of jobs and new consumption patterns. This diffusion process seems also bound to increase inequalities in wage formation. In various ways the development of banking and finance services as well as the reorganization of large networks services, which we discuss next, are instrumental in fostering such new economic trends.

2. INSTITUTIONAL CHANGES IN FINANCE AND INTERMEDIATION SERVICES: FROM DEREGULATION TO THE ASSESSMENT OF NEW PRIORITIES

The deregulation of large network services has been a major development of the late 1980s and 1990s. It started with the financial sector with obvious important consequences for the global functioning of the developed and, more recently, industrializing countries and the economies in transition. Financial sectors were highly and diversely regulated from the 1930s and 1940s. In the 1970s, while international transactions had entered a new, post-Bretton Woods era, the national restrictions were in danger of either being effectively bypassed or blocking the international dynamics of products markets. Deregulation was slow and partial, depending on the initial situation in the various countries. Progressively, and in very different forms among countries, financial capital acquired a new mobility and creativity. It transformed the governance in all activities, on the one hand giving more importance to short-run transparency and profitability and, on the other, helping to finance risky but highly profitable longer-term projects.

By and large, European countries followed in a very diverse fashion the deregulation path taken by the USA. They were in very different positions to do so, depending on the characteristics of their own financial systems. Crucial in that respect were the different national ways in which firms and households financed their projects: for firms, the different ways they financed their investments in the first place, but also their merger and acquisition attempts or their temporary needs for liquidity; for households, how they financed their consumption and investment (of which housing was in most countries the most important component) but also how they saved income for the future, an issue which, as we saw above, makes a big difference for the operation and financial sustainability of welfare states – whether capital is being accumulated, as in the case of pension funds, or whether those currently at work pay the pensions and welfare contributions of those not at work. In both cases external changes, like structural changes in financial markets or in demographic structure, strongly challenge the viability of the systems.

Systems were so different that the deregulation move of the late 1980s did not lead to any real convergence of the systems (on 'national systems for financing innovation', see OECD, 1995b). Still the widespread deregulation moves induced some common trends. For firms, financial markets became a more important means for developing an enlarged range of operations. Banks increased their global reach in developing international networks and linkages. Many more risky projects, most of them innovation projects in a broad sense, could be financed. For households and individuals, new forms of savings developed, especially by means of a larger use of financial markets, whether directly through buying stocks – people capitalism – or indirectly through pension schemes, not to mention much easier forms and management of payments.

This again led to large discrepancies between firms (not only by size and financial strength but also by limited know-how in using the new banking and finance system) but also to discrepancies between individuals (not only in the relative importance of financial gains in their current income but also in their abilities to use the facilities of the system). These discrepancies brought to the forefront, however, the general nature of the move. While the old system had been stressing the principle of universal service through which access was universally guaranteed and every user was in a position to fulfil his basic 'needs' regarding loans and savings as well as payments, the new systems provided a much wider range of services and the issue now became 'prudential', for example in order to make clear the rules under which the systems and services were operating. This concern over the 'transparency' of the services provided is far-reaching and not so easy to solve or calibrate, as it implies knowledge about the capabilities of the users. In this sense, the financial deregulation move is an integral part of the diffusion of a new economy.

What is so clearly identifiable in the case of the financial sectors can also be detected and observed in relation to the evolution of the other large network services which play a role of intermediation in the working of markets such as the communication, transport or distribution sectors. These issues are largely stressed in Chapter 6, which emphasizes the fact that the new technologies of information and communication deeply transformed these activities of intermediation. In effect, the deregulation of these sectors again followed more or less rapidly that of the banking sector under rather similar external pressures, with the added feature that not only were a large proportion of these activities highly regulated but also they were provided by public bodies (whatever their exact status). Privatization thus took more time on average and went through a longer transition period. Telecommunication services give a good example of both the length of time necessary to privatize large, highly profitable, national public enterprises (as in France or in Germany) and the long period of transition it takes for the new structure to adjust to market forces and ongoing technological changes (as became obvious from the earlier transformation of the telecommunications sectors in the USA and in the UK). It follows that, by the end of the 1990s, operators and market regulations still had a long way to go from the old regulatory context to the one which is more likely to prevail in the future. Still the issues are rather similar to the ones mentioned in the case of finance, with the same shift away from universal access, and the existence of discriminations and barriers constituted by differences in know-how and knowledge on the user's side and the same 'prudential' principle of transparency (and privacy) regarding the rules and contents of the services provided.

This state of broad institutional change conditions in different ways the evolution of the various national systems of innovation, as we shall outline below. But the trends accompanied and favoured over the last two decades by the institutional changes in labour markets and financial markets have also had a noticeable effect in terms of prices and distribution. On the one hand, the achievement of a certain level of labour market flexibility helped to get rid, at least partially, of any basic wage inflation pressure; on the other, the reorganization of financial markets appears to have led to some structural inflation pressure of financial assets (of which the unprecedented steady growth of stock market indices is a major sign). The feature may not be so clear-cut. Some financial assets may experience clear downswings or falls (there is of course no indexation of financial assets, as in the case of wages) and wages of some professionals may enjoy long-lasting upswings, as experienced with the salaries of CEOs, most of the time enlarged by stock options. Still the 1990s and the years to come offer a clear, new picture which also changes the context in which technologies are diffusing, innovations are made and employment is created.

Table 16.1 summarizes what we have just outlined regarding the main directions of institutional changes as they appeared to emerge in the developed economies over the last two decades.

Table 16.1   Key issues in the institutional changes of the 1980s and 1990s

	1980s	1990s
Labour markets and educational systems	Flexibility of labour markets and quantity adjustments of educational systems Disindexation of wages	Quality adjustment of education and training, selective adjustment of wage and income formation
Financial markets and banking systems	Deregulation of finance industries	Development of 'prudential' rules and selective risk coverage procedures Limited regulation of international financial flows
	Wage inflation	Financial asset inflation

3. NATIONAL SYSTEMS OF INNOVATION AND THE NEW TECHNOLOGICAL PARADIGM: A PROCESS OF INTERNATIONALIZATION

The evolution of the wage–labour nexus and of the financing conditions, discussed in the two previous sections, has had a direct influence on the diffusion of the new technological system centred on new information and communication technologies. In the first place, as we have already underlined, the institutional changes in the late 1980s and 1990s were in line with the emergence of a new economy where individual knowledge was more directly in a reactive and creative interaction with all kinds of codified knowledge embodied in equipment and organizations. Knowledge management and capital risk financing, which only became major issues in the 1990s, are effectively two key factors driving the development and diffusion of new technologies. Knowledge and finance condition the flow of inventions and the speed of diffusion of new technologies.

As a result, developing and putting to use the potential of new technologies has only been at the forefront of economic development for less than a decade. The 1980s were from this perspective the very first, preliminary phase of restructuring whereby the labour market and finance markets only gained some room for manœuvre in cutting through old, dominant linkages. This view of course fits the historians' general finding that transitions from one technological system to another takes time, as Freeman (1987) and David (1991) in particular stressed, and are accompanied by innumerable institutional mismatches.

Yet, at the same time, despite the valuable insights, such historical comparisons do not allow one to infer that transitions will necessarily and always follow similar patterns. On the contrary, the specificities of each technological paradigm, the space and time in which it develops, all suggest that phasing and sequencing will differ. Consequently, one has to be careful when assessing what the new institutional matching will be, which countries will be leaders on the technological frontier and which will be followers. So far the 1990s have shown a ranking in growth and technological performance even when we limit oneselves to the 'triad' which was not obvious from the start. The 1990s did show the comeback of the US economy at the frontier of the new economy. After the 1980s, when the US leadership appeared to become increasingly challenged by the abilities of some Asian countries to produce IT equipment, their capacity to make wide use of this equipment, combining the new computing and telecommunication potentials, and to develop, on the basis of this know-how, highly intensive knowledge-based activities, has undoubtedly restored the USA to the position of leader. Japan and Europe appear to lag somewhat behind.

But again this is by no means the end of the competition at the technological and economic frontier. The technological paradigm around the ICTs is itself the outcome of the internal speed of development ensured by the ever-increasing storage capacities of semiconductors known as Moore's law (the storage capacity of semiconductors doubles every 18 months) and the constant upgrading of software that this implies, open to a lot of competition. There is intrinsic room for manœuvre here. On top of that there are great potentials for learning how to use efficiently this equipment capacity in complex organizations. It follows that a club of countries remain active in the technological race, beyond the simple hierarchy that one might witness at some point in time, as today with respect to the USA. The weaknesses that one may observe in Europe or in Japan today in some leading technological activities have, in other words, to be seen in a dynamic perspective, allowing for new, unexpected learning potentials to emerge and abilities to diversify, thereby assuming technological leadership in some specific production fields and market niches.

This does not mean that we are unable to make any comparison and prospective assessment on the relative potential of countries to play a major role in the development of the current new technological system, and this volume contains a number of contributions which make such challenging assessments (see, for example, Chapter 1 and Chapter 14). But two lines of investigation have to be distinguished in order make a comparative assessment of the perspectives of technological development open to each country. One is 'diachronic' and has to do with the relative capacity of each national system of innovation to adjust to the challenge of internationalization. The other is more 'synchronic' and concentrates on the intrinsic properties of local adaptation to the new technological system.

The internationalization of national systems of innovation is from this perspective tridimensional. It is channelled by the internationalization of knowledge, of capital and of labour. A priori the internationalization of labour by means of migration plays a relatively small role in the process. The first phase of restructuring, in the 1980s, has even been marked by the relative absence of migration flows, with the exception of the USA, precisely at a time when the internationalization of capital was accelerated. Relatively little changed in that respect in the 1990s, but it has become clear that the migration of qualified labour, of professionals and scientists, will be an important issue for the near future. Already, at the core of the scientific labour force, the number of foreign students and academics is becoming important. It certainly is already the case in the USA, and European countries are starting to compete to attract foreign scientists (Mahroum, 1998).

As a matter of fact, at this level the issue is part of the increased internationalization of knowledge, which is a multidimensional phenomenon. Some knowledge is produced and developed from the start at an international level and circulates via international networks of academics and researchers (see here, for example, the typology of Archibugi and Michie, 1997). A structure like Internet clearly induces further, and expands, such dynamics. At another level, large business services are also major channels for technology transfers, along with all the transactions having a direct bearing on technology sales, from standard patents to all kind of franchising and intellectual property rights. This is also the case with the export of equipment goods, international transactions traditionally considered as passive vectors of technology transfers. Technology can by now be transferred by various means. Beyond this aspect of access to technology, the policy interest lies of course in the capacity of assimilation of each country. As stressed by Nelson and Pack (1999), the challenge of national systems of innovations is not so much to accumulate new technology equipment but to assimilate the technology and get to know how to use it comprehensively. In particular, the learning process may take longer in some countries than in others. Secondly, adjusting the old qualities of the national system of innovation may lead to the development of some niches more than others or to the retention of particular specializations.

Therefore comparing the potential of different national systems of innovation is useful and particularly indicative of the kind of specialization fitting any given country. From this perspective, the contribution by Amable and Boyer to the present volume points to some of the major weaknesses of Europe's innovation capabilities but also hints at the potentials that can advantageously be put to use. Within a certain sub-set of countries, chances to catch up with some developments on the technological frontier, whatever the size of the country, are worth considering. The learning process by which countries acquire the know-how and knowledge in an open environment is from this perspective complex and composed of many alternative ways for countries having reached a certain level of development in terms of wealth and human capital.

Looking at the internationalization of capital and, more precisely, at the internationalization of the national system of financing, one can reach similar conclusions. National systems tend to adjust to the pressure of internationalization in a more or less direct and automatic fashion. Some countries may have to develop entirely new forms of interventions, quite different from old structures. This will take time and the lack of complementarity with the old system may lead to some oppositions between new and old set-ups. Japan is a good example of a country where the old national banking system finds it difficult to follow the new trends to develop risk venture capital induced by the search for short-term, higher profits.

But this comparison should not be confused with the more 'synchronic' perspective which takes into account the specific assets of countries as well as the 'ideal-type' generic properties of the new system. For instance, the new financing system implicit in the new economic context is more open to risks, including some risk of a major systems crisis. However, it does not follow that, in this new context, countries which have gone some way towards this type of adjustment are more specifically prone to financial crises than countries which have not done so. Freeman, whose contribution in this volume emphasizes the various aspects of the relative positions of financial systems in the triad, points quite rightly to the differences between the systems in dealing with risks. The US system appears in this comparison as highly risky and unsteady. Yet, beyond this observation, one has to account for the greater ability of the system to adjust precisely to higher risk levels. The challenge is then to appreciate how far the USA has gone ahead in terms of risk management at all levels, and to what extent they eventually might have gone too far in that direction, taking too many or too huge risks to be dealt with even in a more sophisticated and more risk-monitoring economy.

In such international comparisons one has thus to take into account a 'synchronic' perspective which sufficiently appreciates evolutions in relation to the would-be characteristics of the new economic system. It is certainly an intrinsic feature of the so-called 'new economy' to be more open to risk taking, thus enlarging the spheres of productions and transactions. A real challenge is to assess the limits of this enlargement beyond which the economy is becoming truly unsteady and open to major crises. From this perspective, the present volume invites one to explore further this frontier; it clearly shows that international comparisons and prospective assessments have to be put in a dynamic setting explicitly allowing for interactions of structural and institutional change.

So far we have stressed some broad, qualitative and institutional changes having affected most developed economies in the field of technology and employment over the last two decades. The period, as we have argued, was itself clearly divided into two sub-periods and the concerns tied up with a new growth regime really only emerged in the more recent phase of the 1990s. How the institutional changes on which we have concentrated so far facilitate what we call structural and organizational change is, as we emphasized in the introduction to this book, a key determinant of the relationship between technology and employment. It is to those issues that we turn now.

4. STRUCTURAL CHANGES AND ORGANIZATIONAL ISSUES

Structural changes are here thought of as the ground in which institutional changes are somehow rooted, as well as constituting their soil. Structural changes are thus setting the scene for economic activities. This structural context is marked, in our view, by three interdependent changes. Two of these, internationalization and the diffusion of the new technological system centred upon ICTs, are well recognized and have already been central to our discussions in the previous section. The third dimension remains, at least in most economic analyses, often implicit: the current phase of tertiarization, with a significant rise in new 'business' service activities alongside the more long-term, gradual trend of tertiarization in developed economies associated with the growth of personal and social services.

Reviewing the dynamics of this third 'nexus' helps us to assess in a more explicit way the various trajectories of countries in the building up of their own brand of the new growth regime. This raises complex issues on the internal and external organization of firms, which in turn affects the dynamics of employment in many ways. It manifests itself, for instance, in terms of new relationships between firms and new skill requirements within firms. Many contributions in the present volume precisely address some of these issues. Caroli, in her survey, stressed how the change of technological system induced organizational changes marked not only by the national wage–labour institutional nexus but also by a general trend towards granting more responsibility to workers at lower levels of the hierarchy. Gatti, on the other hand, called attention to similar effects by modelling the microinstitutional context of firms and showing that relevant organizational choices were more important than the increased flexibility of labour markets. Finally, De Loo and Ziesemer tried to model the links between three aspects of structural change, namely trade (internationalization), technological change and labour supply.

We start here, though, with a recapitulation of stylized facts which orders somewhat the various findings of the book. We start with the present phase of tertiarization, keeping for the next section the key issue of labour supply and its determinants, such as changing demographic conditions.[1] We gave in the introduction some stylized facts regarding the new technological system centred on information and communication technologies. One feature is undoubtedly that it does represent only a small amount of the standard notion of installed fixed capital, as most of the financial investment that it corresponds to is intangible (software expenditures, development, training and organizational costs) and therefore difficult to estimate (see Sichel, 1997; Kendrick, 1994) and increasingly much larger than its tangible counterpart. Still one should not underestimate the tangible side of ICT diffusion, for at least three reasons. The first is that the growing capacity to store, to manipulate information and to communicate worldwide and instantaneously, changes the time and space dimensions of economic activities. The second reason is linked to the impetus given to all these tangible equipment flows by Moore's semiconductor performance law (the storage capacity of semiconductors doubles every 18 months) which constantly broadens the range of innovative applications. A third, and more ad hoc, reason is that the diffusion of tangible ICT equipment, whether it relates to the actual use of PCs or to the number of Internet connections, offers indicators of the diffusion process (see OECD, 1999). Such physical use and accumulation trends do not, of course, imply effective assimilation of efficient use. Misuses and inefficiency are also part of the diffusion of the new technological system.

It is clear that the present phase of the process of internationalization fuels further the diffusion of the new technical system, based on ICTs. ICTs can even be considered historically the first technical system, which diffuses, one way or another, so rapidly on a worldwide scale. Of course one has to clarify what is specific to the present phase of internationalization. It can be qualified as a phase of globalization, not so much because the intensity of trade and foreign direct investment (FDI) flows will have reached unprecedented levels – countries such as the UK, the Netherlands or Japan experienced similar levels of internationalization at the end of the 19th century – nevertheless, the fact that trade and FDI patterns now involve an unprecedented range of countries and concern a much more widely diversified range of products, with a particular importance given to the technologically more sophisticated products, is really specific to the current phase of globalization. The present phase of globalization is also characterized by the fact that most economic transactions now take place as part of strategies that account for international conditions and opportunities. This is made possible by the diffusion of information on international opportunities, as well as by the existence of intermediaries, whether in the form of international partnerships, through the internal organization of multinational firms, or via international networks of services. The large flows of information rapidly available on all sorts of topics, as well as the possibilities for cooperation with academics and other non-market institutions, clearly also contribute to the building up of these logistics of international mediation. The development of advanced telecommunications has closely accompanied and enabled this internationalization of firm strategies. Information networks have been gradually developed, slowly leading to the recent explosion of the Internet, which again offers a dramatically new potential for international transactions. This significant enlargement of the scope of actions may not necessarily appear in the balance of payments, as long as it does not lead to effective transactions. Still the new scope to act at international levels and its effects on national economies is a key feature of the present phase of globalization. It leads to a larger capacity for a larger number of (small) firms and even households to develop international transactions.

Central to the enlargement of the international reach of firms is the development of business and intermediary services with a global reach. This is linked to the phenomenon of the increasing tertiarization of economic activities, which has probably not been emphasized enough as a central element in the development of our economies. Tertiarization in a broad sense is not a new phenomenon, and the growth of the share of employment in the services sector follows a long historical trend, not altered by the relative slowdown in economic growth which has affected most developed economies since the mid-1970s. Still the tertiarization trend experienced over the last two decades has some distinctive features, the most marked of which is the unprecedented growth of business services. Their share in total employment rose from a few percentage points to more than 10 per cent today. The business and finance services sector constitutes a 'new' phenomenon, characteristic of a new fabric of relations between firms, and consists of a wide variety of specialized services to business (from audit and research labs to cleaning and surveying). These new service activities mix very diverse types of activities, some using highly qualified labour and others requiring low-qualified labour.

The development of a set of highly qualified service activities in many new market niches appears primarily linked to the diffusion of new information and communication technologies (ICTs). Some of these activities were instrumental in developing a logistics which, together with the spread of large network services in communication, transport and distribution, further induced the enlargement of this global reach capability. On the other hand, the same activities also contributed strongly to the diffusion of the new technological system. In the first place they provided the core of professionals needed to maintain, assist and complement the diffusion of the new information and communication technologies (all business services linked to hardware and software activities). Secondly, they developed new sorts of activities, exploiting for individual firms the data-gathering possibilities of new ICTs in terms of data mining, treatment and fast communication of such increasingly valuable commercial information. The development of such expertise has been an important factor favouring a reshuffling in the internal organization of firms, their outsourcing practices and product turnover. It does not mean though that a new 'optimal' model of firm organization emerged in the process. The latter remains largely a learning process, influenced as much by local traditions and tastes as by the efficiency of local suppliers of qualified services. Thus Schettkat and Russo, in Chapter 4 of this book show that external uses of business services (as shown in input–output statistics) had only a small impact on the dynamics of productivity gains. Beyond the fact that the measures in real terms of these intermediate uses are rather uncertain, the fact that they developed strongly across the board of most countries does not help us to determine their efficiency effect, particularly over a period when productivity gains at the macroeconomic level did, if anything, slow down.

We thus stress the interdependence of a nexus of structural changes more or less common to all countries but difficult to assess in terms of the relative efficiency of the different country experiences. This nexus of transformation is consistent with the microeconomic transformation of activities described in Chapter 6 by Petit and Soete whereby the diffusion of ICTs was shown to enlarge the provisionability and content of both service and manufacturing activities alike.

Thus both the elaboration and diffusion of ICTs are channelled through all kinds of specialized and highly qualified business services (from software producing to all sorts of consultancy), which are at the same time large buyers of ICT equipment. Hence globalization or internationalization has largely drawn on and been enabled by telecommunication and computing facilities, as well as the appearance of business services provided by large multinational firms.

The interdependence of the three aspects of structural change outlined above and in Chapter 3 (see Table 3.3) calls for a multidimensional representation and perspective on the surrounding socioeconomic context. This can be represented as a triangular nexus of structural changes, as in Figure 16.1. The

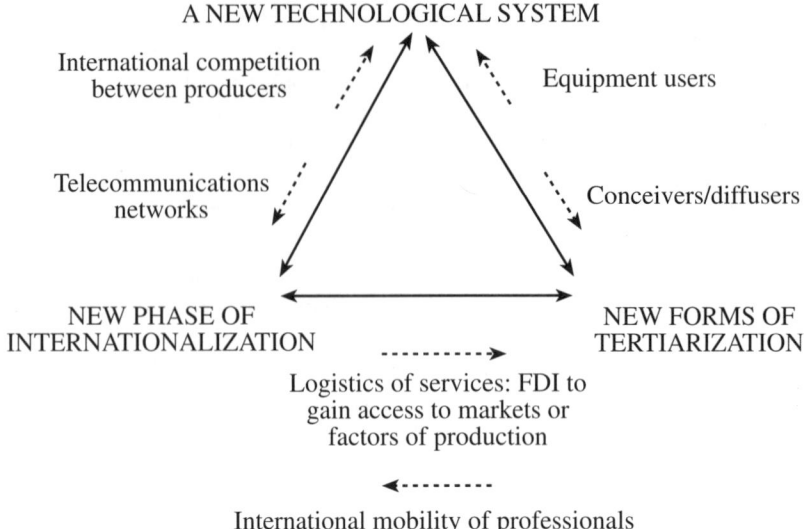

A NEW TECHNOLOGICAL SYSTEM

International competition
between producers

Equipment users

Telecommunications
networks

Conceivers/diffusers

NEW PHASE OF
INTERNATIONALIZATION

NEW FORMS OF
TERTIARIZATION

Logistics of services: FDI to
gain access to markets or
factors of production

International mobility of professionals

Figure 16.1   Three interdependent structural changes

figure illustrates the possible outcomes of the present period of trial and error, where economic agents (essentially firms, but also households) experiment in the form of more or less continuous reorganization in order to try to match the changes taking place. This multidimensional perspective on current structural change stresses the broad organizational challenges with which our economies are being confronted. It concerns firms, how they organize their work in-house or take advantage of external suppliers (especially services). It also concerns households, who can radically reorganize their domestic activities, following or favouring changes occurring at work or with regard to the services they use.

While the above appears more a necessary combination of structural changes, its links to total employment remain largely undetermined. Pianta, in Chapter 5 of the present volume, underlines the fact that innovative activities in manufacturing, because they have been more oriented towards process innovations than towards products innovations, have had no positive impact in Europe on employment. Conversely, Landesmann and Stehrer, in Chapter 7, stress that the late deindustrialization process in many continental European countries can account for the high unemployment observed in the 1980s. It manifested itself in lags in the adjustment of various productive processes with eventually persistent negative effects on structural unemployment. Beyond these structural mismatches, differences in total employment growth remain, by and large, unexplained. Schettkat and Russo in their contribution stress that manufacturing, augmented by the set of services that are carrying the reorganization of the production processes, cannot be considered as an engine of growth which could account for the overall growth rate of GDP and employment. The employment content of growth actually differs widely between countries. A good number of these differences are linked to the expansion of personal, social and community services, as emphasized extensively in the contributions to Part III of this book.

In effect, a steady growth of these services accompanied the development of the nexus of interdependent structural changes mentioned above. This, increasingly massive, component of employment, which benefited from the rise of public budgets in the 1950s and 1960s, continued to grow over the period of slow growth of the 1980s and 1990s. It corresponds to a wide range of good and 'disadvantageous' jobs, and the combination and relative importance of sub-sectors in the field is very much country-specific. One could say that the pattern of services (business or community services) that we currently observe is very much path-dependent, based on services provided somewhere in the past on the basis of some specific initial conditions. This implies the type of welfare state that has prevailed in the immediate, postwar period and developed until recently (see Mjøset's contribution for a discussion of types of welfare states and their evolution).

Having analysed the main features of the differences in institutional and structural changes, we can now draw some conclusions on the trends observed so far. To complement our policy conclusions it will also be important to point to changes which have not yet occurred but are foreseeable in the years to come. They relate to demographic changes and the transformation of household 'services' once the impact of ICTs fully reaches them, whether domestically produced or provided by a public or private organization.

5. GROWTH AND EMPLOYMENT: CONVERGENCE AND DIVERGENCE IN TIMES OF TRANSITION

From a long-term perspective, the last 25 years have been characterized by the widespread slowdown of economic growth by comparison with the 'golden years' of capitalism that represented the 1950s and 1960s. This stagnation followed a marked reduction in productivity gains which has been the centre of many academic, measurement and policy debates for the last 20 years. The productivity slowdown has been a widespread and general phenomenon across both countries and sectors. However, the rates of economic growth have been affected in very different ways, especially in the 1990s (see Table 16.2). On average, countries experienced a prolonged period of stagnation, but some countries did experience extensive economic growth (with growth of GDP

Table 16.2 GDP growth and productivity growth (in real terms, 1990 prices; average growth rates in %)

	EU-15	B	DK	D	E	F	IRL	I	NL	A	P	FIN	S	UK	USA	JP
GDP Growth																
1960–71	4.8	4.9	4.5	4.4	7.3	5.6	4.2	5.7	5.1	4.7	6.4	4.8	4.6	2.9	3.8	10.5
1981–90	2.4	1.9	2.0	2.2	3.0	2.4	3.6	2.2	2.2	2.3	2.9	3.1	2.0	2.6	2.6	4.0
1991–95	1.5	1.2	2.0	2.1	1.3	1.1	5.9	1.1	2.1	2.0	1.4	-0.5	0.5	1.3	2.1	1.4
1995–00	2.4	2.4	2.6	2.2	3.3	2.4	7.9	1.7	3.1	2.4	3.5	4.2	2.1	2.2	2.9	0.6
Productivity Growth																
1960–71	4.6	4.4	3.4	4.2	6.7	4.9	4.2	6.2	3.9	5.2	6.2	4.4	3.9	2.6	1.9	8.9
1981–90	1.9	1.7	1.5	1.7	2.2	2.1	3.8	1.7	1.6	2.2	2.7	2.7	1.3	2.1	0.8	3.1
1991–95	2.0	1.6	2.4	2.5	1.8	1.3	3.9	2.1	1.4	1.6	2.0	3.3	2.7	2.2	0.9	0.7
1995–00	1.7	1.6	1.3	1.9	0.9	1.5	4.0	1.4	0.9	2.1	2.2	2.5	1.9	1.7	1.7	0.6

Sources: *European Economy*, 1998, vol. 65 for the periods 1960–71, 1981–90, 1991–5; *Economic Outlook*, OECD, June 1999 for the period 1995–2000.

being accompanied by a similar growth in employment). This is in particular the case of the USA, which enjoyed a long period of sustained growth through the 1990s (unprecedented in peacetime, but slower than the long expansion period experienced during the Vietnam war), with until recently record low productivity growth gains (measured in terms of GDP per person employed) in comparison with the European countries and Japan.

The three European countries which enjoyed a more rapid economic growth rate in the 1990s benefited from specific opportunities (natural resources, as in the case of Norway, or a rapidly growing logistical base for international transactions, as in the case of Ireland and the Netherlands). Whether they are heralding a general resumption of productivity growth remains an open question. The recent increase in productivity gains in the USA, where GDP per person employed jumped from 1 per cent over the period 1989–95 to 2.5 per cent over the years 1996–9, has also to be treated with caution. The major fact, as underlined in the contribution by Petit in this volume, is that for the last two decades the productivity gains in the USA have been increasingly differentiated between manufacturing and services. Somehow productivity grew at a more rapid rate in sectors heavily investing but losing jobs, while in service sectors, where employment has increased dramatically, investment remained well below the average growth rate and the already weak improvements in productivity growth further slowed down. Even the recent rise in manufacturing productivity follows an upwards move in investment (see Brender and Pisani, 1999). In 1997, 1998 and 1999, for the first time in 30 years, total investment in the USA was higher than in Europe (see Muldur, 1999). The real effect on productivity of the diffusion of the new technological system is hence still in its infancy stages, having only boosted investment in some sectors (with no clear rise in the autonomous trend of technical change), even in a country where the new technologies and the accumulation of knowledge have transformed, and continue to transform, the organization of production and markets, and in particular the distribution of goods and services. The figure is even less clear for the large European countries where productivity gains remain low and where the experience of the last two decades has not, as yet, led to the distinguishing of a 'model country' which, as measured in terms of productivity gains, would have done much better than the other European countries in taking advantage of the new technologies and knowledge accumulated to spur growth and employment. The experience of many of the small European countries (Austria, Denmark, Finland, Ireland, the Netherlands, Norway, Portugal, Sweden) suggests, on the other hand, a variety of models, some of which appear related to the effective exploitation of some comparative advantage in relation to a particular feature of the new technologies, while others are specific to other characteristics of country specialization and locational advantage.

Nevertheless, what we observed in the area of institutional and structural change suggests that, after two decades of trial and error, the preliminary conditions for a change towards a new growth regime are now in place in most countries. The first decade saw a more or less drastic adjustment of practices on the labour markets, while the 1990s were witnessing a difficult process of reorganizing the finance and network-service sectors while increasing the role of qualified labour in production processes. What then could be the future of technology and employment in Europe? This future is, in our view, determined by the answers that one can give to a set of interrelated questions.

The first question bears on the relation between Europe and the USA. Can a 'New Economy' model diffuse to Europe? We more or less acknowledged in our prospective exercise that a new economy has been developing in the USA, but that its development was a long and risky process: long to emerge, when measured in terms of productivity growth, and risky because of the highly speculative nature of many of the new business ventures, and especially the financial ones. We should add that the extensive US boom in the 1990s was accompanied by a large increase in employment in services of both types, qualified and unqualified (see Chapter 6 in this volume). A lot of this last category of jobs in large metropoles fell into the category of working poor. The widening of the distribution of real wages and earnings since the 1970s has been spectacular. For full-time male workers, the average wage of the last 25th percentile was back to its 1960 level, while for the 75th percentile the average wage was 50 per cent above their 1960 level in real terms (see Ellwood, 1999). This is a specific US feature of whatever would be a New Economic Growth Regime that could diffuse to European countries. Another specific US feature is tied to the hegemonic position that gave the USA the technological and intellectual property 'appropriation' leadership forged in the course of the last two decades. The question is how the catching up of Europe and other countries, which over the 1990s experienced some widening of the gap, can take place and to what sharing of rents of innovation it will lead. The extent of restructuring still under way in most European countries and in particular in Germany is such that this remains an open question. So far the USA has enjoyed the bulk of the rents of innovation in the booming intangible new industries, with only a few exceptions (mobile telephony, smart cards, digital television). Norms, standards and regulations established in the meantime tend to strengthen this position, facilitating the huge US deficit in the external trade of standard goods. It goes beyond the objective of this book to assess in more detail the sustainability of such unbalanced growth and how Europe's catching up could take place in this setting. But some elements of the various analyses presented here specify the conditions for European countries to make the most of the opportunities that are available.

As we have already indicated above, not all European countries are in the same position and a dividing line could well separate Europe in two according to the ability of each country to catch up with the technology leader. Not only did European countries start from various levels of development in terms of GDP per head (or persons employed) but the period of transition here under review, with its marked slowdown in overall economic growth, has also experienced a relative slowdown in the process of catching up. When a new technological system is building up and diffusing, it has a double impact on the standard catching-up process, which basically relies on the lower unit costs of the lagging country and its ability to transfer technologies by means of the acquisition of equipment goods.

In a first phase the change to a new technological system occurs through a period of stagnation where the pressure on price differentials is lowered, which in turn reduces the competitive pressure of catching-up countries. In a second phase the transition towards a new technological system is illustrated in the fact that the leading countries are in a position to shape the development of the new system to their own advantage, which in turn reduces the transferability of the new technologies. Agglomeration effects, which are particularly active in the development of the current new technological system, changing in depth the time and space dimension of economic activities, re-enforce the advantage of the leading countries. They may also create specific comparative advantages for some catching-up countries in a strategic position in the diffusion network of the new technological system. The extraordinary expansion of Ireland over the last decade is an example of such an opportunity, but, by and large, under the two effects mentioned above, the growth convergence between European countries and the USA, the technological leading country in the new paradigm, has slowed down over the last decade.

The countries which have been the least affected by this slowdown may seem in the best position to resume the catching-up process. But the process of EU trade convergence may well have itself deeply transformed the process and capacity of European countries to resume catching up growth in new intangible goods and services. The recent collective work on European competitiveness edited by Fagerberg *et al.* (1999) stresses the particular importance of non-price competitiveness in the trade relations of Europe. In our analytical interpretation, this would imply that technology transfer in a world where non-price factors are playing a larger role will not occur to the same extent through the simple acquisition of equipment goods. More important will be the exchange of intangibles, foreign investment and various formal and informal exchanges of codified knowledge and, last but not least, the migration of highly skilled personnel (tacit knowledge). Chesnais *et al.* (1999) insist on the particular role played by multinationals from the leading countries in these processes of technology transfer which condition the diffusion of a new technological system among the follower countries. The development of FDI within the world of developed economies after the mid-1980s is a sign of this technology transfer. However, such traditional 'physical' investment technology transfer channels have a strong sectoral dimension, which limits the learning potential of the transfers.

A potential advantage in this respect is that, contrary to what was experienced in the USA at the beginning of the 20th century, the construction of a united Europe has not led to an extensive specialization of and between countries. The industrial structure of European economies remains diversified (see CGP, 1999). Surprisingly, maybe, this might well represent today a positive factor for a more 'fully fledged access' to the new technological system across Europe and across sectors, and might thereby facilitate innovative adjustment, if only by means of fertilizing spillovers across countries and sectors. The development of business services will help in this adjustment process even through affiliates of foreign, non-European firms, which turn out to be important vectors of technology transfer.

The overriding, present reality remains of course that European countries did not catch up with the development of the core of ICT industries as well as did the USA or Japan. It may be a matter of time, but the importance of the particular ICT production lag may also have been exaggerated if countries prove able to be leaders in sectors using these ICTs and if these technologies are accessible worldwide. A diversity, which may effectively limit the overall capacity that European countries could develop to catch up and be active promoters on the technological frontier, has to do with the national structures of the deregulated large network industries. The relative heterogeneity in the rules and regulating agencies of the old network services, whether finance or telecommunications, limits advantages to be gained at the European Union level. In particular, it limits the synergies and joint ventures that European companies can expect within the large economic space of an integrated Europe and consequently reduces their room for manœuvre in their worldwide strategic operations. It also limits the possibility for countries lagging behind to catch up.

With these caveats in mind, we would conclude that a good majority of countries in Europe can have access to the new technologies. The question behind some Euro-pessimism bears not so much on the capacity of European countries to gain access to the new technologies, but more on their capacity to take advantage of these technologies in terms of economic growth and employment. For this to occur, countries have not only to master the new technologies, but also, as Pianta argues in Chapter 5, to be able to turn them into product innovations. This lack of a crucial capability can also be linked to the delay with which venture capital developed in Europe. It expanded rapidly in the USA in the 1980s, giving birth to a great number of new, today very large, worldwide companies (Microsoft, Federal Express, Digital Equipment, Intel, Cisco and so on), and only developed in the late 1990s in Europe. Again, as with the other large network intermediate services, the segmentation in Europe has hampered their boosting effect.

The above does not mean that all European countries will have an even access to the new technologies. A set of less developed countries will obviously meet some difficulty, as underlined in this catching-up process. They risk being left behind, with a growing gap, if some policy to support the transfer of technology is not given a fair priority on the agenda (and all the more so after the enlargement of Europe by eastern countries). It may also be the case that some regions within countries, with large access and mastering of new technologies, will still need some support for their own development. In effect, if countries within Europe remained diversified, they also developed a tendency towards regional specialization which may lead to some regions forging ahead and others lagging behind. An adequate regional policy will thus certainly remain necessary, but also will require increased and consistent efforts to match what is the real challenge of the new model, that is its content in terms of economic growth and employment.

If a large share of economic growth can follow from the competitiveness of a developed economy in most activities, we saw that employment was much more dependent on the organization of community, social and personal services. Again this organization is closely tied to the choice of welfare state retained by each economy earlier on (as clearly indicated by Scharf, 1997). In Chapter 15, Mjøset underlines the different kinds of welfare states which exist in Europe according to the various roles given to the family, to public services or to market services in caring for our health, education and resources at all ages. These welfare states already underwent some transformations when going through the period of slow growth and mass unemployment. The search for a greater flexibility of labour markets led to some adjustments in their rules and transfers, which can be globally qualified as a turn towards systems keener to preserve incentives to work for all those fit to work (a 'workfare' orientation, to refer to the UK model, which corresponds also to the changes experienced in the USA).

Beyond this transformation, closely linked to the search for some labour market flexibility, demographic factors, namely the increasing effects of the reduction in birth rates across Europe, are challenging even more radically these welfare systems. Esping-Andersen (1999) and Mjøset, in his contribution to this volume, strongly underline that ageing of European populations is a major issue of our present time. This challenge is being largely debated in our societies with regard to the growing question of the future of most pension schemes. There is a pressure to shift to pension funds systems which would not be as vulnerable to drops in participation rates as are the present systems.

We contend that the challenge to the whole organization of the welfare systems is of even greater importance than it appears through this issue of retirement schemes. In effect they are of crucial importance in promoting a reorganization of the whole pattern of personal and collective services which is determinant not only for employment, as we saw, but also for a consistent contribution to economic growth. In other words, the changes in the patterns of household consumption and ways of life which are at the root of fundamental transformations of the welfare systems may, or may not, accordingly, be the supportive missing link to the full expansion of the so-called 'New Economy'. This applies equally to the USA, where the expected rise of a real New Economy may well be marred by too low productivity gains in large service industries. Welfare needs can be provided in many ways, using more or less qualified jobs in marketed or public services. The extent and quality of the coverage vary. Discussions on welfare states address the issue rather indirectly when looking at the propensity of each welfare system to favour or not the employment of low-qualified labour. International comparisons of welfare states thus oppose an Anglo-Saxon model, where low-paid menial services can be developed for such social needs as family care, with a Scandinavian model where similar services are provided by numerous public service schemes. What is at stake at present is precisely the types of services and the ways of life that can be developed with the new technologies and the knowledge that they can mobilize. Indeed, this is a huge programme of redefinition of family care, old age organizations and leisure, not to speak of the huge field of education and health systems.

We have already noted that, strangely enough, most of the signs of the new economy retained in the public debates referred to changes in the supply side of the economy, namely, how new relations to technology and finance will change the production processes. Our contribution in Chapter 6 of this book underlined the potential of the role of demand in a new growth pattern. Therefore a key issue in the transformation of welfare states lies in their ability to adjust the old model (choices of welfare systems are so ethically essential that one cannot imagine countries shifting entirely away from their welfare system with the respective importance it gives to families, public or market provisions of services) to new patterns in accordance with the future of their demography and with patterns of consumption and ways of life, making the best of the new technologies and knowledge to increase their achievements. Such a drive enriching the provision of welfare services is also bound to have cumulative effects on economic growth, while learning processes ensure that the best practices of user–producer relations diffuse throughout the economy. The challenge to this transformation on the demand side is enormous. It is obviously a dimension which has been so far lacking in the US experience of the new model, where the long path of slow growth has generated increases in poverty and inequalities which are not only a failure in terms of welfare but also very likely an important limitation in the future to a full-size recovery of economic growth and GDP per head.

But here again the challenge is quite different from one European country to the other, if only because the demographic futures are quite different. Table 16.3 illustrates the large differences among European countries in both the age structure of the population and the natural rate of growth of the population. It points to the likelihood that labour market situations will increasingly start to differ in the relatively near future. The present low rate of natural growth of the population and the rapid ageing in most member countries may well transform in a decade or so the situation of high unemployment that we now have in Europe into one of low unemployment on average, with some countries experiencing labour shortages and trying to increase the flow of immigration. How countries will ultimately take this turn depends on the type of growth trajectory chosen, and in particular on the favoured adjustment of their welfare system, but also on their internal political situation after more than two decades of slow growth and unemployment.

In redefining models of consumption, access to large systems of education, health and leisure, deeply transformed by the new technologies and knowledge (as suggested in Petit and Soete, 1997) the local dimension will certainly also play a role. It is a well-known paradox of modern changes to transform time and space dimensions in ways which re-enforce local ties. Weak social local networks can be strengthened by the use of these technologies and knowledge, which allow in turn local market and non-market organizations to reap the benefits of synergies in connecting with large networks of services and telecommunications and thus to extend their reach worldwide.

From such a perspective the question of employment and technology in Europe in the years to come is highly contingent on the demographic situations on the one side and on the ability to develop a fully fledged provision of business, intermediate and personal services on the other. This presses for highly differentiated structural policies. The next section recalls these various policies, linking them to the successive issues we have raised as parts of the determination of future economic growth and employment in Europe.

6. POLICY PERSPECTIVES IN A DIVERSE EUROPE

At this stage the general policy perspectives for the various European countries can be broadly indicated. They are centred upon the set of institutional and structural changes that we stressed above taking into account their link-

Table 16.3 Diversified demographic structures in European countries, 1995

Demographic structure	EU-15	B	DK	D	E	F	IRL	I	NL	A	P	FIN	S	UK	USA	JP
Natural growth rate per 1000	0.7	1.0	1.3	−1.5	0.4	3.4	4.7	−0.5	3.5	0.9	0.3	2.7	1.1	1.5	6.0	2.1
Total growth rate per 1000	2.9	1.2	6.8	3.4	1.6	4.1	6.2	1.1	4.5	1.9	0.9	3.5	2.4	3.5	9.3	1.7
Percentage of population under 20 years of age	23.8	24	23.5	21.5	24.6	25.4	33.7	21.2	24.4	23.2	25.8	25.4	24.6	25.3	28.9	22.8

Source: Eurostat surveys for European countries and UN data for others.

ages with the growth regime. Figure 16.2 lists these various links as they have been successively presented in the above sections. The line of argumentation is quite similar for each country but the policy implications turn out to be very country-specific as national policies have to take fully into account the characteristics of each country in terms of wage–labour relationships and services organization. To what extent a European Union policy can help these diverse national structural policies is therefore a central question.

Three areas of structural policies can help in adjusting the European countries to the new conditions of growth and employment: first, policies with respect to labour markets, second with respect to intermediate large network services, and third with respect to welfare systems. These three fields are largely interacting. Actions on the labour markets are obviously linked with interventions on the welfare system; conversely, the forms of actions in these two fields are conditioned by the logistics of intermediate services that are available, whether it is to support local synergies or to facilitate learning processes. Policy interventions should thus take into account these interdependencies when they try to respond to the two basic challenges of (a) ensuring the competitiveness of economies increasingly exposed to external competition, and (b) ensuring a full employment of the labour force, as conventionally defined, in accordance with demographic conditions.

We have already stressed that competitiveness does not imply that the full employment objective will be reached; reciprocally, full employment does not ensure competitiveness, even if in both cases one without the other constitutes an unstable situation. The increasing internationalization of the economies has led in the last two decades to priority being given to competitiveness. Policy interventions on labour markets in Europe, in the first half of the 1980s, focused on the objective of labour market flexibility to accompany the restructuring of manufacturing industries. Deindustrialization has been facilitated in various ways by this increased flexibility. It allowed countries to reallocate their labour force to sectors and niches less exposed to external competition. This restructuring also gave more importance to activities the competitiveness of which relied more on non-price factors. It contributed to shifting the focus of structural policies towards education and training.

This in turn raised in the 1990s two seemingly separate issues: what are the skills needed for such new competitiveness and how to employ the unskilled? On the first question experience has highlighted the fact that the skills required combine formal education, on-the-job training and out-of-work personal experiences. This last factor gained in importance, something which can be expected in a transition period, but also points to the need to restructure the process of training and education along lines of lifelong training schemes, which alternate work experience and diversified training programmes. These

INSTITUTIONS   ORGANIZATIONS &   GROWTH
                      STRUCTURES   REGIME

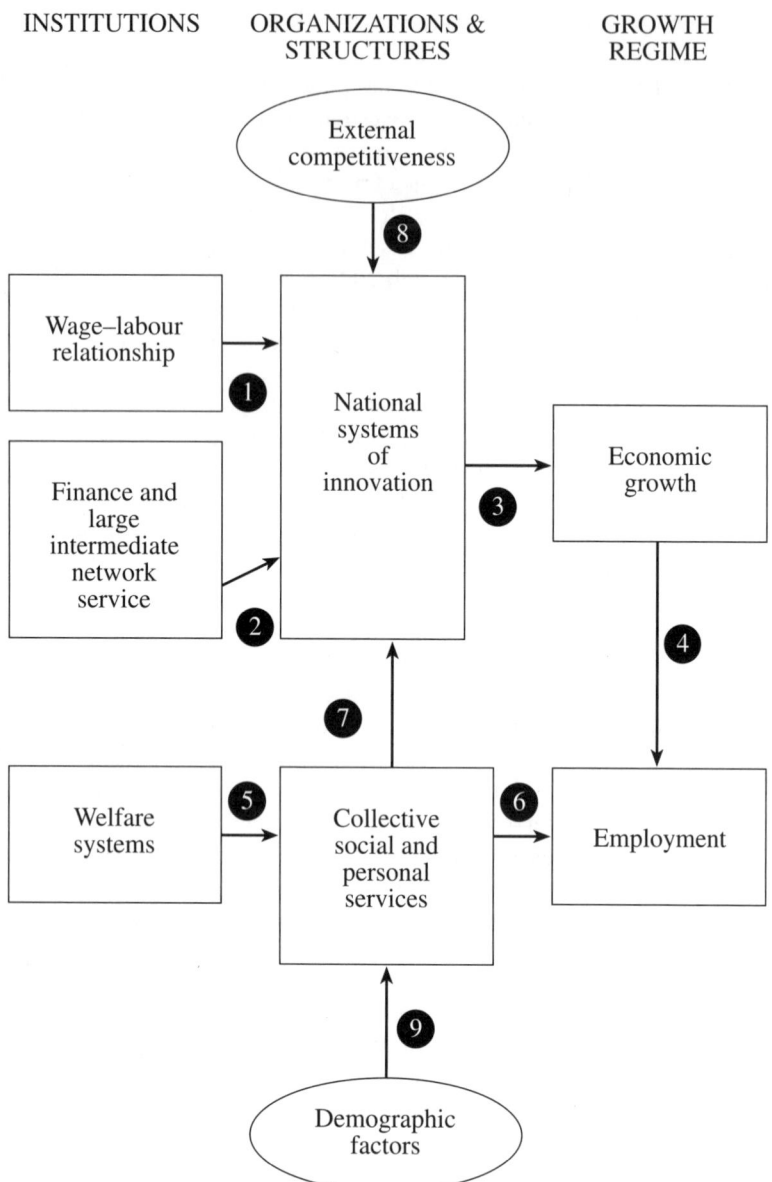

Figure 16.2   Employment and growth: the main issues

'old' recipes have so far rarely been put into practice, the new context is now more favourable to their implementation. European policies can help in launching international schemes in human capital formation and also in setting norms and models of reference, all of which appear necessary and very useful, but the bulk of the effort remains with the individual member countries. A primary condition for the success of these national policies lies in their link with the structural policies implemented in the other areas and especially in the reform of welfare systems with the reorganization of the nexus of community, social and personal services that it implies.

Many contributions to this volume stress the fact that employment in these services was clearly making the difference in the levels of participation rate, whether countries were achieving full employment or not. Furthermore, this was even more the case after the restructuring of manufacturing industries when employment in manufacturing tended to reduce its demand for unskilled labour. This direct employment effect is illustrated in Figure 16.2 by arrow number 6. The magnitude of employment in these services is particularly great, either when there is a fully fledged public provision, as in the Scandinavian countries, or when the labour markets are so 'flexible', with no minimum wage and an increasing number of working poor, leading to the provision of a wide range of 'market services'. Most of the continental European countries display an organization of these community, social and personal services which appears in some way a mix of the two schemes. These organizations stem from the various types of welfare systems developed in different European countries. As the characteristics of these systems mainly depend on the relative role given to the family, the public sector or the market in the provision of basic care, and as one should not expect rapid changes in the hierarchy of values which are grounding the overall organization of social and personal services in each country, it can be changed only in an incremental way (as underlined in Esping-Andersen, 1999). This explains why most continental European countries experienced mass unemployment despite their past commitment to full employment, and only slowly adjusted their pattern of service provision.

The experience of the USA, however, does show how incremental moves may end up with a widely transformed system of income transfer. It has not been noticed sufficiently when assessing the US experience that the US welfare system has been remarkably transformed. The cuts in welfare of the 1970s and 1980s forced more people to work (an extraordinary rise in participation rates) and led to a rise in wage and earning inequality and in the number of working poor. This pressure was finally countered in the 1990s by the implementation of various negative income tax mechanisms, raising the federal spending on low-income families (not on welfare) from six billion dollars in 1984 to 52 billion dollars in 1999. As we have seen, this redistribution was relatively unrelated to the emergence of the 'New Economy'; it might, however, create the conditions for rapid take-off.

It is our strong contention that a deeper change of the nexus of social and personal services is also needed in Europe, which would take advantage of the new technologies and of the vast amount of 'codified' knowledge accumulated. The changes in health systems, in education, as well as in leisure and work organization, that are likely to emerge will strongly influence the way people spend their time throughout their lives. Moreover, the direction of the changes will clearly be influenced by the specific demographic conditions of each country. It could lead to extensive reorganization of the welfare system and of citizens' way of life – the more so if these changes can be presented as some kind of 'new deal'. The impetus to growth from such a transformation is broadly indicated in Figure 16.2 by means of arrow 7.

The bulk of local European experiences that have emerged over the last two decades suggest that such a view is not too utopian. But to consolidate and enlarge such commitments at some national or regional level also requires the support of adequate logistics of intermediation, whether in finance, communications, transport or in specialized business services. A balance has to be found between the adequacy of these supports of specific national needs and the advantages brought by regional harmonization of these logistics. In all cases a real effort has to be made so as to give full political priority in Europe to these issues of large social change. So far the 'New Economy' has been much driven by expectations of the role of technology, mainly implemented by an 'elite' of symbolic professionals, to use Reich's term, and officials, with a much more narrow perception by those who have experienced more hardship in the meantime. Basically, to live up to its promises the 'New Economy', still has to expand its perspectives to the social realm in a much more explicitly debated and enriched form.

NOTE

1. One reason for doing this is that a result common to many chapters in this book is the apparent delinking between an overall growth regime centred on the three structural changes mentioned above and a future determination of overall employment and labour supply which increasingly also involves other variables, and particularly demographic variables. At this final stage from which policies have to take their macroeconomic perspective, the position of European countries differs strongly and imposes in the end a quite differentiated policy diagnosis.

REFERENCES

Archibugi, D. and J. Michie (1995), 'The Globalisation of Technology: a New Taxonomy', *Cambridge Journal of Economics*, 19, 121–40.

Archibugi, D. and J. Michie (1997), 'Technological Globalisation or National Systems of Innovation', in D. Archibugi and J. Michie (eds), *Technology, Globalisation and Economic Performance*, Cambridge: Cambridge University Press.

Brender, A. and F. Pisani (1999), *Le nouvel âge de l'économie américaine*, Paris: Economica.

CGP (1999), *Géographie Economique de L'Europe, Rapport du Commissariat Général du Plan*, Paris: Economica.

Chesnais, F., G. Ietto-Gillies and R. Simonetti (1999), *European Integration and Global Corporate Strategies*, London: Routledge.

David, P. (1991), 'Computer and Dynamo: The Modern Productivity Paradox in a Not-too-distant Mirror', *American Economic Review*, 80 (2).

Ellwood, D.T. (1999), 'The Plight of the Working Poor', Children Roundtable Report, no. 2, November, Brookings Institution.

Esping-Andersen, G. (1999), *Social Foundations of Postindustrial Economies*, Oxford: Oxford University Press.

Fagerberg, J., P. Guerrieri and B. Verspagen (eds) (1999), *The Economic Challenge for Europe: Adapting to Innovation-based Growth*, Cheltenham, UK and Northampton, MA, USA: Edward Elgar.

Freeman, C. (1987), 'Le défi des nouvelles technologies', *Interdépendance et coopération dans le monde de demain*, OECD.

Goux, D. and E. Maurin (1996), 'Changes in the Demand for Labour in France', *STI Review*, no. 18, OECD, Paris.

Heilbroner, R. and W. Milberg (1995), *The Crisis of Vision in Modern Economic Thought*, Cambridge: Cambridge University Press.

Kendrick, J. (1994), 'Total Capital and Economic Growth', *Atlantic Economic Journal*, 22(1), March, 1–18.

Mahroum, S. (1998), 'L'Europe et le défi de l'exode des cerveaux', *The IPTS Report*, no. 29, November, Seville.

Muldur, U. (1999), 'L'allocation des capitaux dans le processus global d'innovation est-elle optimale?', note DGXII-EC, presented to the Conseil d'Analyse Economique, Paris.

Nelson, R. and H. Pack (1999), 'The Asian Miracle and Modern Growth Theory', *Economic Journal*, 109(457), July.

OECD (1995a), *Jobs Study. Taxation, Employment and Unemployment*, Paris: OECD.

OECD (1995b), *National Systems for Financing Innovation*, Paris: OECD.

OECD (1996a), 'Technology, Productivity and Job Creation', vol. 2, Analytical Report.

OECD (1996b), 'Ageing in OECD countries. A critical policy challenge', *Social Policy Studies*, no. 20, Paris: OECD.

OECD (1997), 'Literacy Skills for the Knowledge Society', in *Cooperation with human resources development Canada*, Paris.

OECD (1999), *Science, Technology and Industry Scoreboard: Benchmarking knowledge-based economies*, Paris: OECD.

Petit, P. and L. Soete (1997), 'Is a Biased Technological Change Fuelling Dualism?', working paper, December, CEPREMAP, Paris.

Scharpf, Fritz W. (1997), 'Combating Unemployment in Continental Europe', EUI, Florence, R. Schuman Centre, Policy Papers, RSC, no. 97/3.

Sichel, D. (1997), *The Computer Revolution: an Economic Perspective*, Washington, DC: Brookings Institution Press.

Vendramin, P. and G. Valenduc (1999), 'L'avenir du travail dans la société de l'information', FTU, Namur, FEC Brussels, June.

Index

Abramovitz, M.A. 426, 434
ageing population, implication for
    welfare state 463–5, 486–7
agricultural employment, OECD
    countries 213, 214, 220
Anglo-Saxon model
    employment patterns, services 481
    tax structure 479
    working time pattern 420–21
Aoki, M. 275
Asia, East, economy 28–37
Aston Group 263, 267
Athey, S. 278
Australia, working time patterns 406–11

Bartel, A. 281
Baumol, W. 210, 211, 238
Bell, 481–2, 488
BERD (Business Expenditure on R &
    D) 47–8, 52–3
Berg, P.B. 297
Bowles, S. 300
Boyer, R. 264, 277
Brazil
    employment in manufacturing 32
    investment 31
Bresnahan, T. 273, 283
Britain see UK
British (liberal) welfare state model 454,
    473
Brynjolfsson, E. 268–9, 273, 277
Burns, T. 263–4, 266, 276
business cycles 25–7
Business Expenditure on R & D
    (BERD) 47–8, 52–3
business services, employment 97–8,
    183–5, 522–4

Canada, working time patterns 406–14,
    415–17, 420–21
capital accumulation 127

capital-labour substitution 126–30
capital movements, impact on wages
    352–79
Cappelli, P. 283
Caroli, E. 268, 283, 285
Carr, F. 276
catch-up, Asia 33–6, 39
centralized mode of organization
    298–309
Chandler, A. 260, 275
Chesnais, F. 530
chimney effect 317–48
clientelism, political, influence on
    welfare benefits 493–5
codified knowledge, impact of ICT
    172–3
Community Innovation Survey (CIS)
    164–5
competitiveness, effect on employment
    149–50
continental welfare model 473, 476–92
coordination within firm 302–5
Coriat, B. 297
corruption, government, influence on
    economic collapse 34
cost disease problem 210, 211, 238
Coutrot, T. 266
crowding-out hypothesis, low-skilled
    work 388
cultural environment, and organizational
    change 276–7
Cyert, R. 264, 275

David, P. 434
Davis, L. 272
De Long, J.B. 48
decentralization
    effect of IT 268–9
    of decision making 265, 273–4
decentralized mode of organization
    298–309

deindustrialization 215–18
demand conditions, impact on work
    organization 267–8
demand, North-South model 389–90
demand patterns 143–61
    relationship with employment 88, 90,
        112–13, 115–17, 156–7, 160–61
demand pull model 49, 56
demography, impact on employment
    structure 463–5, 533–5
Denmark
    innovation expenditure 151–7, 164–5
    welfare system 481–3
deregulation, impact on labour market
    513–16
deskilling effect of new technologies
    281–3
domestic demand 112–3
Dosi, G. 27, 297
Dow, C. 25

early retirement 484–7, 497
East Asian economy 28–37
Eastern Europe, patenting activity 33
economic growth engines 183–5
economic trends
    Africa 29
    Europe 78–83
    Japan 78–83
    Latin America 29
    Tiger economies 29
    USA 39–43, 78–83
Economist magazine 42
education and training 285–6, 295–7
    costs 325–7
    effect on employment pattern 511–13
    implementation of new technologies
        279–81
    service sector 198–9
    Southern Europe 493
employees
    competence 295–7
    efficiency, by skill level 334–8
    skills *see* skills
    training cost 325–7
employment
    Brazil 32
    effects of macroeconomic perform-
        ance 160–61
    effects of structural changes 525–6

patterns 497–8
    Europe 160–61, 186–7, 189, 474–6,
        492–3
    Germany, 332–4
    Netherlands 332–4
    service sector 185–97, 481
    relationship to skill levels 295–7,
        317–48, 387–401
    sectoral 213–25
    South Korea 32
    working time patterns 406–11
    see also labour market; employees;
        unemployment
EMU, impact on innovation 445–6
endogenous growth, influence of
    institutions 429–31
equilibrium unemployment 309–13
Ergas, P. 433, 434
error-correction model 54–6
Esping-Andersen welfare state typology
    453–5, 472–500
Esping-Andersen, G. 466–7
EU *see* European Union
Euro, impact on innovation 445–6
Europe *see* individual countries
Europe, Eastern, patenting activity 33
Europe, Southern, welfare state model
    492–6
European integration model 441
European monetary integration, impact
    on innovation 445–6
European Union
    employment trends 186–7, 189
    relationship between wages and
        employment rates 360–78
    working time policy 421–2
Eurosclerosis 86, 89–90, 509
Evangelista, R. 151

Fagerberg, J. 530
family trends, effects on welfare state
    459–65, 492
FDI 97, 102
Feenstra, R.C. 126
female employment, Europe 484, 485–6,
    489, 490
Ferrera, M. 493, 494
final product demand 115–7
final product productivity (FPP) 120–26,
    133, 139–41

financing of welfare system 481–4, 499–500
Finegold, D. 286
Finland, working time patterns 412–14
foreign direct investment 97, 102
France
  early retirement 477
  employment trends 125, 130
    elderly people 477
    service sector 480
    wage-related 360–61, 362–8, 370–71, 378
    working time patterns 406–14, 415–16, 418–20
  productivity 125
Freeman, C. 27 , 434
Freeman, J. 260
Freeman, R. 470, 496

G6 countries, productivity growth 158–61
GDP, relationship with R & D investment 48–53, 56–7, 63–5
GERD (Gross Expenditure on R & D) 47–8, 52, 66, 67
Germany
  employment trends 123, 332–4
    chimney effect 339–44, 347
    elderly people 477–8
    service sector 480–81
    wage-related 360–63
    working time patterns 406–14, 416–20
  innovation expenditure 151–7
  labour force skill level 280, 296–7, 344–8
  productivity and employment changes 123, 159
  worker efficiency 334–8
Gerschenkron, A. 38–9
Gerschenkron uncertainty 38, 39
Gerwin, D. 277
Gilpin, K.G. 40
globalization 191–3, 522–3
Glyn, A. 230
Goffman, E. 272
Gouyette, C. 121, 126
Granger causality tests 362–3
Great Britain *see* UK
Greece, welfare system 492–5

Greenan, N. 273, 283
Greenspan, A. 40
Gregori, T. 126
Griliches, Z. 121
Gross Expenditure on R & D (GERD) 47–8, 52, 66, 67
Guellec, D. 273

Hage, J. 273
Hancké, B. 296
Hannan, M. 260
Hanson, G.H. 126
Heckscher-Ohlin (HO) models 353–4
hierarchical work organisation 297–8
high-skilled labour 393–5
  see also skills
Higher Education Expenditure on R & D (HERD) 48, 53, 68
Hitt, L. 268–9, 273, 277
Hollingsworth, R. 436
horizontal work organization 297–8
Howell, D. 282
human resource management 266, 276

Ichniowski, C. 266, 267, 278
ICT *see* information and communication technologies
imperfect competition model 372–7
incentive schemes 305–9, 512–3
income redistribution, effect on output and employment, Europe 238–48
industrial employment, OECD countries 213–16, 220–21, 224–5
Industrial Growth and World Trade 29–30
industry structure 111–34
information and communication technologies (ICT)
  diffusion 521–6
  equipment exports 30–31
  impact on economy 24
  impact on employment 96, 98–103
  impact on goods and services 170–78
  impact on labour market 508–9, 516–20
  impact on tradeability 174–7, 182
  impact on working time 415
  role in globalization 191–3
  role in innovation 142–3, 179–83
  role in organizational change 268–78

training 198–9
innovation 142–61
    expenditure 151–7, 164–5
    in services 179–83, 196
    relation to growth 425–6
    see also information and communication technologies; technological change
instability of investment 34, 37–44
institutional changes, impact on labour market 508–16
institutions, effects on growth 426–446
interest rates, promoting growth 61
intermediation logistics 101–2
international trade, impact on wages 352–79
internationalization
    of national systems of innovation 518–29
    promotes diffusion of ICTs 521–6
    structural change factor 96–103
investment
    effect on employment 58–9, 87
    instability 34, 37–44
    OECD countries 48–54
    service sector 197–8
Ireland, working time patterns 415–16, 418–19
iso-employment curve 112–3
IT *see* information and communication technologies
Italy
    employment and wages 360–68, 372–8
    innovation expenditure 151–7, 164–5
    productivity 159
    welfare system 492–5
    working time patterns 406–14, 415–19

Japan
    economic trends 28–30, 35, 37, 78–83
    employment trends 84–5, 89–92, 124, 129, 186–7
    equilibrium opportunity rate 307–9
    productivity 124, 158–61
    vocational training 297
    working time patterns 406–14
job finding rate 310–11
job seekers 490, 492–3

job-skill level correspondence 317–46
jobless growth theory 88
Juglar cycle 25, 26

Keynes, J.M. 25–6
knowledge distribution within a firm 297–9, 302–5
Kondratieff waves 25, 26
Korea, South *see* South Korea
Krugman, P.R. 28, 37, 43, 352–3
Krugman North-South model 388–92
Kuznets, S. 26–7

labour absorption 208–11
labour costs 325–7, 409
labour demand 112–3
labour force skill levels 280–86
Labour Force Survey 418–19
labour-management relations 275–6
labour market
    deregulation 312–13
    effects on welfare state 459–65
    Europe 92, 399–400, 469–72, 508–13
    flexibility 293–5
    Japan 92
    mobility 86, 93–4, 295–7, 393–5
    policies 198, 473–87, 537
    service sector 187–97
    UK 472
    USA 92, 469–72
    wage divergence 396–401
labour productivity 95, 114–15, 131
Landau, R. 435
Latin America
    economy 29–31
    patenting activity 33
Lawrence, R.Z. 352, 353
Leamer, E.E. 353
learning asymmetries 325–7
Lichtenberg, P. 281
life expectancy trends 463–5
Lilien, D.M. 87
Lincoln, J. 267
long-run cycle theory 88
long-run growth, relation to technical change 425–6
long wave theory 24–9
low-paid work 93–4
low-skilled labour, wages 387–401
Lücke, M. 353

Lundvall, B.-A. 435

M-firm 260, 275
Machin, S. 268
Maizels, A. 29–30
Malinvaud, E. 87, 127
management organization types 263–5
manufacturing exports 30–31
manufacturing sector employment 57–9,
    127, 117–20, 224–5, 264–5
market-based productive model 438,
    440, 442
Marsden, D. 296
Mason, G. 279, 280, 286
matching workers to jobs 301–2
Maurice, M. 296
Mazzucato, M. 38
mechanistic management structure
    263–4, 273
meso-corporatist productive model
    438–9, 441–2
microinstitutional settings 295–9
Milgrom, P. 277
mobility of workers 86, 93–4, 295–7,
    393–5
mode of organization (MoO) 297–307
monitoring workers 300
Mowery, D. 264, 275, 435
multidimensional firm (M-firm) 260,
    275

national systems of innovation 432–3
nation-welfare state 453
Nelson, R. 27, 435
Netherlands
    employment trends 332–8
        chimney effect 344–7
        wage-related 360–68, 372–6
        working time 412–16, 418–20
    innovation expenditure 151–7, 164–5
Nickel, S. 481–2, 488
Noble, D. 272
Nordic welfare model 454, 472–6, 479,
    481–5, 488–9, 491, 496–8
North-South model 388–92
North 426
Norway
    innovation expenditure 151–7, 164–5
    welfare system 481–3
    working time patterns 406–14

NSI (national systems of innovation)
    432–3

O'Connor, E. 264, 273
OECD countries
    economic performance 48–62
    employment patterns 186–91, 213–37
    low-skilled wages 387–8
    working time patterns 411–14
OECD McCracken Report 23
Ohkawa, K. 434
Oliveira Martins, J. 353
opportunity rate 301
organic management structure 264, 266,
    273
organizational change 259–287
Oscarsson, E. 353
Osterman, P. 266, 267, 276
output growth, Triad countries 80–83
outsourcing 101–2, 113, 120–27, 133
overcapacity in manufacturing 35, 39

paradigm change, theory of 27–8
part-time working 415–20
Pasinetti, L. 205
patent activity 33, 53–4, 69–70, 359
payroll taxation 481–2
pensions, old age 478, 484–5
per capita GDP, relationship with R & D
    investment 48–53, 56–7, 63–5
Perelman, S. 121, 126
Perez, C. 24, 27–8
perfect competition model 360–72
political clientelism, influence on
    welfare benefits 493–5
Portugal, welfare state system 492–5
precarious jobs 93
Prennushi, G. 266, 267
prison population, US 470–71
process innovation 148–9
    industry performance 158–61
    service sector 196
producer durables rate, OECD countries
    48–9
product demand, effect on employment
    demand 112
product innovation 148–9
    industry performance 157–61
    relationship with employment 168–70
    service sector 196

share of R & D 152–5, 164–5
product market, effect on unemployment 91
production regime, North-South model 391–2
productive models 433–46
productivity
    effect of capital-labour ratio 126–31, 133
    effect of outsourcing 120–26
    growth 158–61
    relationship with employment 168–70, 528
    relationship with working time 411–14
    Triad countries 80–83
public institutions-based productive model 439, 441–2

redistribution of income 238–48
regulation theory 432
Reich, R. 43
reorganized firms 266–8
research and development (R & D)
    indicator of technical change 359
    investment, OECD countries 47–54, 56–7, 65–8
retirement, early 477–8
risk structures, impact on welfare state 465–9
Roberts, J. 277
Rosenberg, N. 435
Rosovski, H. 434
Rowthorn, R.E. 87, 230

Sachs, J.D. 34, 36, 353, 377
Sakakibara, Mr. 37
Salais, R. 436
SAM (skill allocation model) 319–48
Scharpf, F.W. 481, 482–3
Schmookler, J. 49
Schmutzler, A. 278
Schumpeter, J.A. 24–6, 29, 88
Schumpeterian structural change 40
Scott, W. 276
search theories 84, 86
sectoral employment, OECD countries 213–25
sectoral unemployment, OECD countries 230–37

sectoral wages 352–79
separation rate 311
service sector 166–7, 171
    employment trends 117–20, 185–97, 214, 215–18, 221, 480–81
    innovation 177–86
    role in economic growth 178–85
    tradeability 173–7
Shatz, H.J. 353, 377
Shaw, K. 266, 267, 278
Shin, J.-S. 39
Siegel, D. 121, 126
Siegenhalter, H. 26
size, impact on work organization 267–8
skill allocation model (SAM) 319–48
skill-biased productivity growth 244–8
skills
    as tacit knowledge 172
    influence on employment opportunities 295–7, 317–48, 387–401
    influence on technical change 279–87
    structure 193–6
Slaughter, M.J. 126, 352
Smithers, A. 42
social capability 426
social democrat productive model 438, 440, 442
socio-economic trends, W. Europe 455–65
Soskice, D. 296
South Korea
    catch-up 39
    economy 31–2
Southern European welfare state model 492–6
Spain
    wages and employment rates 360–71, 374–8
    welfare system 492–5
Stalker, G. 263–4, 266, 276
Steedman, H. 279, 285, 286
Stigler, G.J. 84
Stiglitz, J. 435
Storper, M. 436
Strategy and Structure 260
Strauss, A. 272
Streeck, W. 297
structural breaks in relative employment growth 224–5
structural change 520–26

dynamics 219–23
impact on employment 525–6
relationship to unemployment 205–37
subsidies, effect on demand 238–9,
   242–4
Summers, L.H. 48
supply-demand interaction model 55
supply of goods, North-South model
   390–91
Sweden
  welfare system 480–83
  working time patterns 406–14,
    415–16

tacit knowledge 172–3
Tarondeau, J.C. 277
tax burden, effect of subsidies 242, 244,
   245, 247–8
tax structure, impact on employment
   478–80
Taylor, J. 272
technical change *see* technological
   change
technical skill requirements 283–6
technological change
  impact on economy 530–32
  impact on employment 96–103,
    156–7, 508–9, 511
  impact on growth 142–33
  impact on wages 352–79, 387, 395–6
  indicators 359–60
  North-South model 391
  relation to growth 425–6
  relation to organizational change
    268–78
  service sector 193–6
  see also information and communica-
    tion technologies; innovation
technoeconomic paradigms theory 27–8
technological innovation, influence of
   labour market 516–20
technological unemployment 91
technology
  relation to growth and employment
    168–70
  relation to organization 261–2
Ten Raa, T. 126
tertiarization 97–103, 183–5, 215–18,
   521–6
Tiger economies 28–34

time constraints on service innovations
   177–8
total factor productivity (TFP) 127, 131
trade, international, impact on wages
   352–79
trade flows 96–7
tradeability of services 172, 173–7, 182
trading regime, North-South model
   391–2
training *see* education and training
transaction costs, role in organizational
   structure 260
Treatise on Money 25

U-firm (unified structure) 260, 275
UK
  labour force skill level 280
  productivity growth 159
  relationship between wages and
    employment rates 360–68,
    370–71, 375–6
  technological strategies 279–80
  working time patterns 406–14, 415–20
unemployment
  age distribution 489
  causes 89–91, 481
  effect of structural change 205–37
  equilibrium 309–13
  Europe 84–94, 204, 269–72, 451–3,
    469–72
  impact on welfare state 460–61, 466
  Japan 84–5, 89–91
  job-level related 339–46
  labour market flexibility 293–5
  long-term 491
  Nordic countries 484
  OECD 29, 230–37
  sectoral 230–37
  USA 84–5, 89–91, 469–72
  youth 461, 488–90
  see also employment
unemployment-wage equilibrium 294
unified structure firm (U-firm) 260, 275
upskilling effect of new technologies
   281–3
USA
  economic trends 39–43, 78–83
  employment trends 121–2, 128,
    186–7, 469–72, 480
  equilibrium opportunity rate 307–9

labour market 399–401, 469–72
manufacturing structure 264
productivity 121–2, 158–61
unemployment patterns 84–92,
      469–72
vocational training 297
wages and employment rates 360–61,
      364–78
welfare system 538
working time patterns 406–14,
      416–17, 420–21

vacancy rates, relationship to unemploy-
      ment 309–13
Van Reenen, J. 268, 283, 285
vocational training 296–7
voluntary search unemployment 84
Von Tuzelmann, G.N. 436

Wadhwani, S. 268
wage divergence 396–401
wages
   affecting employment distribution
      318–19
   effects of decreasing working time
      409
   impact on employment composition
      330–31
   low-skilled labour 387–401
   manufacturing employment 59
   optimal level 299–309
   related to employment rates 352–79
   related to productivity 412–14
   related to working time 412–14

sectoral 352–79
subsidies, effect on demand 242–4
welfare services sector 210
Wagner, K. 279, 280, 285
welfare services
   employment absorption 209–11
   structural breaks in employment
      growth 225
welfare systems
   Europe 453–500, 510, 532–3
   impact on employment 532–3
   USA 538
West Germany *see* Germany
Williamson, O. 260, 272
Winter, S.G. 27
Wolfensohn, J. 36
Wolff, E.N. 126, 282
Wood, A. 353
Woodall, P. 24
work organization 297–9
workers *see* employees
workforce matching 206–7
working time
   flexibility 414–20
   patterns 405–22
   reduction, effect on employment
      pattern 510–11
   relationship with productivity 411–14

Young, A. 178
youth unemployment 461, 488–90,
      498–9

Zammuto, R. 264, 273